Addison-Wesley

Consumer Mathematics

Stanley A. Smith
Penelope P. Booth

Program Consultants

Randall I. Charles
John A. Dossey

Addison-Wesley Publishing Company

Menlo Park, California • Reading, Massachusetts
New York • Don Mills, Ontario • Wokingham, England
Amsterdam • Bonn • Sydney • Singapore • Tokyo
Madrid • Bogotá • Santiago • San Juan

ISBN 0-201-21328-1

CDEFGHIJKL-VH-892109

Acknowledgements

Illustration

Art by AYXA
Christensen and Son Design
Marilyn Hill
Marcia McGetrick
Phyllis Rockne
George Samuelson
Sally Shimizu

Typesetting

Typothetae

Design

Design Office

Photography

Page

3 Chris Sorensen/The Stock Market
12 William Strode/Woodfin Camp & Associates
18 Jeffrey D. Smith/Woodfin Camp & Associates
29 Julie Houck/Stock, Boston
35 © R. Hamilton Smith
41 Timothy O'Keefe/Bruce Coleman Inc.
42 Henley & Savage/The Stock Market
44 Elliott Varner Smith*
53 Robert C. Schafer/After-Image
56 Dan McCoy/Rainbow
68 Jim Harrison/Stock, Boston
77 Randy Duchaine/The Stock Market
78 Jeff Zaruba/After-Image
84B Bohdan Hrynewych/Stock, Boston
84T Jerry Howard/Stock, Boston
88 Lois Ellen Frank/West Light
92 Norman Owen Tomalin/Bruce Coleman, Inc.
101 Dan McCoy/Rainbow
104 Edward L. Miller/Stock, Boston
105 Leverett Bradley/After-Image
106 © T & M Grimm/After-Image
108 Peter Menzel/Stock, Boston
110 Dan McCoy/Rainbow
112 Jerry Jones
119 Jon Feingersh/Stock, Boston
125 Paul Barton/The Stock Market
128 Jon Feingersh/Tom Stack & Associates
139 T. J. Florian/Rainbow
149 Coco McCoy/Rainbow
154 Jim Zuckerman/West Light
160 George B. Fry III*
162 Gale Zucker/Stock, Boston
167 William James Warren/West Light
173 David Madison/Bruce Coleman Inc.
177R Robert Beck/Focus West
197 Roman Lezo/After-Image
198 Phil Degginger/Bruce Coleman Inc.
204 D. P. Hershkowitz/Bruce Coleman Inc.
206 © R. Hamilton Smith 1984
212 Martin Rogers/Stock, Boston
221 Ralph B. Pleasant/After-Image
223 Roger Tully/After-Image
226 Bob Daemmrich/Stock, Boston
232 © David Madison
234 Jerry Howard/Stock, Boston
238 Michael Yamashita/West Light
245 Howard Sochurek/Woodfin Camp & Associates
246 © David Madison
254 © David Madison
258 Craig Hammell/The Stock Market
269 Tom Tracy/After-Image
270 Gabe Palmer/After-Image
274 Donald Dietz/Stock, Boston
278 Timothy O'Keefe/Tom Stack & Associates
280 M & T Grimm/After-Image
293 © Harald Sund
294 Gabe Palmer/After-Image
298 Gabe Palmer/The Stock Market
299 Gabe Palmer/The Stock Market
300 Vic Huber/West Light
302 Bob McKeever/Tom Stack & Associates
303 © David Schaefer
304 Henley & Savage/After-Image
306 George W. Gardner/Stock, Boston
307 Vic Huber/West Light
317 Roger Allyn Lee/After-Image
332 A. Edgeworth/The Stock Market
341 Ann Duncan/Tom Stack & Associates
351 Ellis Herwig/Stock, Boston
365 Gabe Palmer/After-Image
389 Ron Sherman/Bruce Coleman Inc.
394 Brian Parker/Tom Stack & Associates
396 Gabor Demjen/Stock, Boston
398 Norman Owen Tomalin/Bruce Coleman Inc.
399 Mike Yamashita/West Light
401 Denis Valentine/The Stock Market
402 Eric Neurath/Stock, Boston
403 Chris McLaughlin/The Stock Market
413 Michael Ventura/After-Image
420 Rene Sheret/After-Image
421 Larry Stevens
427 Gene Stein/West Light
428 Gil Garner/After-Image

Wayland Lee*/Addison-Wesley Publishing Company: Pages 45, 60, 80, 177, 200, 201, 356, 367, 404

*Photographs taken expressly for the publisher

CONTENTS

1 CONSUMER CALCULATIONS

2 SOLVING CONSUMER PROBLEMS

3 USING FRACTIONS

4 USING PERCENT

5 EARNING MONEY

6 PERSONAL FINANCE

7 SPENDING MONEY

8 USING CONSUMER INFORMATION

9 USING PROBABILITY

10 FOOD AND DIET

11 HEALTH AND FITNESS

12 TAXES

13 USING A CAR

14 TRANSPORTATION

15 HOUSING

16 HOME IMPROVEMENTS

17 TRAVEL AND LEISURE

18 INVESTMENTS AND RETIREMENT

Addison-Wesley

Consumer Mathematics

Chapter 1 Overview

Key Ideas

- deciding on the kind of answer needed
- choosing a calculation method
- estimating sums and differences
- calculating sums
- calculating differences
- using mental math
- estimating products and quotients
- calculating products
- calculating quotients

Key Terms

- mental math
- paper and pencil
- calculator
- estimation
- rounded numbers
- clusters
- compatible numbers
- sum
- total
- increased by
- all together
- difference
- remainder
- decreased by
- break apart
- count up/count back
- factor
- product
- quotient

Key Skills

Tell the value of the underlined digit.

1. $\underline{3}2{,}568$	**2.** $108{,}4\underline{1}6$	**3.** $14.7\underline{5}8$
4. $3{,}\underline{6}74{,}035$	**5.** $3.01\underline{5}$	**6.** $159.\underline{4}3$
7. $\underline{5}{,}142{,}768$	**8.** $10\underline{0}.44$	**9.** $6\underline{6}{,}201$

Find the answer.

10. $40.05 - 26.135$	**11.** $85.76 \div 6.4$	**12.** $\$108.27 + \$364.89$
13. 38×79	**14.** $38 - 0.46$	**15.** 60.7×34
16. $7322 \div 14$	**17.** $834 + 478$	**18.** $\$71.68 \div 56$
19. $14 + 37 + 56 + 52$	**20.** 2.5×3.36	**21.** $1123 - 456$
22. $23.5 \div 0.25$	**23.** 217×54	**24.** $7092 - 4884$
25. $284 + 2795$	**26.** $14.3 - 6.52$	**27.** $15 \div 0.6$
28. 15.6×0.34	**29.** $\$0.74 + \$0.58$	**30.** $6136 \div 52$

CONSUMER CALCULATIONS 1

1-1 Choosing a Calculation Method

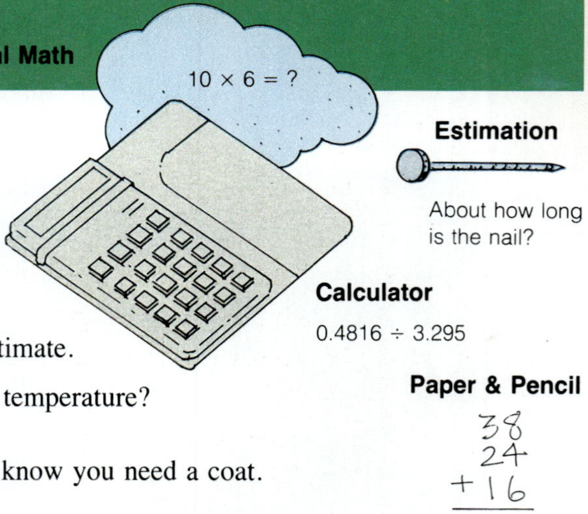

Objective
Decide if an exact answer is needed. Choose the calculation method.

You use mathematics to solve problems every day. An important step in solving problems is deciding what kind of answers are required. Some situations require an exact answer. In other situations an estimate may be more useful.

Examples Decide if you need an exact answer or an estimate.

You are reading a thermometer. What is the temperature?

1. Should you wear a coat today?
You do not need an exact temperature to know you need a coat.

2. Does the patient have a fever?
To provide proper care, you need an exact temperature reading.

Try This Decide if you need an exact answer or an estimate.

You are reading a sales chart. How many sales were made?

a. Find the total sales for the three-month period.

b. Which month had the greatest number of sales?

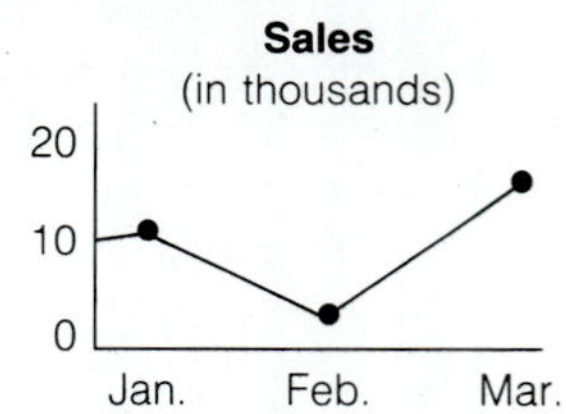

Mental math problems are those that you solve in your head.
You usually use **paper and pencil** to solve more complicated problems.
A **calculator** is fast and accurate when you have a good plan for using it.
You can find an approximate answer using **estimation.**

Examples Choose estimation, mental math, paper and pencil, or a calculator.

3. You are at the movies. Can you afford to buy popcorn and a drink?
Count your money. Use mental math.

4. You are filling out your tax returns. How much do you owe?
Many exact calculations are needed. Use a calculator.

Try This Choose mental math, paper and pencil, a calculator, or estimation.

c. You are balancing your checkbook. How much money is in the account?

d. You are planning a vacation. About how many miles will you drive?

Exercises

Decide if you need an exact answer or an estimate.

You are reading a gauge. How many gallons are in the tank?

1. Do I need to get gas?

2. What is the exact cost to fill the tank?

You are reading a clock. What time is it?

3. Is the time a new world record?

4. Is it time for lunch?

You are reading a chart. What is the score?

5. Which player won the first game?

6. Which player has the best average for three games?

Bowling Scores

Name	Game 1	Game 2	Game 3
Paula	110	128	118
Brian	124	143	152
Emma	130	155	148

You are reading a tape measure.
What is the measurement?

7. Will the bookcase fit between the windows?

8. How tall am I?

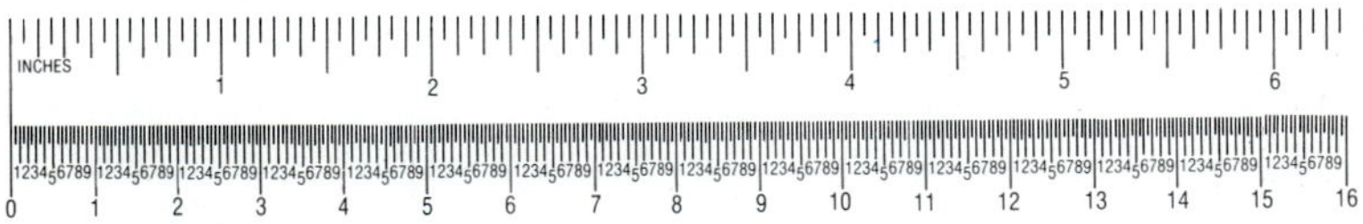

Choose estimation, mental math, paper and pencil, or a calculator.

9. You are painting a room. How many gallons of paint will you need?

10. You are scorekeeper at a basketball game. Who is high scorer?

11. You are the club treasurer. How much money is in the treasury?

12. You are giving a friend a ride to school. What time should you leave?

13. You are preparing a project for the science fair. What was the average daily temperature during the last month?

14. You are buying seed for the lawn. How many pounds do you need?

15. The cafeteria serves 850 students. What is the monthly budget?

16. The public library serves a community of 25,000 people. How many books are borrowed each year?

17. The class has 25 students. What percent wear glasses?

Mixed Skills Review

1. $6 + 3$ **2.** 5×5 **3.** 9×10 **4.** $48 \div 12$

5. $35 \div 7$ **6.** $900 \div 9$ **7.** $4 + 5$ **8.** $21 - 11$

9. $18 - 9$ **10.** $26 \div 2$ **11.** $8 + 10$ **12.** 12×3

1-2 Estimating Sums and Differences

Objective
Estimate using rounding, clustering, and compatible numbers.

Units Handled				September
	Week			
Area	1	2	3	4
East	175	191	185	174
Downtown	253	228	247	279
West	86	112	99	85
Totals				

Metropolitan Courier guarantees overnight delivery. The courier records the number and the weight of items handled.

Understand the Situation

- For which period of time is data given?
- How many items were handled in the East area during Week 1?
- During which week were the most items shipped in the Downtown area?

Example

1. Use rounding to estimate the units handled during the first week.

East	175		200			180	
Downtown	253	or	300	*nearest*	or	250	*nearest*
West	+ 86		+ 100	*hundred*		+ 90	*ten*
			600			520	

Try This Estimate.

a. 5491 + 3924 **b.** \$16.34 + \$43.98 **c.** 956 − 724 **d.** 7.56 − 3.47

Example

2. Estimate the units handled in the East for September.

Since each number **clusters** around 200, a good estimate is 4 × 200.
4 × 200 = 800 The total number of units was about 800.

Try This Estimate. Use clustering.

e. 84 + 78 + 82 **f.** 398 + 406 + 412 + 387

g. 923 + 898 + 906 + 403 + 398 + 412

Example

3. \$12.65 + \$21.62 + \$7.98 \$12.65 + \$21.62 + \$7.98
\$12.65 and \$7.98 are **compatible numbers.** Their sum is about \$20.
\$21.62 is about \$20. Estimate \$40.

Try This Estimate. Look for pairs of compatible numbers.

h. 22 + 36 + 129 **i.** \$29.13 + \$19.54 + \$52.76 **j.** 32 + 18 + 56

Exercises

Round to the indicated place.

nearest ten	**1.** 323	**2.** 4784	**3.** 12,009
nearest hundred	**4.** 478	**5.** 3704	**6.** 56,828
nearest thousand	**7.** 895	**8.** 9530	**9.** 35,254
nearest dollar	**10.** \$2.75	**11.** \$39.50	**12.** \$390.58
nearest cent	**13.** \$1.034	**14.** \$77.197	**15.** \$0.303
nearest tenth	**16.** 0.76	**17.** 0.349	**18.** 6.0955

Estimation

Estimate. Use rounding.

19. $\begin{array}{r} 41 \\ 94 \\ +63 \\ \hline \end{array}$ **20.** $\begin{array}{r} 609 \\ 441 \\ +576 \\ \hline \end{array}$ **21.** $\begin{array}{r} 2859 \\ +3675 \\ \hline \end{array}$ **22.** $\begin{array}{r} 70{,}123 \\ -\quad 352 \\ \hline \end{array}$ **23.** $\begin{array}{r} 6093 \\ -5519 \\ \hline \end{array}$

Estimate. Use clustering.

24. 198 + 212 + 189

25. \$28.98 + \$32.09 + \$35.12

26. \$178 + \$223.12 + \$199 + \$234.06

27. 676 + 702 + 431 + 399 + 689 + 421

Estimate. Look for pairs of compatible numbers.

28. 153 + 48 + 2706 + 322

29. 786 + 144 + 389

30. \$14.98 + \$2.89 + \$5.67

31. \$4.09 + \$5.01 + \$19.87

Calculator

Use a calculator. Use mental math. Compare your answers.

32. Add 9.78 + 2.65 + 5.431. → Round to tenths by rounding up. → Subtract 7.468 from answer. → Round to tenths by rounding down.

Work with a Partner

33. Estimate the total number of units handled by Metropolitan Courier during September. Solve the problem two different ways. Compare your totals.

Write Math

34. Make up a problem whose answer can be estimated using clustering.

Mixed Skills Review

1. 6×7	**2.** $4 - 3$	**3.** $44 \div 11$	**4.** $10 \div 5$
5. $7 + 8$	**6.** $24 - 12$	**7.** 3×10	**8.** $55 \div 11$
9. $8 + 12$	**10.** $6 + 11$	**11.** 8×8	**12.** $45 \div 9$

1-3 Calculating Sums

Objective
Add whole numbers and decimals.

Sam Houston High School is sponsoring a recycling drive. Each class earns money by collecting cans and newspapers.

Class	Pounds Collected		Money Earned	
	Cans	Papers	Cans 0.03/lb	Papers 0.014/lb
Seniors	374	647	$11.22	$9.06
Juniors	416	751	12.48	10.51
Sophomores	299	960	8.97	13.44
Freshmen	441	549	13.23	7.69

Understand the Situation

- How many pounds of cans did the senior class collect?
- Which classes have collected more than 700 pounds of newspapers?
- How much money does a class earn for one pound of cans?
- How much money did the sophomore class earn from newspapers?

Language often suggests which operations are needed to solve problems. The words **sum, total, increased by,** and **all together** usually suggest addition.

Examples

1. How many pounds of cans and newspapers did the junior and senior classes collect all together? *All together* here means to join or add the numbers.

$$\begin{array}{r} 374 \\ 416 \\ 647 \\ +751 \\ \hline 2188 \end{array}$$

The junior and senior classes collected 2188 pounds of cans and newspapers all together.

When adding decimal numbers, line up digits with the same place value.

2. How much money did the freshman class earn?

$$\begin{array}{lr} \text{Cans} & \$13.23 \\ \text{Papers} & +\ 7.69 \\ \hline & \$20.92 \end{array}$$

The freshman class earned $20.92.

If you use a calculator, it lines up the decimal points automatically.

Try This

Add.

a. 378 + 599 + 82 + 104 **b.** 17,914 + 25,637

c. 1936 + 12,789 + 4041 **d.** $54.93 + $65.77

e. 13.844 + 5.09 + 8.347 **f.** 0.5321 + 0.8169

Exercises

Add.

1.
```
  4512
  9704
 +3897
```

2.
```
  60,724
  88,653
 +47,139
```

3.
```
  2937
 +1449
```

4.
```
  $76,901
 +65,098
```

5. 257 + 78 + 129 + 45

6. 19,024 + 4370 + 6499 + 23,108

7. 12,625 + 9814

8. 90,000 + 54,317

9.
```
  89.12
  43.35
 +66.97
```

10.
```
  99.015
  67.339
 +56.48
```

11.
```
  $87.14
  +34.95
```

12.
```
  0.1083
 +0.10462
```

13. 1034.12 + 897.39 + 722.3

14. 0.214 + 1.251 + 0.9975

15. 578.081 + 93.6

16. \$328 + \$64.33 + \$11

17. Which class collected the most cans in the recycling drive?

18. Which class earned the most money?

Calculator 19. Suppose each class stacked the cans they collected. Find the height of each stack to the nearest meter.

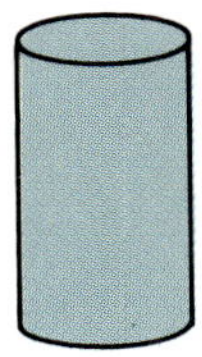

1 can = 0.124 m tall

25 cans per pound

Show Math 20. Many calculators have a display that is 9 digits long. Add 2 numbers that are larger than 500,000,000. Experiment with a few other examples. Show what happens on the display.

Multiply.

21. 4 × 7
22. 3 × 9
23. 8 × 6
24. 5 × 9
25. 7 × 8
26. 3 × 6
27. 9 × 9
28. 7 × 9
29. 5 × 5
30. 6 × 7

Mixed Skills Review

1. 42 ÷ 7
2. 5 × 8
3. 21 − 12
4. 22 − 12
5. 4 + 8
6. 8 + 11
7. 8 × 4
8. 36 ÷ 6
9. 4 × 5
10. 7 + 4
11. 23 − 4
12. 24 ÷ 3

1-4 Calculating Differences

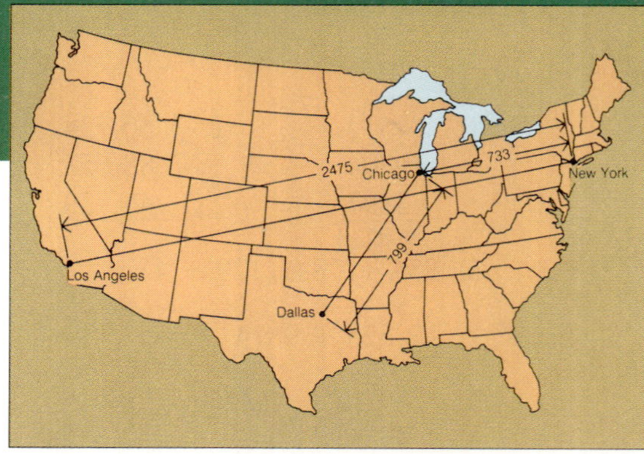

Objective
Subtract whole numbers and decimals.

Don Luepke often travels by airplane on business trips. He is a member of a frequent flyer program and receives discount rates after flying a fixed number of miles. He keeps a record of his trips.

Understand the Situation

- What route did Don take on business trips during the week of April 1?
- What is the distance in air miles between New York and Chicago?
- What is the price of a ticket between Chicago and Dallas?

Date	Flight	Air Miles	Ticket Cost
4/1	Los Angeles to New York	2475	$520.75
4/3	New York to Chicago	733	$285.48
4/4	Chicago to Dallas	799	$301.00

Examples

1. What is the air distance between Chicago and Los Angeles?

You know the distance from New York to Los Angeles and from New York to Chicago. Sometimes it helps to draw a diagram. Subtract to find the distance.

$$\begin{array}{r} 2475 \\ -\ 733 \\ \hline 1742 \end{array}$$

The distance between Chicago and Los Angeles is 1742 air miles.

The words **difference, remainder,** and **decreased by** suggest subtraction.

2. Find the difference in cost of a ticket between Chicago and Dallas and a ticket between New York and Chicago.

Chicago and Dallas	$301.00	
New York and Chicago	−285.48	The ticket for a flight between
	$15.52	Chicago and Dallas costs $15.52 more.

When subtracting decimal numbers, line up the digits that have the same place value. A calculator lines up the decimal points automatically.

Try This

Subtract.

a. 3201 − 945 **b.** 71,000 − 29,154 **c.** $12.85 − $5.99

d. 80.19 − 53.2 **e.** 7.34 − 2.997 **f.** 1.23 − 0.9418

Exercises

Subtract.

1. $\begin{array}{r} 3904 \\ -1679 \\ \hline \end{array}$ 2. $\begin{array}{r} 58{,}043 \\ -47{,}158 \\ \hline \end{array}$ 3. $\begin{array}{r} 2114 \\ -\ \ 989 \\ \hline \end{array}$ 4. $\begin{array}{r} \$72{,}400 \\ -39{,}251 \\ \hline \end{array}$

5. 907 − 35 6. 19,024 − 4730 7. $500 − $346 8. 1003 − 68

9. $\begin{array}{r} \$35.11 \\ -18.35 \\ \hline \end{array}$ 10. $\begin{array}{r} 9.01 \\ -6.339 \\ \hline \end{array}$ 11. $\begin{array}{r} 87.14 \\ -34.9\ \ \\ \hline \end{array}$ 12. $\begin{array}{r} 0.3084 \\ -0.1947 \\ \hline \end{array}$

13. 54.99 − 47.36 14. $32 − 17¢ 15. 9.43 − 5.689

16. 0.354 − 0.19 17. 39 + 45 − 12 18. 723 − 112 + 600

19. 2.15 + 5.75 − 6.31

Number Sense Find the missing digits.

20. $\begin{array}{r} 395 \\ -\square\square\square \\ \hline 68 \end{array}$ 21. $\begin{array}{r} 12\square\square \\ -\ \square 96 \\ \hline 473 \end{array}$ 22. $\begin{array}{r} 4.\square 7 \\ -\square.34 \\ \hline 1.9\square \end{array}$ 23. $\begin{array}{r} 8.\square\square \\ -7.18 \\ \hline \square.76 \end{array}$

24. Carrie made purchases of $9.85, $2.07, and $0.69. Find the amount of change she received from a twenty-dollar bill.

25. Don needs a total of 10,000 miles to earn discount rates on the frequent flyer program. How many more miles does he need?

26. The company budgeted $1500 for Don's business flights during the week of April 1. His ticket from Dallas to Los Angeles cost $425.70. Was the budgeted amount accurate?

Calculator 27. Subtract a larger number from a smaller number. Try 20 − 30. Experiment with a few other examples. Write a sentence to explain what happens.

Divide. Use basic facts.

28. 35 ÷ 7 29. 56 ÷ 8 30. 42 ÷ 6 31. 32 ÷ 4 32. 63 ÷ 9

33. 72 ÷ 6 34. 81 ÷ 9 35. 36 ÷ 4 36. 27 ÷ 3 37. 48 ÷ 8

Mixed Skills Review

1. 24 ÷ 12 2. 51 + 24 3. 54 ÷ 6

4. 28 + 101 5. 33 − 18 6. 72 + 9

7. 9 × 8 8. 63 ÷ 9 9. 41 − 39

10. 29 − 7 11. 4 × 7 12. 12 × 9

1-5 Using Mental Math

Objective
Use mental math by breaking apart. Use mental math by counting. Use mental math to estimate.

On a California vacation trip the Sanchez family traveled from Lone Pine through Bishop and Lee Vining to Yosemite Village. Mr. Sanchez found the distances on his map and used mental math to find the total mileage.

Understand the Situation

- After leaving Lone Pine where did the family go?
- How far is it from Lee Vining to Yosemite Village?

Examples

1. Use mental math to find the distance from Lone Pine to Yosemite Village.

Lone Pine → Bishop → Lee Vining → Yosemite Village

58 miles + 66 miles + 76 miles

Break apart the numbers: 50 + 8 + 60 + 6 + 70 + 6

Think: 50 + 60 = 110; 110 + 70 = 180; 8 + 6 + 6 = 20; 180 + 20 = 200

The distance from Lone Pine to Yosemite Village is 200 miles.

2. Use mental math to find $17.00 − $5.82.

Start with $5.82 and **count up** to $6.00. Then count up to $10 and then to $17. The total counted up is $11.18.

Try This

Use mental math to find the sum or difference.

a. 56 + 39 + 42 **b.** 34 + 26 + 43 + 94 **c.** 87 − 5

Example

3. Estimate the total cost of these purchases. Use mental math and front-end addition.

Program	$3.25
Hot dog and coffee	$2.39
Peanuts	$0.65
Soft pretzel and drink	$1.70

Add the front-end amounts. 3 + 2 + 1 = $6
Group the cents into dollars. 25¢ + 70¢ is about $1.
39¢ + 65¢ is about $1.

Since $6 + $2 = $8, the purchases cost about $8.

Try This

Estimate the sum. Use mental math

e. 205 + 197 + 211 + 184 **f.** $4.98 + $6.29 + $3.80

Exercises

Use mental math to find the answer. Try the break apart strategy.

1. 34 + 45 + 23 **2.** 54 + 17 + 43

3. 25 + 63 + 34 **4.** 19 + 23 + 64

5. $\begin{array}{r} 24 \\ 18 \\ +48 \\ \hline \end{array}$ **6.** $\begin{array}{r} 54 \\ 37 \\ +62 \\ \hline \end{array}$ **7.** $\begin{array}{r} 87 \\ 28 \\ +71 \\ \hline \end{array}$ **8.** $\begin{array}{r} 18 \\ 46 \\ 33 \\ +22 \\ \hline \end{array}$ **9.** $\begin{array}{r} 56 \\ 19 \\ 28 \\ +31 \\ \hline \end{array}$ **10.** $\begin{array}{r} 72 \\ 36 \\ 19 \\ +43 \\ \hline \end{array}$

Use mental math to compute. Try the counting strategy.

11. 45 + 3 **12.** 76 − 4 **13.** 54 + 3 **14.** 64 − 6 **15.** 23 − 5

16. \$6.00 − \$2.87 **17.** \$4.00 − \$3.29

18. \$8.00 − \$3.50 **19.** \$11.00 − \$5.89

20. After leaving Yosemite Village the Sanchez family drove to Merced, then south to Fresno, and on to Sequoia National Park. Use mental math to find the total distance traveled in this part of the trip.

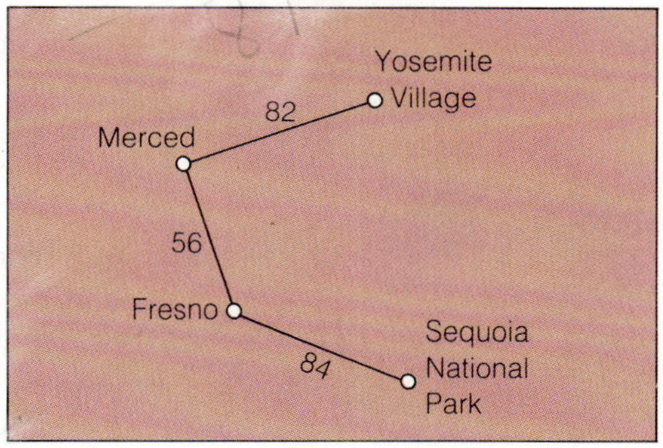

Estimate the sum. Use mental math.

21. 74 + 66 + 81 + 75 **22.** 50 + 490 + 52 + 48

23. 843 + 760 + 594 + 909 **24.** 321 + 435 + 279 + 360

25. \$1.35 + \$2.76 + \$4.58 + \$3.20 **26.** \$9.49 + \$2.08 + \$3.90 + \$6.50

27. 2036 + 1945 + 2177 + 2001 **28.** 7633 + 7612 + 7599 + 7801

Mixed Skills Review

1. \$1.18 + \$4.76 + \$7.99 **2.** 11.6 − 8.7 **3.** 4 × 11

4. 119 − 72 **5.** 4 ÷ 2 **6.** 16 ÷ 4

7. 1724 + 466 + 2 **8.** 87 + 101 **9.** 4.23 − 1.66

10. 3 × 2 **11.** 12 ÷ 3 **12.** 6 × 6

1-6 Estimating Products and Quotients

Objective
Estimate products and quotients using rounding and compatible numbers.

The Yorkshire Cinema has 4 shows on Saturday. To find the receipts for each show, the manager multiplies the number of tickets sold by the price per ticket.

Yorkshire Cinema Saturday Attendance		
	Tickets Sold	Price per Ticket
Adults	263	$4.00
Children	147	$2.25
Senior citizens	65	$2.50

Understand the Situation

- How many people bought an adult ticket?
- What is the price of a child's ticket?
- How much is the discount for senior citizens?

Before counting the cashier's receipts, the manager estimates the expected totals by rounding.

Examples

1. Estimate the total receipts from adult tickets.
263 tickets × $4 per ticket

		Round to nearest 10.	*Round to nearest 100.*
Number of adult tickets	263	260	300
Price per ticket	× $4	× $4	× $4
		$1040	$1200

The manager should expect over $1000 in receipts for adult tickets.

2. Estimate the total receipts from children's tickets.
147 tickets × $2.25 per ticket

		Round to nearest 10.	*Round to nearest 100.*
Number of children's tickets	147	150	100
Price per ticket	×$2.25	×$2.25	×$2.25
		$337.50	$250

The manager should expect about $300 in receipts for children's tickets.

3. On Saturday the Yorkshire Cinema sold a total of 475 tickets. About how many customers attended each show?
475 is about 480. 48 and 4 are compatible. 480 ÷ 4 = 120
So, 475 ÷ 4 is about 120.

About 120 customers attended each show.

Try This

Estimate.

a. 46 × 7 **b.** $3.49 × 95 **c.** 2185 × 64 **d.** 49.29 × 33.1

e. 98 ÷ 9 **f.** $41.33 ÷ 5 **g.** 371 ÷ 61 **h.** 123.4 ÷ 1.9

Exercises

Estimate.

1. 29 × 9	**2.** 64 × 7	**3.** 437 × 5	**4.** 807 × 6
5. 32 × 33	**6.** 78 × 19	**7.** 154 × 89	**8.** 726 × 57
9. 2971 × 32	**10.** 6189 × 65	**11.** 5073 × 441	**12.** 7605 × 167
13. 3.45 × 2.3	**14.** 5.28 × 7.6	**15.** 67.91 × 4.28	**16.** 43.45 × 1.87
17. \$45.90 × 5.34	**18.** \$89.27 × 8.04	**19.** 48 ÷ 7	**20.** 64 ÷ 9

21. 19 ÷ 4 **22.** 43 ÷ 6 **23.** 158 ÷ 8

24. 419 ÷ 7 **25.** \$23.45 ÷ 5 **26.** \$15.88 ÷ 4

27. \$35.21 ÷ 6 **28.** \$71.63 ÷ 8 **29.** 827 ÷ 19

30. 583 ÷ 31 **31.** 9881 ÷ 46 **32.** 4754 ÷ 55

33. \$53.99 ÷ 8.7 **34.** \$61.25 ÷ 3.4 **35.** \$208.45 ÷ 1.8

Suppose

36. Think about the number of students in your class. Suppose every student carries 3 textbooks. Estimate the total number of textbooks.

37. Think about the number of pounds you weigh. Suppose 10 people your size get into an elevator. Estimate the total weight in the elevator.

Work with a Partner

38. Estimate the total receipts from the refreshment stand. Solve the problem in two different ways. Compare.

39. Calculate the actual receipts from the refreshment stand.

Refreshment Stand Sales

	Number Sold	Price per Item
Popcorn	129	\$0.85
Peanuts	45	\$0.45
Ice cream	78	\$0.60
Drinks	151	\$0.75

Mixed Skills Review

1. 10 × 6 **2.** 4419 + 7700 **3.** 14 × 1

4. 1 ÷ 1 **5.** \$800 − \$17.50 **6.** 72,401 − 122

7. 16.748 − 8.9 **8.** 9 × 5 **9.** 148 + 1.6

10. 72 ÷ 9 **11.** \$1.16 + \$2.88 + \$.04 **12.** 88 ÷ 11

1-7 Calculating Products

Objective
Multiply whole numbers and decimals.

The dispatcher keeps statistics for the Metro Bus Line. The numbers in the chart show the daily averages during the month of June.

Metro Bus Line	
Number of buses	65
Passengers per bus	324
Miles per route	9.7
Miles per day	79.3
Bus fare	$0.75
Gallons per mile	0.45

Understand the Situation

- What is the average number of passengers per day carried by one bus?
- What is the average length of a bus route?
- How much does it cost to ride the bus?

To find the average number of passengers served by the Metro Bus Line in one day, multiply 324 × 65.

Examples

1. Multiply. 324 × 65 Multiply the larger factor by the smaller factor.

$$\begin{array}{r} 324 \\ \times\ 65 \\ \hline 1620 \\ 1944\ \ \\ \hline 21060 \end{array}$$

324 and 65 are called **factors.** The **product** of the factors is 21,060.

The Metro Bus Line serves an average of 21,060 passengers per day.

To find the gallons of gasoline used on one bus route, multiply 0.45 gallon × 9.7 miles.

2. Multiply. 0.45 × 9.7 First multiply, then place the decimal point.

$$\begin{array}{r} 0.45 \\ \times\ 9.7 \\ \hline 315 \\ 405\ \ \\ \hline 4365 \end{array}$$

Count the total number of digits to the right of the decimal point in both factors.

$$\begin{array}{r} 0.45 \\ \times\ 9.7 \\ \hline 315 \\ 405\ \ \\ \hline 4.365 \end{array}$$

Count the same number of digits to the left in the product to place the decimal point.

A bus uses 4.365 gallons of gasoline on one route.

Try This

Multiply.

a. 95 × 37 **b.** 1968 × 29 **c.** $1.75 × 38

d. 3067 × $1.25 **e.** 7.5 × 4.9 **f.** 9.8 × 0.15

Exercises

Multiply.

1. $\begin{array}{r} 529 \\ \times\ 7 \\ \hline \end{array}$

2. $\begin{array}{r} 815 \\ \times\ 64 \\ \hline \end{array}$

3. $\begin{array}{r} 4608 \\ \times\ 25 \\ \hline \end{array}$

4. $\begin{array}{r} 7719 \\ \times\ 90 \\ \hline \end{array}$

5. $\begin{array}{r} 683 \\ \times \$45 \\ \hline \end{array}$

6. $\begin{array}{r} \$12.45 \\ \times\ 17 \\ \hline \end{array}$

7. 8924 × 37

8. 1009 × 45

9. 1176 × 21

10. $29.89 × 50

11. $\begin{array}{r} 41.2 \\ \times\ 7 \\ \hline \end{array}$

12. $\begin{array}{r} 6.92 \\ \times\ 23 \\ \hline \end{array}$

13. $\begin{array}{r} 738 \\ \times 4.5 \\ \hline \end{array}$

14. $\begin{array}{r} 24.5 \\ \times\ 3.6 \\ \hline \end{array}$

15. $\begin{array}{r} 7.84 \\ \times 0.58 \\ \hline \end{array}$

16. $\begin{array}{r} \$42 \\ \times 3.6 \\ \hline \end{array}$

17. 3.61 × 2.7

18. 72.9 × 0.46

19. 3.45 × 0.01

20. 7.49 × 0.055

21. Find the average number of miles each bus is driven in 30 days.

22. The cost for gasoline, maintenance, and repairs for a Metro Line bus is $1.54 per mile. Find the total cost per day for the average bus.

Calculator

23. Use a calculator to find each product. Record your results. Predict the product of 11111 × 11111. Write a sentence to describe what happens.

1 × 1
11 × 11
111 × 111
1111 × 1111

Mental Math

24. Carrie rides the Metro Bus Line to and from work 20 days per month. How much should she budget each month for bus transportation?

25. Gail rides the Metro bus to and from the archery range every Saturday. What is the annual cost?

26. The cost for a taxi is $1.75 for the first mile and $1.25 for each additional mile. How much money can you save on a two-mile trip by riding the Metro bus?

27. John rides the bus to and from school five days each week. What is the cost per week?

28. The cost to ride the Metro Subway is 85¢. How much can you save per week by riding the bus to and from work?

Mixed Skills Review

1. 2 × 16
2. 111 + 42.73
3. 60 ÷ 5
4. $9.66 + $11.43
5. 8 × 7
6. 812 + 17 + 48
7. $180.73 − $66.04
8. 1994 − 128
9. 27 ÷ 3
10. 100 ÷ 10
11. 450 − 12.4
12. 9 × 4

1-8 Calculating Quotients

Objective
Divide whole numbers and decimals.

John Hajewski is studying to be a licensed practical nurse. Tuition and fees cost \$1980 for the year. John makes 12 equal monthly payments. How much is each payment?

To find the amount of each monthly payment, divide \$1980 by 12.

Examples

1. Divide. \$1980 ÷ 12

Divide.
Multiply.
Subtract.
Compare.

```
            $ 165  ← quotient
divisor → 12)$ 1980 ← dividend
              12
               78
               72
                60
                60
```

John pays \$165 each month.

2. Divide. 3.24 ÷ 4.8

```
4.8)3.24          48)32.4          0.675
                               48)32.400
                                  288
                                   360
                                   336
                                    240
                                    240
```

Move the decimal point one place to the right in each number until the divisor is a whole number.

Place the decimal point directly above for the quotient. Divide.

3. John paid \$94 for four textbooks in the college bookstore. What is the average cost per book?

\$94 ÷ 4

```
  $ 23.50
4)$ 94.00
    8
    14
    12
     20
     20
```

The average cost of a textbook is \$23.50. A calculator automatically adds a decimal point and zeros to the dividend.

Try This

Calculate.

a. 4248 ÷ 6 **b.** 2491 ÷ 53 **c.** \$18.60 ÷ 24

d. 93.24 ÷ 63 **e.** 1.68 ÷ 0.35 **f.** 9 ÷ 0.18

Exercises

Divide.

1. 846 ÷ 3 2. $6350 ÷ 5 3. 5817 ÷ 7 4. 4324 ÷ 16

5. 255 ÷ 35 6. $89,786 ÷ 7 7. $65.74 ÷ 19 8. $45.60 ÷ 10

9. $9\overline{)241.2}$ 10. $6\overline{)886.8}$ 11. $72\overline{)14.76}$ 12. $44\overline{)55.66}$

13. $0.3\overline{)1.2}$ 14. $0.7\overline{)4.2}$ 15. $1.4\overline{)0.49}$ 16. $0.18\overline{)0.549}$

17. $0.01\overline{)1.8}$ 18. $0.03\overline{)2.4}$ 19. $2.5\overline{)0.112}$ 20. $0.25\overline{)0.625}$

21. $0.35\overline{)\$7}$ 22. $0.5\overline{)\$6}$ 23. $0.4\overline{)8}$ 24. $0.3\overline{)9}$

25. Each student nurse wears a uniform to class. The cost of 3 uniforms is $83.94. Find the cost of 1 uniform.

26. During the first semester John earned 52.5 quality points for 15 course credits. How many quality points per credit did he earn?

Mental Math

Is the answer more than 1 or less than 1?

27. 8 ÷ 5 28. 5 ÷ 8 29. 0.3 ÷ 6 30. 6 ÷ 0.3

Calculator

31. Use a calculator to find each answer. Describe what happens.
    a. 4 ÷ 9 b. 5 ÷ 6 c. 2 ÷ 3
    d. 5 ÷ 12 e. 7 ÷ 33 f. 11 ÷ 9

32. John uses a calculator to work problems in his chemistry class. Round each answer to the nearest tenth.
    a. 0.7 ÷ 8 b. 2.6 ÷ 3.9
    c. 1.23 ÷ 4.95 d. 9 ÷ 0.8

Find each answer. Use a calculator or paper and pencil.

33. Subtract $4.95 from the quotient of $378 ÷ 60.

34. Add $8.75 to the quotient of $16.45 ÷ 5.

35. Divide 79.2 by the difference between 7.8 and 6.9.

Mixed Skills Review

1. 16.7 × 84 2. 160 ÷ 5 3. 77 ÷ 11

4. $1874 + $690 + $11 5. $12.55 × 18 6. 101 + 1001 + 16

7. 864 − 42.78 8. 68.6 − 15 9. 16.88 + 2.444

10. $111 − $11.10 11. 48 ÷ 6 12. 77 × 104

1-9 Using Mental Math: Multiplication and Division

Objective
Use mental math by breaking apart.

The popularity of spectator sports in the United States is increasing. Here are the attendance records for five seasons of the National Basketball Association.

NBA Attendance		
Season	Total	Average per Game
1981–82	9,964,919	10,567
1982–83	9,637,614	10,220
1983–84	10,014,543	10,620
1984–85	10,506,355	11,141
1985–86	11,214,888	11,893

Understand the Situation

- Did attendance increase every year?
- When was total attendance about 700,000 more than the previous season?
- When was average attendance about 500 more than the previous season?

Example

1. The average price of a ticket during the 1982–83 season was about \$8. Use mental math to find the receipts for one game.

Estimate: 10,220 is about 10,000. $10{,}000 \times \$8 = \$80{,}000$

Think: Break apart the larger number. $10{,}220 = 10{,}000 + 200 + 20$

$$10{,}220 \times \$8 = (10{,}000 \times \$8) + (200 \times \$8) + (20 \times \$8)$$
$$= \$80{,}000 - \$1600 + \$160 = \$81{,}760$$

The total receipts for one game were about \$81,760.

Try This

Use mental math to find the product.

a. 36×5 **b.** 4021×9 **c.** $\$1.40 \times 3$ **d.** $\$5460 \times 6$

Example

2. About 51,300,000 spectators attended NBA games during the five seasons listed above. What was the average attendance each season?

Estimate: 51,300,000 is about 50,000,000. $50{,}000{,}000 \div 5 = 10{,}000{,}000$

Think: Break apart the larger number.

$$51{,}300{,}000 = 50{,}000{,}000 + 1{,}000{,}000 + 300{,}000$$
$$51{,}300{,}000 \div 5 = (50{,}000{,}000 \div 5) + (1{,}000{,}000 \div 5) + (300{,}000 \div 5)$$
$$= 10{,}000{,}000 + 200{,}000 + 60{,}000 = 10{,}260{,}000$$

The average attendance each season was 10,260,000.

Try This

Use mental math to find the quotient.

e. $26{,}120 \div 4$ **f.** $21{,}012 \div 3$ **g.** $36{,}108 \div 12$ **h.** $71{,}428 \div 7$

Exercises

Mental Math

Find the product. Use mental math.

1. 79 × 4	**2.** 51 × 7
3. 204 × 9	**4.** 416 × 3
5. \$5.20 × 6	**6.** \$7.09 × 8
7. 3004 × 12	**8.** 5016 × 20
9. \$60.25 × 40	**10.** \$103.04 × 16

Sports on Television	Number of Viewers
Professional football	63,200,000
Baseball	62,700,000
College football	48,900,000
Boxing	37,200,000
College basketball	36,200,000
Professional basketball	34,700,000
Professional wrestling	28,800,000
Bowling	28,600,000
Tennis	26,300,000
Auto racing	25,800,000

Use the chart for Exercises 11 and 12.

11. Is the total number of viewers in the study more or less than 500,000,000?

Suppose

12. Suppose you had to pay \$25 for a season ticket to watch each sport on TV. Which sports would make more than \$1,000,000,000?

Mental Math

Find the quotient. Use mental math.

13. 1460 ÷ 4	**14.** 9021 ÷ 3
15. 3426 ÷ 6	**16.** 1325 ÷ 5
17. 28,714 ÷ 7	**18.** 5427 ÷ 9
19. 16,824 ÷ 8	**20.** 20,144 ÷ 4
21. 48,600 ÷ 12	**22.** 22,515 ÷ 15

To multiply by 5 you can multiply by 10 and then take half.

Example: 5 × 32 → 10 × 32 = 320 Half of 320 is 160.
So, 5 × 32 = 160.

Multiply.

23. 5 × 26 **24.** 5 × 16 **25.** 5 × 42 **26.** 5 × 86

Show Math

27. Give a mental math tip for multiplying by 50.

28. Give a mental math tip for multiplying by 9.

Mixed Skills Review

1. 168 × 4.6	**2.** 114 + 12.3 + 1.41	**3.** \$460 ÷ \$72.9
4. \$44,688 + \$69,111	**5.** \$16 − \$4.23	**6.** 727 + 146 + 16.21
7. 1808 × \$11.75	**8.** 28 ÷ 12.5	**9.** 4292 − 622
10. 17.9 × 4.11	**11.** 168.4 − 12	**12.** 199 ÷ 2.5

1-10 Enrichment: More Mental Math Strategies

Multiplicative Property of Zero
- For every real number a, $a \times 0 = 0$.

Commutative Property of Multiplication
- For any real numbers a and b, $a \times b = b \times a$.

Distributive Property of Multiplication Over Addition
- For any real numbers a, b, and c, $(a \times b) + (a \times c) = a \times (b + c)$.

Associative Property of Multiplication
- For any real numbers a, b, and c, $a \times (b \times c) = (a \times b) \times c$.

Objective
Use mental math and the properties of multiplication to solve problems.

The properties of multiplication can sometimes help you solve problems faster with mental math than with a calculator.

The **Multiplicative Property of Zero** says any number times 0 equals 0.
$68 \times 374 \times 0 \times 943 = 0$

The **Commutative Property of Multiplication** shows that you can multiply numbers in any order without changing the answer.
$258 \times 643 = 643 \times 258$

The **Distributive Property of Multiplication Over Addition** shows that you can first add multipliers of the same number.
$(13 \times 4) + (13 \times 6) = 13 \times (4 + 6) = 13 \times 10 = 130$

The **Associative Property of Multiplication** shows that you can group numbers any way you want to for multiplication.
$20 \times 5 \times 6 = 100 \times 6 = 600$ or $20 \times 5 \times 6 = 20 \times 30 = 600$

Exercises

Tell what property you can use to find the missing terms.

1. $(84 \times 6) + (84 \times 4) = \square$
2. $128 \times 536 \times \square = 0$
3. $72 \times 41 \times 18 = \square \times 41 \times 72$
4. $847 \times 916 \times 709 = 847 \times \square$

Solve. Use mental math. Check your answer with a calculator.

5. $743 \times 0 \times 618 \times 587 \times 22 = \square$
6. $3185 \times 5427 \times 685 = 685 \times \square \times 5427$
7. $37 \times (20 + 80) = \square$
8. $(45 \times 3) + (45 \times 7) = \square$
9. $5 \times 43 \times 2 = \square$
10. $25 \times 73 \times \square = 0$
11. $\square \times 46 \times 91 = 91 \times 46 \times 78$
12. $(98 \times 7) + (\square \times 3) = 980$
13. $60 \times \square \times 5 = 900$
14. $312 \times 684 \times \square = 9 \times 684 \times 312$

1-11 Calculator: Choosing a Calculation Method

Objective
Find products using mental math and a calculator.

You can calculate the answers to some multiplication problems faster with mental math than with a calculator, especially if one of the factors is a power of ten.

Multiply 38×100.

Using a calculator you may need seven keystrokes.

[3] [8] [×] [1] [0] [0] [=]

Using mental math, move the decimal as many places to the right as there are zeros as placeholders in the multiple of ten.

$38 \times 100 = 38.0 \times 100 = 3800.0$ or 3800

Multiply 295×0.001.

Using a calculator you may need nine keystrokes.

[2] [9] [5] [×] [.] [0] [0] [1] [=]

Using mental math, move the decimal point to the left as many times as there are places to the right of the decimal.

$295 \times 0.001 = 295.0 \times 0.001 = 0.295$

Exercises

Solve. Use a calculator.

1. 52×100	**2.** 52×10	**3.** 52×1	**4.** 52×0.1	**5.** 52×0.01
6. 8.7×1000	**7.** 71×0.001	**8.** 2.33×10	**9.** 9.2×0.001	**10.** 20×0.01

Solve. Use mental math.

11. 73×100	**12.** 73×10	**13.** 73×1	**14.** 73×0.1	**15.** 73×0.01
16. 3.8×0.1	**17.** 6×0.001	**18.** 1.5×1000	**19.** 0.4×0.1	**20.** 13.2×10

Time yourself. Solve using a calculator.

21. 24×100	**22.** 5×0.1	**23.** 92×0.01	**24.** 150×10	**25.** 41×1000
26. 2.78×1	**27.** 5.2×0.1	**28.** 94.5×100	**29.** 6.7×0.01	**30.** 7.69×0.001

Time yourself. Solve using mental math.

31. 38×100	**32.** 7×0.1	**33.** 45×0.01	**34.** 320×10	**35.** 69×1000
36. 82.5×1	**37.** 6.8×0.1	**38.** 3.6×100	**39.** 2.9×0.01	**40.** 9.14×0.001
41. 39.2×10	**42.** 93×1000	**43.** 62×0.01	**44.** 80×0.001	**45.** 0.32×0.001

CHAPTER 1 REVIEW

Complete each sentence. Use this list of words.

quotient	remainder
product	factor
sum	multiplication
mental math	approximate
estimation	division
decreased by	difference

1. Sum is to addition as _______ is to subtraction.
2. Calculator is to exact answer as estimate is to _______ answer.
3. Multiplication is to product as division is to _______.
4. Addend is to addition as _______ is to multiplication.
5. Rounding is to estimation as breaking apart is to _______.
6. Addition is to "increased by" as subtraction is to _______.

Find the word that doesn't belong.

7. mental math
   compatible numbers
   break apart
   calculator
8. total
   remainder
   all together
   sum
9. difference
   factor
   product
   multiply
10. estimation
    quotient
    mental math
    paper and pencil
11. decreased by
    mental math
    difference
    subtract
12. estimation
    front end
    rounding
    mental math
13. division
    quotient
    divisor
    difference
14. approximate answer
    exact answer
    rounded number
    estimate
15. remainder
    sum
    quotient
    mental math
16. divide
    product
    multiply
    subtract
17. decide
    understand
    solve
    quotient
18. remainder
    quotient
    sum
    difference

CHAPTER 1 TEST

Choose estimation, mental math, paper and pencil, or a calculator.

1. You are using a measuring cup to make pancake batter.
2. You are a clerk in a food store. How much should you charge the customer for 7.3 oz of mushrooms?
3. Alex's clothing store is having a sale. All items are selling for 50% off the regular price. How much will a $20 pair of slacks cost?

Round to the indicated place.

ten	**4.** 4687	**5.** 734	dollar	**6.** $67.63	**7.** $1.65
thousand	**8.** 749	**9.** 8936	tenth	**10.** $82.39	**11.** 7.392

Estimate. Use rounding, clustering, or compatible numbers.

12. 56 + 74 + 49 **13.** 5963 − 3905 **14.** 15.49 − 12.23
15. 59 + 62 + 60 + 61 **16.** 302 + 299 + 305 **17.** 77 + 31 + 82 + 28
18. $124.56 + $32.09 + $74.33 **19.** 127 + 848 + 74 + 251 + 52

Add.

20. 483 + 194 + 659 **21.** 4.582 + 12.45 + 0.632

Subtract.

22. 6942 − 759 **23.** 0.6195 − 0.3936

Use mental math to compute. Try the break apart strategy or the count up/back strategy.

24. 32 + 46 + 78 **25.** 23 + 47 + 68
26. 82 − 7 **27.** 75 + 9 **28.** $8.00 − $7.73 **29.** $20.00 − $14.49

Estimate.

30. 49 × 7 **31.** 4.23 × 2.9 **32.** 311 ÷ 29 **33.** 52.13 ÷ 5.2

Multiply.

34. 634 × 8 **35.** $54.65 × 47 **36.** 84.6 × 43.9

Divide.

37. 693 ÷ 7 **38.** $4272 ÷ 89 **39.** 607.5 ÷ 0.75

Find the product or quotient. Use mental math.

40. $4.15 × 4 **41.** 3010 × 5 **42.** $2.20 × 9
43. 4450 ÷ 25 **44.** 7218 ÷ 9 **45.** 763 ÷ 7

CHAPTER 1 CUMULATIVE REVIEW

Decide if you need an exact answer or an estimate.

1. You are a bank teller. What is the customer's account balance?
   A. exact
   B. estimate

2. You are making punch for a party. How much will you make?
   A. exact
   B. estimate

3. You are mixing some expensive chemicals. How much of each should be added?
   A. exact
   B. estimate

Estimate. Use rounding.

4. $\begin{array}{r} 760 \\ 302 \\ +465 \\ \hline \end{array}$
   A. 1000
   B. 1400
   C. 1600
   D. 2000

5. $\begin{array}{r} 51{,}598 \\ -\ \ 620 \\ \hline \end{array}$
   A. 5000
   B. 600
   C. 60,000
   D. 51,000

6. $44 \div 5$
   A. 9
   B. 5
   C. 4
   D. 90

7. Multiply.
   $\begin{array}{r} 736 \\ \times\ \ 8 \\ \hline \end{array}$
   A. 5648
   B. 5888
   C. 5924
   D. 5788

8. Divide.
   $8\overline{)360.8}$
   A. 45.1
   B. 451
   C. 4.51
   D. 40.1

9. Add.
   $\begin{array}{r} 87.026 \\ 56.228 \\ +43.41\ \ \\ \hline \end{array}$
   A. 176.654
   B. 186.664
   C. 186.285
   D. 176.664

10. Subtract.
    $\begin{array}{r} 67{,}104 \\ -36{,}245 \\ \hline \end{array}$
    A. 31,109
    B. 31,141
    C. 30,869
    D. 30,859

11. Add.
    $\begin{array}{r} \$65.15 \\ +\$42.09 \\ \hline \end{array}$
    A. $17.24
    B. $107.14
    C. $100.04
    D. $107.24

12. Multiply.
    3007×52
    A. 15,964
    B. 156,364
    C. 21,364
    D. 150,964

Use mental math.

13. $47 + 32 + 21$
    A. 79
    B. 83
    C. 150
    D. 100

14. 305×5
    A. 1525
    B. 155
    C. 15,025
    D. 25,015

15. $5.00 − $3.65
    A. $2.65
    B. $2.35
    C. $1.35
    D. $1.65

16. Divide.
7928 ÷ 8
A. 901
B. 890
C. 991
D. 990

17. Use mental math.
8124 ÷ 4
A. 231
B. 2031
C. 206
D. 2301

18. Estimate.
$$\begin{array}{r} 3783 \\ \times \quad 29 \\ \hline \end{array}$$
A. 9000
B. 1200
C. 60,000
D. 120,000

19. Subtract.
7.21 − 4.578
A. 2.632
B. 3.368
C. 2.638
D. 2.732

20. Multiply.
8.57 × 0.033
A. 28.281
B. 0.28281
C. 0.028281
D. 2.8281

21. Divide.
$0.27\overline{)0.837}$
A. 4.1
B. 3.1
C. 0.31
D. 0.041

22. Subtract.
$$\begin{array}{r} 0.4105 \\ -0.2858 \\ \hline \end{array}$$
A. 0.2753
B. 0.0357
C. 0.1247
D. 0.2957

23. Estimate.
Use clustering.
$$\begin{array}{r} \$19.02 \\ 21.19 \\ +\ 18.07 \\ \hline \end{array}$$
A. 40.00
B. 50.00
C. 60.00
D. 70.00

24. Add.
$$\begin{array}{r} 97.396 \\ 34.978 \\ 10.652 \\ \hline \end{array}$$
A. 158.926
B. 158.815
C. 143.026
D. 148.016

25. Use mental math.
58 + 41 + 18
A. 100
B. 117
C. 137
D. 96

26. Estimate.
$$\begin{array}{r} 4963 \\ \times \quad 19 \\ \hline \end{array}$$
A. 100,000
B. 10,000
C. 80,000
D. 40,000

27. Use mental math.
14994 ÷ 7
A. 2128
B. 2440
C. 214
D. 2142

Chapter 2 Overview

Key Ideas

- using problem-solving guidelines
- using graphs and tables
- finding data and advice
- solving multi-step problems
- using formulas
- deciding on the reasonableness of an answer
- choosing problem-solving strategies
- applying problem-solving strategies

Key Terms

- situation
- data
- plan
- answer
- check
- table
- graph
- advise
- profit
- formula
- strategy
- problem solving
- reasonable

Key Skills

Find the answer.

1. 15,074 − 10,675

2. 812 × 34

3. 1736 ÷ 28

4. 26,075 + 37,128

5. 703 ÷ 19

6. 99.6 + 8.43

7. 8.6 × 4.9

8. 1186 − 1098

9. 7.11 ÷ 0.9

10. \$12.35 − 56¢

11. \$6.29 + \$108.74

12. 74 × 83

13. 5.98 ÷ 1.3

14. \$15.34 × 17

15. 12 − 4.15

16. 603 − 378

17. 53 + 26 + 76

18. 59 × 312

19. 6 × 0.48

20. 335.8 ÷ 23

21. \$200 − \$137.14

22. 1316 ÷ 47

23. 0.68 + 7.5 + 8

24. 145 + 16,386

Solve.

25. 18 + □ = 34

26. □ − 23 = 16

27. 98 ÷ □ = 7

28. □ + 82 = 123

29. 9 × □ = 126

30. □ ÷ 14 = 14

31. \$18.63 − □ = \$7.26

32. □ × 2 = \$17

SOLVING CONSUMER PROBLEMS 2

2-1 Problem-Solving Guidelines

Objective
Use problem-solving guidelines.

According to the National Yellow Pages Service Association, about 6000 different Yellow Pages directories are published each year. The average adult uses the directory three times a week. The rates for a standard entry in the Yellow Pages are often based on the number of households receiving the directory.

Yellow Pages Advertising Rates

Location	Number of Households	Rates per Household
Hancock	24,600	32¢
Millers	9,340	24.5¢
Salem	36,400	37¢
Seaford	15,300	29¢

Example Find the cost of an advertisement in the Seaford Yellow Pages Directory.

A set of guidelines can help us to solve the problem.

Understand the Situation

How much does an ad in the Seaford directory cost?

What question is asked?
Should you read the problem again?
What information do you need to know?

Find the needed Data

There are 15,300 households.
The cost is \$0.29 per household.

What information is needed?
Is all the data present?

Plan what to do

Multiply. 15,300 × \$0.29

Is an exact answer needed?
What operation should you choose?
Should you use a calculator?

Find the Answer

Estimate: 15,000 × \$0.30 = \$4500
Exact: 15,300 × \$0.29 = \$4437
The cost of an ad is \$4437.

Is the data accurate?
Is the computation correct?

Check the Answer

The answer is close to the estimate.

Is the answer reasonable?
Is the answer stated clearly?

Try This

a. Find the cost of an advertisement in the Salem Yellow Pages.

b. A local company delivers the Millers directory for a cost of \$0.15 per household. Find the delivery cost.

Exercises

Solve.

1. Find the cost of an advertisement in the Hancock Yellow Pages.
2. In Lansing the cost to advertise in the Yellow Pages is $300 plus $0.10 per household. There are 8360 households. What is the total cost for a Yellow Pages ad?
3. Conowingo Telephone distributes 20,000 Yellow Pages directories. The cost to print the directory is $4.95 each. Find the total printing cost.
4. In Seaford the Junior Achievement Club delivered directories for a cost of $0.12 per household. How much did the club gross on the project?
5. The Blue Dolphin Seafood House hired an artist to prepare a new layout for a Yellow Pages ad. The artist charged $132 and worked 8 hours on the design. How much did the artist charge per hour?
6. An airport limousine service lists its rates in the Yellow Pages. What is the cost of a ride to the Holiday Inn?
7. How much will the limousine service earn by taking three passengers to Harbor Towers and two passengers to the Town Court?

Chariot Limousines

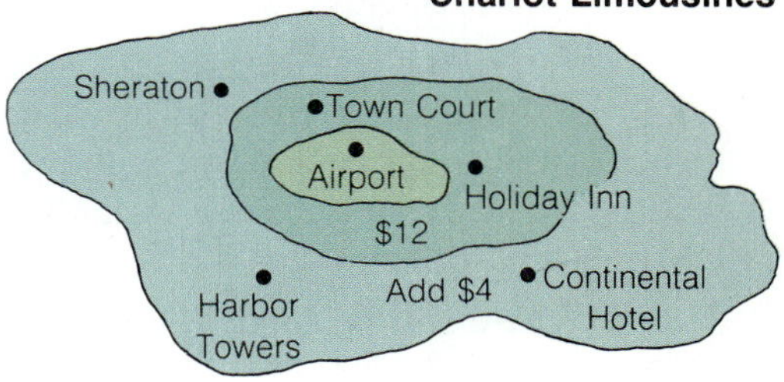

Use the Annual References chart to answer Exercises 8–11.

8. In one year how many times do people look in the directory for information about doctors?
9. What do most people look for in the Yellow Pages?
10. How many times a year do people look for information about cars?
11. How many more people look for information about restaurants than for information about doctors?
12. How many times a year does the average adult use the directory?

Annual References	Millions
Restaurants	929
Doctors	882
Auto parts	845
Auto repairs	585
Auto dealers	407
Appliances/electronics	402

Mixed Skills Review

1. 144 × 218
2. 64 ÷ 10
3. $4.18 − $3.03
4. $27.48 ÷ 10
5. 118 + 624 + 436
6. 18.6 × 44.4
7. 125 ÷ 5
8. 886 − 116
9. 1001 + 225 + 6.7
10. $642 × 12.2
11. 1.4 + 22.48 + 6
12. $4258 − $2223

2-2 Using Tables and Graphs

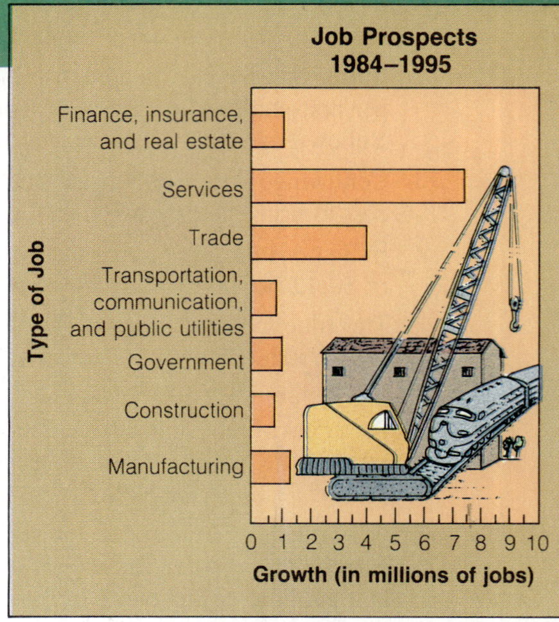

Objective
Read data from tables and graphs.

Donna Novack is an employment counselor. Donna uses this graph to show projected growth in job prospects.

Understand the Situation

- What is the title of the graph?
- What types of jobs are shown?
- What does each bar show?

Example

1. In 1984 there were about 6.3 million jobs in finance, insurance, and real estate. About how many jobs will there be in 1995?

To find the increase go to the end of the bar and look down to the scale. The increase is about 1.1 million Add. 6.3 + 1.1 = 7.4

There will be about 7.4 million jobs in finance, insurance, and real estate in 1995.

Try This

a. What is the increase in manufacturing?

b. Which type of job will have the third highest increase?

Theresa uses Gro-Fast fertilizer on her African violets. Each packet has a table that tells how much fertilizer to use.

Gro-Fast Fertilizer

Container Diameter	Amount per Feeding	Feedings per Packet
10″–12″	2 level tsp	12
8″	1 level tsp	24
6″	$\frac{1}{2}$ level tsp	48
4″	$\frac{1}{4}$ level tsp	96

Example

2. How much fertilizer should you use in an 8-inch pot?

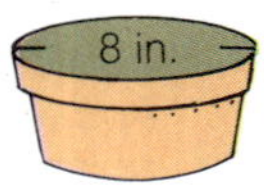

Look at the table in the row for 8-inch containers. You will need 1 level teaspoon of fertilizer.

Try This

c. How much fertilizer should you use to fertilize a plant in a 10-inch pot?

d. Suppose you have only 6-inch pots for your houseplants. How many feedings can you expect from a packet of fertilizer?

Exercises

Solve. Use the job prospects graph.

1. What is the increase in job prospects for construction?
2. About how many new jobs will be available in services?
3. In 1995 the number of jobs in services is projected to be 31.2 million. About how many people held service jobs in 1984?

This graph shows the increase in job prospects for scientific and technical types of jobs.

4. What was the increase in the number of jobs for computer programmers?
5. In 1984 there were about 237,000 jobs for mechanical engineers. About how many jobs will there be in 1995?
6. In 1995 the number of jobs for electrical and electronics technicians is projected to be 607,000. About how many jobs were there in 1984?

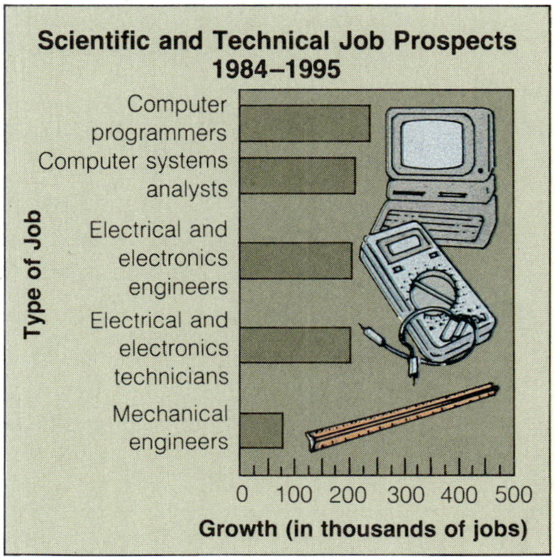

The table below shows how to mix a powdered fruit drink.

No. of Glasses (with Ice)	Amount of Drink Mix	Amount of Water
1	1 tbsp	$\frac{2}{3}$ cup
3	$\frac{1}{4}$ cup	1 pint
6	$\frac{1}{2}$ cup	1 quart
12	1 cup	2 quarts
24	2 cups	1 gallon

7. How much powdered drink mix should you use to make 12 glasses of fruit drink?
8. How much water should you use to make 6 glasses of fruit drink?
9. When you add $\frac{1}{4}$ cup of drink mix to 1 pint of water how many glasses of drink do you get?

Mixed Skills Review

1. $45 + 56 + 14$
2. $910 - 87$
3. $\$14.87 + \$0.91$
4. 1.5×36.98
5. $19.10 + 7.73 + 1.37$
6. $48 \div 12$
7. 370×979
8. 48.25×16
9. $196 \div 14$
10. $45.98 + 23 + 7.65$
11. $14.911 + 65.9$
12. $909 \div 6$

2-3 Finding Data and Advice

Objective
Find and use sources of information.

- QUESTION
- **DATA**
- PLAN
- ANSWER
- CHECK

Finding the needed data is an important step in solving a consumer problem. In the back of this book is a Data Bank which can be used to solve certain problems. The Data Bank contains charts, tables, graphs, formulas, and lists of data.

Example

1. What is the difference in daily calorie needs between an 18-year-old boy and an 18-year-old girl?

The table on p. 437 in the Data Bank lists Recommended Daily Dietary Allowances. The daily required calorie intake for an 18-year-old boy is 2800, while the required intake for an 18-year-old girl is 2100.

The question asks for a difference. It makes sense to subtract. Subtract 2100 from 2800. This can be done using mental math.
2800 − 2100 = 700

An 18-year-old boy needs 700 more calories than an 18-year-old girl.

Try This

Use the Data Bank on p. 437 to help solve this problem.

a. How many fewer calories does a 22-year-old woman need daily than a 10-year-old girl?

To solve some consumer problems, you will need to find data from sources other than this book. These are called **Data Hunt** problems.

Examples

2. What is the cost of a new washing machine? Data to solve this problem can be found in a newspaper ad or appliance store catalog. You would need to know the size and type of washing machine.

3. Whom would you consult for advice when seeking an auto loan? A credit counselor at a bank or credit union could give advice on how to get an auto loan.

Try This

Where would you look to find data needed to solve these problems?

b. How much more electricity is used by a 100-watt bulb than by a 75-watt bulb?

c. What kind of trees should you buy when landscaping your yard?

Exercises

Use the Data Bank on p. 437 to help solve these problems.

1. How many more daily calories does a 33-year-old man need than a 5-year-old child?
2. How many fewer daily calories does a 36-year-old woman need than a 12-year-old girl?

Data Hunt Where would you look to find data needed to solve these problems?

3. What part-time job opportunities are available?
4. How much will a new food processor cost?
5. What is the average starting salary for a beginning carpenter?
6. How much will a used car cost?

Whom would you consult for advice in each of these situations?

7. What type of insulation should you install to decrease heat loss in your home?
8. What scholarship opportunities are available for apprenticeship programs?
9. What factors should you consider when buying homeowners insurance?

Work with a Partner

10. Make a list of questions that you would ask if you consulted an expert about scholarship opportunities for apprenticeship programs.

Estimate

11. Estimate the total number of calories needed in a week by a 19-year-old woman.

Number Sense

12. Find the total yearly calorie requirement for a 25-year-old woman and a 25-year-old man.

Mixed Skills Review

1. 89.56 + 67.3 + 43.123	2. 250 × 32	3. 360.75 ÷ 55.5
4. 760 − 579	5. $34.98 + $78.00	6. 98 × 12.22
7. 33.33 ÷ 11	8. 909 − 199	9. 1248 ÷ 16
10. 56 + 87 + 27	11. 12.22 ÷ 2	12. 765 × 657

2-4 Solving Multi-Step Problems

Objective
Solve multi-step problems.

While shopping at the Music House, Nina Rankin bought 5 videotapes and 2 compact discs. What was the total cost for all of her purchases?

Understand the Situation

- How many videotapes did Nina buy?
- How many compact discs did she buy?
- What is the cost of each item that she bought?

Sometimes complete solutions to problems require more than one step. A plan helps you define the steps and check your progress.

Example Find the total cost for Nina Rankin's purchases.

The solution requires three steps.

A. Find the cost of the 5 videotapes.
Estimate: $5 \times 4 = \$20$
or
Calculate: $5 \times \$3.98 = \$19.90$

B. Find the cost of the 2 compact discs.
Estimate: $2 \times \$13 = \$26$
or
Calculate: $2 \times \$12.99 = \$25.98$

C. Add to find the total.
Estimate: $\$20 + \$26 = \$46$
or
Calculate: $\$19.90 + \$25.98 = \$45.88$

The total cost for Nina's purchases was $45.88, or an estimated $46.

Try This Solve.

a. At the basketball game, Janine sold 7 cartons of peanuts. Each carton contained 48 bags of peanuts. A bag of peanuts costs $0.75. How much did Janine receive for the peanuts?

b. The Sports Boosters Club bought eight cartons of popcorn at $20.40 per carton. Each carton contained 24 boxes of popcorn. The Club sold the popcorn for $0.95 a box. What was their **net profit** after selling all the popcorn?

Exercises

Solve.

1. Larry runs 9 kilometers on Sundays, Tuesdays, and Thursdays. On Monday, Wednesdays, and Saturdays he runs 15 kilometers. How many kilometers does he run each week?
2. Toni sold 22 push-up packs of tomato plants at the school fair. Each pack contained twelve plants. The cost of each plant was $0.12. How much did Toni receive for the tomato plants?
3. Jessica bought a calculator and a sweater. She paid with $25 cash. What was her change?
4. Nick bought a cordless drill on sale. He also bought a set of drill bits for $6.98. He paid with a check for $50. How much cash did he receive in change?

Suppose 5. You have $50. Do you have enough money to purchase one of each item? If not, list the combination of items that you can buy spending as close to, but not exceeding, $50.

Estimate 6. Estimate the total cost of all the items that are on sale.

Write Math 7. Write a problem that requires three steps for solution.

Data Bank 8. Fran is 23 years old. Did she meet her daily requirement for calorie intake with this diet? Use the Calorie Count Table on page 438 of the Data Bank.

Breakfast	Bagel and cream cheese, half grapefruit, bran flakes, 1 cup skim milk
Lunch	Grilled American cheese on 2 slices of wheat bread, 1 cup skim milk, apple
Dinner	Fish sticks with tartar sauce, baked potato with 1 pat butter and cheddar cheese, broccoli, cranberry juice, strawberries, 1 banana, and $\frac{1}{2}$ pint (16 tbsp) of half-and-half cream

Mixed Skills Review

1. $29.80 + 21¢	2. 374 × 98	3. 87.65 − 7.65
4. 55 ÷ 15.5	5. 570 + 18 + 68	6. 72.8 × 93
7. 240 ÷ 12	8. $27.43 × 0.8	9. 935 − 25
10. $43 ÷ 12.70	11. 4.5 + 1.6 + 2.2	12. 36 × 48

2-5 Using Formulas

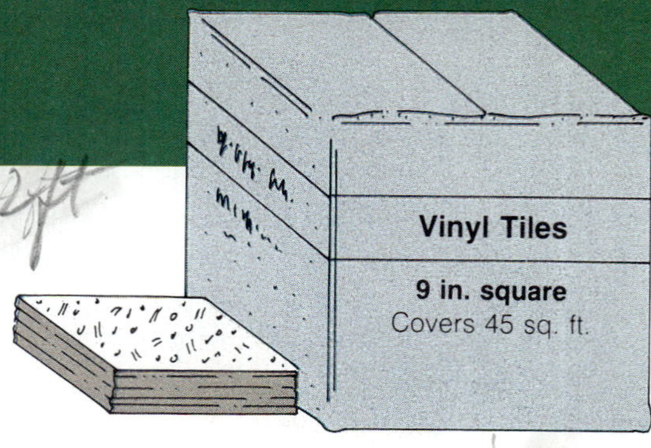

Objective
Solve problems by using formulas.

Frank is planning to install new vinyl tile in his kitchenette. The kitchenette is 8 feet by 9 feet. How many boxes of tile should Frank buy?

Understand the Situation

- What is the size of a tile?
- What is the size and shape of the kitchenette?
- How many square feet of floor will a box of tiles cover?

Some consumer problems can be solved with the use of a **formula**. A formula is a rule or recipe for doing a certain calculation. For some problems you may remember the needed formula. For other problems you will have to find the formula or search for data.

Example Find how many boxes of tile Frank should buy.

A. Make a drawing of the room and label the dimensions.
The kitchenette is shaped like a rectangle.

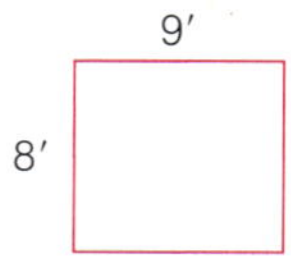

B. Find the area of the rectangle.
The formula for the area of a rectangle is $A = lw$.
9 ft long × 8 ft wide = 72 square feet or 72 ft^2

C. Divide the area by 45 to find the number of boxes of tile needed.
Estimate: $80 \div 40 = 2$
or
Calculate: $72 \div 45 = 1.6$

Since tile is sold by the box, Frank will need to buy two boxes. The answer is reasonable. One box of tiles covers only 45 square feet. Two boxes will cover 90 square feet, more than enough for Frank's 72 ft^2 kitchenette.

For other problems you may need to look up the formula. The Data Bank on page 441 gives a list of useful formulas.

Try This Jean Motyer is 45 years old. Calculate her maximum pulse rate for exercising. Use the formula on page 441 in the Data Bank.

Exercises

Use a formula to solve these problems.

1. Roofing shingles are sold by the **square.** A square of shingles covers 100 square feet. How many squares of shingles should you buy to put a new roof on the garage shown below?

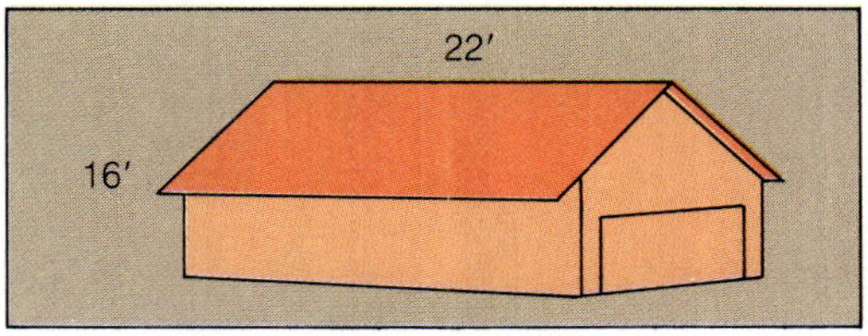

2. You are planning a vacation trip. You estimate that you can average 45 mph and can drive 6 hours per day. How far can you expect to drive in a day?
3. How many feet of fencing is needed to enclose a square-shaped garden that is 17 feet on a side?

Data Bank Solve these problems by looking up the needed formula on p. 441 in the Data Bank. You may wish to use a calculator.

4. An electric toaster in a 120-volt circuit draws 6.7 amps. Find the number of watts of power required by the toaster.
5. Cyrilla Fenhagen is 40 years old and is 146 cm tall. Find her **vital capacity.**
6. The time elapsed between a lightning flash and the sound of thunder is 10 seconds. Find the distance to the storm.
7. A cylindrical tank with a 4-foot radius holds gasoline that is 20 inches deep. Find the number of gallons of gasoline in the tank.

Suppose

8. If the length and width of Frank's kitchenette were each doubled, how many boxes of tile would he need?

Write Math

9. Write a problem that can be solved using one of the formulas on page 436.

Mixed Skills Review

1. 4 − 2.5	2. 6 + 66.6	3. 42 ÷ 16
4. 15 × 19	5. 6.32 ÷ 2.8	6. 101 × 102
7. 8 + 4.3 + 11	8. \$9 + \$16 + \$3.94	9. 406 ÷ 14
10. 16.9 × \$86	11. 177 − 12	12. 4.3 − 2.8

2-6 Reasonable Answers

Objective
Decide if an answer is reasonable.

Lesa ordered 4 tickets for a trip to Orlando. She calculates the total cost of her tickets is $631.80.

Understand the Situation

- Where is Lesa's family going on vacation?
- How many people will be going on the trip?
- What is the special rate for a plane ticket?

When solving problems, estimation and mental math can help you decide if an answer is reasonable. This is part of checking your answer.

Examples

1. Find the cost of Lesa's 4 tickets. Answer: $631.80

To determine if Lesa's calculation is reasonable, make estimates greater and less than the given answer.

First round $157.95 up to $160, which is compatible with 4. $4 \times 160 = 640$

Now round down to 150, which is also compatible with 4. $4 \times 150 = 600$

The answer of $631.80 is reasonable since it lies between each estimate.

Calculation will show the 4 tickets cost $631.80.

2. The ticket price for infants is half the adult fare, or $90.15. If the given answer is not reasonable, find the correct answer.

$157.95 is about $160. $\$160 \div 2 = \$80$

The given answer is not reasonable.

Calculation shows the correct answer is $78.98.

Try This

Estimate the answer to each problem. If the given answer is not reasonable, find the correct answer.

a. A car driven at 55 miles per hour went 24.6 miles on a gallon of gasoline. The same car driven at 45 miles per hour went 27.3 miles on a gallon of gasoline. How much farther per gallon did it go at the slower speed? Answer: 124.5 miles

b. 442 people signed up for a museum trip and tour. The people were divided into groups of 25. How many groups were formed? Answer: 18

Exercises

Estimate the answer to each problem. If the given answer is not reasonable, find the correct answer.

1. Freda bought a sweater for $29.65, a blouse for $18.65, a skirt for $42.85. How much did she spend in all? Answer: $101.15
2. Denise bought a packet of 3 videotapes for $14.69. She paid with a $20 bill. How much change did she receive? Answer: $44.07
3. In California's Humboldt Redwood State Park there is a coast redwood tree that is 362 feet tall. In Washington's Gifford Pinchot National Forest there is a noble fir that is 278 feet tall. How much taller is the redwood? Answer: 94 ft
4. Ivan owns 3 Brittany spaniels. He pays $14.23 for a 25-kilogram bag of dog food each month. What is his annual cost for dog food? Answer: $170.76

5. In a business education classroom there are 15 word processors. Each costs $6523. What is the total cost? Answer: $9784.50
6. Shannon bought 3 pairs of jeans costing $29.99 each. What was the total cost? Answer: $89.97
7. The average human hair grows about 3.25 millimeters per week. If left uncut, how many millimeters will a human hair grow in a year (52 weeks)? Answer: 143.04 mm

Suppose

8. If Lesa cancelled her Florida reservations and there was a 25% cancellation penalty, how much of a penalty would she pay?

Talk Math

9. Carolyn wrote a check for $29.76. She recorded the check in her checkbook and quickly subtracted it from her previous balance of $618.20. She entered $647.96 as her new balance. Use logical reasoning to explain why her new balance is not reasonable.

Number Sense

10. Instead of buying the four airline tickets at the special rate, Lesa could have used the family plan. In the family plan, one ticket is full fare at $352 and the remainder cost $98 each. Which is the better deal?

Mixed Skills Review

1. 428 + 161.2	2. $31 \div 11$	3. \$604 − \$110
4. $55 \div 9.1$	5. 6.4 + 8.9 + 102.4	6. \$17 × 74
7. 18 − .06	8. 39 − 15	9. 10 + 445 + 615
10. 1.1 × 34.1	11. 12 × 62	12. \$61 ÷ 7

2-7 Choosing Problem-Solving Strategies

Objective
Identify problem-solving strategies.

The landscaper needs to buy 19 pounds of seed for several projects. Grass seed comes in 5-pound bags for $5.15 and in 3-pound bags for $3.10. How much of each size bag should she buy to have at least 19 pounds at the lowest cost?

Understand the Situation

- How much grass seed is needed?
- In what size bag is grass seed sold?
- What is the cost per bag of grass seed?

It is helpful to select a strategy to solve the problem. Many problems can be solved in more than one way.

Problem-Solving Strategies

- Make a drawing.
- Write an equation.
- Choose an operation.
- Find a pattern.
- Solve a simpler problem.
- Make a model.
- Make an organized list.
- Guess and check.
- Work backward.
- Make a chart or table.
- Use logical reasoning.
- Act it out.

Examples

Guess and check.

1st guess

1. four 5-pound bags = 20 lb
$4 \times \$5.15 = \$20.60$

2nd guess

2. two 5-pound bags = 10 lb $\quad 2 \times \$5.15 = \$10.30$
three 3-pound bags = + 9 lb $\quad 3 \times \$3.10 = + \$9.30$
19 lb $\quad \$19.60$

Make a chart.

3.

5-lb Bags	Cost @ $5.15	3-lb Bags	Cost @ $3.10	Total Amount	Total Cost
4	$20.60			20 lb	$20.60
2	$10.30	3	$9.30	19 lb	$19.60

Buy two 5-pound bags and three 3-pound bags to have at least 19 pounds at the lowest cost.

Try This

The 13 members of the Bobby Sox softball team need new socks. The socks are sold for $2.50 a pair or $6.25 for 3 pairs. How should the manager purchase the socks to get at least 13 pairs at the lowest cost?

Exercises

Work with a Partner

Work with a partner to pick a strategy. Solve the problem.

1. A landscaper has the following kinds of tulips, daffodils, and crocuses: early blooming tulip, yellow tulip, white daffodil, late-blooming crocus, and yellow crocus. For a job she wants to have every type and color of each type of bulb. What does she need? (Hint: **Make a table** like the one on the right and place an X for every type and color that she has. An X has been placed for the yellow tulip.)

	Early	Late	Yellow	White
Tulip			X	
Daffodil				
Crocus				

2. At the Bay View High School homecoming parade there were twice as many business-sponsored floats as student floats. There were one-third as many civic floats as business floats. There were 6 civic floats. How many of each type of float were there? How many floats were there in all? (Hint: Try the strategy **work backward**.)

3. A small garden needs 72 feet of new fencing. The new fence is to be placed near an old fence as shown below. The old fence will serve as one of the sides. What should be the dimensions of the fence so that the largest area is enclosed? (Hint: Try various possibilities and **find a pattern**. Remember that the area of a rectangle can be found by multiplying length by width.)

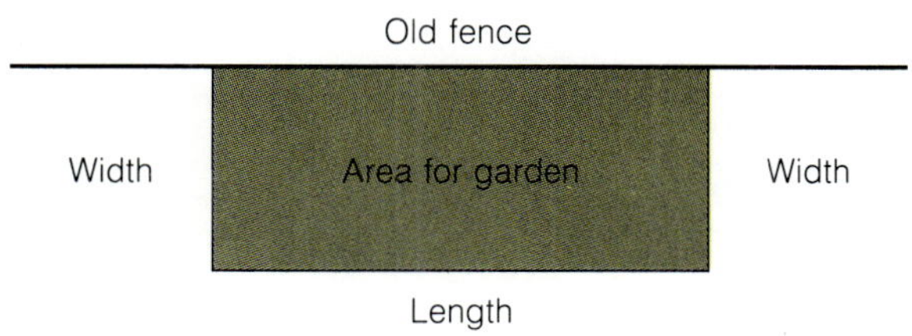

4. In the cross-country bicycle race Hector came in first. Tom came in last. Gloria came in just after Anna. Bill came in after Gloria. Mary Lou did not finish. Who came in third? (Hint: Try the strategy **make a drawing**.)

Mixed Skills Review

1. 234 − 148	**2.** 85.1 × 58.7	**3.** 87 − 33.33
4. 432 ÷ 4	**5.** 91 + 11 + 12	**6.** 864 × 46.3
7. 99.9 + 3.3	**8.** \$56.80 − \$51.40	**9.** 10.8 ÷ 2.1
10. \$17.80 × 22	**11.** 128 ÷ 13	**12.** 738.1 − 720.8

2-8 Applying Problem-Solving Strategies

Objective
Identify problem-solving strategies.

Dave Salla works as a ticket agent. Tickets are on sale for a concert. When the ticket counter opens there are 2 people in line. Dave knows that it takes about 2 minutes to serve a customer. The first new customer arrives 1 minute after Dave opens the counter. After that, a new customer arrives every 5 minutes. How much time will pass before a customer arrives and does not find a waiting line?

Understand the Situation

- How many people are in line when the ticket counter opens?
- How long does it take to serve a customer?
- When does the first new customer arrive?
- How often does a new customer arrive?

Select a strategy to help solve the problem.

Examples

1. Make a drawing.

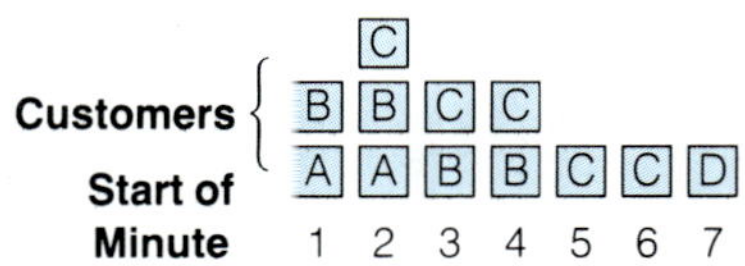

2. Make a table.

	Minutes						
	0–1	1–2	2–3	3–4	4–5	5–6	6–7
Customer 1	1st min	2nd min					
Customer 2	wait	wait	1st min	2nd min			
Customer 3		new	wait	wait	1st min	2nd min	
Customer 4							new
Customer 5							

It will take 6 minutes before a customer arrives and finds no waiting line.

Try This

When a hot dog stand opens there are 2 customers waiting. The vendor takes about 2 minutes to serve a customer. After 3 minutes a new customer arrives. After that, a new customer arrives every 4 minutes. How long will it be before the first "no wait" customer arrives?

Exercises

Pick a strategy. Solve the problem.

1. In a production line an automatic welder takes 3 minutes to make a weld. When the production line is turned on in the morning there are 2 pieces waiting to be welded. The first new piece arrives 2 minutes after the operation starts. After that a new piece arrives every 6 minutes. How many minutes after the production line is turned on will the first "no wait" piece arrive?
2. Willy worked three jobs during the summer. He made a total of $1245. The first job paid twice as much as the second and third jobs combined. The second job paid one third of the third job. How much did each job pay?
3. At the annual Maple Valley fall race there are 6 check points 2.5 miles apart. The first and last check point are 2 miles each from the start and end of the race. How long is the race?
4. Herman had $100 to spend on clothes for school. What are the different ways he could spend his money? Use the price list shown.

Clothing			
Shoes	$32 a pair	T-shirts	3 for $14
Shirt	$15 each	Sweatshirt	$12
Pants	$25 each	Jacket	$46
Socks	$1.75 a pair	Sweater	$16

5. Janet and Holly raced to see who could finish eating a 6-ft long string of spaghetti first. Janet took 1-inch bites every 1.5 seconds, while Holly took $\frac{3}{4}$-inch bites every 1.1 seconds. Who finished first?

6. Marta Fisk paid $20,800 for her new car. While she owns it, her car devaluates (loses value) at the rate of $1,200 per year. For every 10,000 miles on the odometer, the car loses another $1000 in value. The dealer will not accept the car as a trade-in if it has lost more than half of its original worth. Marta drives 23,000 miles per year. To the nearest year, how long can she keep the car before trading it in?

Mixed Skills Review

1. 428 + 161.2	2. 31 ÷ 11	3. 18 × 64
4. 178 − 24	5. $4.02 × 187	6. 12 + 68 + 1022
7. 2.2 × 0.5	8. 8.6 + 4.1 + .009	9. $16 ÷ 2.8
10. 148 ÷ 19	11. $418 − $99.07	12. 4.45 − 2.02

2-9 Enrichment: Clever Reasoning

Objective
Solve problems using inductive and deductive reasoning.

Some problems take very careful reading or clever reasoning to be solved.

Solve.

1. If you take 2 apples from 3 apples, how many apples do you have?
2. Seven months have exactly 31 days. Four months have exactly 30 days. How many months have 28 days?
3. A man wearing a green shirt found that it took him 80 minutes to read an 80-page book. The same man found that, while wearing a blue shirt, he took an hour and 20 minutes to read the same book. Can you explain this?

For each of the following, a rule has been devised that determines whether a letter should be written above or below the line. See if you can figure out each rule. Then decide in each case whether the letter *z* should be written above or below the line.

4.
a	e	i	o	u	y
bcd	fgh	jklmn	pqrst	vwx	y

5.
g	j	pq	y
abcdef	hi	klmno	rstuvwx

6.
A	EF	HI	KLMN	T	VWXY
BCD	G	J	OPQRS	U	

Solve the following problems. You may choose to draw a diagram.

7. In the 800-meter dash Delbert finished 0.53 seconds before Arnold. Arnold finished 0.81 seconds before Joe. Harry finished 0.48 seconds before Delbert. Who finished first, second, and third? How many seconds faster was Harry than Arnold?
8. In the 600-meter race Rose finished 2.16 seconds behind Heather. Heather finished 1.04 seconds behind Kim. Kim finished 1.04 seconds behind Betty. Who finished first, second, and third? What was the difference between Betty's time and Rose's time?
9. In the 5000-meter race Rick finished 15.47 seconds after Charlie and 22.56 seconds before Manuel. Tom finished 37.29 seconds after Rick. Who finished first, second, and third? What was the difference between Manuel's time and Tom's time?
10. In the 1500-meter race Trisha finished 6.99 seconds after Connie and 7.10 seconds before Jane. Sally finished 4.12 seconds before Trisha. Who finished first, second, and third? What was the difference between Sally's time and Connie's?

2-10 Computer: Finding Costs

Objective
Use a computer program to find the cost of items.

Many stores use computers to find the cost when you buy several of the same item. The computer also calculates the total cost of all items that you buy at one time. The computer uses more complicated versions of a program (set of instructions) like the one given below.

```
10  REM This program will find the total cost of purchased items.
20  PRINT "Name the item."
30  INPUT I$
40  PRINT "How much did one " I$ " cost?"
50  INPUT C
60  PRINT "How many " I$ "s did you buy?"
70  INPUT N
80  T = N * C
90  PRINT "The cost of " N" " I$ "s is $"T"."
100 PRINT "Do you want to try another? (Y/N)"
110 INPUT Q$
120 IF Q$ = "Y" OR Q$ = "y" THEN 20
130 END
```

Mr. Arnold is buying a bicycle for each of his three children. Each bicycle costs $129.95. Use mental math to estimate the cost of the three bicycles. Use a calculator to find the exact cost. Compare your answers to the computer solution given.

```
Name the item.
bicycle
How much did one bicycle cost?
129.95
How many bicycles did you buy?
3
The cost of 3 bicycles is $389.85.
Do you want to try another? (Y/N)
N
```

Exercises

1. Type the program into the computer.
2. Run the program using the bicycle problem above.
3. Use the program to find the total cost of the following items.
   A. 24 cans of dog food @ $0.59 B. 6 bars of soap @ $0.85
4. Mrs. Elkwood is having a picnic and plans to buy the following items.

   Use the computer program to find the cost of each group of items.

Item	Cost	Number
Chili mix	$0.49	24
Burger buns	$0.65	12
Punch concentrate	$1.59	8

CHAPTER 2 REVIEW

1. Consider the following list of questions. Classify each question by the number of the problem-solving guideline to which it relates.

SITUATION	**1. Understand the SITUATION**
DATA	**2. Find the needed DATA**
PLAN	**3. PLAN what to do**
ANSWER	**4. Find the ANSWER**
CHECK	**5. CHECK the answer**

A. Is all the data present?
B. Is the answer reasonable?
C. What question is asked?
D. Should I use a calculator?
E. Are my computations correct?
F. Can I draw a picture?
G. Have I stated the answer clearly?
H. Should I read the problem again?
I. Is there extra data?
J. What operation should I use?
K. Can I make a table?
L. Is the answer close to my estimate?
M. Is an exact answer needed?
N. What data do I need?

Choose the one that does not belong.

2. Guess and check
Find a pattern
Clustering
Make a drawing

3. $A = lw$
$A = \frac{1}{2}bh$
$V = S^3$
$A = S^2$

4. Plan
Understand
Check
Estimate

Move the underlined words around so that these sentences make sense.

5. Data can often be found in a <u>strategy</u>
6. Some problems can be solved using <u>reasonable</u>.
7. We may need to seek <u>estimating</u> to solve some problems.
8. A <u>multi-step problem</u> helps solve a <u>advice</u>.
9. <u>Formulas</u> can help decide if an answer is <u>plan</u>.
10. It is helpful to select a <u>table</u> to solve a problem.

CHAPTER 2 TEST

1. An ad in the Lincoln Yellow Pages costs \$0.09 per household for each 20 words in the ad. About 5760 households receive the Yellow Pages. How much does a 100-word ad cost?

The table at the right shows how to make pancakes.

2. How many eggs should be used to make 13 pancakes?
3. How many cups of mix are needed to make 55 pancakes?
4. How many pancakes can we make using $10\frac{1}{2}$ cups of milk?

Ingredients				
No. of Pancakes	Cups Mix	Cups Milk	No. Eggs	Tbsp Oil
6–7	$\frac{1}{2}$	$\frac{1}{2}$	1	1
12–14	1	1	1	1
25–30	2	2	2	2
40–44	$3\frac{1}{2}$	$3\frac{1}{2}$	3	4

Use the Data Bank on p. 437 to help solve this problem.

5. How many more calories per day does a 7-year-old child need than a 21-year-old woman?
6. Bill bowled in a local charity tournament. Eight people pledged 5¢ per pin and 12 people pledged 10¢ per pin for his highest game. He scored 127, 136, and 159. How much money did he raise for the local charity?
7. Johanna lives 25 minutes from the campus where the PSAT is given. The registration booklet recommended that students allow 50 minutes each for the verbal and math sections of the test, 30 minutes for instructions, and 10 minutes for a break. Johanna left home at 8:35 a.m. What time should her mother leave home to pick her up after the test?
8. How long does it take to drive 100 miles at 40 mph? 50 mph? 60 mph?
9. Jack intends to build a patio using 3 feet × 4 feet slabs of concrete. His patio will be 9 feet on each side. How many concrete slabs does he need? Will he have to cut any of them?
10. Nina bought a gallon of milk for \$1.69 and 3 pounds of hamburger at \$1.19 a pound. She paid with a \$10 bill and received \$5.20 in change. Is this reasonable? If not, what is the correct change?

Choose a strategy and solve the problem.

11. Martin must mail a package that requires 92¢ in postage. He has 13¢ stamps and 22¢ stamps. How many of each type should he use?
12. Workers earn \$3 for every bookcase they assemble. Bret assembles 2 bookcases every 15 minutes. Perry completes 3 of them every 20 minutes. Alysia finishes 4 bookcases every 30 minutes. What is the total they earn together in 4 hours?

CHAPTER 2 CUMULATIVE REVIEW

1. Add.
   18,035 + 5480 + 4377
   A. 27,982
   B. 27,892
   C. 27,782
   D. 28,882

2. Taxi fare from downtown to the airport is $20. How much will the driver earn from 12 trips to the airport?
   A. $360
   B. $240
   C. $260
   D. $220

3. Joe bought a notebook for $1.76 and a dictionary for $20.52. He paid with $25 cash. What was his change?
   A. $5
   B. $3.72
   C. $2.72
   D. $4.28

4. Estimate. Use clustering.
   289 + 313 + 295
   A. 900
   B. 600
   C. 700
   D. 1200

5. The Casey family's playroom is 6 ft wide by 7 ft long. What is the area of the room?
   Use the formula $A = lw$.
   A. 13 ft^2
   B. 49 ft^2
   C. 36 ft^2
   D. 42 ft^2

6. Multiply.
   $38.79 × 60
   A. $2227.40
   B. $1884.20
   C. $2327.40
   D. $2238.20

7. Find the product. Use mental math.
   521 × 6
   A. 3126
   B. 3026
   C. 3136
   D. 3226

8. Oranges are 2 lb for 89¢. How much do 5 lb cost?
   Answer: $2.20
   Is the answer reasonable?
   A. reasonable
   B. unreasonable

9. Add.
   $$\begin{array}{r} 1826 \\ +2559 \\ \hline \end{array}$$
   A. 4375
   B. 3375
   C. 4385
   D. 4285

10. **Hot Cocoa Mix**

Cups	Mix	Hot Water
1	3 tsp	$\frac{3}{4}$ cup

   How much cocoa mix should you use for 6 cups of cocoa?
   A. 1 cup
   B. 6 tsp
   C. 2 tsp
   D. 18 tsp

11. Sam bought 3 packs of fruit drink. Every pack contained 6 drinks. The drinks were $0.34 each. How much did he pay in all?
   A. $5.04
   B. $6.12
   C. $1.02
   D. $2.04

12. Divide.
   $56.20 ÷ 10
   A. $0.56
   B. $562
   C. $5.62
   D. $0.06

13. Add.

$$\begin{array}{r} 0.2072 \\ +0.20351 \\ \hline \end{array}$$

A. 0.5071
B. 0.41071
C. 0.4171
D. 0.401071

14. Maria made $258 working at Office Inc. She worked 40 hours. How much did she make per hour?

A. $5.95
B. $6.20
C. $6.12
D. $6.45

15. How much ribbon binding is needed for a blanket 6.5 ft long by 4 ft wide?

A. 21 ft
B. 14.5 ft
C. 104 ft
D. 17 ft

16. Decide if you need an exact answer or an estimate. You are driving to camp. How long will it take?

A. exact answer
B. estimate

17. Estimate.

$$\begin{array}{r} 8.31 \\ \times\ 5.8 \\ \hline \end{array}$$

A. 32
B. 48
C. 54
D. 40

18. Where would you look to find the cost of car accessories?

A. sports catalog
B. auto parts store
C. hardware store
D. encyclopedia

19. Estimate the answer. If the given answer is not reasonable, find the correct answer. Helen is 18 years old. Irene is 10 years old. How old will Helen be when Irene is 30 years old? Answer: 40 years old

A. reasonable
B. unreasonable, 36 years old
C. unreasonable, 68 years old
D. unreasonable, 54 years old

20. Subtract.

$$\begin{array}{r} 5921 \\ -4050 \\ \hline \end{array}$$

A. 1871
B. 971
C. 1931
D. 871

21. Solve.
Kris is standing in line at the bank. There are 5 people ahead of her. There are 2 tellers. Each customer takes about 2 minutes. How long will Kris wait?

A. about 2 min
B. about 4 min
C. about 10 min
D. about 15 min

Chapter 3 Overview

Key Ideas

- meaning of fractions
- estimating fractions
- comparing fractions
- using fractions in a number sentence
- addition, subtraction, multiplication, and division with fractions
- mental math with fractions

Key Terms

- fraction
- numerator
- denominator
- mixed number
- lowest terms
- reciprocal
- lowest common denominator (LCD)
- equivalent fractions
- rename
- invert

Key Skills

Find the answer.

1. $8 + 8 + 8 + 8$

2. $25 + 25 - 25$

3. $256 + 256$

4. $\frac{1}{8} + \frac{1}{8} + \frac{1}{8} + \frac{1}{8}$

5. $\frac{4}{25} + \frac{9}{25} + \frac{7}{25}$

6. $\frac{29}{256} + \frac{99}{256}$

7. $49 - 7$

8. $35 - 7$

9. $28 - 14$

10. $1 - \frac{3}{7}$

11. $\frac{11}{35} - \frac{4}{35}$

12. $\frac{5}{28} - \frac{1}{28}$

13. $4 \times 12\frac{1}{2}$

14. $3 \times 3\frac{1}{3}$

15. $\frac{2}{5} \times 100$

16. $\frac{1}{2} \times \frac{1}{4}$

17. $\frac{1}{3} \times \frac{3}{4}$

18. $\frac{1}{7} \times \frac{21}{23}$

19. $\frac{3}{4} \div 3$

20. $3 \div \frac{3}{4}$

21. $\frac{1}{2} \div 8$

Arrange in order from least to greatest.

22. $\frac{1}{9}, \frac{1}{4}, \frac{3}{7}, \frac{4}{5}$

23. $\frac{3}{17}, \frac{5}{34}, \frac{9}{68}, \frac{11}{68}$

24. $\frac{11}{10}, \frac{7}{5}, 1, \frac{7}{9}$

25. $4\frac{1}{3}, \frac{29}{3}, \frac{5}{24}, \frac{9}{5}$

USING FRACTIONS

3

3-1 Understanding Fractions

Objective
Write fractions.

These road signs must be understood by all drivers. Warning signs in the shape of a diamond are usually yellow with black letters.

Understand the Situation

- How many signs are there?
- How many signs give information about road conditions?
- How many signs are shaped like a diamond?

We use fractions to represent parts of a whole.

numerator → 3
denominator → 12

$$\frac{3}{12} = \frac{3 \div 3}{12 \div 3} = \frac{1}{4}$$

$\frac{3}{12}$, or $\frac{1}{4}$ of the signs are yellow.

$\frac{3}{12}$ and $\frac{1}{4}$ are **equivalent** fractions.

To write a fraction in **lowest terms**, divide the numerator and the denominator by their **greatest common factor**.

A fraction is in lowest terms when 1 is the greatest common factor.

Examples Write in lowest terms.

1. $\frac{15}{18} = \frac{15 \div 3}{18 \div 3} = \frac{5}{6}$ **2.** $\frac{16}{24} = \frac{16 \div 8}{24 \div 8} = \frac{2}{3}$ **3.** $\frac{72}{81} = \frac{72 \div 9}{81 \div 9} = \frac{8}{9}$

Try This Write in lowest terms.

a. $\frac{6}{15}$ **b.** $\frac{8}{12}$ **c.** $\frac{10}{35}$ **d.** $\frac{14}{28}$ **e.** $\frac{25}{40}$

To write equivalent fractions, multiply the numerator and the denominator by the same number.

$$\frac{4}{9} = \frac{4 \times 3}{9 \times 3} = \frac{12}{27}$$

Example Write the missing numeral.

4. $\frac{3}{8} = \frac{\square}{32}$ Think: Since $8 \times 4 = 32$, multiply 3×4.
The missing numeral is 12.

Try This Write the missing numeral.

f. $\frac{3}{5} = \frac{\square}{15}$ **g.** $\frac{7}{9} = \frac{21}{\square}$ **h.** $\frac{5}{8} = \frac{\square}{48}$ **i.** $\frac{13}{12} = \frac{\square}{84}$ **j.** $\frac{5}{6} = \frac{60}{\square}$

Exercises

Write in lowest terms.

1. $\frac{6}{9}$	**2.** $\frac{7}{28}$	**3.** $\frac{9}{45}$	**4.** $\frac{14}{16}$	**5.** $\frac{12}{30}$
6. $\frac{24}{30}$	**7.** $\frac{27}{45}$	**8.** $\frac{35}{40}$	**9.** $\frac{18}{33}$	**10.** $\frac{28}{42}$
11. $\frac{36}{49}$	**12.** $\frac{25}{64}$	**13.** $\frac{27}{81}$	**14.** $\frac{40}{60}$	**15.** $\frac{45}{54}$
16. $\frac{56}{72}$	**17.** $\frac{49}{98}$	**18.** $\frac{275}{500}$	**19.** $\frac{240}{840}$	**20.** $\frac{324}{432}$

Write the missing numeral.

21. $\frac{4}{7} = \frac{\square}{28}$	**22.** $\frac{3}{5} = \frac{\square}{40}$	**23.** $\frac{9}{10} = \frac{27}{\square}$	**24.** $\frac{8}{9} = \frac{32}{\square}$	**25.** $\frac{3}{8} = \frac{18}{\square}$
26. $\frac{11}{14} = \frac{\square}{42}$	**27.** $\frac{3}{2} = \frac{24}{\square}$	**28.** $\frac{10}{9} = \frac{40}{\square}$	**29.** $\frac{7}{16} = \frac{\square}{48}$	**30.** $\frac{15}{13} = \frac{45}{\square}$
31. $\frac{7}{6} = \frac{49}{\square}$	**32.** $\frac{13}{8} = \frac{\square}{40}$	**33.** $\frac{11}{12} = \frac{44}{\square}$	**34.** $\frac{5}{3} = \frac{\square}{36}$	**35.** $\frac{16}{21} = \frac{32}{\square}$

Write an equivalent fraction with a denominator of 84.

36. $\frac{1}{2}$ **37.** $\frac{2}{3}$ **38.** $\frac{11}{14}$ **39.** $\frac{20}{21}$ **40.** $\frac{8}{7}$

Show Math

41. Complete the following steps for each pair of fractions. Then write a rule about equivalent fractions.

Multiply the numerator of the first fraction by the denominator of the second. → Multiply the denominator of the first fraction by the numerator of the second. → Compare the results.

$\frac{5}{8} = \frac{20}{32}$ $5 \times 32 = 160$ $8 \times 20 = 160$ $5 \times 32 = 8 \times 20$

$160 = 160$

A. $\frac{10}{12} = \frac{15}{18}$ **B.** $\frac{28}{32} = \frac{49}{56}$ **C.** $\frac{40}{30} = \frac{16}{12}$

Mixed Skills Review

1. 397×654	**2.** $\frac{8}{11} = \frac{24}{\square}$	**3.** $26.3 - 17.34$
4. $29.61 \div 1.8$	**5.** $2734 + 418 + 5448$	**6.** $\frac{9}{12} = \frac{3}{\square}$
7. \$27.18 + \$56.84 + \$45	**8.** $\frac{5}{8} = \frac{\square}{48}$	**9.** $10{,}948 \div 46$
10. $4062 - 2665$	**11.** 7.4×0.85	**12.** $\frac{20}{35} = \frac{\square}{7}$

3-2 Fractions and Mixed Numbers

Objective
Write mixed numbers as fractions. Write fractions as mixed numbers and whole numbers.

Leilani Ho makes and sells shell jewelry on the island of Hawaii. She bought $1\frac{3}{4}$ lb of shells from a diver. How many packages did she buy?

Understand the Situation

- How much does each package of shells weigh?
- How many packages are needed to weigh 1 lb?

The **mixed number** $1\frac{3}{4}$ is the sum of $1 + \frac{3}{4}$.

$1 + \frac{3}{4} = 1\frac{3}{4}$ $\quad \frac{4}{4} + \frac{3}{4} = \frac{7}{4}$ $\quad$ Leilani bought 7 packages of shells.

Example

1. Write as a fraction. $\quad 4\frac{2}{3}$

Multiply the whole number by the denominator. $\quad 4 \times 3 = 12$
Add the numerator. $\quad 12 + 2 = 14$

Write the sum over the denominator. $\quad \frac{14}{3}$

Try This

Write as a fraction.

a. $1\frac{3}{5}$ **b.** $2\frac{1}{2}$ **c.** $4\frac{1}{6}$ **d.** $1\frac{4}{7}$ **e.** $3\frac{5}{8}$

Example

2. Leilani bought 14 more $\frac{1}{4}$-lb packages of shells, or $\frac{14}{4}$ lb of shells.

Write as a mixed number or a whole number. $\quad \frac{14}{4}$

Divide the numerator by the denominator. $\quad 14 \div 4 = 3$ R2

Write the remainder over the divisor. $\quad 3\frac{2}{4}$

Write the fraction in lowest terms. $\quad 3\frac{2}{4} = 3\frac{1}{2}$

Try This

Write as a mixed number or a whole number.

f. $\frac{9}{5}$ **g.** $\frac{12}{7}$ **h.** $\frac{42}{6}$ **i.** $\frac{32}{20}$ **j.** $\frac{16}{2}$

Exercises

Write as a fraction.

1. $3\frac{1}{2}$
2. $2\frac{1}{4}$
3. $1\frac{4}{5}$
4. $1\frac{4}{7}$
5. $4\frac{2}{3}$
6. $3\frac{5}{6}$
7. $2\frac{2}{9}$
8. $1\frac{11}{12}$
9. $6\frac{1}{4}$
10. $2\frac{5}{8}$
11. $2\frac{3}{16}$
12. $8\frac{1}{2}$
13. $12\frac{3}{4}$
14. $10\frac{2}{3}$
15. $7\frac{4}{5}$
16. $3\frac{9}{10}$
17. $6\frac{3}{8}$
18. $11\frac{1}{6}$

Write as a mixed or a whole number.

19. $\frac{9}{4}$
20. $\frac{7}{3}$
21. $\frac{17}{5}$
22. $\frac{24}{6}$
23. $\frac{16}{10}$
24. $\frac{20}{8}$
25. $\frac{8}{6}$
26. $\frac{10}{8}$
27. $\frac{18}{4}$
28. $\frac{34}{17}$
29. $\frac{42}{14}$
30. $\frac{56}{12}$
31. $\frac{17}{6}$
32. $\frac{84}{7}$
33. $\frac{26}{10}$
34. $\frac{52}{12}$
35. $\frac{36}{15}$
36. $\frac{64}{18}$

Number Sense

Use the clues to find the number.

37. A. The number is a fraction.
    B. The numerator is 4 more than the denominator.
    C. The denominator is a multiple of 4.
    D. The number is equal to $1\frac{1}{2}$.

38. A. The number is a fraction.
    B. The numerator is a multiple of 6.
    C. The denominator is less than 10.
    D. The number is equal to $2\frac{2}{3}$.

Calculator

Use a calculator to write the following fractions in order from the least to the greatest, as shown at right.

$\frac{5}{3} = 5 \div 3 = 1.666666$ $\quad \frac{6}{4} = 6 \div 4 = 1.5$

$\frac{5}{3} > \frac{6}{4}$ $\quad \frac{6}{4}, \frac{5}{3}$

39. $\frac{7}{6}, \frac{8}{7}, \frac{9}{8}$
40. $\frac{7}{3}, \frac{8}{5}, \frac{9}{7}$
41. $\frac{13}{10}, \frac{23}{20}, \frac{33}{30}$
42. $\frac{10}{33}, \frac{20}{23}, \frac{30}{13}$

Mental Math

43. Margarine is sold in $\frac{1}{4}$-lb sticks. How many sticks together weigh $3\frac{1}{4}$ lb?

Mixed Skills Review

1. $588 + 167 + 304$
2. $\frac{17}{4} = 4\frac{\square}{4}$
3. $6 + 2.08 + 0.96$
4. $\frac{2}{9} = \frac{16}{\square}$
5. 46×804
6. $2\frac{3}{4} = \frac{\square}{4}$
7. $11 - 2.358$
8. $48.8 \div 0.16$
9. $9111 - 1652$
10. $27{,}909 \div 63$
11. $\frac{12}{15} = \frac{4}{\square}$
12. 19.3×2.08

3-3 Multiplying Fractions and Mixed Numbers

Objective
Multiply fractions.

A go-kart track is $\frac{7}{8}$ mile long. In one race drivers make $1\frac{1}{3}$ laps around the track. How long is the race?

To find the length of the race, multiply the length of the track by the number of laps.

Examples

1. Multiply. $\frac{7}{8} \times 1\frac{1}{3}$

Write each number as a fraction. $1\frac{1}{3} = \frac{4}{3}$

Multiply the numerators. Multiply the denominators. $\frac{7}{8} \times \frac{4}{3} = \frac{28}{24}$

Write as a mixed number in lowest terms. $\frac{28}{24} = 1\frac{4}{24} = 1\frac{1}{6}$ miles

2. A short method helps to make some problems easier. Divide numerator and denominator by a common factor before multiplying.

Long Method

$\frac{9}{10} \times \frac{5}{6} = \frac{45 \div 15}{60 \div 15} = \frac{3}{4}$

Short Method

$\frac{\overset{3}{\cancel{9}}}{\underset{2}{\cancel{10}}} \times \frac{\overset{1}{\cancel{5}}}{\underset{2}{\cancel{6}}} = \frac{3}{4}$

Multiply.

3. $1\frac{1}{8} \times 6 = \frac{9}{\underset{4}{\cancel{8}}} \times \frac{\overset{3}{\cancel{6}}}{1} = \frac{27}{4} = 6\frac{3}{4}$ Write the whole number as a fraction.

4. $1\frac{3}{4} \times 2\frac{2}{7} = \frac{\overset{1}{\cancel{7}}}{\underset{1}{\cancel{4}}} \times \frac{\overset{4}{\cancel{16}}}{\underset{1}{\cancel{7}}} = 4$ Write each mixed number as a fraction.

Try This

Multiply.

a. $\frac{4}{5} \times \frac{1}{2}$ **b.** $14 \times \frac{3}{7}$ **c.** $\frac{3}{8} \times 1\frac{2}{30}$ **d.** $2\frac{9}{16} \times 8$ **e.** $3\frac{3}{4} \times 1\frac{7}{10}$

Exercises

Multiply.

1. $\frac{1}{2} \times \frac{3}{4}$ **2.** $\frac{2}{5} \times \frac{4}{7}$ **3.** $\frac{3}{2} \times \frac{5}{8}$ **4.** $\frac{1}{6} \times \frac{7}{4}$

5. $\frac{5}{8} \times \frac{7}{10}$ **6.** $\frac{4}{3} \times \frac{9}{16}$ **7.** $\frac{12}{7} \times \frac{21}{8}$ **8.** $\frac{9}{5} \times \frac{7}{6}$

9. $\frac{8}{9} \times 9$ **10.** $\frac{5}{7} \times 14$ **11.** $16 \times \frac{7}{12}$ **12.** $28 \times \frac{5}{21}$

13. $3\frac{1}{3} \times \frac{3}{5}$ **14.** $4\frac{2}{3} \times \frac{9}{14}$ **15.** $\frac{2}{5} \times 3\frac{3}{4}$ **16.** $\frac{1}{2} \times 3\frac{2}{5}$

17. $1\frac{3}{8} \times 1\frac{2}{3}$ **18.** $4\frac{1}{2} \times 1\frac{3}{5}$ **19.** $3\frac{3}{4} \times 1\frac{7}{8}$ **20.** $1\frac{5}{6} \times 2\frac{1}{4}$

21. $\frac{2}{3}$ of \$63 **22.** $\frac{5}{8}$ of 20 **23.** $\frac{1}{4}$ of \$13.56 **24.** $\frac{5}{6}$ of \$12.40

Mental Math

Find the missing numeral. Use mental math and the Data Bank on p. 436.

25. $\frac{1}{2}$ ft = □ in. **26.** $\frac{1}{3}$ h = □ min **27.** $\frac{1}{4}$ yr = □ mo

28. $\frac{1}{3}$ yd = □ ft **29.** $\frac{1}{4}$ lb = □ oz **30.** $\frac{1}{2}$ gal = □ qt

31. The go-kart track at Stephens City is $1\frac{1}{4}$ miles long. How long is the third race?

32. A go-kart travels at a rate of 12 miles per hour. How far can the kart travel in 15 minutes?

33. Juan Gan's go-kart averages 20 miles per gallon of gasoline. How many gallons of gasoline will he need for the fourth race at Stephens City?

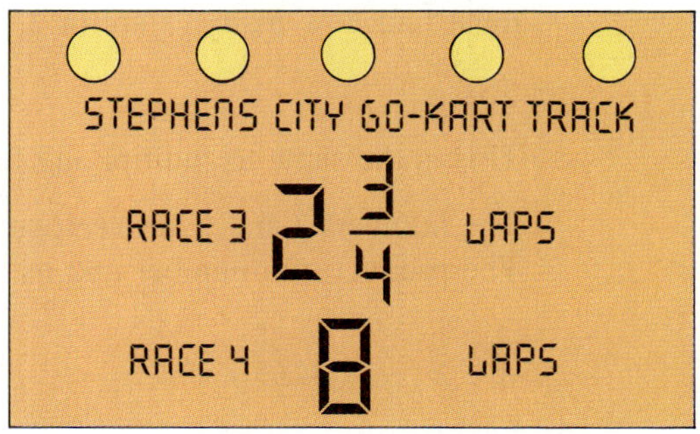

34. Dave Williams built his go-kart for about \$240. He earned three fourths of this amount mowing lawns and borrowed the rest from his dad. How much did Dave borrow?

Mixed Skills Review

1. $\frac{4}{9} \times \frac{3}{8}$ **2.** $0.62 + 0.8 + 0.008$ **3.** $3\frac{1}{4} \times 3\frac{1}{3}$

4. $11{,}232 \div 39$ **5.** 156×6037 **6.** $32.005 - 18.646$

7. $15 \times 1\frac{4}{5}$ **8.** $2819 - 1757$ **9.** 2.934×45

10. $123.152 \div 8.6$ **11.** $\frac{1}{2}$ of \$11.42 **12.** $6473 + 2651 + 718$

3-4 Dividing Fractions and Mixed Numbers

Objective *Divide fractions.*

Each year the United States Post Office issues special stamps to commemorate famous people and events in history. Collectors buy envelopes postmarked the day the stamp was first issued. Mark Smith mounts his first-day covers in a scrapbook.

Understand the Situation

- What part of a page does each envelope cover?
- How many pages will Mark use for 9 envelopes?

Examples

1. Mark used 2 pages in his book for stamps issued in January. How many first-day covers were issued in January?

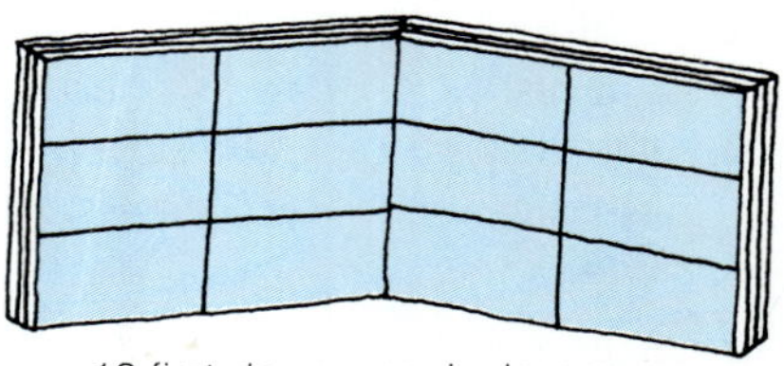

12 first-day covers in January

To find the number of covers issued, we must find how many $\frac{1}{6}$s are in 2.

$2 \div \frac{1}{6} = 12$

This is the same as multiplying 2×6.

The numbers $\frac{1}{6}$ and 6 are **reciprocals.** The product of a number and its reciprocal is 1.

$\frac{2}{1} \boxtimes \frac{1}{2}$ $\qquad$ $\frac{5}{8} \boxtimes \frac{8}{5}$

Dividing by a number is the same as multiplying by its reciprocal.

2. Divide. $1\frac{1}{4} \div \frac{2}{3}$

Write each number as a fraction. $\quad 1\frac{1}{4} = \frac{5}{4}$

Write the reciprocal of the divisor. $\quad \frac{2}{3} \boxtimes \frac{3}{2}$

Multiply. $\quad \frac{5}{4} \times \frac{3}{2} = \frac{15}{8} = 1\frac{7}{8}$

Try This Divide.

a. $\frac{7}{8} \div \frac{7}{8}$ **b.** $\frac{5}{6} \div \frac{1}{3}$ **c.** $\frac{8}{7} \div 4$ **d.** $10 \div 1\frac{5}{9}$ **e.** $3\frac{1}{2} \div 2\frac{3}{4}$

Exercises

Divide.

1. $\frac{1}{2} \div \frac{1}{4}$
2. $\frac{1}{6} \div \frac{1}{3}$
3. $\frac{6}{7} \div \frac{6}{7}$
4. $\frac{9}{4} \div \frac{9}{4}$
5. $\frac{3}{2} \div \frac{4}{3}$
6. $\frac{5}{2} \div \frac{1}{4}$
7. $\frac{1}{3} \div \frac{8}{9}$
8. $\frac{3}{8} \div \frac{2}{3}$
9. $\frac{1}{2} \div 3$
10. $\frac{2}{3} \div 4$
11. $6 \div \frac{3}{4}$
12. $9 \div \frac{3}{5}$
13. $1\frac{3}{4} \div \frac{1}{8}$
14. $2\frac{1}{2} \div \frac{1}{3}$
15. $3\frac{1}{5} \div \frac{8}{7}$
16. $4\frac{2}{3} \div \frac{9}{4}$
17. $3\frac{1}{2} \div 7$
18. $1\frac{2}{3} \div 4$
19. $3 \div 1\frac{3}{5}$
20. $10 \div 2\frac{1}{12}$
21. $2\frac{3}{8} \div 1\frac{3}{4}$
22. $7\frac{1}{2} \div 1\frac{1}{2}$
23. $4\frac{1}{3} \div 2\frac{1}{4}$
24. $1\frac{9}{16} \div 1\frac{5}{8}$
25. $1\frac{7}{12} \div \frac{5}{6}$
26. $3\frac{3}{7} \div 6$
27. $3\frac{2}{3} \div 1\frac{1}{3}$
28. $9 \div 4\frac{1}{2}$

Number Sense

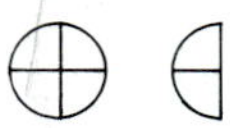 shows the number sentence $1\frac{1}{2} \div \frac{1}{4} = 6$.

Write a number sentence for each diagram.

29. 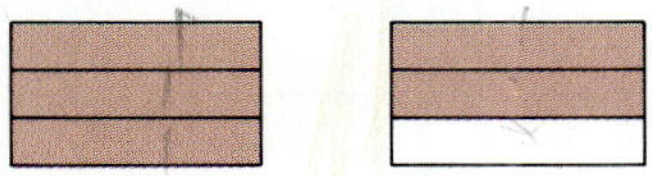

30. 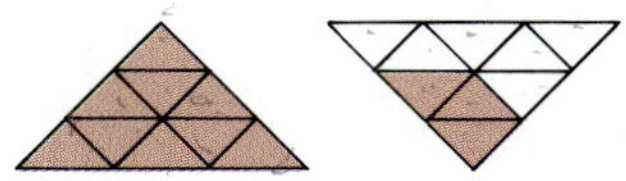

Mental Math

Mark records the number of pages used for stamps issued in a series. How many stamps were issued for each series?

	Stamp Series	Year Issued	Pages Used	Stamps Issued
31.	Great Americans	1982	$\frac{2}{3}$	
32.	Transportation	1985	$1\frac{1}{2}$	
33.	U.S. Presidents	1986	$5\frac{5}{6}$	
34.	American Wildlife	1987	$8\frac{1}{3}$	

Mixed Skills Review

1. $\$0.26 + \$0.59 + \$2.34$
2. $\frac{7}{10} \div 2\frac{4}{5}$
3. $86,147 + 153,624$
4. $10,712 - 4727$
5. 49×2185
6. 5.6×184.7
7. $50,084 - 17,346$
8. $0.402 - 0.38$
9. $9\frac{1}{6} \times 4\frac{4}{11}$
10. $3\frac{5}{12} \times 2\frac{1}{7}$
11. $54 \div 0.027$
12. $2142 \div 126$

3-5 Adding Fractions and Mixed Numbers

Objective
Add fractions.

A chess set contains 32 pieces, 16 white and 16 black. Many people collect chess sets as works of art. Each wooden piece is carefully made by a master craftsman.

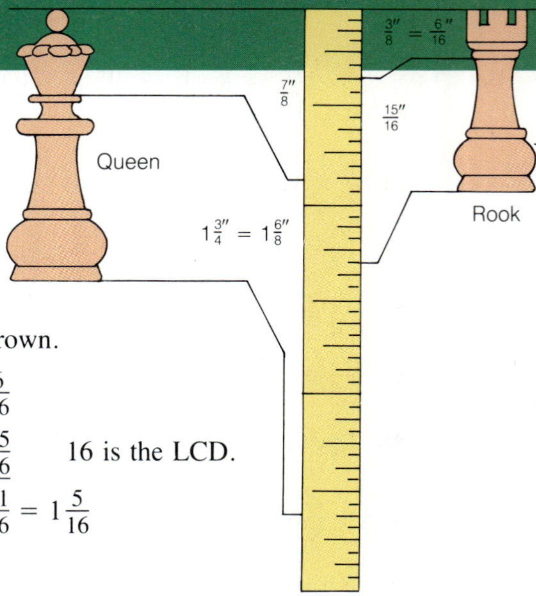

Examples

1. What is the total height of the rook?

Add the dimensions of the base and the crown.

$$\frac{3}{8} = \frac{3 \times 2}{8 \times 2} = \frac{6}{16}$$

$$\begin{array}{rcr} \frac{3}{8} & = & \frac{6}{16} \\ +\frac{15}{16} & = & +\frac{15}{16} \\ \hline & & \frac{21}{16} = 1\frac{5}{16} \end{array}$$

16 is the LCD.

To add fractions, we write equivalent fractions with the **least common denominator**.

The rook is $1\frac{5}{16}$ in. tall.

2. Find the total height of the queen.

Write equivalent fractions with the LCD. → Add the numerators. Write the sum over the common denominator. → Add the whole numbers. Rename.

$$\begin{array}{rcr} 1\frac{3}{4} & = & 1\frac{6}{8} \\ +\ \frac{7}{8} & = & \frac{7}{8} \\ \hline \end{array} \qquad \begin{array}{rcr} 1\frac{3}{4} & = & 1\frac{6}{8} \\ +\ \frac{7}{8} & = & \frac{7}{8} \\ \hline & & \frac{13}{8} \end{array} \qquad \begin{array}{rcr} 1\frac{3}{4} & = & 1\frac{6}{8} \\ +\ \frac{7}{8} & = & \frac{7}{8} \\ \hline & & 1\frac{13}{8} = 2\frac{5}{8} \end{array}$$

The queen is $2\frac{5}{8}$ in. tall.

Add.

3. $$\begin{array}{rcr} \frac{1}{6} & = & \frac{2}{12} \\ +\frac{3}{4} & = & \frac{9}{12} \\ \hline & & \frac{11}{12} \end{array}$$

4. $$\begin{array}{rcr} 2\frac{2}{5} & = & 2\frac{4}{10} \\ +\ \frac{1}{10} & = & \frac{1}{10} \\ \hline & & 2\frac{5}{10} = 2\frac{1}{2} \end{array}$$

5. $$\begin{array}{rcr} 3\frac{1}{2} & = & 3\frac{3}{6} \\ +4\frac{2}{3} & = & 4\frac{4}{6} \\ \hline & & 7\frac{7}{6} = 8\frac{1}{6} \end{array}$$

Try This

Add.

a. $\frac{5}{8} + \frac{1}{8}$ **b.** $\frac{2}{3} + \frac{1}{4}$ **c.** $\frac{7}{10} + 1\frac{2}{5}$ **d.** $3\frac{11}{16} + 1\frac{3}{8}$ **e.** $4\frac{1}{9} + 2\frac{5}{6}$

Exercises

Add.

1. $\frac{3}{5} + \frac{1}{5}$
2. $\frac{5}{6} + \frac{1}{6}$
3. $\frac{5}{8} + \frac{5}{8}$
4. $\frac{7}{10} + \frac{9}{10}$
5. $\frac{11}{12} + \frac{7}{12}$
6. $\frac{1}{2} + \frac{1}{6}$
7. $\frac{2}{3} + \frac{1}{8}$
8. $\frac{3}{5} + \frac{3}{10}$
9. $\frac{2}{3} + \frac{5}{6}$
10. $\frac{2}{7} + \frac{9}{14}$
11. $\frac{4}{15} + \frac{1}{3}$
12. $\frac{5}{6} + \frac{1}{9}$
13. $\frac{9}{10} + \frac{3}{4}$
14. $2\frac{1}{3} + \frac{2}{3}$
15. $\frac{7}{8} + 3\frac{3}{8}$
16. $9\frac{3}{7} + \frac{5}{7}$
17. $\frac{3}{10} + 2\frac{4}{5}$
18. $3\frac{1}{9} + \frac{2}{3}$
19. $4\frac{1}{6} + \frac{5}{9}$
20. $5\frac{1}{3} + 1\frac{1}{2}$
21. $1\frac{3}{4} + 2\frac{5}{6}$
22. $4\frac{3}{5} + 2\frac{1}{6}$
23. $4\frac{3}{4} + 2\frac{9}{16}$
24. $6\frac{1}{5} + 5\frac{1}{3}$
25. $3\frac{11}{12} + 7\frac{3}{8}$
26. $\frac{1}{2} + \frac{1}{4} + \frac{1}{3}$
27. $\frac{1}{8} + \frac{3}{4} + \frac{11}{16}$
28. $2\frac{2}{3} + 1\frac{3}{4} + \frac{5}{6}$

Estimation Estimate each distance. Then measure with a ruler to check your estimate.

29. AB
30. BC
31. CD
32. DE

Show Math Draw a line 6 inches long. Mark off the following segments on the line. Measure the distance from F to J. Check your work by adding the lengths.

33. FG $\frac{1}{4}''$
34. GH $\frac{7}{8}''$
35. HI $1\frac{3}{4}''$
36. IJ $\frac{1}{2}''$

Find the total height of each chess piece.

37. 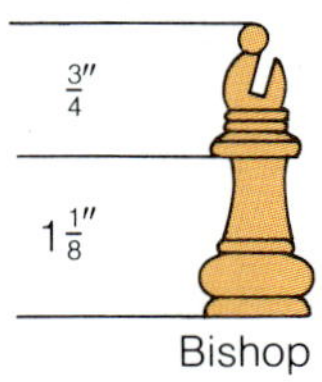

38. 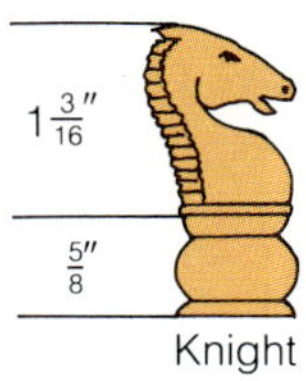

39. 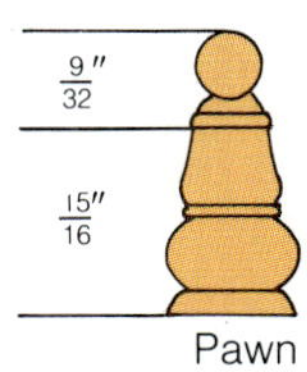

40. 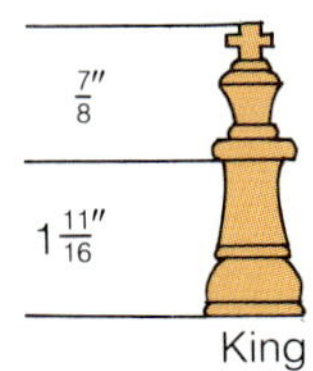

Mixed Skills Review

1. $\frac{5}{8} + \frac{7}{8}$
2. \$906 + \$87.63 + \$174.82
3. 21,523 − 6761
4. 11.79 × 0.62
5. $8 \div 3\frac{11}{15}$
6. 94.1 − 87.634
7. 93 + 2762 + 194
8. 35.376 ÷ 0.201
9. $3\frac{2}{5} + 1\frac{3}{4}$
10. 78 × 382
11. $6\frac{3}{10} \times 3\frac{3}{4}$
12. 35,554 ÷ 58

3-6 Subtracting Fractions and Mixed Numbers

Objective
Subtract fractions.

The Ranger Club has planned a hike around Hidden Lake. Sarah and Tom are trying to draw a map for new members. They know the lake is in the shape of a peanut.

Examples

1. What is the distance between Echo Glen and Rocky Ridge?

Are the denominators the same? → If so, can the fractions be subtracted? If not, rename. → Subtract the numerators. Subtract the whole numbers. Simplify.

$$\begin{array}{r} 3\frac{1}{8} \\ -1\frac{5}{8} \\ \hline \end{array} \quad \text{Yes.}$$

No, $\frac{1}{8}$ is less than $\frac{5}{8}$.

$$3\frac{1}{8} = 2 + \frac{8}{8} + \frac{1}{8} = 2\frac{9}{8}$$

$$\begin{array}{rr} 3\frac{1}{8} = & 2\frac{9}{8} \\ -1\frac{5}{8} = & -1\frac{5}{8} \\ \hline & 1\frac{4}{8} = 1\frac{1}{2} \end{array}$$

The distance between Echo Glen and Rocky Ridge is $1\frac{1}{2}$ miles.

2. What is the distance between Grist Mill and Arrowhead Point?

$$\begin{array}{rrr} 1\frac{1}{2} = & 1\frac{3}{6} = & \frac{9}{6} \\ -\ \frac{2}{3} = & \frac{4}{6} = & \frac{4}{6} \\ \hline & & \frac{5}{6} \end{array}$$

The denominators are not the same. Write equivalent fractions using the LCD.

Rename. $1\frac{3}{6} = \frac{9}{6}$

The distance between Grist Mill and Arrowhead Point is $\frac{5}{6}$ mile.

Subtract.

3. $$\begin{array}{rr} \frac{4}{7} = & \frac{8}{14} \\ -\frac{1}{2} = & -\frac{7}{14} \\ \hline & \frac{1}{14} \end{array}$$

4. $$\begin{array}{rr} 5 = & 4\frac{4}{4} \\ -2\frac{3}{4} = & -2\frac{3}{4} \\ \hline & 2\frac{1}{4} \end{array}$$

5. $$\begin{array}{rrr} 6\frac{2}{5} = & 6\frac{8}{20} = & 5\frac{28}{20} \\ -3\frac{3}{4} = & -3\frac{15}{20} = & -3\frac{15}{20} \\ \hline & & 2\frac{13}{20} \end{array}$$

Try This Subtract.

a. $\frac{15}{16} - \frac{7}{16}$ **b.** $\frac{2}{3} - \frac{2}{9}$ **c.** $4\frac{5}{6} - 2\frac{1}{6}$ **d.** $3\frac{1}{2} - 1\frac{3}{4}$ **e.** $5\frac{1}{3} - 3\frac{3}{5}$

Exercises

Subtract.

1. $\frac{11}{12} - \frac{5}{12}$
2. $\frac{5}{8} - \frac{3}{8}$
3. $\frac{3}{4} - \frac{1}{2}$
4. $\frac{2}{3} - \frac{4}{9}$
5. $\frac{3}{5} - \frac{7}{15}$
6. $\frac{9}{10} - \frac{5}{6}$
7. $\frac{13}{16} - \frac{5}{8}$
8. $\frac{3}{4} - \frac{3}{5}$
9. $8\frac{2}{5} - 6$
10. $3\frac{5}{9} - 3$
11. $4\frac{1}{2} - \frac{1}{2}$
12. $5\frac{9}{10} - \frac{7}{10}$
13. $6\frac{7}{8} - \frac{1}{8}$
14. $4\frac{11}{18} - \frac{5}{18}$
15. $7\frac{2}{3} - \frac{1}{6}$
16. $1\frac{3}{4} - \frac{5}{12}$
17. $3\frac{7}{8} - \frac{1}{4}$
18. $5\frac{1}{4} - \frac{3}{4}$
19. $3\frac{1}{3} - \frac{2}{3}$
20. $9\frac{5}{11} - \frac{7}{11}$
21. $1 - \frac{3}{5}$
22. $2 - \frac{7}{9}$
23. $9\frac{1}{4} - 3\frac{5}{8}$
24. $4\frac{1}{3} - 1\frac{3}{4}$
25. $5\frac{1}{4} - 2\frac{3}{16}$
26. $2\frac{3}{5} - 1\frac{7}{10}$
27. $4\frac{1}{6} - 3\frac{5}{9}$
28. $9\frac{2}{3} - 5\frac{7}{8}$

Number Sense

Continue the pattern.

29. $10\frac{1}{5}$, $9\frac{3}{5}$, 9, $8\frac{2}{5}$, ___, ___, ___, 6
30. $12\frac{2}{7}$, $11\frac{4}{7}$, $10\frac{6}{7}$, ___, ___, ___, 8
31. $9\frac{1}{2}$, $8\frac{7}{8}$, $8\frac{1}{4}$, ___, ___, ___, $5\frac{3}{4}$
32. Anne Banta compared the effects of different fertilizers on the growth of plants. The numbers indicate the percent of nitrogen, phosphorus, and potash in the plant food. Find the difference in total growth at the end of each week for a three-week period.

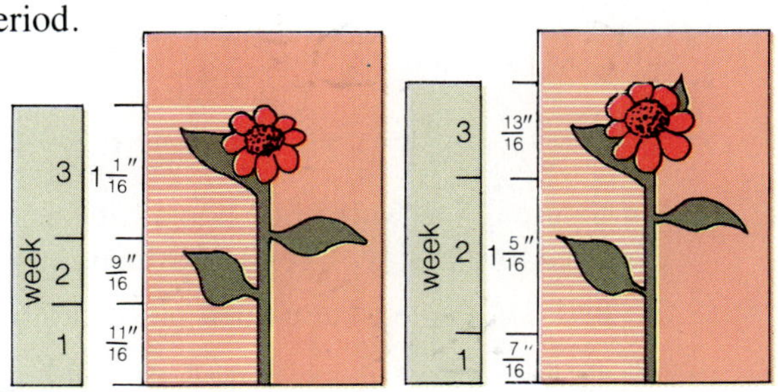

Fertilizer 5-10-5 Fertilizer 10-10-5

Mixed Skills Review

1. 3.6 + 273.54 + 18.28
2. 197 × 346
3. $\frac{2}{5} - \frac{1}{3}$
4. 10,712 − 4727
5. 5.008 × 55
6. $2\frac{4}{9} \div 11$
7. 25,252 ÷ 107
8. $2\frac{5}{8} - 1\frac{3}{4}$
9. \$100.06 − \$63.39
10. $10\frac{1}{2} \times 5\frac{1}{3}$
11. 29.376 ÷ 6.12
12. 6158 + 11,313 + 754

3-7 Mental Math and Fractions

Objective
Use mental math with fractions.

Pat works in a flower shop on Saturdays and after school. She must work 20 hours each week to qualify for vacation time. Pat checks her time card to see how many hours she has worked during the week.

Examples

1. Add. Use mental math. $2\frac{1}{4} + 4 + 2\frac{3}{4} + 3\frac{1}{2}$

Notice that two of the fractions fit together to make a whole number.

$2\frac{1}{4} + 2\frac{3}{4} = 4 + 1 = 5$ Add the compatible numbers. Use mental math.

$5 + 4 + 3\frac{1}{2} = 12\frac{1}{2}$ Add the other numbers to the sum.

Pat has worked $12\frac{1}{2}$ hours this week.

2. $5\frac{7}{8} + 1\frac{1}{3} + 3\frac{1}{8}$ *Think*: $5\frac{7}{8} + 3\frac{1}{8} = 9$ $9 + 1\frac{1}{3} = 10\frac{1}{3}$

Try This

Add. Use mental math.

a. $3\frac{1}{6} + 2 + 1\frac{5}{6}$ **b.** $2\frac{1}{2} + 4\frac{1}{2} + 1\frac{1}{2}$ **c.** $5\frac{4}{7} + 3\frac{1}{7} + \frac{3}{4}$

Examples

3. On Saturday Pat worked $6\frac{1}{2}$ hours. She earned \$4 per hour. Find her total earnings for the day. Use mental math.

To find the amount Pat earned on Saturday, multiply $\$4 \times 6\frac{1}{2}$.

Think: $6\frac{1}{2} = 6 + \frac{1}{2}$ Break apart the mixed number.

$\$4 \times 6\frac{1}{2} = (\$4 \times 6) + \left(\$4 \times \frac{1}{2}\right) = \$24 + \$2 = \$26$

Pat earned \$26 on Saturday.

4. $9 \times 3\frac{1}{8} = (9 \times 3) + \left(9 \times \frac{1}{8}\right) = 27 + 1\frac{1}{8} = 28\frac{1}{8}$

Try This

Multiply. Use mental math.

d. $6 \times 4\frac{1}{3}$ **e.** $10 \times 3\frac{3}{5}$ **f.** $5 \times 3\frac{1}{4}$

Exercises

Add. Use mental math.

1. $2\frac{1}{4} + 5 + 3\frac{3}{4}$ **2.** $1\frac{3}{7} + 4 + 2\frac{4}{7}$ **3.** $9\frac{1}{3} + 5\frac{1}{3} + \frac{1}{3}$

4. $1\frac{1}{6} + 3\frac{5}{6} + 4$ **5.** $4\frac{1}{12} + 2 + 3\frac{7}{12}$ **6.** $1\frac{5}{8} + 6 + 3\frac{1}{8}$

7. $\frac{6}{11} + 1 + 4\frac{5}{11}$ **8.** $8\frac{3}{4} + 3\frac{1}{4} + \frac{3}{4}$ **9.** $5\frac{1}{9} + \frac{2}{9} + 1\frac{6}{9}$

10. $\frac{9}{16} + 2\frac{7}{16} + 1\frac{3}{8}$ **11.** $5\frac{1}{4} + 3\frac{2}{3} + 2\frac{3}{4}$ **12.** $5\frac{11}{18} + \frac{4}{18} + 1\frac{3}{18}$

Multiply. Use mental math.

13. $9 \times 2\frac{1}{3}$ **14.** $10 \times 3\frac{1}{2}$ **15.** $6 \times 1\frac{5}{6}$ **16.** $15 \times 4\frac{2}{5}$

17. $\$12 \times 1\frac{1}{2}$ **18.** $\$16 \times 2\frac{3}{8}$ **19.** $\$20 \times 1\frac{7}{10}$ **20.** $\$24 \times 3\frac{7}{12}$

21. $2\frac{2}{3} \times 18$ **22.** $1\frac{1}{2} \times 14$ **23.** $3\frac{1}{4} \times 8$ **24.** $6\frac{3}{7} \times 7$

25. $7 \times 1\frac{1}{5}$ **26.** $9 \times 2\frac{1}{4}$ **27.** $4 \times 3\frac{1}{3}$ **28.** $8 \times 1\frac{1}{6}$

Mental Math

Find the cost of the flowers. Use mental math.

29. $\frac{3}{4}$ doz roses

30. $1\frac{1}{2}$ doz roses

31. $1\frac{1}{4}$ doz carnations

32. $2\frac{3}{4}$ doz carnations

	Dozen		
Flowers	$\frac{1}{4}$	$\frac{1}{2}$	1
Roses	\$3.75	\$7.50	\$15
Carnations	\$2.25	\$4.50	\$9

33. Pat needed to measure fertilizer. She filled a $\frac{1}{2}$-c container and poured the fertilizer into a $\frac{1}{3}$-c container. How much was left in the $\frac{1}{2}$-c container?

Mixed Skills Review

1. 42,186 ÷ 79 **2.** $6\frac{2}{3} + 8 + 4\frac{1}{3}$ **3.** 4751 × 84

4. 38,670 + 64,780 **5.** $9\frac{5}{6} - 5\frac{7}{8}$ **6.** 1051.84 ÷ 3.04

7. $3\frac{6}{7} \times 5\frac{5}{6}$ **8.** 12.36 + 17.8 + 35.67 **9.** 8216 − 449

10. 15.9 × 2 × 3.6 **11.** $7\frac{3}{11} \div 2\frac{4}{33}$ **12.** 300 − 258.037

3-8 Estimating with Fractions

Objective
Estimate with fractions.

A bolt of fabric usually contains 25 yards of material. A yard goods store sells fabric in fractions of a yard in the lengths shown on the yardstick.

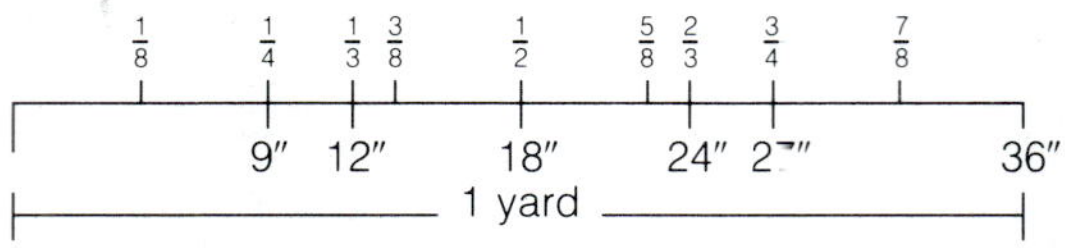

Understand the Situation

- Can you purchase exactly $\frac{5}{8}$ yd of fabric?
- Can you purchase exactly $\frac{1}{5}$ yd of fabric?
- Can you purchase exactly 9 in. of fabric?

Examples

1. Julia needs 19 in. of a fabric that costs \$5.65 per yd. Estimate the cost.

Since there are 36 inches in a yard, express the amount of fabric as $\frac{19}{36}$ yd.
We need to estimate the product. $\frac{19}{36} \times \$5.65$
Write compatible numbers.

$\frac{19}{36}$ is about $\frac{1}{2}$. \$5.65 is about \$6.

$\frac{1}{2} \times \$6 = \$3$ The fabric will cost about \$3.

2. A bow takes 26 in. of ribbon. About how many bows can be made from 5 yards of ribbon?

We need to estimate the quotient $5 \div \frac{26}{36}$
Write as a compatible number.

$\frac{26}{36}$ is about $\frac{27}{36}$ $\frac{27}{36} = \frac{3}{4}$

$5 \div \frac{3}{4} = 5 \times \frac{4}{3} = 6\frac{2}{3} \approx 7$

About 7 bows can be made from 5 yd of ribbon.

Try This

Estimate the answer. Use compatible numbers.

a. $\frac{15}{19} \times 12$ **b.** $\frac{3}{31} \times \$19.95$ **c.** $5 \div \frac{31}{36}$ **d.** $9.9 \div \frac{7}{13}$

e. $\frac{19}{24} \times \$9.67$ **f.** $\frac{3}{16} \times 27$ **g.** $34 \div \frac{9}{17}$ **h.** $13.6 \div \frac{8}{21}$

Exercises

Estimate the answer. Use compatible numbers.

1. $\frac{17}{36} \times \$44$ **2.** $\frac{29}{36} \times \$18$ **3.** $\frac{16}{21} \times 40$ **4.** $\frac{10}{48} \times 25$

5. $\frac{11}{36} \times \$9.11$ **6.** $\frac{8}{36} \times \$15.84$ **7.** $\frac{40}{62} \times 5.49$ **8.** $\frac{20}{39} \times 7.5$

9. $10 \div \frac{4}{25}$ **10.** $30 \div \frac{31}{40}$ **11.** $45 \div \frac{26}{36}$ **12.** $2 \div \frac{10}{39}$

13. $\$6.03 \div \frac{12}{17}$ **14.** $29.7 \div \frac{51}{60}$ **15.** $7.9 \div \frac{29}{35}$ **16.** $\$15.99 \div \frac{30}{39}$

17. Claire needs 29 inches of fabric that costs $4.59 per yard. Estimate the cost.

18. A large pillow case requires 31 inches of fabric that costs $6.75 per yard. Estimate the cost.

19. A costume requires 17 inches of lace. About how many costumes can be prepared from 8 yards of lace?

20. A child's pair of pajamas requires 25 inches of elastic. About how many pairs of pajamas can be prepared from 10 yards of elastic?

Mental Math

A yard goods store only sells fabric in halves, thirds, fourths, and eighths of a yard. What fractions of a yard must a customer buy for each of the following lengths of fabric? Use the yardstick on page 68.

21. 21 inches **22.** 8 inches **23.** 14 inches **24.** 28 inches

25. 42 inches **26.** 49 inches **27.** 64 inches **28.** 55 inches

Data Bank

Suppose you find a remnant which is 44 inches wide and $3\frac{1}{2}$ yards long. Use the sewing pattern on p. 440 of the Data Bank to decide whether you can make the following items.

29. size 12 shorts and shirt **30.** size 10 pants and shirt

31. three pairs size 14 shorts **32.** size 14 shorts and shirt

Mixed Skills Review

1. $13\frac{1}{3} \div 2\frac{7}{9}$ **2.** $\$1.98 + \$0.36 + \$7.23$ **3.** 627×364

4. $6\frac{5}{12} - 4\frac{5}{8}$ **5.** $30{,}552 \div 38$ **6.** $15.6 - 9.236$

7. $5632 + 2765 + 184$ **8.** $8.987 \div 0.19$ **9.** $2\frac{3}{4} + 4\frac{1}{2} + 2\frac{3}{8}$

10. $\$28.43 \times 0.7$ **11.** $6073 - 4268$ **12.** $9\frac{3}{5} \times 7\frac{1}{2}$

3-9 Enrichment: Fractions and Figures

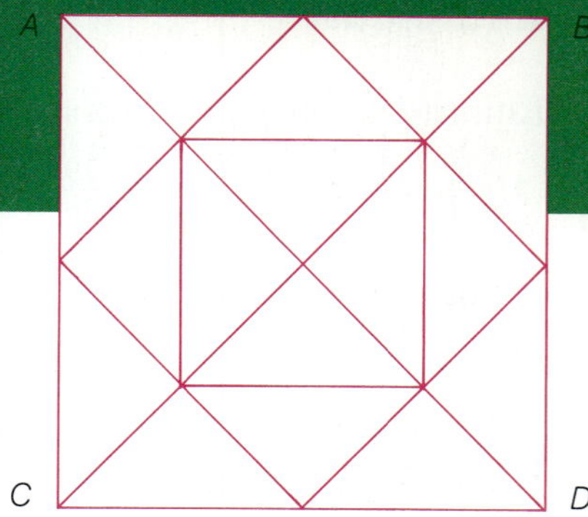

Objective
Write fractions for parts of figures.

You can use drawings to represent a fractional part of a whole. Figure *ABCD* is 1 unit. It is divided into several equivalent triangles. Find what fractional part a figure is of *ABCD* by counting how many triangles from *ABCD* fill the figure.

Write what fraction of *ABCD* this figure represents.

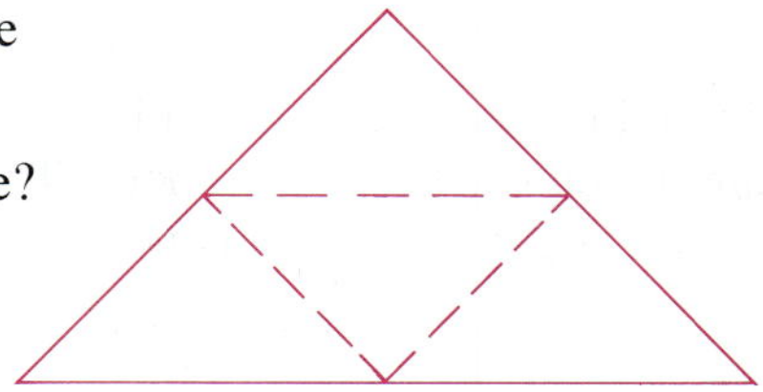

How many small triangles fill the figure?
How many small triangles fill *ABCD*?

$$\frac{\text{Triangles in the figure}}{\text{Triangles in } ABCD} = \frac{4}{16} = \frac{1}{4}$$

The figure is $\frac{1}{4}$ of *ABCD*.

You can determine what fraction of a figure is shaded by using this formula:

$$\text{Fraction} = \frac{\text{Shaded segments}}{\text{Total segments}}$$

Write the fraction for the shaded part.

$$\frac{\text{Shaded segments}}{\text{Total segments}} = \frac{12}{24} = \frac{1}{2}$$

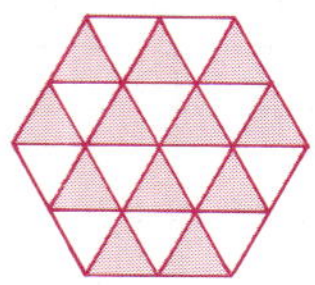

$\frac{1}{2}$ of the segments are shaded.

Exercises

Write what fraction of *ABCD* each of these figures represents.

1. 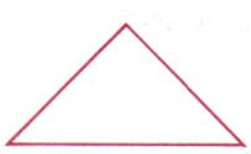

2. 

3. 

4. 

Write the fraction for the shaded part.

5. 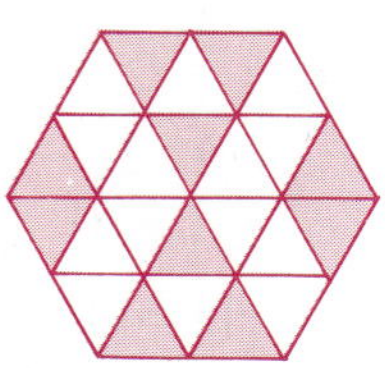

6. 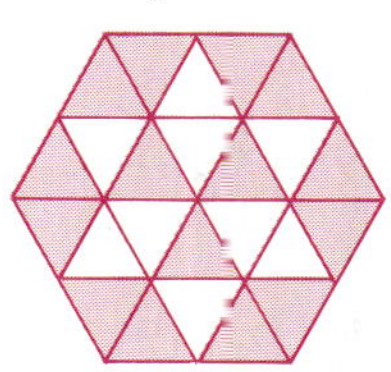

7. 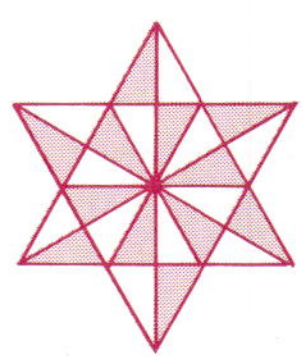

8. 

3-10 Calculator: Operations with Fractions

Objective
Use a calculator to add, subtract, multiply, and divide fractions.

Memory functions on calculators can help add and subtract fractions.

Press [M+] to add a number into the memory.

Press [M−] to subtract a number from the memory.

Press [MRC] to recall and display a number in the memory.

Press [MRC] twice to clear the memory.

To add fractions on a calculator, divide the first numerator by its denominator and store the quotient in the memory. Then divide the second numerator by its denominator, and add the quotient to the memory.

Examples

1. Add. $\frac{3}{4} + \frac{5}{6}$ 3 [÷] 4 [M+] 5 [÷] 6 [M+] [MRC] 1.5833333

To subtract fractions, use the [M−] key.

2. Subtract. $\frac{3}{8} - \frac{1}{4}$ 3 [÷] 8 [M+] 1 [÷] 4 [M−] [MRC] 0.125

To multiply fractions, perform each operation from left to right.

3. Multiply. $\frac{1}{5} \times \frac{4}{7}$ 1 [÷] 5 [×] 4 [÷] 7 [=] 0.1142857

To divide fractions, remember to invert the divisor and multiply.

4. Divide. $\frac{7}{8} \div \frac{2}{5}$ *Think*: $\frac{7}{8} \times \frac{5}{2}$. 7 [÷] 8 [×] 5 [÷] 2 [=] 2.1875

Try This

a. $\frac{6}{7} + \frac{14}{15}$ **b.** $\frac{2}{3} - \frac{1}{6}$ **c.** $\frac{4}{5} \times \frac{3}{4}$ **d.** $\frac{3}{4} \div \frac{3}{8}$

Exercises

Use a calculator.

1. $\frac{3}{7} + \frac{1}{6}$ **2.** $\frac{5}{8} + \frac{3}{5}$ **3.** $\frac{4}{5} + \frac{5}{12}$ **4.** $\frac{1}{2} + \frac{1}{3}$ **5.** $\frac{3}{8} + \frac{8}{9}$

6. $\frac{4}{5} - \frac{1}{2}$ **7.** $\frac{2}{3} - \frac{2}{5}$ **8.** $\frac{8}{9} - \frac{3}{10}$ **9.** $\frac{7}{6} - \frac{5}{7}$ **10.** $\frac{11}{12} - \frac{3}{8}$

11. $\frac{4}{5} \times \frac{3}{4}$ **12.** $\frac{5}{12} \times \frac{3}{8}$ **13.** $\frac{1}{9} \times \frac{2}{3}$ **14.** $\frac{7}{8} \times \frac{2}{13}$ **15.** $\frac{5}{12} \times \frac{1}{2}$

16. $\frac{4}{11} \div \frac{1}{5}$ **17.** $\frac{7}{8} \div \frac{1}{3}$ **18.** $\frac{11}{12} \div \frac{3}{7}$ **19.** $\frac{8}{9} \div \frac{5}{7}$ **20.** $\frac{4}{5} \div \frac{2}{5}$

CHAPTER 3 REVIEW

Match each $\square$ in 1–10 with A–J.

1. $\frac{15}{20} = \square$

2. $\frac{2}{3} \times \frac{9}{10} = \square$

3. $\frac{1}{6} + \frac{1}{2} = \square$

4. $1\frac{\square}{7} = \frac{10}{7}$

5. $\frac{7}{8} \div \frac{8}{7} = \square$

6. $4\frac{1}{3} - 3\frac{5}{6} = \square$

7. $\frac{21}{20}$ is about $\square$

8. $\frac{6}{8} \div \frac{3}{8} = \square$

9. $2\frac{1}{8} - 1\frac{7}{8} = \square$

10. $\frac{1}{2} + \frac{1}{3} + \frac{1}{4} = \square$

A. 2

B. $\frac{1}{4}$

C. $\frac{3}{4}$

D. 1

E. $1\frac{1}{12}$

F. $\frac{3}{5}$

G. $\frac{2}{3}$

H. 3

I. $\frac{1}{2}$

J. none of the above

Which is incorrect and does not belong in the group?

11. **A.** $\frac{6}{4} = 1\frac{1}{2}$ **B.** $\frac{9}{6} = 1\frac{1}{2}$ **C.** $\frac{7}{5} = 1\frac{1}{2}$

12. **A.** $\frac{1}{2} = \frac{3}{6}$ **B.** $\frac{2}{3} = \frac{5}{6}$ **C.** $\frac{3}{2} = \frac{12}{8}$

13. **A.** $\frac{51}{7} = 7\frac{2}{7}$ **B.** $\frac{51}{9} = 5\frac{1}{3}$ **C.** $\frac{51}{11} = 4\frac{7}{11}$

14. **A.** $\frac{1}{5}$ of 240 is 48. **B.** $\frac{1}{240}$ of 5 is $\frac{1}{48}$. **C.** $\frac{4}{5}$ of 10 is 240.

15. **A.** $\frac{3}{4}$ of 75 is about 50. **B.** $\frac{2}{3}$ of 14 is about 10. **C.** $\frac{5}{8}$ of 50 is about 30.

16. **A.** $\frac{1}{3} \times \frac{3}{2} = \frac{1}{2}$ **B.** $\frac{1}{3} \times \frac{2}{5} = \frac{2}{15}$ **C.** $\frac{1}{3} \times \frac{1}{2} = \frac{1}{5}$

17. **A.** $1\frac{3}{5} \div \frac{1}{5} = 8$ **B.** $1 \div \frac{1}{5} = 5$ **C.** $\frac{1}{5} \div 2 = 10$

CHAPTER 3 TEST

Write in lowest terms.

1. $\frac{6}{8}$ **2.** $\frac{5}{35}$ **3.** $\frac{28}{64}$ **4.** $\frac{112}{168}$

Write the missing numeral.

5. $\frac{2}{5} = \frac{8}{\square}$ **6.** $\frac{7}{5} = \frac{56}{\square}$ **7.** $\frac{14}{17} = \frac{84}{\square}$ **8.** $\frac{35}{9} = \frac{175}{\square}$

Write as a fraction.

9. $3\frac{4}{5}$ **10.** $4\frac{7}{9}$ **11.** $12\frac{5}{8}$ **12.** $9\frac{10}{11}$

Write as a mixed number or a whole number.

13. $\frac{8}{5}$ **14.** $\frac{26}{8}$ **15.** $\frac{44}{13}$ **16.** $\frac{89}{7}$

Multiply.

17. $\frac{2}{3} \times \frac{5}{8}$ **18.** $28 \times \frac{9}{16}$ **19.** $4\frac{4}{9} \times \frac{3}{4}$ **20.** $\frac{7}{8}$ of \$82.96

Divide.

21. $\frac{3}{4} \div \frac{1}{2}$ **22.** $\frac{5}{8} \div \frac{5}{16}$ **23.** $3\frac{5}{7} \div \frac{13}{14}$ **24.** $5\frac{5}{6} \div 3\frac{1}{9}$

Add.

25. $\frac{16}{23} + \frac{7}{23}$ **26.** $\frac{5}{12} + \frac{11}{18}$ **27.** $3 + 2\frac{5}{9}$ **28.** $\frac{3}{4} + 2\frac{5}{16}$

Subtract.

29. $\frac{6}{7} - \frac{2}{7}$ **30.** $\frac{3}{8} - \frac{5}{24}$ **31.** $3\frac{7}{18} - 1\frac{8}{9}$ **32.** $5 - 3\frac{14}{15}$

Add. Use mental math.

33. $4\frac{5}{7} + 3 + 2\frac{1}{7}$ **34.** $1\frac{2}{5} + 3\frac{3}{10} + 5\frac{3}{10}$

35. $8\frac{2}{5} + 1 + \frac{2}{5}$ **36.** $\frac{3}{4} + \frac{3}{8} + \frac{2}{16}$

Multiply. Use mental math.

37. $3\frac{1}{4} \times 16$ **38.** $\$25 \times 2\frac{1}{5}$ **39.** $33 \times 2\frac{3}{11}$ **40.** $\frac{3}{23} \times 69$

Estimate the answer. Use compatible numbers.

41. $\frac{11}{24} \times 50$ **42.** $\frac{13}{40} \times \$15$ **43.** $\frac{24}{97} \times \$39.96$ **44.** $\frac{7}{37} \times 5.5$

CHAPTER 3 CUMULATIVE REVIEW

1. Write as a fraction.
$8\frac{3}{5}$
- A. $\frac{24}{5}$
- B. $\frac{43}{8}$
- C. $\frac{17}{5}$
- D. $\frac{43}{5}$

2. Multiply.
$78[illegible]3 \times 26$
- A. 203,138
- B. 151,138
- C. 193,138
- D. 203,128

3. Add.
$\frac{1}{3} + \frac{5}{6}$
- A. 1
- B. $1\frac{1}{6}$
- C. $1\frac{1}{3}$
- D. $\frac{2}{3}$

4. The playing field is 40 yards by 100 yards. What is the area?
Use $A = l \times w$.
- A. 250 yd^2
- B. 400 yd^2
- C. 4000 yd^2
- D. 2000 yd^2

5. Subtract.
$54 − 21¢
- A. $54.79
- B. $53.21
- C. $53.79
- D. $33

6. Divide.
$\frac{7}{5} \div \frac{7}{3}$
- A. $1\frac{1}{15}$
- B. $\frac{15}{49}$
- C. $\frac{49}{15}$
- D. $\frac{3}{5}$

7. Keith is 5 years older than Darryl. Darryl is 2 years younger than Erin. Erin is 16. How old is Keith?
- A. 21
- B. 14
- C. 19
- D. 18

8. Add. Use mental math.
$2\frac{3}{5} + 5 + 1\frac{2}{5}$
- A. 8
- B. 9
- C. 10
- D. 11

9. Write in lowest terms.
$\frac{28}{32}$
- A. $\frac{4}{7}$
- B. $\frac{1}{2}$
- C. $\frac{2}{3}$
- D. $\frac{7}{8}$

10. Divide.
$6482 \div 14$
- A. 463
- B. 648
- C. 462
- D. 420

11. Multiply.
$\frac{5}{7} \times \frac{14}{15}$
- A. $2\frac{1}{3}$
- B. $\frac{2}{3}$
- C. $\frac{3}{7}$
- D. $1\frac{3}{7}$

12. Add.
$$\begin{array}{r} 2786 \\ 8693 \\ +3401 \\ \hline \end{array}$$
- A. 13,770
- B. 14,880
- C. 13,780
- D. 14,870

13. Add.

$1\frac{2}{3} + 5\frac{1}{9}$

A. $6\frac{1}{3}$

B. $7\frac{1}{3}$

C. $6\frac{7}{9}$

D. $7\frac{5}{9}$

14. Multiply. Use mental math.

$4 \times 3\frac{3}{4}$

A. 12

B. $9\frac{1}{4}$

C. $12\frac{3}{4}$

D. 15

15. Sophie bikes 27 km on Monday and Tuesday. She bikes 15 km on Thursday and Saturday. How many kilometers does she bike in a week?

A. 54 km

B. 30 km

C. 84 km

D. 57 km

16. Subtract.

$3\frac{7}{12} - \frac{5}{12}$

A. $3\frac{1}{6}$

B. $2\frac{1}{4}$

C. $2\frac{1}{6}$

D. $3\frac{1}{12}$

17. Is the answer reasonable? The school lunch special is $2.50. Sal buys it 4 days a week. How much does it cost him a month?

Answer: $75

A. Reasonable

B. Unreasonable

18. Find the product. Use mental math.

85×5

A. 435

B. 415

C. 405

D. 425

19. Estimate. Use compatible numbers.

$\frac{30}{63} \times 77$

A. about 20

B. about 80

C. about 60

D. about 40

20. Round to nearest hundred.

45,717

A. 45,700

B. 45,800

C. 45,600

D. 46,000

21. Divide.

$8 \div \frac{2}{7}$

A. 32

B. $\frac{1}{28}$

C. 28

D. $2\frac{2}{7}$

Chapter 4 Overview

Key Ideas

- meaning of percents
- estimating percents
- writing percents as decimals and fractions
- writing fractions as decimals and percents
- making comparisons with percents
- using percents in equations
- finding a percent of a number
- finding what percent one number is of another
- finding the percent of increase or decrease

Key Terms

- percent
- percent sign
- circle graph
- probability
- statistics
- comparisons
- greater than (>)
- less than (<)
- translate
- number sentence
- compatible numbers
- increase
- decrease
- original amount

Key Skills

Find the answer.

1. $100 - 15\frac{1}{2}$ **2.** $100 - 64$ **3.** $100 - 83\frac{1}{3}$

4. $37\frac{1}{2} + 18\frac{1}{5}$ **5.** $25\frac{1}{10} + 38\frac{1}{5}$ **6.** $27\frac{3}{4} + 36\frac{7}{8} + 64\frac{1}{2}$

7. $\frac{3}{7} \times 100$ **8.** 19×100 **9.** $\frac{8}{7} \times 100$

10. 8.2×0.01 **11.** 18.5×0.01 **12.** 29.6×0.01

13. $\frac{1}{100} \times 46$ **14.** $38 \times \frac{1}{100}$ **15.** $59 \times \frac{1}{100}$

16. $0.345 \div 0.01$ **17.** $0.695 \div 0.01$ **18.** $7.85 \div 0.01$

19. 26.9×37.4 **20.** 19.9×27.8 **21.** 47.2×13.3

Solve.

22. $n \times 240 = 60$ **23.** $34 = n \times 102$ **24.** $17 \times n = 7$

25. $n \times 350 = 50$ **26.** $42 = n \times 84$ **27.** $18 \times n = 2$

Compare. Use $<$, $>$, or $=$ for each □.

28. 0.38 □ 0.45 **29.** $\frac{66}{100}$ □ 0.57 **30.** $\frac{11}{2}$ □ 1.5 **31.** 0.9 □ 0.91 **32.** 0.53 □ $\frac{53}{100}$ **33.** $\frac{12}{5}$ □ 2.6

USING PERCENT

4

4-1 Understanding Percent

Objective
Recognize percents.

Practical problems often involve percent. Many numbers we use each day are expressed as percents. Percent means **per hundred.**

Example

1. In a recent poll at Alamo High School 65% of the students said Carl La Fong was their favorite recording artist. What percent of the students polled did not pick Carl La Fong?

Think of the total number of students as 100%. Subtract using mental math. 100% − 65% = 35%

Try This

Solve. Use mental math.

a. Only 41% of the students in the junior class had summer jobs. What percent of the students did not have summer jobs?

b. A national poll reported that $16\frac{2}{3}\%$ of the people surveyed attended a movie at least once a month. What percent of those surveyed did not?

Information is frequently presented in the form of a circle graph.

Example

2. Find the missing percent.

Add the known percents.

$$37\frac{1}{2} = 37\frac{5}{10}$$
$$19\frac{3}{5} = 19\frac{6}{10}$$
$$+26\frac{7}{10} = 26\frac{7}{10}$$
$$82\frac{18}{10} = 83\frac{4}{5}$$

Subtract the sum from 100%.

$$100 = 99\frac{5}{5}$$
$$-83\frac{4}{5} = 83\frac{4}{5}$$
$$16\frac{1}{5}$$

Favorite Sherbet Flavors
Orange $37\frac{1}{2}\%$
Rainbow $19\frac{3}{5}\%$
Raspberry $26\frac{7}{10}\%$
Lime

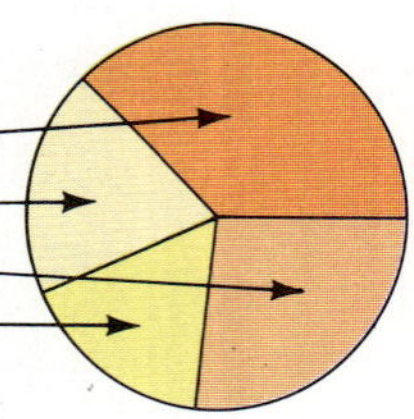

$16\frac{1}{5}\%$ of the people surveyed preferred lime.

Try This

Find the missing percent.

c. Student Body
Van Horn High School
Seniors 26.6%
Juniors 25.25%
Freshmen 24.9%
Sophomores

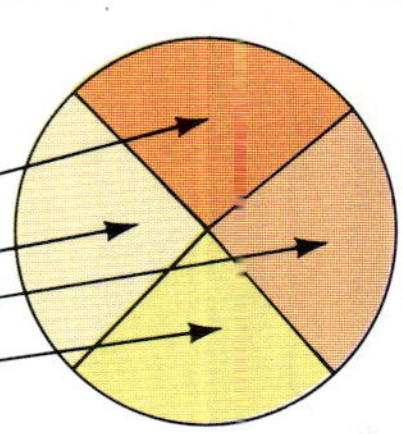

d. U.S. Family Incomes
Under \$5000 7%
\$5000–\$25,000
\$25,000–\$50,000 $29\frac{1}{2}\%$
Over \$50,000 $5\frac{1}{10}\%$

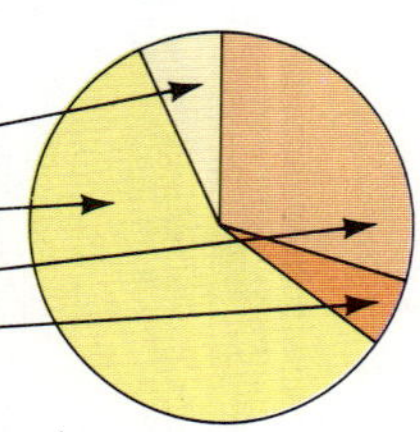

Exercises

Estimation Estimate the missing percent.

1. 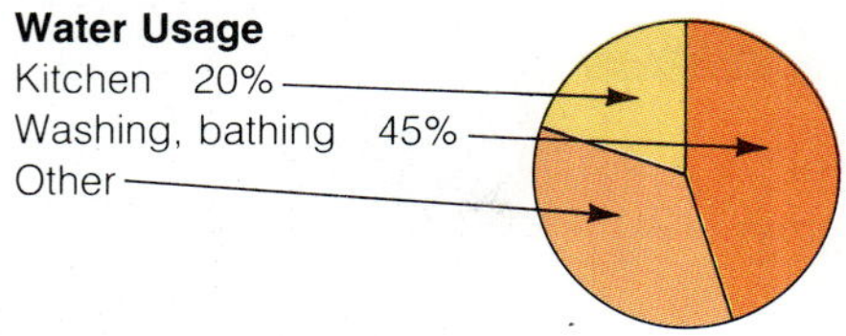

2. 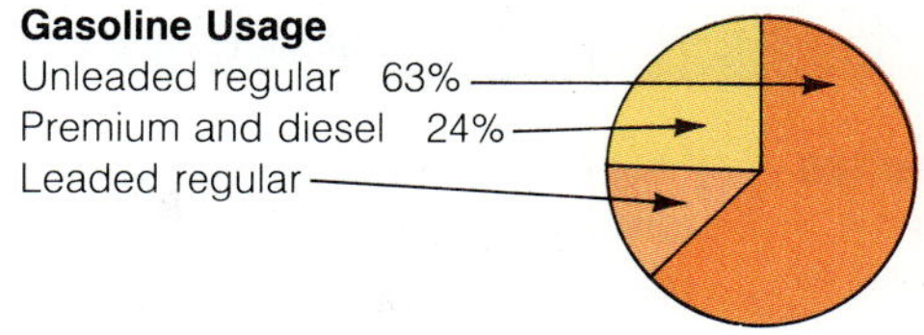

3. 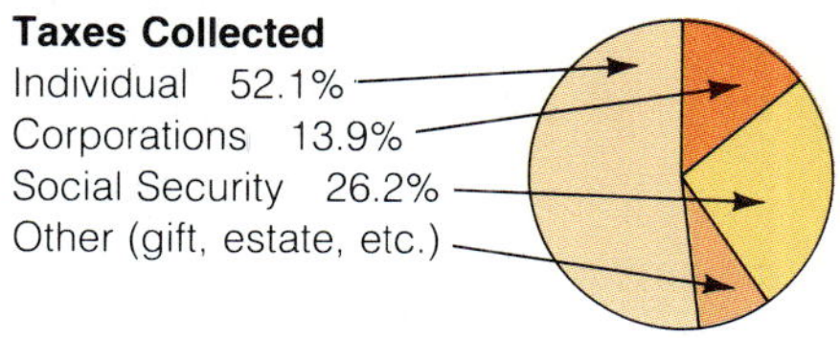

4. 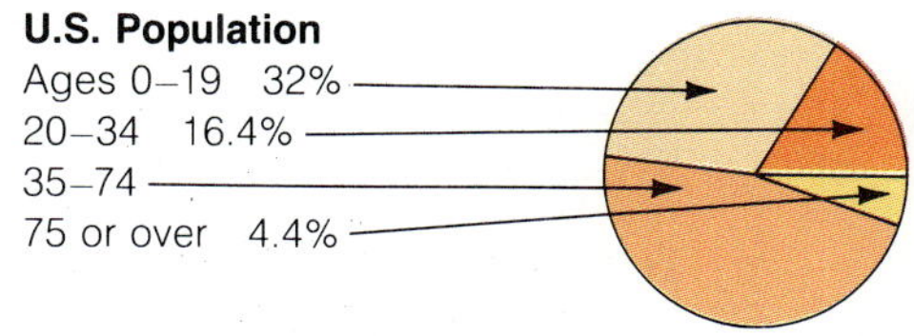

5. 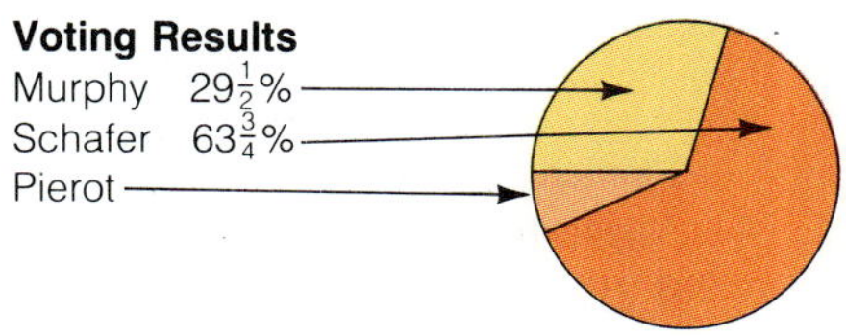

6. 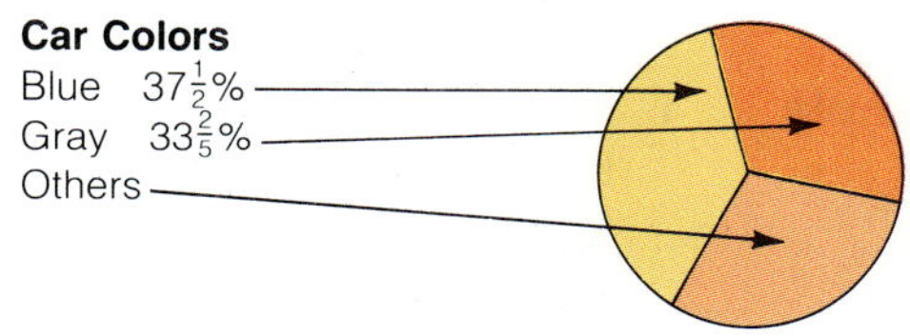

Mental Math Solve. Use mental math.

7. At Kingston Airport 57% of the tickets were purchased for business trips. What percent of the tickets were purchased for other reasons?
8. The center on the Cougars basketball team missed $12\frac{3}{4}\%$ of his free-throw attempts. What percent of his free throws did he make?

Estimation Estimate what percent of each circle is shaded.

9. 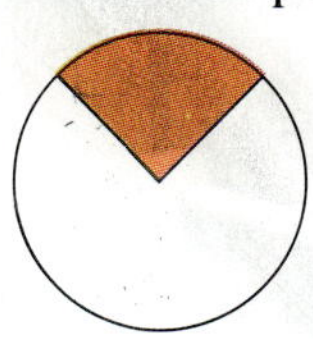

10. 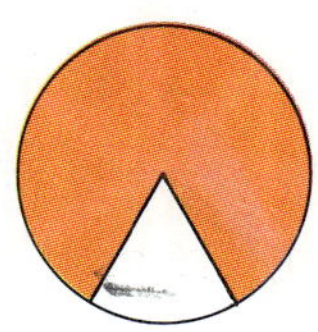

11. 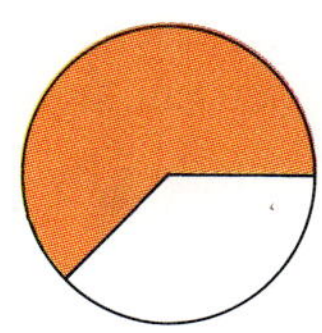

12. 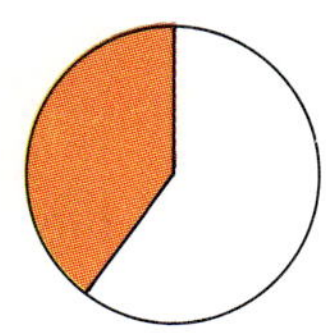

Mixed Skills Review

1. $8026 - 6538$
2. 846×367
3. $3\frac{7}{8} + 6\frac{1}{3}$
4. $216{,}568 \div 506$
5. 6.35×148.5
6. $16.2 - 12.54$
7. $6472 + 469 + 783$
8. $3.16388 \div 43.7$
9. $\frac{6}{7} \times \frac{3}{4} \times \frac{2}{9}$
10. $9\frac{3}{5} - 3\frac{1}{2}$
11. $0.38 + 12.6 + 18.07$
12. $\frac{3}{5} \div \frac{7}{8}$

4-2 Writing Percents as Decimals and Fractions

Objective
Write percents as decimals and fractions.

One hundred discs are stacked in piles of ten. Thirty discs are green. We can express the portion of the discs that are green in three ways.

Fraction	*Decimal*	*Percent*
$\frac{30}{100}$ or $\frac{3}{10}$	0.30 or 0.3	30%

To write a percent as a decimal, remove the percent sign, then multiply by 0.01 or move the decimal point two places to the left. 30% means 30 per hundred, or 30 hundredths, or 0.30.

Example

1. Write 9.2% as a decimal.

Remove the percent sign.
Multiply by 0.01.
$9.2\% = 9.2 \times 0.01 = 0.092$

or

Remove the percent sign.
Move the decimal two places to the left.
$9.2\% = 0.09.2 = 0.092$

Try This Write as a decimal.

a. 51% **b.** 7% **c.** 275% **d.** 83.2% **e.** 0.1%

1% means $\frac{1}{100}$. 30% means $30 \times \frac{1}{100}$ or $\frac{30}{100}$.
Writing $\frac{30}{100}$ in simplest terms shows that $30\% = \frac{3}{10}$.

Examples

2. Write 56% as a fraction.

Remove the percent sign. $56\% \rightarrow 56$

Multiply the number by $\frac{1}{100}$. $56 \times \frac{1}{100} = \frac{56}{100}$

Write in simplest terms. $\frac{56}{100} = \frac{14}{25}$

Sometimes we use percents that are equal to or greater than 1.

3. Write 125% as a whole number or a mixed number.

$125\% = 100\% + 25\% = 1 + \frac{25}{100} = 1\frac{1}{4}$

Try This Write as a fraction, whole number, or mixed number.

f. 83% **g.** 16% **h.** $\frac{1}{5}\%$ **i.** $16\frac{2}{3}\%$ **j.** 275%

Exercises

Write as a decimal.

1. 17%
2. 91%
3. 35%
4. 20%
5. 54%
6. 88%
7. 9%
8. 4%
9. 42.3%
10. 8.3%
11. 1.9%
12. 0.8%
13. 100%
14. 500%
15. 235%

Write as a fraction, whole number, or mixed number.

16. 19%
17. 37%
18. 75%
19. 60%
20. 95%
21. 1%
22. 6%
23. $\frac{1}{4}\%$
24. $\frac{2}{5}\%$
25. $6\frac{1}{4}\%$
26. $8\frac{1}{3}\%$
27. $62\frac{1}{2}\%$
28. 200%
29. 425%
30. $112\frac{1}{2}\%$

Mental Math

Find the equivalent fraction. Use mental math.

31. If $10\% = \frac{1}{10}$, then $30\% = \square$
32. If $25\% = \frac{1}{4}$, then $75\% = \square$
33. If $40\% = \frac{2}{5}$, then $80\% = \square$
34. If $16\% = \frac{4}{25}$, then $32\% = \square$
35. If $12\frac{1}{2}\% = \frac{1}{8}$, then $37\frac{1}{2}\% = \square$
36. If $8\frac{1}{3}\% = \frac{1}{12}$, then $16\frac{2}{3}\% = \square$
37. If $72\% = \frac{18}{25}$, then $36\% = \square$
38. If $24\% = \frac{6}{25}$, then $4\% = \square$
39. If $60\% = \frac{3}{5}$, then $20\% = \square$
40. If $66\frac{2}{3}\% = \frac{2}{3}$, then $33\frac{1}{3}\% = \square$

Show Math

Draw the figure on a separate piece of paper. Shade the indicated percent.

41. 25% of a square
42. $33\frac{1}{3}\%$ of a rectangle
43. 40% of a pentagon
44. $62\frac{1}{2}\%$ of a circle

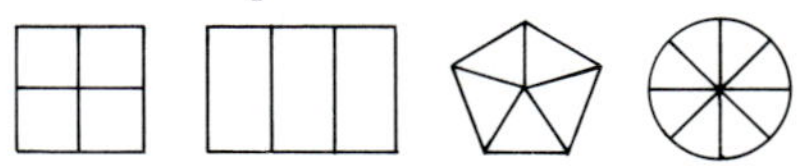

Think Math

What percent of the time will the spinner land on blue?

45.

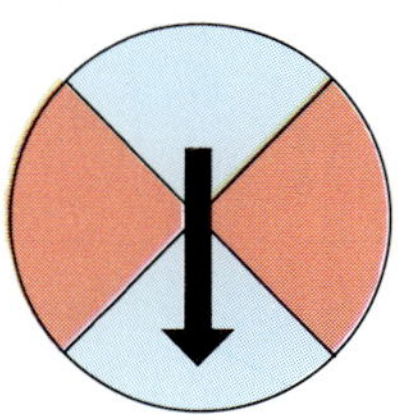

46.

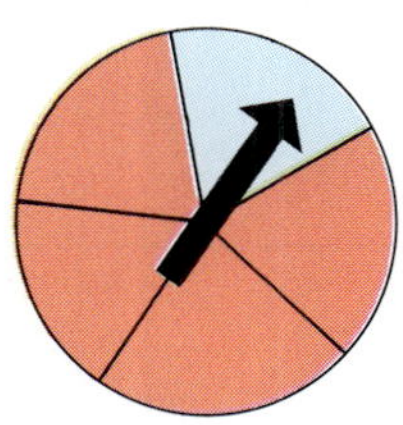

Mixed Skills Review

1. $72.8 - 34.067$
2. $197{,}064 \div 84$
3. $6\frac{3}{4} \times 3\frac{3}{9}$
4. $16.132 \div 2.18$
5. $93 + 1284 + 688$
6. $5\frac{1}{4} - 3\frac{3}{8}$
7. $\frac{16}{21} \div 4\frac{4}{7}$
8. $\$15.36 + \$1.66 + \$2.93$
9. 126.4×2.83
10. $10{,}634 - 2596$
11. 428×638
12. $7\frac{2}{3} + 4\frac{2}{3} + 2\frac{1}{6}$

4-3 Writing Fractions and Decimals as Percents

Objective
Write fractions and decimals as percents.

Magazines, newspapers, and almanacs often represent statistics in picture form. Charts like this help us to learn more about ourselves.

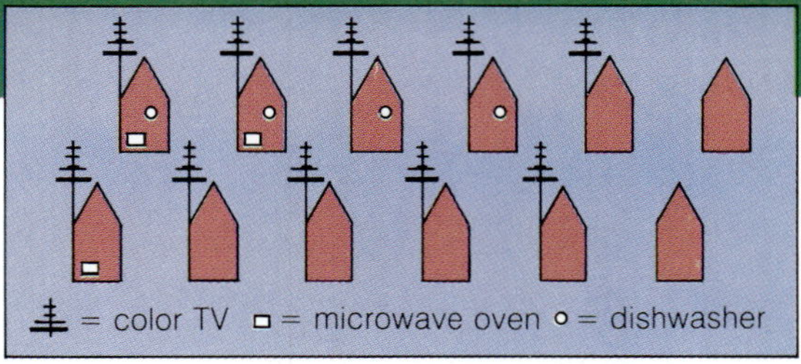

Understand the Situation

- What fraction of homes contain a dishwasher?
- What percent of American families own a color television?

Write $\frac{1}{4}$ as a percent.

Multiply the fraction by 100. → Write in simplest terms. → Add a percent sign.

$\frac{1}{4} \times 100 = \frac{100}{4}$ $\qquad \frac{100}{4} = 25$ $\qquad 25\%$

Examples Write as a percent.

1. $\frac{5}{6}$ $\quad \frac{5}{6} \times 100 = \frac{500}{6} = 83\frac{1}{3}\%$

2. $\frac{9}{8}$ $\quad \frac{9}{8} \times 100 = \frac{900}{8} = 112\frac{1}{2}\%$

3. 4 $\quad 4 \times 100 = 400 = 400\%$

4. $1\frac{3}{4}$ $\quad$ Think $1\frac{3}{4} = 1 + \frac{3}{4}$
$100\% + 75\% = 175\%$

Try This Write as a percent.

a. $\frac{9}{10}$ **b.** $\frac{4}{5}$ **c.** $\frac{2}{3}$ **d** $\frac{13}{6}$ **e.** 2 **f.** $1\frac{3}{8}$

Examples **5.** Write 0.245 as a percent.

Multiply by 100.
$0.245 \times 100 = 24.5\%$

or

Move the decimal two places to the right.
$0.245 = 0.24.5 = 24.5\%$

6. $0.16 = 16\%$ **7.** $0.09 = 9\%$ **8.** $1.5 = 150\%$

9. $3 = 300\%$ **10.** $0.0425 = 4.25\%$ **11.** $0.06\frac{1}{5} = 6\frac{1}{5}\%$

Try This Write as a percent.

g. 0.94 **h.** 0.03 **i.** 2.2 **j.** 1 **k.** 0.0325 **l.** $0.12\frac{1}{2}$

Exercises

Write each fraction as a percent.

1. $\frac{31}{100}$ **2.** $\frac{17}{50}$ **3.** $\frac{18}{25}$ **4.** $\frac{13}{20}$ **5.** $\frac{3}{5}$ **6.** $\frac{1}{10}$

7. $\frac{3}{4}$ **8.** $\frac{5}{16}$ **9.** $\frac{7}{8}$ **10.** $\frac{9}{32}$ **11.** $\frac{2}{3}$ **12.** $\frac{5}{12}$

13. $\frac{1}{6}$ **14.** $\frac{4}{9}$ **15.** $\frac{7}{12}$ **16.** $\frac{5}{2}$ **17.** $\frac{9}{4}$ **18.** $\frac{6}{5}$

Write each decimal as a percent.

19. 0.42 **20.** 0.18 **21.** 0.67 **22.** 0.3 **23.** 0.1

24. 0.77 **25.** 0.65 **26.** 0.31 **27.** 0.4 **28.** 0.7

29 0.125 **30.** 0.028 **31.** $0.33\frac{1}{3}$ **32.** $0.16\frac{2}{3}$ **33.** $0.37\frac{1}{2}$

Write each number as a percent.

34. 1 **35.** 3 **36.** $2\frac{4}{5}$ **37.** $1\frac{1}{2}$ **38.** $1\frac{5}{6}$

39. $2\frac{2}{3}$ **40.** $1\frac{11}{12}$ **41.** $1\frac{3}{8}$ **42.** $3\frac{2}{9}$ **43.** $4\frac{7}{10}$

44. 4.78 **45.** 2.4 **46.** 1.0084 **47.** $1.66\frac{2}{3}$ **48.** $1.16\frac{2}{3}$

Mental Math

Which fractions are greater than 75%?

49. $\frac{4}{5}$ **50.** $\frac{2}{3}$ **51.** $\frac{7}{10}$

52. $\frac{5}{6}$ **53.** $\frac{19}{25}$

Strategy Practice

Write a percent to indicate the number of American homes that contain the following appliances. Use the figure on p. 82.

54. a dishwasher and a microwave oven

55. either a dishwasher or a microwave

56. all three appliances

Mixed Skills Review

1. $18 \div \frac{5}{8}$ **2.** $0.4 - 0.132$ **3.** $157 + 864 + 529$

4. 3.46×61.7 **5.** $6\frac{5}{12} + 3\frac{3}{4}$ **6.** $114{,}504 \div 156$

7. $5034 - 3673$ **8.** 2786×25 **9.** $9\frac{1}{2} - 7\frac{2}{3}$

10. $144.088 \div 0.83$ **11.** $\$0.63 + \$0.48 + \$0.74$ **12.** $\frac{9}{20} \times 1\frac{3}{5}$

4-4 Using Percents to Compare

30% off
Dupree's Cycles

Objective
Make comparisons using percents.

Dupree's Cycles and Barclay Wheel are both having a sale on racing bikes. Which store offers the larger discount?

We can compare the discounts by comparing percents.

Compare 30% and $\frac{1}{3}$.

Write each number as a percent.	→	Compare the percents.	→	Substitute the original fraction.
$\frac{1}{3} \times 100 = 33\frac{1}{3}$ $\frac{1}{3} = 33\frac{1}{3}\%$		$30\% < 33\frac{1}{3}\%$		$30\% < \frac{1}{3}$

Barclay Wheel offers the larger discount.

$\frac{1}{3}$ off
Barclay Wheel

Examples

Compare. Use <, >, or =.

1. $\frac{3}{5}$ and 55%

$\frac{3}{5} \times 100 = 60$

$\frac{3}{5} = 60\%$

$60\% > 55\%$

$\frac{3}{5} > 55\%$

2. $116\frac{2}{3}\%$ and $\frac{7}{6}$

$\frac{7}{6} \times 100 = \frac{700}{6}\%$

$\frac{700}{6}\% = 116\frac{2}{3}\%$

$116\frac{2}{3}\% = \frac{7}{6}$

3. 0.85 and $83\frac{1}{2}\%$

$0.85 = 85\%$

$85\% > 83\frac{1}{2}\%$

$0.85 > 83\frac{1}{2}\%$

4. 89.6% and 0.9

$0.9 = 90\%$

$89.6\% < 90\%$

$89.6\% < 0.9$

Try This

Compare. Use <, >, or =.

a. 34% and $\frac{3}{4}$ **b.** $\frac{2}{5}$ and 25% **c.** 66% and $\frac{2}{3}$ **d.** $\frac{12}{7}$ and 127%

e. 0.49 and 49% **f.** 32% and 0.333 **g.** 0.7 and 71% **h.** $8.3\frac{1}{3}$ and 0.83

Exercises

Compare. Use <, >, or =.

1. $\frac{1}{4}$ □ 25%
2. $\frac{1}{5}$ □ 20%
3. 33% □ $\frac{1}{3}$
4. 16% □ $\frac{1}{6}$
5. $\frac{2}{3}$ □ 75%
6. $\frac{4}{5}$ □ 80%
7. 22% □ $\frac{2}{9}$
8. 44% □ $\frac{4}{11}$
9. $\frac{4}{7}$ □ 60%
10. $\frac{6}{25}$ □ 15%
11. 21% □ $\frac{2}{15}$
12. 89% □ $\frac{8}{9}$
13. $\frac{11}{16}$ □ 70%
14. $\frac{5}{6}$ □ 56%
15. 100% □ $\frac{8}{8}$
16. 150% □ $\frac{12}{8}$
17. 0.23 □ 24%
18. 0.95 □ 95%
19. 6.25% □ 0.62
20. 4.3% □ 0.34
21. $\frac{3}{4}$ □ 0.75
22. $\frac{1}{8}$ □ 0.18
23. 0.23 □ $\frac{2}{3}$
24. 0.72 □ $\frac{7}{20}$
25. $\frac{3}{7}$ □ 0.37
26. $\frac{4}{9}$ □ 0.49
27. 0.12 □ $\frac{7}{12}$
28. 0.815 □ $\frac{8}{15}$
29. $\frac{6}{5}$ □ $\frac{1}{2}$
30. $\frac{5}{2}$ □ 2.5
31. 0.625 □ $\frac{6}{5}$
32. 0.875 □ $\frac{8}{75}$
33. $\frac{3}{10}$ □ 0.310
34. $\frac{5}{8}$ □ 0.58
35. $0.31\frac{1}{4}$ □ $\frac{5}{16}$
36. $0.58\frac{1}{3}$ □ $\frac{7}{12}$

Estimation

37. In Japan there is one bicycle for every two people. The estimated population in Japan in 1984 was 119,896,000. About how many bicycles are there in Japan?

38. The Tour de France is a 2542-mile bicycle endurance race. In 1986 Greg Lemond of the U.S. won the three-week race in 110 hours 35 minutes 19 seconds. Estimate his average speed per hour.

39. A recent survey found the five most popular American sports. List the activities in order of popularity from the least to the most popular.

Bicycling	Camping	Fishing	Jogging	Swimming
0.337	23 3/10%	0.213	16.7%	23/50

Mixed Skills Review

1. $3\frac{4}{7} \times 2\frac{4}{5}$
2. 168,954 ÷ 29
3. 8.253 − 7.876
4. $6\frac{7}{10} - 4\frac{3}{4}$
5. 0.6 + 41.83 + 8.542
6. $4\frac{2}{7} \div 1\frac{1}{2}$
7. 873 × 604
8. 23.232 ÷ 0.726
9. $1\frac{3}{8} + 2\frac{1}{4} + 2\frac{1}{2}$
10. 8394 + 616 + 758
11. 23,058 − 17,495
12. 9237 × 48

4-5 Translating to a Number Sentence

Objective
Use percents in equations.

The Recording Industry Association presents awards for singles and albums.

Recording Awards, 1983–1985				
		1983	1984	1985
Singles	Gold (1,000,000 sold)	17	25	13
	Platinum (2,000,000 sold)	2	2	1
Albums	Gold (500,000 sold)	0	0	0
	Platinum (1,000,000 sold)	49	58	65

In 1984, 56% of the Gold Awards were presented for records by a solo artist. How many winners were solo artists?

You can solve many problems by translating the given information into a number sentence.

Let n stand for the number you are trying to find.
What number is 56% of 25?
$n = 56\% \times 25$

Examples Translate to a number sentence.

1. What percent of 36 is 9?
$n\% \times 36 = 9$

2. 84 is 28% of what number?
$84 = 28\% \times n$

Try This Translate to a number sentence.

a. 105% of 50 is what number?

b. 92 is what percent of 200?

c. 14.7 is 7% of what number?

d. Find $33\frac{1}{3}\%$ of 66.

Decimal or fractional equivalents are often used in number sentences.

Examples Translate to a number sentence. Write each percent using the given notation.

3. $66\frac{2}{3}\%$ of what number is 12?
as a fraction

$\frac{2}{3} \times n = 12$

4. 60% of what number is 4.8?
as a decimal

$0.6 \times n = 4.8$

Try This Translate to a number sentence. Write each percent using the given notation.

e. What number is $83\frac{1}{3}\%$ of 18?
as a fraction

f. $62\frac{1}{2}\%$ of what number is 12?
as a decimal

g. 50% of what number is 9.37?
as a fraction

h. $\frac{2}{3}$ is $16\frac{2}{3}\%$ of what number?
as a fraction

Exercises

Translate to a number sentence. Write each percent as a decimal.

1. 40% of 25 is what number?
2. 50 is 25% of what number?
3. 60 is what percent of 80?
4. Find 3% of $65.
5. Find 7% of $92.
6. 8.3% of what number is 13?
7. 90 is what percent of 63?
8. What number is 200% of 7.5?

Translate to a number sentence. Write each percent as a fraction.

9. What number is $33\frac{1}{3}\%$ of 27?
10. $12\frac{4}{7}\%$ of what number is 1?
11. $66\frac{2}{3}\%$ of what number is 24?
12. $24 is what percent of $9?
13. $\frac{4}{5}$ is 75% of what number?
14. $\frac{2}{3}$ is $83\frac{1}{3}\%$ of what number?
15. 498 is what percent of 575?
16. 625 is what percent of 450?

Translate to a number sentence. Choose the better notation.

17. Find $37\frac{1}{2}\%$ of 8.
18. 20 is 73% of what number?
19. What percent of 45 is 90?
20. What number is 250% of 6?
21. 40 is $62\frac{1}{2}\%$ of what number?
22. $37\frac{1}{2}\%$ of what number is 11?
23. 80% of what number is 25?
24. What number is 10% of 2.9?

Estimation

25. The average cost of a single record is $1.59. Estimate the gross receipts from sales of Gold Record winners in 1984.

Think Math

26. Twelve of the Gold Single Awards presented in 1983 went to songs from Walt Disney movies. Explain why these songs won so many awards.
27. A long-playing record makes $33\frac{1}{3}$ revolutions per minute. How many revolutions does it make in $2\frac{1}{2}$ minutes?
28. A single record makes 45 revolutions per minute. How many minutes does it take to make 72 revolutions?
29. A CD makes from 300 to 500 revolutions per minute. How many minutes does a CD take to make 3000 revolutions?

Mixed Skills Review

1. $2\frac{1}{6} + 1\frac{1}{4} + 4\frac{2}{3}$
2. $56,248 \div 178$
3. $26.08 - 13.472$
4. 697×548
5. $\frac{5}{11}$ of 176
6. $83,174 - 79,196$
7. $310.34 \div 11.8$
8. $1926 + 3854 + 7498$
9. $2\frac{2}{9} \div \frac{8}{21}$
10. 83% of 1056
11. $2.641 + 4.25 + 4.109$
12. $16\frac{1}{3} - 6\frac{1}{2}$

4-6 Finding a Percent of a Number

Objective
Find a percent of a number.

A restaurant makes 30 gallons of coffee each day. The restaurant sells 25% of the coffee before 9:00 a.m. How many gallons of coffee does it sell before 9:00 a.m.?

Find 25% of 30 gallons.
$n = 25\% \times 30$
Multiply to find 25% of 30. Use either a decimal or a fraction for the percent.

Decimal $25\% = 0.25 \rightarrow n = 0.25 \times 30 \quad 0.25 \times 30 = 7.50$

Fraction $25\% = \frac{25}{100} = \frac{1}{4} \rightarrow n = \frac{1}{4} \times 30 \quad \frac{1}{4} \times \frac{30}{1} = \frac{30}{4} = 7\frac{1}{2}$

The restaurant sells 7.5 or $7\frac{1}{2}$ gallons of coffee before 9:00 a.m.

Examples Give the percent of each number.

1. 4.8% of 70

$0.048 \times 70 = 3.36$

2. $62\frac{1}{2}\%$ of \$32

$\frac{5}{8} \times \$32 = \$20$

3. 150% of 8.7

$1.5 \times 8.7 = 13.05$

Try This Give the percent of each number. Use a decimal for the percent.

a. 28% of 72 **b.** 3% of 49 **c.** 110% of \$16
d. 5.6% of 30 **e.** 16.4% of 90 **f.** 0.3% of 2.5

Give the percent of each number. Use a fraction for the percent.

g. 75% of 36 **h.** 40% of \$51 **i.** 125% of 64

j. $16\frac{2}{3}\%$ of 108 **k.** $33\frac{1}{3}\%$ of 20 **l.** $37\frac{1}{2}\%$ of 400

Example

4. The restaurant served 321 customers during the lunch-hour rush. Of these customers, 48% ordered their food to carry out. About how many customers ordered carry-out food?

48% is about 50%.
321 is about 300.

$\frac{1}{2} \times 300 = 150$

Try This Estimate the percent of each number. Use compatible numbers.

m. 21% of 49 **n.** 73% of 442 **o.** 9.9% of 54

Exercises

Give the percent of each number. Use a decimal for the percent.

1. 30% of 50 **2.** 40% of 92 **3.** 18% of 45 **4.** 23% of 12

5. 65% of $134 **6.** 81% of $257 **7.** 6% of 93 **8.** 5% of 78

9. 15% of 548 **10.** 35% of 672 **11.** 175% of 19 **12.** 200% of 718

13. 50% of 85.3 **14.** 90% of 2.58 **15.** 1.5% of 2 **16.** 8.3% of 718

Give the percent of each number. Use a fraction for the percent.

17. 40% of 20 **18.** 90% of 80 **19.** 29% of 800 **20.** 75% of 280

21. 25% of $72 **22.** 15% of $600 **23.** 56% of 125 **24.** 16% of 150

25. 5% of 260 **26.** 4% of $2450 **27.** 120% of 24 **28.** 175% of 94

29. $33\frac{1}{3}\%$ of 21 **30.** $12\frac{1}{2}\%$ of 68 **31.** $87\frac{1}{2}\%$ of 16 **32.** $16\frac{2}{3}\%$ of 18

Estimation Estimate the percent of each number. Use compatible numbers.

33. 33% of 19 **34.** 29% of 41 **35.** 24% of 323 **36.** 88% of 691

37. 76% of 123 **38.** 57% of 230 **39.** 16% of $150 **40.** 13% of $639

41. 19.8% of 78 **42.** 34.1% of 199 **43.** $24\frac{3}{4}\%$ of 37 **44.** $99\frac{1}{2}\%$ of 52

45. Which country consumes the greatest number of cups of coffee per person per day?

46. What is the coffee consumption in the United States?

47. What is the difference between the amounts of coffee consumed in France and Canada?

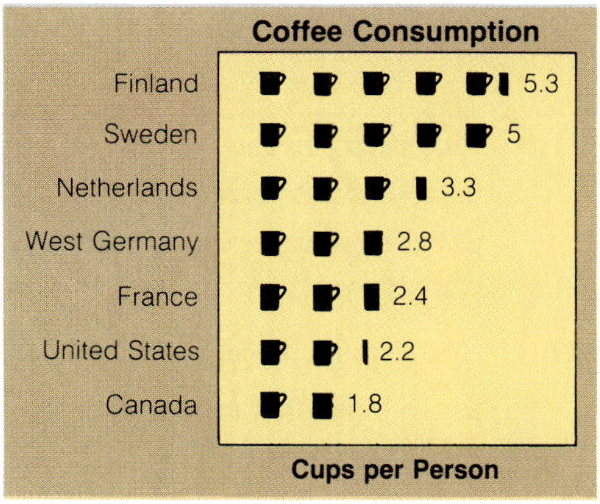

Mixed Skills Review

1. $172.137 \div 49$ **2.** $102.6 - 57.063$ **3.** $5\frac{3}{4} \div \frac{5}{8}$

4. $6.2 + 8.46 + 3.78$ **5.** 7416×47 **6.** $8563 + 729 + 5437$

7. $12\frac{1}{2} \times 7\frac{3}{5}$ **8.** $11,763 - 8888$ **9.** $6\frac{2}{5} - 4\frac{1}{4}$

10. $138,380 \div 340$ **11.** $4\frac{5}{8} + 9\frac{5}{16}$ **12.** 16.5% of 180

4-7 Finding What Percent One Number Is of Another

Objective
Find what percent a number is of another number.

Percents are often used to compare part of a number to the whole number.

> **"29/100, or 29% of American car buyers purchase imports"**
> **"143 people per 1000, or 14.3% of world population speak English"**

In the United States 3 people in every 25 travel abroad. What percent of Americans travel to other countries?

To solve the problem, compute what percent one number is of another. We can translate to a number sentence.

$3 = n\% \times 25 \qquad \frac{3}{25} = n\% \qquad \frac{3}{25} = 0.12 = 12\% \qquad n = 12\%$

Examples Find each percent.

1. 16 is what percent of 20?

$16 = n\% \times 20$

$\frac{16}{20} = n\%$

$\frac{16}{20} = 0.8 = 80\%$

2. What percent of 16 is 20?

$n\% \times 16 = 20$

$n\% = \frac{20}{16}$

$\frac{20}{16} = 1.25 = 125\%$

Try This Find each percent.

a. 15 is what percent of 60?
b. 8 is what percent of 24?
c. 36 is what percent of 30?
d. What percent of 150 is 2?

Example Estimate the percent. Use compatible numbers.

3. Two hundred twelve students attend the International School in Vienna. Nineteen of the students are from the United States. What percent of the students are from the United States?

$\frac{19}{212}$ is about $\frac{20}{200}$

$\frac{20}{200} = \frac{1}{10} = 10\%$ About 10% of the students are from the United States.

Try This Estimate the percent.

e. 39 is what percent of 49?
f. 13 is what percent of 22?
g. 63 is what percent of 89?
h. 97 is what percent of 24?

Exercises

Find each percent.

1. 30 is what percent of 60?
2. 20 is what percent of 50?
3. 17 is what percent of 25?
4. 27 is what percent of 30?
5. What percent of 140 is 35?
6. What percent of 250 is 50?
7. 18 is what percent of 24?
8. 18 is what percent of 81?
9. 24 is what percent of 72?
10. 16 is what percent of 96?
11. What percent of 16 is 2?
12. What percent of 12 is 5?
13. 7.2 is what percent of 8?
14. 1.2 is what percent of 24?
15. What percent of 3.5 is 7?
16. What percent of 7.2 is 9?

Estimation Estimate the percent. Use compatible numbers.

17. 19 is what percent of 39?
18. 34 is what percent of 346?
19. What percent of 52 is 41?
20. What percent of 99 is 67?

Calculator Use your calculator to find the percents. Round to the nearest tenth.

21. 77 is what percent of 94?
22. What percent of 77 is 94?
23. What percent of 132 is 54?
24. 23.9 is what percent of 12.7?
25. There are 325 students in the freshman class at Hereford High School. To the nearest tenth, what percent of the freshmen take each of the following languages?
    a. French **b.** Spanish
    c. German **d.** Chinese

Number of Freshmen Enrolled in Language Classes	
Chinese	19
German	24
French	64
Spanish	78
None	140

26. What percent of the freshmen do not take a foreign language?
27. What percent of the freshmen take either French or Spanish?
28. What percent of the freshmen take German, Chinese, or French?
29. What percent of the freshmen take neither Chinese nor French?

Mixed Skills Review

1. $946 + 1066 + 273$
2. $15.7168 \div 30.4$
3. 634×256
4. $20{,}060 - 1974$
5. $18.6 \times 4 \times 3.18$
6. $3\frac{8}{9} + 1\frac{1}{3}$
7. $3\frac{3}{4} \times 4\frac{1}{2}$
8. $11 - 3.783$
9. $34{,}394 \div 58$
10. $3\frac{3}{5} \div 20$
11. $3.163 + 2.084 + 0.915$
12. $12\frac{3}{16} - 7\frac{3}{8}$

4-8 Finding the Percent of Increase or Decrease

Objective
Find the percent of increase or decrease.

The cost for a local call on a pay telephone in the 1960s was 10¢. In the 1980s the price was raised to 25¢. What was the percent of increase?

To find the percent of increase or decrease, we begin by finding the difference in the two amounts.

Amount of increase $\rightarrow$ 25¢ − 10¢ = 15¢

Write a fraction using the difference as the numerator and the original amount as the denominator.

Amount of increase $\rightarrow$ / Original amount $\rightarrow$ $\frac{15}{10} = 150\%$

The cost of a local call from a pay phone increased 150%.

During the 1960s the cost of a three-minute call from Newport to Rockford was $2.50. During the 1970s the same call cost $2.25. What was the percent of decrease?

Amount of decrease $\rightarrow$ $2.50 − $2.25 = $0.25

Amount of decrease $\rightarrow$ / Original amount $\rightarrow$ $\frac{0.25}{2.50} = 10\%$

The cost of a three-minute call from Newport to Rockford decreased 10%.

Examples Find the percent of increase or decrease.

1. from 24 to 16
$24 - 16 = 8$
$\frac{8}{24} = \frac{1}{3} = 33\frac{1}{3}\%$ *decrease*

2. from 4 to 9
$9 - 4 = 5$
$\frac{5}{4} = 125\%$ *increase*

Try This Find the percent of increase.

a. from 15 to 20 **b.** from 5 to 8 **c.** from 100 to 200

Find the percent of decrease.

d. from 20 to 15 **e.** from 8 to 5 **f.** from 25 to 10

Find the percent of increase or decrease.

g. A teen-age babysitter in the 1960s earned 50¢ an hour. In the 1980s the average amount earned per hour rose to $2.

h. The cost of a black-and-white television in the 1960s was $400. In the 1980s the same television cost $150.

Exercises

Mental Math

Find the percent of increase or decrease. Use mental math whenever possible.

1. from 12 to 15
2. from 6 to 9
3. from 10 to 8
4. from 4 to 1
5. from 95 to 19
6. from 37 to 51
7. from 25 to 16
8. from 100 to 99
9. from 75 to 90
10. from 10 to 20
11. from 9 to 27
12. from 5 to 11
13. from 600 to 150
14. from 150 to 225
15. from 560 to 224
16. from 21 to 28
17. from 64 to 56
18. from 18 to 21
19. from 9 to 15
20. from 12 to 13
21. from 8 to 3
22. from 12 to 22
23. from 42 to 28
24. from 80 to 130

Calculator

Use your calculator to find the percent of increase or decrease. Round to the nearest tenth.

25. Between 1970 and 1980 the amount of farm products grown in the United States and exported to other countries rose from \$8 billion to \$42 billion.
26. In 1978, 12 minutes of commercials were shown each hour on Saturday mornings. In 1983 the number of commercial minutes dropped to 7.
27. Advertisements for candies and snacks decreased from 32% to 19% between 1978 and 1983.
28. The number of women selecting computer programming as a career is expected to change from 99,000 in 1980 to 308,000 in 1990.
29. Between 1983 and 1984 the average annual income in the United States rose from \$16,740 to \$17,544.
30. A new automobile purchased for \$12,000 depreciates to a value of \$8,640 after the first year.
31. In 1986, about 52 million women were in the labor force. By the year 2000, about 65 million women will be in the labor force.
32. Capital spending by top electronics companies was about \$17 billion in 1985 and \$14 billion in 1986.

Mixed Skills Review

1. $54{,}995 \div 647$
2. $468.708 \div 56.2$
3. $2\frac{5}{21} - 1\frac{4}{7}$
4. $3\frac{2}{5} + 6\frac{1}{6}$
5. $105.6 - 77.861$
6. $2\frac{3}{8} \times 24$
7. $1110 - 354$
8. 8084×63
9. $2\frac{13}{25} \div 1\frac{4}{5}$
10. 72.4% of 285
11. $784 + 329 + 53$
12. $1021.47 + 6.285 + 16.168$

4-9 Enrichment: Percent Equivalents

Objective
Write fractions as percents using unit equivalents.

Here is a simple way to write fractions as percents. Any fraction that has a numerator of 1 is called a unit fraction. Multiply the percent equivalent of a unit fraction by any numerator larger than 1. The result is the fraction written as a percent.

Examples

1. Write $\frac{5}{8}$ as a percent. Use the unit equivalent.

The fraction is the product of the numerator and a unit fraction. $\frac{5}{8} = 5 \times \frac{1}{8}$

Recall the percent equivalent for the unit fraction. $\frac{1}{8} = 12\frac{1}{2}\%$

Multiply the numerator by the percent. $5 \times 12\frac{1}{2}\% = 62\frac{1}{2}\%$

$$\frac{5}{8} = 62\frac{1}{2}\%$$

2. Write $\frac{13}{12}$ as a percent. Use the unit equivalent.

Think of the fraction as a product. $\frac{13}{12} = 13 \times \frac{1}{12}$

Recall the percent unit equivalent. $\frac{1}{12} = 8\frac{1}{3}\%$

Multiply. $13 \times 8\frac{1}{3}\% = 108\frac{1}{3}\%$

Try This

Write as percents. Use unit equivalents. Recall that $\frac{1}{8} = 12\frac{1}{2}\%$.

a. $\frac{7}{8}$ **b.** $\frac{9}{8}$ **c.** $\frac{11}{8}$ **d.** $\frac{3}{8}$

Exercises

Write as percents. Use unit equivalents.

1. If $\frac{1}{15} = 6\frac{2}{3}\%$, find $\frac{7}{15}$.

2. If $\frac{1}{7} = 14\frac{2}{7}\%$, find $\frac{6}{7}$.

3. If $\frac{1}{9} = 11\frac{1}{9}\%$, find $\frac{4}{9}$.

4. If $\frac{1}{13} = 7\frac{9}{13}\%$, find $\frac{10}{13}$.

5. If $\frac{1}{8} = 12\frac{1}{2}\%$, find $\frac{15}{8}$.

6. If $\frac{1}{11} = 9\frac{1}{11}\%$, find $\frac{12}{11}$.

4-10 Calculator: Keying Percent

Objective
Use a calculator to work with percent.

To write a fraction as a decimal using a calculator, divide the numerator by the denominator.

Write $\frac{1}{4}$ as a decimal. 1 [÷] 4 [=] 0.25 $\frac{1}{4} = 0.25$

To write a fraction as a percent using a calculator, press [%] instead of [=].

Write $\frac{1}{4}$ as a percent. 1 [÷] 4 [%] 25 $\frac{1}{4} = 25\%$

Use your calculator to complete the tables.

	Fraction	Decimal	Percent
1.	$\frac{6}{8}$		
2.	$\frac{1}{2}$		
3.	$\frac{3}{4}$		

	Fraction	Decimal	Percent
4.	$\frac{2}{3}$		
5.	$\frac{5}{6}$		
6.	$\frac{7}{8}$		

Find the percent of each number.

Write the percent as a decimal and multiply. Use a calculator.

35% of 124 0.35 [×] 124 [=] 43.4

Or, press [%] instead of [=] without changing the percent to a decimal.

35% of 124 35 [×] 124 [%] 43.4

7. 58% of 685 = _______ **8.** 21% of 900 = _______

9. 16% of 98 = _______ **10.** 79% of 180 = _______

11. 24.8% of 139 = _______ **12.** 6.3% of 41 = _______

Find the number when a percent of it is known.

Write the percent as a decimal and divide.

60% of ? = 300 300 [÷] 0.60 [=] 500

Or, press [%] instead of [=] without changing the percent to a decimal.

60% of ? = 300 300 [÷] 60 [%] 500

13. 37% of _______ = 148 **14.** 72% of _______ = 482.4

15. 95% of _______ = 23.75 **16.** 40.3% of _______ = 604.5

17. 5.83% of _______ = 1.6324 **18.** 3.9% of _______ = 1.755

19. 85.7% of _______ = 479.92 **20.** 16.2% of _______ = 14.904

21. 182% of _______ = 134.68 **22.** 50.9% of _______ = 108.926

23. 0.4% of _______ = 8.112 **24.** 67.5% of _______ = 10.8

CHAPTER 4 REVIEW

Which is incorrect and does not belong in the group?

1. A. $25\% = \frac{1}{4}$
B. $66\frac{2}{3}\% = \frac{2}{3}$
C. $12\frac{1}{2}\% = \frac{25}{2}$

2. A. $\frac{3}{4} = 75\%$
B. $1\frac{3}{4} = 1.75\%$
C. $\frac{1}{2} = 50\%$

3. A. $200\% = 2$
B. $175\% = 1\frac{3}{4}$
C. $1.5\% = 1\frac{1}{2}$

4. A. $\frac{3}{5} > 55\%$
B. $\frac{7}{6} > 117\%$
C. $\frac{7}{8} > 80\%$

5. A. $0.66 = \frac{2}{3}$
B. $0.375 = \frac{3}{8}$
C. $0.16\frac{2}{3} = \frac{1}{6}$

6. A. 21% of 49 is about 10
B. 9.9% of 54 is about 5
C. 50% of 20 is about 40

7. A. $11\frac{1}{9}\% = \frac{9}{11}$
B. $16\frac{2}{3}\% = \frac{1}{6}$
C. $33\frac{1}{3}\% = \frac{1}{3}$

8. A. $51\% = 0.51$
B. $34\% = 0.34$
C. $117\% = 0.117$

9. A. $0.89 < \frac{9}{10}$
B. $0.66 > \frac{2}{3}$
C. $0.11 < \frac{1}{9}$

10. A. If $60\% = \frac{3}{5}$, then $20\% = \frac{1}{15}$
B. If $40\% = \frac{2}{5}$, then $80\% = \frac{4}{5}$
C. If $24\% = \frac{6}{25}$, then $4\% = \frac{1}{25}$

11. A. 200% of 15 is 30
B. 150% of 20 is 30
C. 100% of 5 is 500

12. A. 39 is about 80% of 49
B. 22 is about 50% of 11
C. 11 is about 25% of 44

13. A. $33\frac{1}{3}\%$ of 25 is about 8
B. $66\frac{2}{3}\%$ of 50 is about 33
C. $16\frac{2}{3}\%$ of 50 is about 10

Match.

14. 16 is what percent of 20?
15. From 5 to 8 is what percent increase?
16. 15 is what percent of 60?
17. Estimate: 9.9% of 250
18. Estimate: $16\frac{2}{3}\%$ of 100
19. From 8 to 3 is what percent decrease?
20. Estimate: 125% of 23

A. 60%
B. 80%
C. 17
D. $62\frac{1}{2}\%$
E. 30
F. 25
G. 25%

CHAPTER 4 TEST

Find the missing percent.

1. During a recent election, 38.2% of the votes were for Art Jones, 44.7% were for Arlene Summers, and the rest were for Shelly Thomas. What percent of the votes were cast for Shelly Thomas?

Write as a decimal.

2. 23% **3.** 8% **4.** 67.3% **5.** 157% **6.** 0.7% **7.** 349%

Write as a fraction, whole number, or mixed number.

8. 40% **9.** 94% **10.** 87.5% **11.** $\frac{3}{4}\%$ **12.** $166\frac{2}{3}\%$ **13.** 250%

Write each as a percent.

14. $\frac{5}{18}$ **15.** 0.582

Compare. Use <, >, or =.

16. $\frac{2}{3}$ and $66\frac{2}{3}\%$ **17.** $\frac{3}{7}$ and 44% **18.** 0.097 and 97%

Translate to a number sentence. Write each percent as a decimal.

19. 16 is $33\frac{1}{3}\%$ of what number? **20.** Find 2.5% of 124.

Translate to a number sentence. Write each percent as a fraction.

21. Find $\frac{1}{2}\%$ of 200. **22.** $\frac{4}{5}$ is what percent of $\frac{5}{8}$?

Give the percent of each number. Use a decimal for the percent.

23. 32% of 45 **24.** 120% of 75 **25.** 0.75% of 40 **26.** 5.6% of 129

Give the percent of each number. Use a fraction for the percent.

27. $66\frac{2}{3}\%$ of 333 **28.** 87.5% of $194.40 **29.** $233\frac{1}{3}\%$ of 891

Estimate the percent of each number. Use compatible numbers.

30. 34% of 181 **31.** 81% of 120 **32.** 76% of $4.88 **33.** 123% of 83

Find each percent.

34. 81 is what percent of 350? **35.** What percent of 279 is 93?

Estimate the percent. Use compatible numbers.

36. What percent of 148 is 51? **37.** 68 is what percent of 417?

Find the percent of increase or decrease. Use mental math whenever possible.

38. from 25 to 30 **39.** from 66 to 44 **40.** from 1 to 5

CHAPTER 4 CUMULATIVE REVIEW

1. Translate to a number sentence. Write each percent as a decimal.
7.2% of what number is 25?
A. $n = 0.072 \times 25$
B. $0.072 = 25 \times n$
C. $25 = 0.072$
D. $0.072 \times n = 25$

2. Write the missing numeral.
$\frac{5}{12} = \frac{20}{\square}$
A. 40
B. 100
C. 48
D. 60

3. Nellie rode 3 rides that cost 25¢ and 2 rides that cost 75¢. How much were her rides?
A. $2.25
B. $0.45
C. $1.50
D. $1.95

4. Find the percent of increase or decrease. Use mental math.
from 48 to 32
A. $-33\frac{1}{3}\%$
B. $+50\%$
C. $-66\frac{2}{3}\%$
D. $+150\%$

5. Multiply. Use mental math.
$15 \times 3\frac{1}{3}$
A. 48
B. 50
C. 45
D. 55

6. Write as a percent.
$2\frac{5}{8}$
A. $2\frac{5}{8}\%$
B. 02625%
C. 2625%
D. 262.5%

7. The All Sport Shop discounted basketballs $23\frac{1}{3}\%$. What percent of the original price did Joe pay for one?
A. $77\frac{1}{3}\%$
B. $76\frac{2}{3}\%$
C. $76\frac{1}{3}\%$
D. $77\frac{2}{3}\%$

8. Write as a mixed number.
$\frac{39}{18}$
A. $3\frac{1}{18}$
B. $2\frac{1}{9}$
C. $2\frac{1}{6}$
D. $1\frac{17}{18}$

9. How much time will it take to drive $437\frac{1}{2}$ miles if you average 50 miles per hour?
Use the formula $T = \frac{d}{s}$.
A. $8\frac{3}{4}$ hours
B. 22 hours
C. 8 hours 37 minutes
D. $4\frac{1}{2}$ hours

10. Compare.
Use >, <, or =.
$\frac{5}{7} \square 70\%$
A. <
B. =
C. >

11. Estimate. Use compatible numbers.
$6.5 \div \frac{7}{24}$
A. 21
B. 42
C. 4
D. 2

12. Divide.
$75.66 ÷ 13
A. $6.29
B. $5.82
C. $7.37
D. $5.99

13. Find the percent.
What percent of 15 is 8?

A. $1\frac{7}{8}\%$
B. $4\frac{2}{3}\%$
C. $87\frac{1}{2}\%$
D. $53\frac{1}{3}\%$

14. Divide.
$\frac{5}{8} \div 10$

A. $\frac{1}{18}$
B. $2\frac{1}{4}$
C. $1\frac{1}{2}$
D. $\frac{1}{16}$

15. Estimate.
$$\begin{array}{r} \$63.47 \\ \times\ 3.82 \\ \hline \end{array}$$

A. \$180
B. \$240
C. \$280
D. \$300

16. Write as a mixed number.
750%

A. $7\frac{1}{2}$
B. 75
C. $7\frac{5}{7}$
D. $75\frac{1}{2}$

17. Add.
$\frac{5}{6} + 2\frac{1}{3} + \frac{7}{6}$

A. $3\frac{1}{3}$
B. $4\frac{1}{6}$
C. $4\frac{1}{3}$
D. $3\frac{1}{6}$

18. Give the percent. Use a decimal for the percent.
32% of 76

A. 14.08
B. 24.32
C. 237.5
D. 2.432

19. Subtract.
$4\frac{3}{4} - \frac{1}{8}$

A. $3\frac{5}{8}$
B. $3\frac{1}{2}$
C. $4\frac{1}{2}$
D. $4\frac{5}{8}$

20. A rectangular plot 120 ft by 84 ft is being divided into 3 equal plots. How much fencing will be needed to fence all 3 plots?

A. 408 ft
B. 500 ft
C. 576 ft
D. 792 ft

21. Snake Abe is twice as long as Snake Ben. Snake Ben is $\frac{1}{2}$ as long as Snake Cal. Snake Cal is 3 feet long. How long is Snake Abe?

A. 3 ft
B. $1\frac{1}{2}$ ft
C. 6 ft
D. 5 ft

22. Estimate the difference. Use mental math.
$$\begin{array}{r} \$184.00 \\ -\$80.37 \\ \hline \end{array}$$

A. \$97.68
B. \$80.00
C. \$103.60
D. \$135.75

23. Estimate. Use clustering.
278 + 314 + 311 + 282

A. 1000
B. 1185
C. 1570
D. 798

24. Find the quotient. Use mental math.
2727 ÷ 3

A. 1,020
B. 909
C. 440
D. 336

Chapter 5 Overview

Key Ideas

- finding a job
- comparing different pay rates
- comparing job benefits
- finding overtime and gross pay
- finding income from tips
- finding income from commissions
- finding income from piecework
- comparing future earnings

Key Terms

Here are some of the key terms you will use in this chapter.

- white-collar job
- blue-collar job
- service job
- farm job
- resume
- weekly pay
- hourly pay
- yearly pay
- benefit
- overtime
- gross pay
- tip
- commission
- straight commission
- piecework

Key Skills

Add.

1. \$45.90 + \$6.30 + \$19.83
2. \$31.09 + \$15.98 + \$0.76
3. \$109.84 + \$23.73
4. \$32.68 + \$21.99 + \$1.29
5. \$65.32 + \$19.73 + \$54.87
6. \$76.04 + \$43.21 + \$302

Subtract.

7. \$292.06 − \$186.78
8. \$34.02 − \$18.64
9. \$402.12 − \$18.93
10. \$676.10 − \$199.29
11. \$59.05 − \$18.77
12. \$90.03 − \$17.54

Multiply.

13. 42.5 × \$6.75
14. 88.6 × 33.7
15. 21.7 × 63.19
16. 54.9 × 73.9
17. 19.73 × 38.02
18. 22.6 × 76.54

Solve.

19. What is $33\frac{1}{3}\%$ of \$600?
20. What is 50% of \$864?
21. What is 28% of \$64.12?
22. What is 19% of \$75.44?
23. Find 32% of \$32,980.
24. Find 45% of \$43,860.
25. 3% of \$12,500 is what number?
26. 8% of \$19,600 is what number?
27. Find 10% of \$45.67.
28. Find 10% of \$19.63.
29. Find 5% of \$23.98.
30. Find 5% of \$37.88.

EARNING MONEY

5

5-1 Finding a Job

Objective *Make career-related decisions.*

More than 125 million people in the United States hold jobs. Nearly 2 million new jobs are created each year. The chart shows the working population in 4 basic categories.

- A **white-collar** job involves clerical or professional skills.
- A **blue-collar** job involves manual skills in crafts or industry.
- A **service** job involves selling or delivering directly to the customer
- A **farm** job involves agricultural skills, usually outdoors.

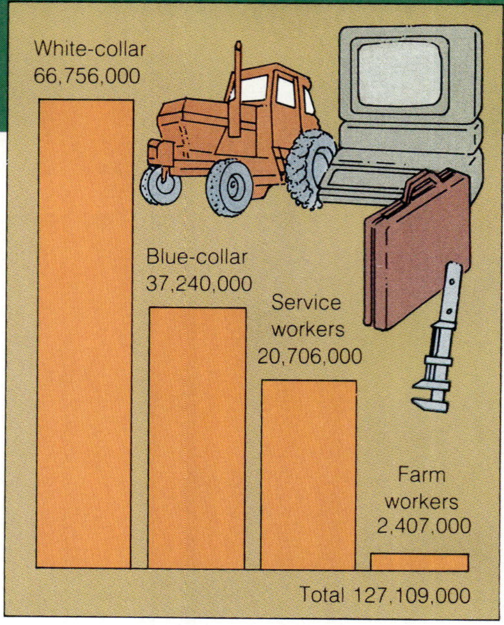

Example Many high school students work part-time or temporary jobs. Tom Stevens is an eleventh grader at Alamo High School. His guidance counselor helps him to identify some important steps in finding an after-school job.

Understand the Situation

What jobs are available?
What jobs interest you?
When can you work?
What jobs are you qualified to do?
How many hours per week can you work?
How much money do you expect to make?

Find the Data

- employment agencies
- family, friends, and counselors
- classified ads
- help-wanted signs

Plan what to do

- Obtain a Social Security number.
- Prepare a **resume** of your qualifications and experience.
- Apply for a position. This may require a test and an interview.

Find the Answer

You and the prospective employer should know if you can do the work.

Check the Answer

Did you get the job? If not, write a follow-up letter.

Try This Work with a partner to determine each job category. Name two more occupations in each category. Decide which jobs you might like.

a. welder in automobile plant **b.** tree farm worker
c. data processing clerk **d.** nursery school aide

You are an employer. Interview a classmate for one of these jobs.

e. department store salesperson **f.** gas station attendant
g. baby sitter **h.** florist's helper

Exercises

Many high school students have after-school jobs. Use this chart to answer the following questions.

1. What percent of high school students worked between 11 and 19 hours per week after school?
2. What percent worked more than 30 hours per week?
3. What percent worked at least 11 hours per week?
4. What percent of students did not work?

Data Bank Use the resources on pp. 441–446 of the Data Bank to complete the following.

5. Apply for a Social Security number.
6. Write a resume.
7. Complete the job application form.
8. Take the aptitude test.

Data Hunt

9. Find a help-wanted ad. Answer the following questions about it.
   a. Are you qualified for the job?
   b. Do you need an appointment?
   c. How much does it pay?
   d. What hours will you work?
   e. Where do you apply?
   f. How will you get to work?

Mixed Skills Review

1. 134,000 + 87,000
2. 206 × 521
3. 7196 ÷ 28
4. \$7086.11 − \$6570.46
5. 731 + 1294 + 18
6. $4\frac{2}{3} + 3\frac{1}{2} + \frac{3}{4}$
7. $4\frac{3}{8} \times 3\frac{1}{5}$
8. 6.5% of \$24
9. 23.6 ÷ .008
10. 126% of 70
11. $3\frac{1}{5} \div \frac{1}{2}$
12. 9.4 − 0.534

5-2 Comparing Pay

Objective
Compare different pay rates.

An employee may be paid by the hour, day, week, month, or year. It is important to consider the same time period when comparing pay for two different jobs.

Examples

1. Job A pays \$8400 per year. Job B pays \$4.35 per hour for a 36-hour week. Compare the weekly pay.

Find the weekly pay.

Job A	$\$8400 \div 52 \approx \$161.54$	Divide the annual pay by 52 and round.
Job B	$36 \times \$4.35 = \$156.60$	Multiply the hourly rate by hours worked.

Find the difference.

Job A	\$161.54	
Job B	− 156.60	
	\$4.94	Job A pays \$4.94 more per week than Job B.

2. Compare the pay.

Job A pays \$5.20 per hour for a 36-hour week.
Job B pays \$200 per week for a 40-hour week.

Understand the Situation How can you decide which job pays more?

Data The number of hours worked for each job per week is different.

Plan You should compare both the weekly pay and the hourly pay.

Answer

Week
Multiply the hourly pay for Job A by 36 hours per week.
$36 \times \$5.20 = \$187.20$ per week

Subtract to find the difference.

Job B	\$200.00
Job A	− 187.20
	\$ 12.80

Job B pays \$12.80 more per week.

Hour
Divide the weekly pay for Job B by 40 hours per week.
$\$200 \div 40 = \$5$ per hour

Subtract to find the difference.

Job A	\$5.20
Job B	− 5.00
	\$0.20

Job A pays \$0.20 more per hour.

Check Job A pays more per hour Job B pays more per week. How many hours per week do you prefer to work?

Try This

Compare the pay for the indicated period.

a. Yearly Job A \$1109 per month Job B \$263 per week
b. Weekly Job A \$15,750 per year Job B \$7.65 per hour for a 40-hour week

Exercises

Compare the pay for the indicated period.

1. Yearly
Job A $19,700 per year
Job B $1806 per month

2. Yearly
Job A $243.10 per week
Job B $12,650 per year

3. Yearly
Job A $183.50 per week
Job B $800 per month

4. Weekly
Job A $13,600 per year
Job B $265 per week

5. Weekly
Job A $14,500 per year
Job B $73 per day for a 4-day week

6. Weekly
Job A $16,276 per year
Job B $8 per hour for a 40-hour week

Compare the pay for both time periods.

7. Job A $5 per hour for a 38-hour week
Job B $180 per week for a 36-hour week

8. Job A $330 per week for a 40-hour week
Job B $8.75 per hour for a 36-hour week

9. Job A $45 per day for a 6-day week
Job B $260 per week for a 5-day week

10. John Turnbaugh has been offered two similar jobs. Transportation costs could be important in his decision. Compare the annual pay.

Job A pays $1590 per month. Riding the train costs $1.80 per day for 22 working days per month.

Job B pays $367.50 per week. Driving to work costs $2.20 per day for 5 working days per week.

Data Hunt

11. Find the starting salary for each of the following occupations in your area.

a. police officer **b.** firefighter **c.** teacher **d.** mail carrier

Mixed Skills Review

1. $17,684 + $22,371

2. $\frac{3}{5} \times \frac{5}{8}$

3. $16 − $7.49

4. 22 × $1.50

5. $3\frac{1}{4} \div 6\frac{1}{2}$

6. 16% of $2500

7. $9\frac{1}{2} - 7\frac{5}{8}$

8. $17,208 ÷ 50

9. 38 × $12.71

10. $\frac{1}{4}$ of 296

11. $2406.44 − $1920.50

12. $22,136.16 ÷ 12

5-3 Comparing Benefits

Objective
Compare job benefits.

Employers often provide benefits to attract qualified employees. Benefit packages typically include health care and insurance programs, stock purchase plans, child care services, and discounts.

Some employees receive benefits in the form of services.

Example

1. Karen Bishop is a teacher's aide at a private school. Tuition is \$48.50 per month per student. Karen's two children may attend for free.

Find Karen's annual benefit in free tuition.

The annual tuition cost is \$48.50 per month per child for 9 months.

Multiply to find the cost per year
Estimate: \$48.50 is about \$50. 9 months × \$50 × 2 children = \$900

Exact: 9 × \$48.50 × 2 = \$873. The annual benefit is \$873.

The exact answer is reasonable.

Try This

Find the annual employee benefit.

a. John Flanigan flies from Richmond to Dallas 6 times a year. A round-trip ticket costs \$338. John is a ticket agent and flies for free.

b. Sue Thompson works at First and Merchants Bank. The bank pays her a parking allowance of \$1.75 per day. Sue works 220 days per year.

Some companies give their employees a discount on purchases.

Example

2. Books in Print offers its employees a $33\frac{1}{3}\%$ discount on the first \$25 of purchases each month. Find the annual employee benefit.

Multiply to find the annual amount discounted. 12 × \$25 = \$300
Find the amount saved. Use mental math.
$33\frac{1}{3}\% = \frac{1}{3}$ $\frac{1}{3} \times \$300 = \$100$ The annual benefit is \$100.

Try This

Find the annual employee benefit.

c. Carrie spends \$600 per year on clothes. The department store where she works allows its employees a 25% discount on clothing.

d. A small telephone company offers its employees a 50% discount on their local service. The cost of service is \$13.25 per month.

Exercises

Find the annual employee benefit.

1. Lotell Labs purchases 8 season box seats at the ballpark for employee use. A box seat costs $11.25 per game for 81 home games. What is the total annual cost to the company?
2. Each year the Titan Die Company purchases two uniforms and one pair of shoes for each employee. Each uniform costs $23.75. Each pair of shoes costs $39.90.
3. Martha Donovan works the late shift at Southern Restaurant five days a week. The restaurant provides her dinner. The average cost of a meal is $5.95.
4. Drivers for the Metropolitan Transit Company are given 60 bus tokens worth 75¢ each every month.
5. A wholesale distributor pays salespersons 80% of monthly gasoline purchases. Each salesperson averages $125 per month in gasoline costs.
6. Wilmington Dry Goods offers employees a 60% discount on the first $50 of fabric purchased each month.
7. Karyl Luck earns $187.50 per week as a bookkeeper and participates in her company's stock purchase plan. The company adds 2% of Karyl's annual salary to her contribution.
8. Crawford Manufacturing has a child care service at the company plant for working parents. The cost per week for child care is $65. The company pays $\frac{1}{2}$.
9. Brighton Bottling Company encourages fitness by paying $\frac{1}{3}$ of its employees' fees at a local health club. An individual membership at the health club is $29.70 per month.

Number Sense

10. The Wicomico Public Schools pay $\frac{2}{3}$ of the costs of teachers who take courses at local colleges. June Danaher took 3 courses during the summer for $195 per course plus a $15 registration fee.

Mixed Skills Review

1. $0.26 + 82 + 4.2 + 27.8$
2. 45% of $129.60
3. $8772 \div 86$
4. $9\frac{4}{7} - 6\frac{1}{2}$
5. $7\frac{1}{4} + 5 + \frac{3}{5} + 8\frac{1}{2}$
6. $\frac{1}{3}$ of $174.36
7. $0.186 \div 0.3$
8. $136.49 + $181.37
9. $10{,}006 - 8544$
10. $6 - 3\frac{5}{8}$
11. 80% of $646
12. $4\frac{3}{4} \div 12\frac{1}{2}$

5-4 Overtime Rates

Objective *Find regular, overtime, and gross pay.*

Terry Wilson works as an orderly at Memorial Hospital. When Terry works more than 37.5 hours per week, he is paid at a higher rate for **overtime** hours. The sum of regular pay and overtime pay is **gross pay.**

Example

1. Find the gross pay.

Reg. Hr	Reg. Rt.	Reg. Pay	OT Hr	OT Rt.	OT Pay	Gross Pay
37.5	$6.26		10	$9.39		

Find the regular pay. $37.5 \times \$6.26 = \$234.75$
Find the overtime pay. $10 \times \$9.39 = \$93.90$
Add to find the gross pay. $\$234.75 + \$93.90 = \$328.65$

Try This

Find the gross pay.

	Reg. Hr	Reg. Rt.	Reg. Pay	OT Hr	OT Rt.	OT Pay	Gross Pay
a.	40	$4.95		8	$9.90		
b.	37.5	$6.50		7.5	$9.75		

Example

2. Joan's regular rate is $8.50 per hour for 36 hours per week. Her overtime rate is 1.5 times her regular rate. Use the time card to find Joan's gross pay for the period ending April 10.

Period End	M	Tu	W	Th	F	S	Total	Reg. Hr	OT Hr
4-10-89	9	10	10	10	6		45	36	9
4-17-89		9	9	11	11	9	49	36	13
4-24-89	8	6.5	7.5	8	9.5	4	43.5	36	7.5

Find the regular pay. $36 \times \$8.50 = \$306$
Find the overtime pay. $1.5 \times \$8.50 = \$12.75$ $9 \times \$12.75 = \$114.75$
Add to find the gross pay. $\$306 + \$114.75 = \$420.75$

Joan's gross pay for the period ending April 10 is $420.75.

Try This

c. Find Joan's gross pay for the period ending April 24.

Exercises

Find the gross pay.

	Reg. Hr	Reg. Rt.	Reg. Pay	OT Hr	OT Rt.	OT Pay	Gross Pay
1.	25	$7.53		0	$11.30		
2.	35	6.40		3	9.60		
3.	36	4.50		8	6.75		
4.	38	9.90		0	14.85		
5.	80	5.85		10	11.70		
6.	75	8.75		7	13.13		
7.	37.5	4.90		6	6.90		
8.	36.5	7.70		5	19.25		

Joe Gleaton repairs outside cable for the telephone company. His schedule is based on the repairs required. Joe is paid $12.60 per hour for the first 40 hours of work each week. His overtime rate is 1.5 times his regular rate. Use the time card to calculate his gross pay for each period listed.

	Period End	M	Tu	W	Th	F	Sa	Su	Total	Reg. Hr	OT Hr
9.	2-01-89	9	10	10	6	5			40	40	0
10.	2-08-89	8	8	8	8	8	2		42	40	2
11.	2-15-89	6	9	8	11	8	3	1	46	40	6
12.	2-22-89	8	9	7.5	8	9.5	4.5		46.5	40	6.5

Estimation Estimate each answer.

13. Gwen wants to purchase a stereo for $49.90. She baby-sits for $2.25 per hour. Each job averages 3 hours. How many jobs does she need to buy the stereo?

14. The Booster Club is selling singing birthday grams to earn $500 for a ski trip. Each birthday gram is 75 cents. The club averages 42 sales per week. About how many weeks will it take to earn the money for the trip?

15. Ondine Seville, a tour bus driver, earns a $35 bonus for every 100 miles she drives over 8000 miles each month. In October, she traveled 11,358 miles. How much was her bonus?

Mixed Skills Review

1. 14.8×6.22 **2.** $10{,}260 \div 27$ **3.** $3\frac{7}{8} + 1\frac{1}{4} + 2\frac{1}{2}$

4. $9\frac{1}{3} \times 3\frac{3}{7}$ **5.** $28.3 - 16.905$ **6.** $3.4 \div 170$

7. $4781 + $326 + $3364 **8.** 2.8% of 906 **9.** $17{,}260 - 15{,}274$

10. $12\frac{1}{3} - 3\frac{5}{9}$ **11.** $16.08 + 0.7 + 12 + 3.75$ **12.** $10\frac{1}{2} \div 4\frac{3}{8}$

5-5 Tips

Objective
Find income from tips.

Tips are a large part of earnings in many service jobs. Tips help limit the employer's expenses and are an incentive for the employee to do a good job.

Most customers tip a certain percent of their bill. Good service is usually rewarded by a tip from 15% to 20%. Calculating a tip can be done using mental math.

Example

1. The bill was \$15.90. Use mental math and rounding to estimate a 15% tip.

\$15.90 is about \$16.

10% of \$16 = \$1.60 → 5% is $\frac{1}{2}$ of 10% → $\frac{1}{2} \times \$1.60 = \$0.80$

\$1.60 + \$0.80 = \$2.40 The tip is \$2.40.

Try This

Estimate the amount of the tip. Use mental math.

a. Dinner bill, \$65.60, tip 20% **b.** Taxi ride, \$9.50, tip 15%

Example

2. Nicholas is a waiter at the Blue Dolphin Restaurant. He works a 6-day, 30-hour week. Nicholas earns \$4.95 an hour and averages \$28 a day in tips. Find the total weekly earnings.

Find the regular pay.	$30 \times \$4.95 = \$148.50$
Find the tip income.	$6 \times \$28 = \$168$
Find the total.	$\$148.50 + \$168 = \$316.50$

The total earnings for the week are \$316.50.

Use estimation to check the results.

\$4.95 is about \$5.	$30 \times \$5 =$	\$150
\$28 is about \$30.	$6 \times \$30 =$	+\$180
		\$330

An answer of \$316.50 seems reasonable.

Try This

Find the total weekly earnings.

c. Bix is a bellhop at the Mayflower Hotel. He works 5 days a week, 8 hours a day. He is paid \$5.45 per hour and averages \$42 a day in tips.

d. Toni works in a beauty shop. She averages \$350 in sales each week. She is paid one half of what she sells and averages 10% of her sales in tips.

Exercises

Estimate the amount of the tip. Use mental math.

1. Lunch bill, \$8, tip 20%
2. Haircut, \$12, tip 15%
3. Taxi ride, \$7.65, tip 15%
4. Room service, \$24.35, tip 25%
5. Parking valet, \$6.50, tip 10%
6. Ski instruction, \$18, tip 30%

Find the total weekly earnings.

7. Larry Kimmel is a parking lot attendant. He works 40 hours in a 5-day week. He is paid \$3.45 per hour and averages \$21 per day in tips.
8. Jake Bochinski is a skycap at the airport. He works 4 days a week, 9 hours a day for \$4.90 per hour and about \$30 per day in tips.
9. Tom Walters bags groceries and loads at Graul's Supermarket. He earns \$3.35 per hour for a 4-day, 20-hour week. He averages \$15.40 per day in tips.
10. Mai-Ling works 3 days a week as a waitress in her family's restaurant. She earns \$4.75 per hour, 6 hours per day, and averages \$23.50 per day in tips.
11. Megan Jackson is a door attendant at a large hotel. She helps guests by opening doors, carrying luggage into the hotel lobby, and calling for taxis. Megan works 6 hours a day, 6 days a week. She is paid \$4.85 per hour and averages \$18 per day in tips.
12. Andrew Katz is the host at an elegant restaurant. He takes reservations and guides people to their seats. Andrew works 7 hours a day, 6 days a week. He earns \$8.90 per hour and averages \$35 per day in tips.
13. Josie Conners is a taxi driver. Her salary is two-thirds of the total she collects in fares. Her tips average one-fifth of the total in fares. One week she collected \$768 in fares.

What percent of the total weekly earnings is from tips? Use mental math.

14. Hotel maid
Tips \$25
Wages \$100
15. Barber
Tips \$60
Wages \$300
16. Waiter
Tips \$125
Wages \$125
17. Parking valet
Tips \$120
Wages \$80

Mixed Skills Review

1. $58 \div 4\frac{5}{6}$
2. $\frac{1}{2} + \frac{2}{3} + \frac{3}{4}$
3. $1560 \div 1.5$
4. 15% of \$46
5. 1924×255
6. $10 - 7\frac{2}{3}$
7. $230 - 0.761$
8. 1.62×16.2
9. $874 \div 19$
10. $7 + 26{,}026 + 148$
11. $0.006 + 27.34 + 8.257$
12. $12{,}345 - 3456$

5-6 Commission

Objective
Find income from commissions.

A **commission** is a percent of the total amount of goods or services sold. Commissions are used by employers as incentives to employees. Both the employer and the employee benefit. The more they sell, the more they make.

Some people in sales jobs earn a **straight commission.** This means they do not receive wages or tips.

Examples

1. Bonnie Kabara gives home cosmetics demonstrations. She receives a 45% commission on sales. Find Bonnie's commission on sales of \$385.75.

   Find 45% of \$385.75. Round to the nearest cent.
   $0.45 \times \$385.75 = \$173.5875 \approx \$173.59$

   Bonnie's commission is \$173.59.

Some people are paid **wages plus commission** on sales or labor.

2. George Tong delivers bottled water. He earns \$4.75 an hour plus a 4% commission on sales. In one week he worked 40 hours and had sales of \$950.75. Find the total earnings.

Find the wages.	$40 \times \$4.75 = \$190$
Find the commission.	4% of \$950.75
	$0.04 \times \$950.75 = \$38.03$
Find the total earnings.	$\$190 + \$38.03 = \$228.03$

   George's earnings were \$228.03.

Try This Find the total earnings.

a. A real estate agent sold a home for \$86,900. She earned a commission of 6% on the sale.

b. Joe Watkins is an auctioneer. He earned a commission of 15% on an antique chest which sold for \$3250.85.

c. Ed Chaput installs mufflers. He works a 40-hour week and is paid \$5.75 per hour. He earns a 40% commission on sales of \$605.

Exercises

Find the total earnings.

1. Real estate agent
   7% straight commission
   $54,000 in sales

2. Modeling agent
   15% straight commission
   $29,800 in sales

3. Cookware salesperson
   40% straight commission
   $195.60 in sales

4. Magazine subscription salesperson
   30% straight commission
   $670.82 in sales

5. Auto mechanic
   38 hours, $10.75 per hour
   12% commission on labor
   $848 in labor

6. Apartment cleaner
   36 hours, $6.50 per hour
   15% commission on labor
   $785 in labor

7. Costume designer
   40 hours, $4.90 per hour
   10% commission on sales
   $975.81 in sales

8. Tour guide
   36 hours, $4.95 per hour
   6% commission on sales
   $665 in labor

Calculator Stockbrokers work for commission. Find the total amount of each sale and the amount of commission.

	Stock	Number of Shares Sold	Price per Share	Amount of Sale	Rate of Commission	Amount of Commission
9.	General Electric	20	$58.25		4%	
10.	Maytag	25	54.50		4.5%	
11.	Sears Roebuck	40	33.75		5.1%	
12.	Westinghouse	24	38.00		4.8%	
13.	Whirlpool	80	44.375		5.3%	

Number Sense Choose a calculation method. Decide which amount is greater.

14. 25% of ($95 − $35), or
    20% of ($50 + $45)

15. 50% of (30 × 40), or
    75% of (1000 ÷ 50)

16. $\frac{3}{4}$ of ($80 + $16), or
    $\frac{5}{6}$ of ($75 − $14)

17. $\frac{3}{10}$ of (450 ÷ 9), or
    $\frac{2}{3}$ of (57 × 6)

Mixed Skills Review

1. $5\frac{5}{16} \div 4\frac{3}{8}$
2. 1564 − 1377
3. 256 × 34
4. 12,864 + 19,376
5. 7011 ÷ 36
6. 101,255 − 89,638
7. 1853.1 ÷ 0.18
8. 20,247 + 9844
9. 16.03 × 6.4
10. $8\frac{2}{5} - 3\frac{5}{8}$
11. $25\frac{1}{3} \div 3\frac{4}{5}$
12. 28% of 1760

5-7 Piecework

Objective
Find income from piecework.

Some people are paid for each unit of work they complete or produce. This is called **piecework.** People who work for **piece rates** may be **self-employed** or may work for someone else.

Michael Warner charges piece rates for word processing and copying.

		Double Spaced	Single Spaced	Space and One-Half
Word Processing	letter legal	$1.35 $2.25	$2.50 $4.15	$1.80 $3.10
Duplicating copies	letter legal	$0.10 per page $0.15 per page		
Typing mailing labels		$0.03 per label		
Preparing envelopes		$0.025 per envelope		
Minimum charge is $3.50.				

Examples

1. Michael typed a 4-page, legal-size, double-spaced report and made 12 copies of each page. Find the amount earned.

Find the word processing charge. Multiply the rate by the units of work.
$4 \times \$2.25 = \$9$

Find the duplicating charge. Multiply the rate by the units of work.
$4 \times 12 \times \$0.15 = \$7.20$

Add the word processing charge and duplicating charge.
$\$9.00 + \$7.20 = \$16.20$ Michael earned $16.20.

2. Mowen Opticians ordered 70 mailing labels and 35 envelopes stuffed for mailing. Find the amount earned.

Find the mailing labels charge. $70 \times \$0.03 = \$2.10$

Find the envelopes charge. Round to the nearest cent. $35 \times \$0.025 \approx \$0.88$

Find the total piece-rate cost. $\$2.10 + \$0.88 = \$2.98$

Michael will earn the minimum charge of $3.50.

Try This

Find the amount earned. Use the piece-rate chart.

a. 94-page, letter-size, space-and-one-half term paper typed, 3 copies of each page

b. Single-spaced, one-page letter typed, 400 copies, mailing labels, and envelopes prepared

Exercises

Find the total earnings for each job. Use the Four Seasons Condo Service piece-rate chart.

1. 4 large lawns, 3 small lawns mowed
2. 84 medium windows washed, 9 large shrubs pruned
3. 7 cars washed, 5 waxed
4. 3 small pools cleaned, 5 small lawns mowed
5. 14 bags of leaves, 19 small windows washed
6. 6 rooms cleaned, 14 medium windows washed
7. 7 large sidewalks, 9 small sidewalks shoveled
8. 4 small lawns mowed, 13 bags of leaves
9. Ginny Hanson operates a sewing machine in a shirt factory. She is paid \$0.21 for each sleeve sewn on a shirt. During one shift Ginny sewed 332 sleeves. Find Ginny's total earnings for the shift.
10. In the same factory Jim Stone is paid \$0.43 for each collar sewn. Find the total earned for 187 collars.
11. Another employee earns \$0.085 for each button sewn. Find the total earned for 676 buttons sewn during one shift.
12. An inspector checks, irons, and packages shirts. The piece rate for each shirt completed is \$0.145. Find the total earnings for 158 shirts.

Four Seasons Condo Service

	Size	Rate
Lawn care	large	\$12.00
	small	8.50
Shrub service	large	3.25
	small	2.25
Pool cleaning	large	35.00
	small	24.50
Snow removal	small	6.25
	large	9.75
Window wash	small	.85
	medium	1.40
	large	3.00
Car wash		3.75
Car wax		17.50
Leaf rake and bag		\$0.75 per bag
Apartment cleaning		\$5.25 per room

Look for a Pattern

Find the missing numbers.

13. 1, 4, 7, ____, ____, ____
14. 13, 28, 43, ____, ____
15. \$9, \$27, ____, \$63, ____, ____
16. \$1.35, \$3, ____, ____, \$7.95
17. $\frac{2}{11}, \frac{4}{11}, \frac{6}{11}$, ____, ____, ____
18. $\frac{1}{8}, \frac{1}{4}$ ____, $\frac{1}{2}$ ____

Mixed Skills Review

1. 20,502 ÷ 72
2. $2\frac{4}{5} + 2\frac{1}{2} + \frac{7}{8}$
3. $4\frac{2}{5} \times \frac{8}{11}$
4. 3002 − 1648
5. \$12.85 + \$16.27 + \$17.53
6. 175 × \$0.225
7. 3 + 0.078 + 2.29
8. 24 is what percent of 16?
9. $6\frac{5}{9} - 4\frac{7}{12}$
10. $2\frac{7}{10} \div 4\frac{1}{5}$
11. 16.5% of 104
12. 0.426 − 0.348

5-8 Comparing Future Earnings

Objective
Compare future earnings.

Before accepting a position, it is important to consider future earnings as well as the initial salary. Employers reward merit and longevity with raises.

Starting	6 months	1 year	18 months	2 years	30 months	3 years
\$12,500	\$12,675	\$12,850	\$13,025	\$13,200	\$13,375	\$13,550

Example

1. The starting salary at Peterson Manufacturing is \$12,500 with raises of \$175 every 6 months. Find the annual salary after 3 years.

The annual salary after 3 years is \$13,550.

You can also use the formula $S = a + nd$ to solve this problem.

S is the salary after a given period — a is the starting annual salary
n is the number of increases — d is the amount of each increase

$S = a + nd$
$S = \$12{,}500 + 6(\$175)$ — 3 years → 6 increases
$S = \$12{,}500 + \$1050 = \$13{,}550$ — Multiply. Then add.

Try This

Solve.

a. The starting salary for an operations assistant at Smith Transfer is \$13,700. Increases of \$650 are given each year. Find the annual salary after 5 years.

Many companies give raises by paying employees a percent of the existing salary.

Example

2. Tom Murphy is an apprentice mechanic. He started at \$13,200 per year with 3% increases every 6 months. Find his salary after 18 months.

Plan Add a 3% increase to Tom's salary in six-month intervals.

Add 3% to find the salary after 6 months. — $13{,}200 + (0.03 \times 13{,}200) = 13596$
Add 3% to find the salary after 12 months. — $13{,}596 + (0.03 \times 13{,}596) \approx 14004$
Add 3% to find the salary after 18 months. — $14{,}004 + (0.03 \times 14{,}004) \approx 14424$

The salary after 18 months is about \$14,424 rounded to the nearest dollar.

Try This

Solve. Round to the nearest dollar.

b. Dick Nair is the track coach for a small college. His starting salary 2 years ago was \$14,900. He has received increases of 3% every 6 months. Find Dick's current salary.

Exercises

Calculator Find the annual salary at the end of the given period. Round to the nearest dollar.

1.	Butcher	$ 9800	$250 every year	3 years
2.	Records clerk	8400	$175 every 6 months	2 years
3.	Retail manager	12,600	$325 every 6 months	3 years
4.	Teacher	13,800	$540 every year	10 years
5.	Dispatcher	9500	3% every 6 months	3 years
6.	Secretary	11,900	4% every 6 months	2 years
7.	Manufacturing manager	15,900	3.5% every year	4 years
8.	Summer lifeguard	3600	4.5% every year	3 years

Ron Harris and Doug Tyree were hired with starting salaries of $10,000 each. Ron was given annual increases of $500. Doug's salary was increased 4.7% per year. Use the bar graph to compare their salaries for eight years.

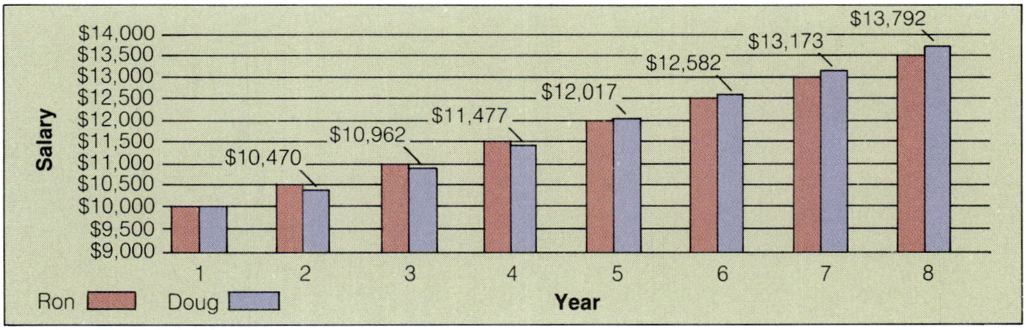

9. Which employee made more money during each of the following years?
a. second **b.** fourth **c.** sixth **d.** eighth

10. In what year did Doug first make more money than Ron?

11. Which employee had the greater total income for the eight years shown?

12. Using these two plans, which method of salary increases would you prefer if you planned to change jobs every three years?

Mixed Skills Review

1. $17,920 + $1245 **2.** 19.011 − 8.27 **3.** 1060 × 427

4. $6\frac{3}{8} + 3\frac{3}{4} + 5\frac{1}{2}$ **5.** $6\frac{1}{10} - 4\frac{4}{5}$ **6.** 19.5% of 385

7. $1\frac{3}{20} \times 4$ **8.** 60% of what number is 45? **9.** 25,218 − 19,326

10. 22,560 ÷ 96 **11.** 2.63 + 7.8 + 2.064 **12.** $\frac{7}{8} \div 2\frac{5}{8}$

5-9 Enrichment: Consumer Price Index

Objective
Use the Consumer Price Index to compare changing costs.

The **Consumer Price Index (CPI)** shows how prices in a given year compare with prices for the same items during a base year.

Year		Index %
1967	Base	100.0
1968		104.2
1969		109.8
1970		116.3
1971		121.3
1972		125.3
1973		133.1
1974		147.7
1975		161.2
1976		170.5
1977		181.5
1978		195.4
1979		217.4
1980		246.8
1981, May		269.0

Consumer Price Index 1967–1981

The Websters bought a home for \$24,000 in 1967. How much would it have cost in 1976?

Use 1967 as the base year. The CPI table shows the index for 1976 was 170.5% of the 1967 prices. Write the percent as a decimal. Then multiply it times the base year price of \$24,000.

170.5% = 1.705   1.705 × \$24,000 = \$40,920

A similar home would have cost the Websters \$40,920 in 1976.

A bicycle cost \$169 in 1981. How much would a similar bicycle have cost in 1967?

The CPI for 1981 is 269% of 1967 prices. Write the CPI as a decimal and divide into the given price.

269.0% = 2.69   \$169 ÷ 2.69 = \$62.83

A similar bicycle would have cost \$62.83 in 1967.

Exercises

Use 1967 as the base year. Round answers to the nearest cent.

1. In 1967, Bob and Sarah paid \$8.30 for dinner at their favorite restaurant. How much would dinner have cost in 1970?
2. Ms. Forest bought a car for \$2100 in 1967. How much would a similar new car cost in 1977?
3. Mr. Snow paid \$75 for a new suit in 1967. How much would he have paid for a similar new suit in 1973?
4. A pair of jeans cost \$24.50 in 1981. How much would a similar pair of jeans have cost in 1967?
5. Mr. Black paid \$150 a month for rent in 1967. How much rent would he have paid for a similar apartment in 1980?
6. The Clemmons family spent \$165 a month for food in 1967. How much would the same amount of food have cost in 1969?

5-10 Computer: Future Earnings

Objective
Use a computer to find future earnings.

Sonya started as a telemarketing agent at \$18,400 per year. She gets a 3.2% increase every 6 months. What will be her salary after 5 years?

Example The following computer program can be used to solve such problems:

```
10  REM   This program will calculate future earnings
20  PRINT   "What is the starting salary?"
30  INPUT A
40  T = A
50  PRINT   "How many increases will be given per year?"
60  INPUT I
70  PRINT   "What is the rate (in %) of each increase?"
80  INPUT R
90  PRINT   "How many years?"
100 INPUT Y
110 N = I*Y
120 FOR X = 1 TO N
130 A = A + R/100*A
140 NEXT X
150 PRINT   "Starting with a salary of $";T;","
160 PRINT   "with increases given" . . . ; I "times a year for "; Y "years,"
170 PRINT   "at a rate of" . . . ; R; "%, your salary will be $"; A; "."
180 END
```

Try This Solve the problem using your calculator. Then compare your results with the computer solution given below.

Starting salary is \$18,400. With increases given 2 times a year for 5 years, at a rate of 3.2%, your salary will be \$25,163.57.

Exercises

1. Type the program into the computer.
2. Run the program and enter the data for the example above.
3. Starting salary is \$15,000. Increases of 1% are given every 3 months. What is the salary after 3 years?
4. Starting salary is \$25,000. Increases of 6% are given once a year. What is the salary after 5 years?

CHAPTER 5 REVIEW

The following summer jobs are used for questions 1–5.

A. Yardwork—$3.95/hr 25 hr/week
B. Housepainting—$5.30/hr 20 hr/6-day week
C. Lifeguard—$75 per weekend
D. Daycare—$4/hr, 5 hr/day 5 days/week

1. Which job pays the most per week? How much?
2. Which job pays the least per week? How much?
3. Joaquin wants to buy a new bicycle and must work to get the money. He practices soccer 2 afternoons a week. Which of the 4 jobs would be best for Joaquin? Why?
4. To maximize her summer income Ilene plans to work on weekends as well as weekdays. At which jobs can she work? What will be her maximum weekly income?
5. Glenda is considering a job selling food at the ballpark. The job pays a commission of 12% of gross sales. What information does Glenda need to help decide whether to take this job or one of those listed above.

For questions 6–11, match the number of hours with the correct pay. Assume a 40-hour regular work week.

	Hours	Regular Rate	Overtime Rate	Pay		
6.	40	$4.76	$ 7.14		A.	$455.00
7.	48	$5.95	$ 8.9		B.	$309.20
8.	55	$6.50	$13.00		C.	$182.40
9.	38	$4.80	$ 9.60		D.	$370.50
10.	45	$7.80	$11.70		E.	$440.00
11.	60	$5.50	$11.00		F.	$190.40

Which does not belong in the group? Explain.

12. A. straight commission
B. wages plus commission
C. stock purchase plan
D. salary

13. A. health care
B. tips
C. child care service
D. discounts

14. A. paycheck stub
B. application form
C. resume
D. social security number

15. A. waiter
B. factory worker
C. skycap
D. bellhop

16. Why is it less expensive for an employer to provide benefits such as health insurance than it is to pay employees enough to buy their own?

CHAPTER 5 TEST

Use the chart to estimate items 1–3.

1. How many seniors from Barclay High have part-time jobs?
2. How many juniors have part-time jobs?
3. How many sophomores and freshmen all together have part-time jobs?

Compare the pay for the indicated period.

4. Annual: Job A pays $22,500 per year; Job B pays $432.92 per week.
5. Monthly: Job A pays $237.50 per week; Job B pays $18,450 per year.

Find the annual employee benefit.

6. Jonathan works at Bill's Burger Palace three days a week. He gets two hamburgers ($0.69 each), an order of french fries ($0.49), and a soft drink ($0.54) each night that he works.
7. Schofield Copiers pays $0.18 per mile to its salespersons. Martha averaged 342 miles per week last year for a 50-week work year.

Find the gross pay.

8. Al Martin drives a delivery truck for $8.50 per hour for the first 40 hours each week. His overtime rate is 1.5 times his regular rate. Find his monthly gross pay if he worked as follows: 1st week—42 hours; 2nd week—48 hours; 3rd week—52 hours; 4th week—51 hours.

Find the total weekly earnings.

9. Karen Daly is a beautician who works 40 hours a week at $6.35 per hour. She averages $2.50 each in tips from 25 customers per week.

Find the total earnings.

10. Marketing agent; $600 per week; 12% commission on average weekly sales of $1200
11. Insurance agent; $600 per week; 4% commission on sales over wages; $2400 in sales

Find the annual salary at the end of the given period.

12. Sales agent; starting at $15,000 a year with a 2% increase every three months, period 1 year

CHAPTER 5 CUMULATIVE REVIEW

1. Find the total earnings. Encyclopedia salesperson, 15% straight commission, \$3,672 in sales
   - A. \$520.80
   - B. \$219.32
   - C. \$550.80
   - D. \$547.70

2. Multiply. Use mental math.
   $5 \times 2\frac{3}{5}$
   - A. $10\frac{3}{5}$
   - B. 13
   - C. $13\frac{1}{5}$
   - D. 12

3. Write as a decimal.
   2.5%
   - A. 0.025
   - B. 0.25
   - C. 0.205
   - D. 0.0025

4. Help Wanted: Delivery person. Must lift 30 lb. Valid driver's license. Call for appointment.

   Which statement is true about this job?
   - A. You do not need to drive.
   - B. You can walk in for an application.
   - C. It is an office job.
   - D. It requires certain physical strength.

5. Add. Use mental math.
   $7\frac{1}{6} + 5 + 4\frac{5}{12}$
   - A. $16\frac{7}{12}$
   - B. $17\frac{1}{6}$
   - C. $16\frac{1}{2}$
   - D. $17\frac{7}{12}$

6. In the Tune Town Band there are one-third as many tubas as drums. There are half as many piccolos as tubas. There is one piccolo. How many drums are there?
   - A. 3
   - B. 2
   - C. 6
   - D. 9

7. Find the annual salary at the end of the given period. Office manager, \$12,500, 3% every 6 months, 3 years.
   - A. \$14,750
   - B. \$12,875
   - C. \$13,625
   - D. \$14,925.65

8. Write as a percent.
   0.8
   - A. 8%
   - B. 80%
   - C. $\frac{8}{10}$%
   - D. 0.8%

9. Write as a fraction.
   $4\frac{5}{16}$
   - A. $\frac{20}{16}$
   - B. $\frac{64}{16}$
   - C. $\frac{69}{16}$
   - D. $\frac{80}{16}$

10. Compare the pay for the indicated period.
    Monthly
    Job A \$13,110 a year
    Job B \$1105 a month
    - A. A pays \$12.50 more.
    - B. B pays \$206 more.
    - C. A pays \$10 more.
    - D. B pays \$12.50 more.

11. **Hot Cocoa Mix**

Cups	Mix	Hot Water
1	3 tsp	$\frac{3}{4}$ cup

    How much water should you use for 6 cups of cocoa?
    - A. $4\frac{1}{2}$ cups
    - B. 8 cups
    - C. 5 cups
    - D. $3\frac{1}{2}$ cups

12. Find the total earnings. Lupe decorates cakes at a bakery. She earns \$3.50 per cake. How much does she make for 17 cakes?
    - A. \$59.50
    - B. \$56.50
    - C. \$25.00
    - D. \$37.45

13. Estimate the amount of tip. Use mental math.
Dinner bill \$25.75, tip 15%.
A. \$5
B. \$4
C. \$7.50
D. \$1.50

14. Write in lowest terms.
$\frac{27}{81}$
A. $\frac{1}{3}$
B. $\frac{7}{8}$
C. $\frac{4}{5}$
D. $\frac{13}{19}$

15. Divide.
$306 \div 6.12$
A. 20
B. 50
C. 5
D. 2

16. Find the annual employee benefit. A sporting goods store offers employees 25% discount on the first \$75 purchased each month.
A. \$56.25
B. \$18.75
C. \$225
D. \$675

17. Find the percent.
What percent of 75 is 15?
A. 5%
B. 40%
C. 20%
D. 8%

18. Find the gross pay.

Regular hours	37
Regular rate	\$10.34
Overtime hours	4
Overtime rate	\$12

A. \$430.58
B. \$382.58
C. \$48
D. \$320.58

19. Multiply.
$3\frac{1}{3} \times 1\frac{5}{8}$
A. $6\frac{1}{4}$
B. 1
C. $3\frac{1}{4}$
D. $5\frac{5}{12}$

20. Find the percent of increase or decrease.
From 455 to 273.
A. +40%
B. $-66\frac{2}{3}\%$
C. −40%
D. −60%

21. Translate to a number sentence.
575 is what percent of 1050?
A. $n\% = 1050 \times 575$
B. $1050 = 575 \times n\%$
C. $575 = n\% \times 1050$
D. $n = 575 \div 1050$

22. What law or property can you use to solve $(49 \times 7) + (49 \times 9) = \square$?
A. Distributive Law of Multiplication over Addition
B. Multiplicative Property of Zero
C. Commutative Law of Multiplication
D. Associative Law of Multiplication

23. Dan takes the train 7 miles to work Monday through Friday. Last week he and his family rode the train to visit relatives 45 miles away. How many miles did Dan travel by train last week?
A. 100 mi
B. 160 mi
C. 52 mi
D. 80 mi

24. Multiply.
$$\begin{array}{r} \$412.91 \\ \times \quad 13 \\ \hline \end{array}$$
A. \$5011.93
B. \$5367.83
C. \$412.93
D. \$1086.44

Chapter 6 Overview

Key Ideas

- filling out a deposit slip
- writing checks
- balancing a checking account
- reconciling a monthly statement
- calculating simple interest
- finding the amount to be repaid
- finding compound interest
- working with a budget

Key Terms

Here are some of the key terms you will use in this chapter.

- checking account
- deposit
- net deposit
- check notation
- payee
- check stub
- check register
- duplicate check
- reconcile
- cancelled check
- balance
- service charge
- bank fee
- interest
- principal
- rate
- promissory note
- cosigner
- secured loan
- compound interest
- effective annual rate
- semiannually
- quarterly
- budget
- gross income
- net income
- terminating debts

Key Skills

Find the answer.

1. \$35.72 + \$94.06 + \$1.87

2. \$22.87 + \$198.08 + \$54.17

3. \$9.02 + \$0.87 + \$328.65

4. \$82.99 + \$12.03 + \$2.13

5. \$65.28 − \$41.99

6. \$203.21 − \$128.65

7. \$51.83 − \$19.90

8. \$55.87 − \$32.98

9. \$15.69 − \$2.70

10. \$200.16 − \$67.65

11. $72 \div 3$ **12.** $72 \div 12$ **13.** $72 \div 7.5$ **14.** $72 \div 8.2$

15. $\$800 \times 0.05 \times \frac{1}{2}$

16. $\$918 \times 0.08 \times \frac{1}{2}$

Write in words.

17. 54 **18.** 346.8 **19.** 287.32 **20.** 2.871 **21.** 998.087

Solve.

22. What is 27% of 87?

23. What is 14% of 823?

24. What is 7% of \$230.98?

25. What is 9% of \$432.00?

26. 8.5% of \$613.65 is how much?

27. 10.8% of \$4098 is how much?

28. \$350 = n% of \$1400

29. \$450 = n% of \$1500

30. n% of \$3600 = \$1200

31. n% of \$1850 = \$250

PERSONAL FINANCE 6

6-1 Checking Accounts: Using Deposit Slips

Objective
Fill out a checking account deposit slip.

A checking account is useful for paying bills and is more secure than carrying cash. You must fill out a **deposit slip** when you deposit money in your checking account.

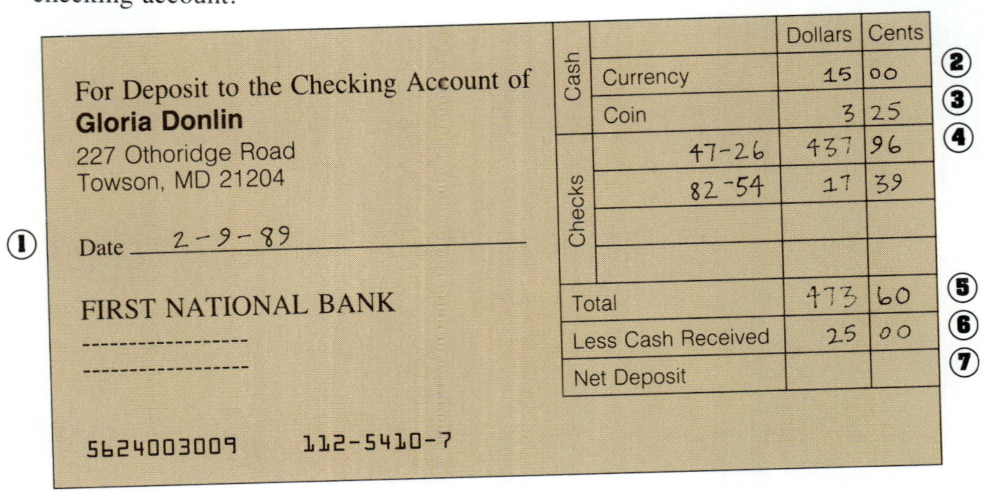

For Deposit to the Checking Account of
Gloria Donlin
227 Othoridge Road
Towson, MD 21204

① Date 2-9-89

FIRST NATIONAL BANK

5624003009 112-5410-7

		Dollars	Cents	
Cash	Currency	15	00	②
	Coin	3	25	③
Checks	47-26	437	96	④
	82-54	17	39	
Total		473	60	⑤
Less Cash Received		25	00	⑥
Net Deposit				⑦

1. Write the date.
2. Write the amount to be deposited in paper money.
3. Write the amount to be deposited in coins.
4. Write the bank number and amount of each check.
5. Add the cash and the check totals.
6. Write the amount of cash you want back.
7. Subtract cash received. The remainder is the amount deposited.

The **net deposit** is the difference between the total deposit and the cash given back. Subtract to find the difference.

Example Find the net deposit from the deposit slip above.

$473.60 − $25.00 = $448.60

The net deposit is $448.60.

Try This Find the net deposit. Make and use a deposit slip like the one above.

Deposit 1 twenty-dollar bill, 1 ten-dollar bill, 2 one-dollar bills, 47 nickels, a check for $45.98 (bank number 27-46), and a check for $132.00 (bank number 825-4). Subtract $50.00 for cash received.

Exercises

Find the net deposit. Make and use a deposit slip like the one shown on page 126.

1. Deposit 2 five-dollar bills, 7 one-dollar bills, 22 quarters, 7 dimes, 11 pennies, and a check for \$38.95 (bank number 14-164).
2. Deposit checks for \$485.17 (bank number 15-17), \$75.16 (bank number 71-109), and \$136.27 (bank number 87-4). Subtract \$25 for cash received.
3. Deposit 1 twenty-dollar bill, 2 ten-dollar bills, 2 five-dollar bills, 6 one-dollar bills, 34 quarters, 27 dimes, 59 nickels, 59 pennies, and a check for \$53.89 (bank number 124-15).
4. Deposit checks for \$426.87 (bank number 17-215), \$86.49 (bank number 9-18), and \$758.65 (bank number 24-10). Subtract \$50 for cash received.

Number Sense

5. Here are some amounts of money written in money notation and check notation. Arrange them in order from least to greatest.
\$0.54, Fifty four and $\frac{00}{100}$ dollars, Zero and 98/100 dollars, \$32.00, \$506.00

Checking Account Costs

Average costs of interest-bearing checking accounts

	Banks/S&Ls	Credit unions
Have no routine charges	5%	75%
Require minimum balance	38%	19%
Charge both flat, per-check fees	29%	1.2%
Avg. required minimum monthly balance	\$880.00	\$300.00
Avg. monthly fee if balance below minimum	\$5.28	\$3.00

6. What percent of banks and savings and loans require a minimum balance?
7. What is the average required minimum monthly balance at credit unions?
8. Where are you most likely to find the least expensive checking account—at a bank, savings and loan, or credit union?

Mixed Skills Review

1. $2\frac{2}{7} + 4\frac{3}{4}$
2. 16% of what number is 18?
3. $128 \div 3.2$
4. \$478.29 + \$194.64
5. $16 - 3.042$
6. 42% of 38
7. $6\frac{1}{3} - 2\frac{3}{8}$
8. \$2900 − \$843.17
9. $4\frac{3}{5} \times 15$
10. $0.7 + 12 + 13.015$
11. $2\frac{1}{3} \div 1\frac{3}{4}$
12. 0.86×1.7

6-2 Checking Accounts: Writing Checks

Objective *Write checks.*

Up to 50 billion checks are written in the U.S. every year.

Example Write a check dated October 23, 1990, from Kathryn Lopez to Griffin Pharmacy in the amount of $17.86 for a prescription.

Kathryn Lopez
624 Marsh Road
San Antonio, TX 78212

132

10/23 19 90 ①

11-7000/3210

② Pay to the Order of Griffin Pharmacy $ 17.86 ③

④ seventeen and $\frac{86}{100}$ Dollars

Memo for prescription Kathryn Lopez ⑤

11 386:48053:06372:29206

Understand the Situation

- What is the number of this check?
- To whom is the check written?
- On what date was the check written?
- For what amount is the check written?
- Whose checking account is involved?
- For what was the check written?

When writing checks it is important to write them very carefully. Errors can sometimes lead to service charges.

1. Write the correct date. Never date ahead.
2. Write the payee's name. Never leave this line blank.
3. Write the check amount. Use money notation.
4. Write the check amount in check notation. Fill in the entire line.
5. Sign the check after filling in all the other blanks.

Try This Make or use a form like the one in the example.

a. Write a check from Joe Clarkson to Ace Electric in the amount of $28.59 for motor repair on January 10, 1990.

b. Write a check to Bob's Supermart on March 9, 1990, in the amount of $60.05 from LaToya Greene.

Exercises

1. Tell the payee, the account number, and the amount the bank will pay.

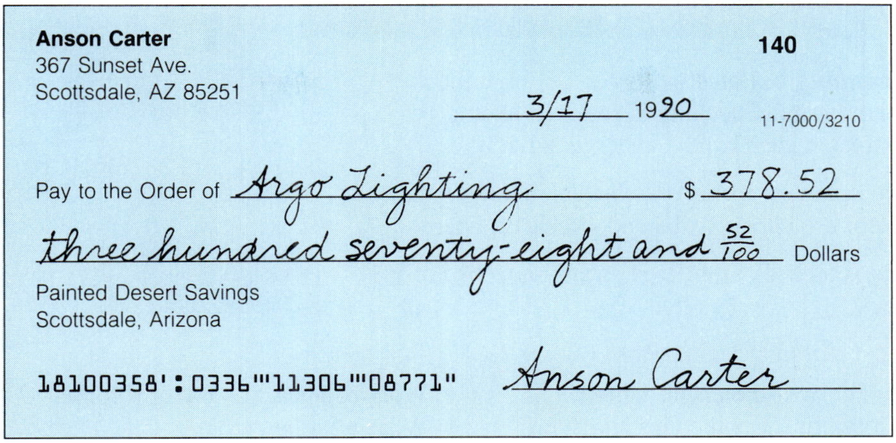

Anson Carter
367 Sunset Ave.
Scottsdale, AZ 85251

140

3/17 1990

11-7000/3210

Pay to the Order of Argo Lighting $ 378.52

three hundred seventy-eight and $\frac{52}{100}$ Dollars

Painted Desert Savings
Scottsdale, Arizona

18100358':0336"11306"08771" Anson Carter

Write checks using the given information.

2. Sam Chin paid C & P Telephone Company \$45.38 on September 15, 1989.

3. Alicia Howard paid \$575.45 for the May home mortgage payment to First Mortgage Company on May 5, 1990.

4. Mike Wollensak made a quarterly tax payment on April 12, 1989 to Internal Revenue Service in the amount of \$450.00.

5. American Express received a monthly payment dated March 7, 1989 for \$234.98 from Pietr Hitzig.

Examine the following check entries carefully. Find the mistakes.

6. \$146.97: Forty-Six and $\frac{97}{100}$ Dollars

7. \$2339.00: Two Thousand Three Hundred Thirty-Nine Dollars

8. \$17.35: Seventy and $\frac{35}{100}$ Dollars

9. \$376.61: Three Hundred Seventy-Six and $\frac{66}{100}$ Dollars

Mixed Skills Review

1. $5\frac{3}{8} \times 5\frac{1}{3}$

2. 30% of what number is 240?

3. 4.5% of \$12.95

4. $\frac{1}{3} - \frac{2}{15}$

5. $5406 \div 53$

6. $1 - 0.163$

7. $8\frac{3}{8} + 2\frac{7}{12}$

8. $19.29 \div 64.3$

9. \$12.56 + \$7.29 + \$0.73

10. 225×48

11. \$1935.07 − \$1258.25

12. $0.93 + 0.6 + 0.07$

6-3 Keeping a Running Balance

Objective
Keep a checking account running balance.

A **running balance** is a record of how much money is in a checking account. It should be completed each time a check is written.

Two common forms on which a running balance is kept are **check stubs** and **check registers**.

This is a check stub. It is usually attached to the check.

Date 6-20 1990
To Tom's Hardware
For Plumbing Supplies

Balance Forward	$ 498	20
Amount of Deposit		
Amount of Check	53	95
Other ATM withdrawal	20	00
Balance Carried Forward		

Understand the Situation

- When was the check written?
- To whom was the check written?
- What was purchased?
- What balance was brought forward?
- What was the amount of the check?
- Was there also a withdrawal?

Examples

1. Find the balance carried forward from the check stub above. Subtract the amount of the check from the balance forward.
$498.20 − $53.95 = $444.25

$20 was withdrawn from an ATM. $444.25 − $20.00 = $424.25

The balance carried forward is $424.25.

2. Another way of keeping a running balance is with a **check register**, shown below. Find the balance forward after the deposit on 4-19.

The balance forward was $675.94. Check 485 was $438.00.
$675.94 − $438.00 = $237.94

The deposit of 4-19 was $296. $237.94 + $296 = $533.94

The balance after the deposit of 4-19 is $533.94.

Number	Date	Description	Payment/Debt (−)		Fee (−)		Deposit/Credit (+)		Balance $675.94	
485	4-14	IRS for taxes	$ 438	00	$		$		237	94
	4-19	deposit					296	00		
486	4-20	Dr. Cassidy for checkup	65	00						
487	4-27	Andy's Auto Repair	49	18						
488	4-30	ABC Dept. Store	78	25						

Try This Find the balance. **a.** after check 486 **b.** after check 487 **c.** after check 488

Exercises

Find the balance carried forward.

1.

Date 4-1 1990
To
For

	$	
Balance Forward	630	50
This check	15	85
Total		
Other	0	
Balance Carried Forward		

2.

Date 6/27 19 90
To
For

	$	
Balance Forward	819	48
This check	138	80
Total		
ATM Withdrawal	50	00
Balance Carried Forward		

	Number	Date	Description	Payment/Debt (−)	Fee (−)	Deposit/Credit (+)	Balance $1465.87
3.	707	8-14	Fairview Apartments	$365.00			
4.	708	8-14	A-1 Market	$ 48.19			
5.		8-14	Checking fee		$6.90		
6.	709	8-16	Bob's Motors	$187.80			
7.		8-18	Deposit			$385.00	

8. What percent of people in the United States use an ATM most of the time?

9. Out of 1000 people, how many would rarely use an ATM?

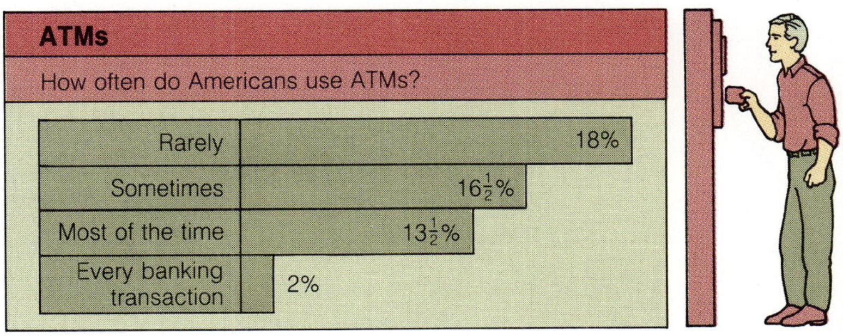

Mixed Skills Review

1. $9 \times 8\frac{1}{3}$
2. $16.2 - 13.34$
3. 16 is what percent of 24?
4. $1\frac{5}{8} - \frac{7}{9}$
5. $0.83 − $0.68
6. $\frac{3}{4}$ of 32
7. $147 \div 4.9$
8. $11,682 + 13,926$
9. 8 × $56.73
10. $3\frac{1}{7} \div \frac{11}{14}$
11. $\frac{2}{5} + \frac{3}{8} + \frac{3}{10}$
12. $0.06 + 3 + 0.075$

6-4 Reconciling Monthly Statements

Objective
Reconcile monthly bank statements.

Each month banking institutions send statements to their customers. They often include the cancelled checks. By accounting for outstanding checks and deposits, you can **reconcile** your current balance with the statement.

BANKERS TRUST COMPANY
Canton, Ohio

Ardele Reese
2625 Market St.
Canton, Ohio 44703

All Items Are Received by This Bank for Purposes of Collection.
All Credits for Items Are Provisional.

	Account Number	Disp. Code	Period Ending	Number of Enclosures	Page
	025-1299-8		10-02-92	5	1

Previous Balance		− Checks and Other Debits Number	Amount		+ Deposits and Other Credits Number	Amount		− Service Charge		= New Balance	
426	92	5	255	92	2	405	81	4	50	572	31

Checks and Other Debits		Credits		Date		Balance	
		135	17	9	10	622	09
16	15			9	11	605	94
28	34			9	12	577	60
95	10			9	14	482	50
104	17			9	16	378	33
		210	64	9	21	588	97
12	16			9	24	576	81
4	50			10	02	572	31

1. Put the cancelled checks in numerical order.
2. Match the checks with the stubs or register.
3. Add deposits not shown on the statement.
4. Subtract checks not shown on the statement.
5. Subtract any bank fees and charges.
6. Compare the balances.

Example Reconcile the statement. Write agree or disagree.

Balance on Bank Statement	Deposits Not Shown	Total of Checks Not Shown	Customer Balance	Bank Fees
\$392.12	\$147.80	\$36.25	\$517.80	\$7.65

Add the deposits not shown. \$392.12 + \$147.80 = \$539.92

Subtract the checks not shown. \$539.92 − \$36.25 = \$503.67

Subtract any fees from the customer's balance. \$517.80 − \$7.65 = \$510.15

\$503.67 ≠ \$510.15 The bank statement and the customer's statement do not agree.

Try This Reconcile the statement. Write agree or disagree.

	Balance on Bank Statement	Deposits Not Shown	Total of Checks Not Shown	Customer Balance	Bank Fees
a.	$498.37	$186.75	$48.17	$629.65	$7.50
b.	$1148.70	$398.42	$575.00	$972.12	0

Exercises

Reconcile the statements. Write agree or disagree.

	Balance on Bank Statement	Deposits Not Shown	Total of Checks Not Shown	Customer Balance	Bank Fees
1.	$785.41	None	$18.75	$741.81	$5.25
2.	$875.39	$187.25	None	$1061.64	None
3.	$1148.70	$398.15	$473.42	$1073.43	$6.75
4.	$2487.87	$560.18	$1137.18	$1910.87	None
5.	$1948.10	$187.75	$310.10	$1822.45	$3.00
6.	$1486.19	$347.00	$198.53	$1628.36	$6.30

7. Jackie has two deposits not shown by the bank: $235.98 and $56.32. She has three checks not shown: $142.67, $22.13, and $10.00. The bank balance on the statement is $556.64. Jackie's check register shows a balance of $674.14. Reconcile the statement.

8. Ken has three deposits not shown by the bank: $500.00, $43.18, and $67.98. All his checks are shown. The bank balance on his statement is $12.75. Ken's check register balance is $633.91. Reconcile the statement.

Write Math

9. Make a list of some possible reasons why the bank statement balance and the customer's balance may disagree.

Data Hunt

10. What must be done if you can't reconcile your checking account?

11. What happens if you write a check for more than is in your account?

Mixed Skills Review

1. $1712 - 238$	**2.** $\frac{19}{24} \times \frac{3}{8}$	**3.** $7.6 + 3.08 + 6.56$
4. $2160 \div 18$	**5.** $2.414 - 1.64$	**6.** $\frac{1}{2} - \frac{5}{12}$
7. 27.3% of 21	**8.** $7 \div 0.35$	**9.** $12 \times \$2064.12$
10. $83 + $27 + $64	**11.** $3\frac{2}{3} + 8\frac{2}{9}$	**12.** $12\frac{1}{2} \div 4$

6-5 Borrowing and Interest

Objective
Calculate simple interest.

In a recent year Americans had borrowed so much money that their debts totaled \$7.145 trillion. People borrow money for many reasons. A major reason is to buy high-priced items such as homes and cars.

Interest Rates	
Home Equity Loan	9%
Signature Loan	10%
Passbook Savings	5.5%
Money Market	5.85%
IRA	6.0%

When people borrow money they must pay **interest** to the lender. Interest is money paid for the use of money. If you borrow money, you pay interest. If you put money in a savings account you receive interest.

One type of interest is called **simple interest.** Each year an agreed-upon percent of the **principal** (the amount borrowed) is charged to the borrower.

interest = principal × rate × time or $i = prt$

The **rate** is given as a percent. The **time** is given in years.

Examples Find the simple interest.

1. Jo borrowed \$700 for 2 years at 9% interest.
Use the interest formula $i = prt$

$$\begin{aligned} i = p \times r \times t &= \$700 \times 9\% \times 2 \\ &= \$700 \times 0.09 \times 2 \\ &= \$700 \times 0.18 \\ i &= \$126 \end{aligned}$$

Estimate: $700 \times 0.10 \times 2 = 140$

Jo will pay \$126 interest on the loan.

2. Rosalie borrowed \$300 for 6 months at 11.5% interest.

$$\begin{aligned} i = p \times r \times t &= \$300 \times 11.5\% \times \tfrac{1}{2} \\ &= \$300 \times 0.115 \times \tfrac{1}{2} \\ &= \$34.50 \times \tfrac{1}{2} \\ i &= \$17.25 \end{aligned}$$

Estimate: $300 \times 0.10 \times \frac{1}{2} = 15$

Rosalie will pay \$17.25 interest on the loan.

Try This Find the simple interest.

a. Carey borrowed \$6000 for 4 years at 11% interest.

b. Patti borrowed \$2500 for 9 months at 12.6% interest.

c. Pier Luigi borrowed \$1000 for 18 months at 10.9% interest.

Exercises

$10,000 Loan–5 Years		
Loan Rates	Monthly Payment	Total Interest
11%	$217.43	$3046
12%	$222.45	$3347
13%	$227.53	$3652
14%	$232.68	$3961
15%	$237.90	$4274
16%	$243.18	$4591

Find the simple interest.

1. $450 loan, 1 year, 10%
2. $700 loan, 2 years, 9%
3. $675 loan, 6 months, 11%
4. $750 loan, 9 months, 12%
5. $900, 3 years, 9.9%
6. $1000 loan, 2 years, 10.8%
7. $1825 loan, 2 years, 12.6%
8. $2345 loan, 3 years, 14.7%
9. $12,000 loan, 3 years, 16%
10. $12,000 loan, 5 years, 16%
11. $65,000 loan, 20 years, 12.8%

Compare. Use the loan chart.

12. What is the difference in monthly payments between an 11% loan and a 12% loan?
13. What is the difference in total interest between an 11% and a 12% loan?
14. What is the difference in monthly payments between a 12% loan and a 16% loan?
15. What is the difference in total interest between an 11% loan and a 16% loan?
16. How much will be paid in monthly payments on a $10,000 loan for 5 years at 11%?
17. How much will be paid in monthly payments on a $10,000 loan for 5 years at 15%?

Data Hunt

18. Make a list of consumer guidelines for borrowing money. Why is prompt repayment important? Should you shop for a loan? Who can you ask for advice?

Mixed Skills Review

1. $113,650 + $436,288
2. $2\frac{1}{10} \div \frac{3}{14}$
3. $27\frac{1}{2} - 16\frac{3}{4}$
4. $19.026 - 13.475$
5. $3\frac{1}{2} + \frac{3}{5} + 4\frac{5}{6}$
6. $22{,}000 - 16{,}485$
7. 264×38
8. $3.423 \div 21$
9. 65% of 113
10. $5.8 + 11.83 + 16.24$
11. $8\frac{1}{8} \times 1\frac{3}{5}$
12. What percent is 17 of 85?

6-6 More on Borrowing

Objective
Find the total principal plus interest on loans.

A **personal loan** can be obtained by signing a **promissory note.** With a good credit rating no security is required. To qualify, sometimes the borrower needs a **cosigner.**

$ 1500.00 Oshkosh, Wisconsin May 5, 19 89

9 months *after date, for value received,* We, the undersigned

promise to pay to Wisconsin Federal Bank, *or order,*

at Wisconsin Federal Bank

the sum of ***One Thousand Five Hundred*** *dollars*

in lawful money of the United States of America, with interest thereon,

payable in like lawful money, at the rate of 15% *percent*

per year *from date until paid.*

Joshua Randall

Tony Walters

No. 5629 *Date* February 5, 1980

Principal

Rate of interest

Signature

Cosigner

Due date

Examples

1. How much will Joshua Randall have to repay for this loan?

Find the interest. $i = prt$
$i = \$1500 \times 15\% \times 9$ months
$i = \$1500 \times 0.15 \times \frac{3}{4}$
$i = \$168.75$

Find the total.
$\$1500 + \$168.75 = \$1668.75$
Joshua will pay $1668.75 for $1500 loan.

Another type of loan is the **secured loan**, obtained by pledging something valuable, called **collateral**, to guarantee repayment.

2. Stephanie borrows $4800 at 14.7% interest for 3 years to buy a used car. The car is security for the loan. How much money must she repay?

Find the interest.
$i = \$4800 \times 0.147 \times 3$
$i = \$2116.80$

Find the total.
Add the principal and interest.
$\$4800 + \$2116.80 = \$6916.80$

Try This

a. Freda obtains a secured loan of $8000 at 9.5% for 2 years. How much money must she repay?

b. Jake signs a promissory note to repay a $2500 loan in 18 months at the rate of 10.5%. How much must he repay?

Exercises

How much money must be repaid on the following promissory notes?

1.

$ 900.00 June 17, 19 89

9 months *after date, for value received,* I

promise to pay to First National Bank*, or order,*

at First National Bank, 497 Market St.,

Lexington, KY

the sum of ***********Nine hundred*********** *dollars*

in lawful money of the United States of America, with interest thereon, payable in like lawful money, at the rate of 7 1/2% *percent per* year *from date until paid.*

Mary Anne Wilkenson

No. 5627 *Date* May 17, 1990

2.

$ 1500.00 January 7, 19 88

2 years 6 months *after date, for value received,* I

promise to pay to First Federal Bank*, or order,*

at First Federal Bank, 2202 East 12th Street,

Chico, CA

the sum of ****One Thousand Five Hundred**** *dollars*

in lawful money of the United States of America, with interest thereon, payable in like lawful money, at the rate of 11% *percent per* year *from date until paid.*

David R. Smith

No. 8356 *Date* July 7, 1990

For each of the following simple interest loans, tell how much must be repaid.

	Principal	Rate	Time		Principal	Rate	Time
3.	\$14,000	16.5%	2 years	**4.**	\$7500	16.75%	$2\frac{1}{2}$ years
5.	\$23,000	14%	$1\frac{1}{2}$ years	**6.**	\$12,750	18%	3 years
7.	\$2500	15.25%	9 months	**8.**	\$15,000	17.5%	4 years

9. Nina Ramblin borrows \$10,500 to buy a car. She agrees to repay the loan in five years at 10.5% interest. How much must she repay?

10. Steven Monet borrows \$2800 at 11% interest to buy a snowmobile. He has three years to repay the loan. How much money must he repay?

11. Jose Cruz obtains a secured loan of \$850 at 10.5% interest for 2 years to buy a stereo system. How much money must Jose repay?

12. Kim Hargett borrows \$1789 at 12.5% interest for 1 year to buy new furniture. How much money will she repay?

Mixed Skills Review

1. $1093 - 694$ **2.** $\$800 \times 0.11 \times \frac{1}{2}$ **3.** $6 \div \frac{1}{8}$

4. $1023 - 0.68$ **5.** $0.161 + 3 + 2.763$ **6.** $346 + 183 + 274$

7. $3\frac{2}{3} + 6 + 4\frac{5}{8}$ **8.** $13.6 \div 17$ **9.** 0.115×0.62

10. $9 - 4\frac{3}{11}$ **11.** 27% of 3750 **12.** $2\frac{1}{4} \div 10\frac{1}{2}$

6-7 Savings and Interest

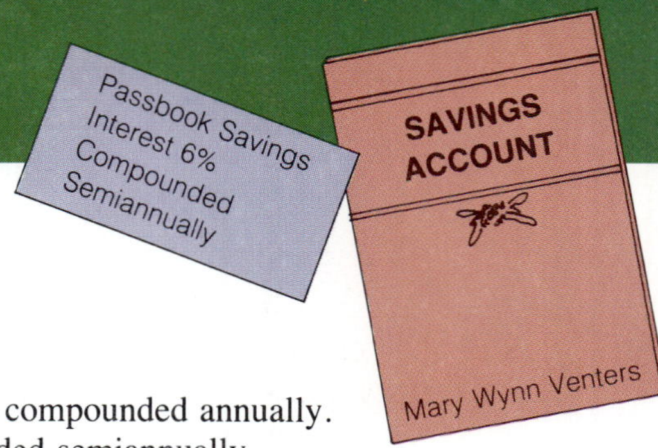

Objective
Find compound interest.

Money in a savings account earns interest. Interest that is earned and added to your account also earns interest. This is called **compound interest.**

Interest compounded once a year is said to be compounded annually. Interest compounded twice a year is compounded semiannually. Interest can be compounded quarterly (four times a year) or even daily.

Examples

1. Mary Wynn Venters has \$800 in the savings account advertised above. Find the amount in her account after one year.

   Interest is compounded semiannually. Find the interest after six months.

   $i = prt = \$800 \times 0.06 \times \frac{1}{2} = \$48 \times \frac{1}{2} = \$24$

   Add the interest to the principal to find the new principal. $\$800 + \$24 = \$824$

   Find the interest on the new principal.

   $i = prt = \$824 \times 0.06 \times \frac{1}{2} = \$24.72$

   Add the interest to find the new principal. $\$824 + \$24.72 = \$848.72$

   Mary Wynn will have \$848.72 in her account at the end of one year.

Many savings institutions express the compound rate as a simple interest rate. 6% compounded semiannually is an **effective annual rate** of 6.09%.

How long does it take savings to double? Use the **Rule of 72.** Divide 72 by the effective annual rate. The answer is the number of years it will take to double the savings.

2. How long will it take Mary Wynn's money to double, if the effective annual rate is 6.09%?

   $72 \div 6.09 = 11.82$

   It will take about 11 years and 10 months for Mary Wynn's \$800 to double.

Try This

a. Find the interest on \$1750 at 9% compounded semiannually after one year.

b. Find the interest on \$2500 at 8% compounded quarterly after one year.

c. The effective yield on a compound interest account is 9.16%. How long will it take for a deposit of \$1000 to double?

Exercises

Find the amount in each account after the given time.

1. \$500 principal, 9% interest, compounded annually for 2 years

2. \$1000 principal, 7% interest, compounded semiannually for 1 year

3. \$500 principal, 10% interest, compounded quarterly for 1 year

4. \$1500 principal, 12% interest, compounded annually for 3 years

5. \$2225 principal, 6% interest, compounded semiannually for 1 year

6. \$3000 principal, 12.5% interest, compounded quarterly for 1 year

7. \$3000 principal, 12.5% interest, compounded semiannually for 1 year

8. \$3000 principal, 12.5% interest, compounded annually for 2 years

9. \$5000 principal, 10.7% interest, compounded annually for 2 years

10. \$5000 principal, 10.7% interest, compounded semiannually for 2 years

Mental Math

Use the *Rule of 72*. About how long will it take to double your money?

11. 6%	**12.** 8%	**13.** 9%	**14.** 10%
15. 11%	**16.** 12%	**17.** 7.5%	**18.** 8.8%

Find the new principal.

19. \$600 at 10% compounded quarterly for 2 years

20. \$2500 at 12% compounded semiannually for 5 years

21. \$5000 at 8% compounded semiannually for 10 years

22. \$525 at 9% interest compounded semiannually for 5 years

Mixed Skills Review

1. $5\frac{4}{9} \times 1\frac{2}{7}$	**2.** 16.3 − 12.086	**3.** 26 + 3846 + 293
4. 9.7% of \$2900	**5.** 12,003 − 726	**6.** 30% of what number is 12?
7. 0.26 × 43.8	**8.** 927 × 43	**9.** What percent of 42 is 7?
10. $7\frac{1}{10} - 4\frac{5}{6}$	**11.** 13,872 ÷ 204	**12.** $1\frac{2}{3} + 4\frac{5}{6} + 7\frac{8}{9}$

6-8 Budgeting

Objective
Use a budget.

A plan for managing your money is called a **budget.** A good budget can be the key to spending wisely.

This circle graph shows the budget for the Pestridge family. Their **gross income** is $1756 per month.

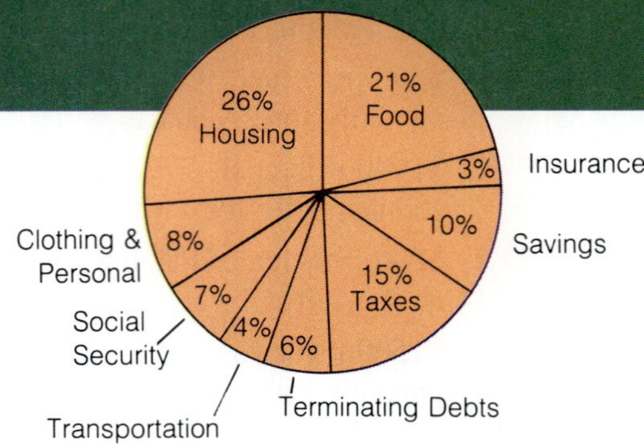

Understand the Situation

- What percent is for housing?
- What part has the least amount allocated?
- What percent is for taxes?
- What percent is for transportation?
- What has 23% allocated for it?

Example

1. **Terminating debts** are debts that end, such as car or appliance payments. Find the amount that the Pestridges budgeted each month for terminating debts.

The circle graph shows 6% budgeted monthly for terminating debts.
6% of the gross income $= 0.06 \times \$1756 = \$105.36$

The Pestridges budgeted $105.36 monthly for terminating debts.

Try This

Find the amount budgeted for each of the following:

a. housing **b.** insurance **c.** transportation **d.** food

Net income is income after taxes. The chart shows how Walt Preston budgets his $1100 net income.

Expense	Amount	Percent
Utilities and Rent	$450	41%
Food	170	
Clothing	90	
Transportation	85	
Savings	110	
Entertainment	75	
Terminating Debts	120	
Total	$1100	

Example

2. Find the percent of net income budgeted for utilities and rent.

$\$450 = n\% \text{ of } \$1100$

$\frac{\$450}{\$1100} = 0.409 = 41\%$

Walt Preston budgeted 41% of his net income for utilities and rent.

Try This

Find the percent of Walt's income budgeted for each of the following:

e. clothing **f.** food **g.** entertainment **h.** terminating debts

Exercises

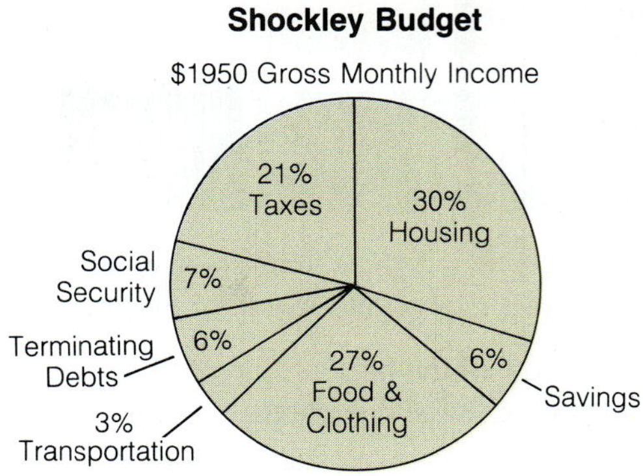

Find the amount budgeted by the Shockleys for each of the following.

1. housing
2. savings
3. food and clothing
4. taxes
5. transportation
6. terminating debts

Find the percent of Sheri's net income budgeted for each of the following.

7. personal
8. savings
9. contributions
10. transportation
11. terminating debts
12. food and clothing

Expense	Amount	Percent
Utilities and rent	$500	
Food and clothing	180	
Savings	80	
Transportation	90	
Personal	110	
Contributions	75	
Terminating debts	150	
Total	$1185	

Write Math 13. Make a list of your gross income from all sources. Prepare a budget of your expenses. Make a circle graph of your budget.

Data Hunt 14. Find a circle graph of your local government's budget. Order the expenses from greatest to least. Do you think that the size of the expenses matches their importance?

15. Obtain a list of the expenses for your school's athletic program. Make a circle graph of the athletic budget.

Mixed Skills Review

1. $10\frac{1}{5} \div 3\frac{2}{5}$
2. $4\frac{2}{3} \times 3\frac{1}{2}$
3. $6.003 - 4.276$
4. 1.203×16
5. $8\frac{5}{12} - 6\frac{3}{4}$
6. $17{,}388 \div 27$
7. $32{,}764 - 24{,}685$
8. $2.8 + 3.5 + 2.6$
9. 704×35
10. $2206 + 4795$
11. $136.32 \div 9.6$
12. $6\frac{3}{8} + 6\frac{3}{7}$

6-9 Enrichment: Saving on Interest

Objective
Compare the interest on loans.

Before you borrow money, shop around for the best rates. Differences in interest rates could mean saving money on loan interest payments.

Decide which option represents the least amount of interest. Find the savings in interest for this choice.

New Car
Option A: \$5800 at 12% per year
Option B: \$5500 at 11% per year
t = 1 year

Calculate the interest for both loan options. Remember $i = prt$.
Option A: $i = \$5800 \times 0.12 \times 1 = \$696$ Option B: $i = \$5500 \times 0.11 \times 1 = \$605$

Subtract to find the savings on interest payments.
$\$696 - \$605 = \$91$

Option B saves \$91 in interest.

Exercises

Decide which option represents the least amount of interest. Find the savings in interest for this choice.

1. **Refrigerator**
Option A: \$1230 at 9% per year
Option B: \$1020 at 12% per year
t = 18 months

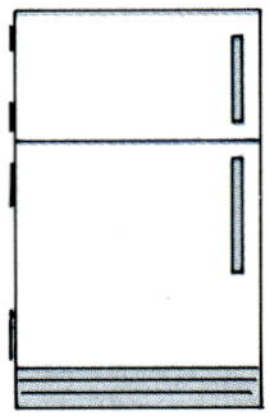

2. **Rowboat**
Option A: \$450 at 10% per year
Option B: \$550 at 9% per year
t = 2 years

3. **Motorcycle**
Option A: \$2500 at 12% per year
Option B: \$2400 at 14% per year
t = 24 months

4. **Home Improvement Loan**
Option A: \$12,000 at 13%
Option B: \$14,000 at 11%
t = 30 months

6-10 Calculator: Compound Interest

Objective
Use a calculator to find compound interest.

If your calculator has memory keys, you can do compound interest problems easily. With one memory key you can store a number for later use. You can also add a new number to the one already stored. With a second key you can recall the sum in memory and display it on the screen.

Experiment with your calculator to learn exactly how the memory keys operate. To do compound interest problems, you'll need to add a number into memory and to **recall** the number in memory and display it. Keys are labeled differently on different calculators. Here, [M+] will indicate *add to memory* and [MR] will indicate *recall and display.*

Find the compound interest.
$2000 principal, 8% interest compounded annually, 2 years

Store 2000 in memory.
2000 [M+] 2000
It may help to think of memory as the bank where $2000 is saved.

Find the first year's interest.
2000 [×] 0.08 [=] 160
You may prefer to use your [%] key to do this step.
The first year's interest is $160.

Add the interest to the amount in the bank. 160 [M+] 160

Display the total amount saved after 1 year. [MR] 2160

Find the second year's interest.
2160 [×] 0.08 [=] 172.8
The second year's interest is $172.80.

Add the new interest to the amount in memory and display the total.
172.8 [M+] 172.8, [MR] 2332.8

There is $2332.80 in the bank after 2 years.

Exercises

Find the compound interest using your calculator.

1. $5246 principal, 9% interest compounded annually, 2 years
2. $63,129 principal, 11% interest compounded annually, 2 years
3. $14,777 principal, 8.3% interest compounded annually, 3 years
4. $6299 principal, 12% interest compounded annually, 5 years
5. $26,914 principal, 12% interest compounded quarterly, 2 years

CHAPTER 6 REVIEW

Which new balance is incorrect in each set?

1.

	Balance Forward	Paid	Deposit	New Balance
A.	255	35		220
B.	444	46	56	434
C.	328		72	400
D.	527	33	46	540

2.

	Balance Forward	Paid	Deposit	New Balance
A.	340	50		290
B.	295		35	330
C.	151	68	34	117
D.	213	14		201

3.

	Balance Forward	Paid	Deposit	New Balance
A.	257.64	57.63		200.01
B.	165.98		34.03	200.02
C.	400.01	222.22	22.22	200.01
D.	207.47	103.57	96.11	200.01

4.

	Balance Forward	Paid	Deposit	New Balance
A.	146.07	111.07	57.00	102.00
B.	232.33	43.44	55.55	244.44
C.	199.50	27.55	50.00	221.95
D.	347.27	95.99	102.11	353.39

Match the following.

5. Interest paid on \$500 at 10% (1 year) — A. \$85
6. Interest paid on \$500 at 20% (1 year) — B. \$510
7. Interest paid on \$500 at 15% $\left(\frac{1}{2}\text{ year}\right)$ — C. \$100
8. Interest paid on \$500 at 17% (1 year) — D. \$1110
9. Interest paid on \$3000 at 17% (1 year) — E. \$50
10. Interest paid on \$3000 at 20% $\left(\frac{1}{2}\text{ year}\right)$ — F. \$300
11. Interest paid on \$3000 at 18.5% (2 years) — G. \$37.50

Find the amount in each savings account after the given time. Compare.

	Principal	Interest	Time	Characteristic	Total Amount
12.	\$500	8%	1 yr	Simple interest	________
13.	\$500	10%	$\frac{1}{2}$ yr	Simple interest	________
14.	\$500	7%	2 yrs	Compounded annually	________
15.	\$500	6%	1 yr	Compounded annually	________
16.	\$500	5%	2 yrs	Compounded semiannually	________

17. Use the "rule of 72" to determine how long it will take to have your savings of \$500 double at 8%, compounded semiannually.

CHAPTER 6 TEST

Find the net deposit.

1. Deposit 2 fifty-dollar bills, 2 twenty-dollar bills, 7 one-dollar bills, 15 quarters, 66 dimes, and a check for $192.58 (bank number 6-137).
2. Deposit currency of $212.83, checks for $8.94 (bank number 17-14), $54.76 (bank number 11-138). Subtract $50 for cash received.

Write a check using the given information.

3. Sam Bomback paid Speedy Electric $63.89 on April 30, 1987.

Find the balance forward.

4.

Balance forward	$345.72
Deposit	$125.56
Check #130	$ 64.55
ATM withdrawal	$ 40.00
Check #131	$ 98.45

5.

Balance forward	$352.68
Checking fee	$ 7.20
Deposit	$ 43.65
ATM withdrawal	$ 80.00
Check #133	$264.72

Reconcile the statements. Write agree or disagree.

	Balance on Bank Statement	Deposits Not Shown	Total of Checks Not Shown	Customer's Balance	Bank Fees
6.	$2376.74	$458.24	$ 549.03	$2292.15	$6.20
7.	$1008.03	$ 42.20	$ 8.42	$1038.72	$3.10

Find the simple interest.

8. $500 loan, 1 year, 5%
9. $2500 loan, 3 years, 10.7%
10. Tell how much money must be repaid on a loan of $45,000 at 14.7% for 5 years.

Find the amount in the account after the given time.

11. $1400 principal, 7.5% interest, compounded semiannually for 2 years

Find Harold's budget for each of the following to the nearest percent:

12. transportation
13. rent and utilities
14. car loan
15. food and clothing

Expense	Amount
Rent & utilities	$ 625
Food & clothing	$ 225
Transportation	$ 85
Car loan	$ 179
Total	$1114

CHAPTER 6 CUMULATIVE REVIEW

1. Estimate. Use clustering.
   $37.86 + $41.07 + $45.33
   A. $120
   B. $90
   C. $240
   D. $40

2. Is the answer reasonable? One calculator costs $8.95. What is the cost for 37? Answer: $351.50
   A. Reasonable
   B. Unreasonable

3. Compare the pay for both time periods.
   Job A $290 per week for a 40-hour week. Job B $8.25 per hour for a 35-hour week.
   A. Job A pays $1 more per week.
   B. Job A pays $1.25 more per week.
   C. They are equal.
   D. Job B pays $1.25 more per week.

4. Write $\frac{36}{54}$ in lowest terms.
   A. $\frac{4}{9}$
   B. $\frac{11}{18}$
   C. $\frac{12}{27}$
   D. $\frac{2}{3}$

5. Find the simple interest.
   $1750 loan, 2 years, 10.7%. Use the formula $i = prt$.
   A. $350
   B. $187.25
   C. $374.50
   D. $175.50

6. Find the total earnings.
   Larry Lui is paid 75% of receipts. His tips average 15% of receipts. One week's receipts were $750.
   A. $675 B. $562.50
   C. $112.50 D. $674

7. Tell how much must be repaid.

Principal	Rate	Time
$13,560	12.5%	3 years

   Use the formula
   Repayment $= p + prt$.
   A. $1695
   B. $5085
   C. $18,645
   D. $15,255

8. Find the equivalent fraction. Use mental math.
   If 45% $= \frac{12}{27}$, then 15% = □.
   A. $\frac{3}{7}$
   B. $\frac{1}{3}$
   C. $\frac{3}{27}$
   D. $\frac{4}{27}$

9. Find the balance carried forward.

Date 6-2 19 91
To Food Mart
For Groceries

	Dollars	Cents
Balance Forward	261	56
This check	19	38
Total		
Other Deposit	50	00
Balance Carried Forward		

   A. $242.18
   B. $261.56
   C. $330.94
   D. $292.18

10. Find the net deposit.
Deposit checks for
$251.62 (#18-135)
$105 (#24-12)
$5.56 (#8-30)

A. $361.08
B. $362.18
C. $362.08
D. $352.18

11. Find the percent of increase or decrease from 12 to 17.

A. $-41\frac{2}{3}\%$
B. +5%
C. $+41\frac{2}{3}\%$
D. −29.4%

12. Brad Gold earns $1250 per month. He budgets $500 for housing. What percent of his budget is used for housing?

A. 40%
B. 16%
C. 10%
D. $62\frac{1}{2}\%$

13. Agusto baby-sits children under 3 for $1.50 per child per hour. He charges $1.75 per child per hour for children over 3 years old. In one month he sat 12 hours for 6 children under 3 and 20 hours for 8 children over 3. Find his total earnings.

A. $108
B. $388
C. $280
D. $475

14. Multiply.

$$\begin{array}{r} 849 \\ \times\ 5.6 \\ \hline \end{array}$$

A. 475.44
B. 4554.4
C. 474.44
D. 4754.4

15. Harvey has a $752 deposit not shown by the bank. He has checks for $56.20 and $134.75 not shown. The balance on his statement is $152.63. The balance in his register is $713.68. Reconcile the statement.

A. Agree
B. Disagree

16. Estimate the percent. Use compatible numbers.
74% of 233

A. 180
B. 60
C. 120
D. 200

17. Find the total earnings. An insurance agent worked 40 hours at $8 per hour plus 7% commission on $1256 in sales.

A. $320
B. $87.92
C. $407.92
D. 563.91

18. Find the amount in this account after the given time.
$1200 principal, 8.5% interest, compounded quarterly for 1 year. Use the formula $i = prt$.

A. $1329.92
B. $1358.18
C. $1302.25
D. $1305.30

Chapter 7 Overview

Key Ideas

- comparing food costs
- buying on sale
- finding costs of installment buying
- finding the cost of using credit cards
- buying from a catalog
- finding the cost of renting items
- finding the cost of local telephone services
- finding the cost of long-distance calls

Key Terms

- unit price
- down payment
- installment price
- carrying charge
- average daily balance
- finance charge
- catalog order
- shipping charges
- handling charges
- day rate
- night rate
- direct dial
- operator-assisted calls

Key Skills

Find the answer.

1. 0.9 ÷ 0.06	**2.** $20 × 0.045
3. 21 − 16.73	**4.** 14 × 28.37
5. $0.07 + $37.16 + $0.25	**6.** $39.40 × 0.065
7. $50.75 ÷ 1.25	**8.** $183 ÷ 2.5
9. 19¢ + $1.38 + 53¢	**10.** $18.05 − $17.67
11. 3.08 + 27.65	**12.** $43.45 × 16

Write as ounces.

13. 3 lb	**14.** 2 lb 5 oz	**15.** 6 lb 15 oz
16. 4 lb 9 oz	**17.** 13 lb 11 oz	**18.** 21 lb 2 oz

Solve.

19. 6% of $7.50	**20.** $\frac{1}{3}$ of $23.10
21. 84% of 860	**22.** $\frac{3}{4}$ of $70
23. 50% of $37	**24.** $\frac{3}{8}$ of $16.40
25. 12% of $28	**26.** $\frac{2}{5}$ of $13.65
27. 3% of $129	

SPENDING MONEY 7

7-1 Grocery Shopping

Objective
Compare food costs using unit pricing.

In a recent year Americans spent $251 billion on groceries. Statistics show that larger families spend less per person.

Understand the Situation

- What is the weekly cost of groceries for a person who lives alone?
- What is the weekly cost of groceries for a family of 4?
- A family of 6 budgets $100 per week for groceries. Is this reasonable?

The **unit price** is the cost of one unit of an item, such as 1 oz of cereal or 1 oz of detergent. Unit prices are usually calculated to the nearest tenth of a cent. If two products are similar in quality, the item with the lower unit price is the better buy.

Example

1. Find the better buy in Bran Flakes.

Divide the cost by the number of units to find the unit price to the nearest 0.1¢.
$\$1.78 \div 21 \text{ oz} = \$0.0848$, or 8.5¢ per oz
$\$2.39 \div 29 \text{ oz} = \$0.0824$, or 8.2¢ per oz

8.2¢ < 8.5¢ Brand Y is the better buy.

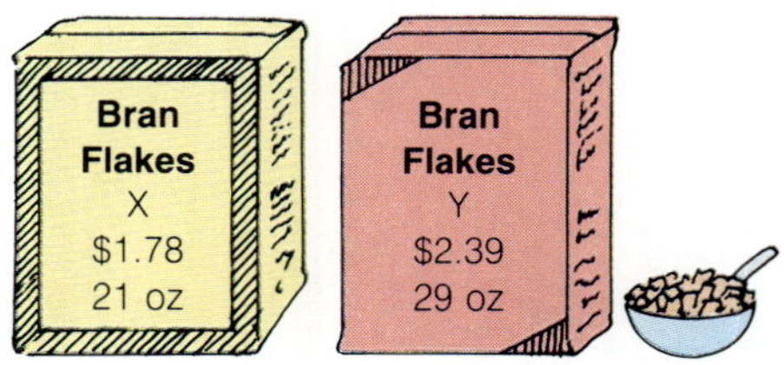

Coupons save money on the actual cost of an item and on the unit price.

2. You have a 25¢ coupon on the Brand X box of Bran Flakes. Is Brand Y still the better buy?

Find the actual cost of the coupon item.
$\$1.78 - \$0.25 = \$1.53$

Find the adjusted unit price.
$\$1.53 \div 21 \text{ oz} = \$0.0729$ 7.3¢ per oz

Since 7.3¢ < 8¢, Brand X is the better buy.

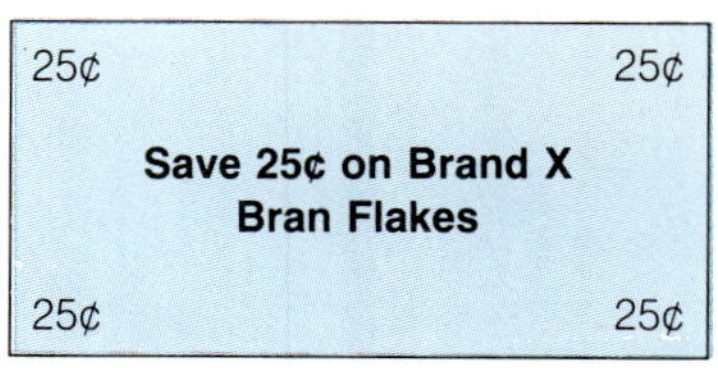

Try This

Find the better buy.

a. peanut butter

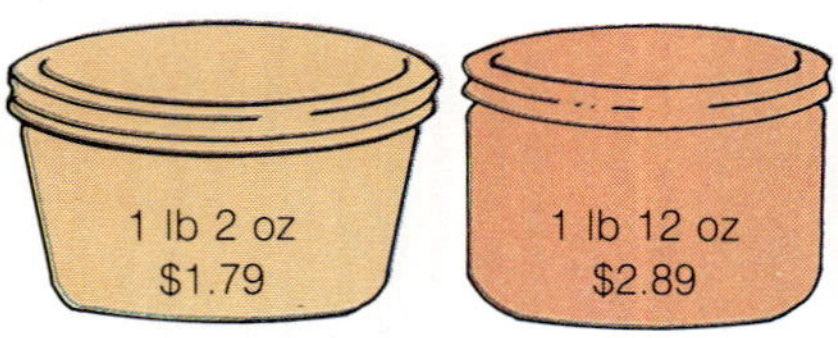

b. paper towels with a 15¢ coupon

Exercises

Find the better buy.

1.	Dog food	15 oz 35¢	1 lb 6 oz 49¢
2.	Detergent	32 fl oz $2.19	64 fl oz $3.79
3.	Soup	2 for 51¢	3 for 74¢
4.	Margarine	1 lb in tubs $1.45	1 lb in sticks $1.39
5.	Toothpaste	Tube 6.5 oz $1.89	Pump 5 oz $1.89 15¢ coupon
6.	Apples	89¢ per lb	3 lb bag $1.99
7.	Frozen vegetables	Box 2 c $1.32	Bag 6 c $2.29

Use a calculator to find the missing numbers.

8. Trail Mix
Net Wt.
.75 lb
Unit Price
1.60
Total Price

9. Macaroni
Net Wt.
1.40 lb
Unit Price

Total Price
0.91

10. Cheese
Net Wt.

Unit Price
1.75
Total Price
2.25

11. Flour
Net Wt.
6.5 lb
Unit Price

Total Price
1.56

Estimation Which items cost less than $3? Use mental math.

	Thrifty Market
12.	6 oranges @ $0.35 ....
13.	8 tomato soup @ 2/$0.49 ....
14.	4 doz eggs @ $0.89/doz ....
15.	6 lb bananas @ $0.32/lb ....
16.	3 lb potatoes @ $0.79/lb ....
17.	4 can pears @ 3/$2 ....

Mixed Skills Review

1. 436×215

2. $8452.82 \div 26.3$

3. $2\frac{5}{9} + 4\frac{2}{3}$

4. $1376 + 762 + 4557$

5. $8\frac{1}{2} - 6\frac{3}{4}$

6. $2\frac{4}{5} \div 2\frac{7}{10}$

7. $8050 - 4763$

8. $1\frac{7}{8} \times \frac{5}{6}$

9. $812.14 − $699.37

10. $37{,}600 \div 160$

11. $2.63 + 16.54 + 8.7$

12. 3.16×24.8

7-2 Buying on Sale

Objective *Find savings on sale items.*

Gail Vesely is shopping for a VCR. The same model is on sale in two different stores.

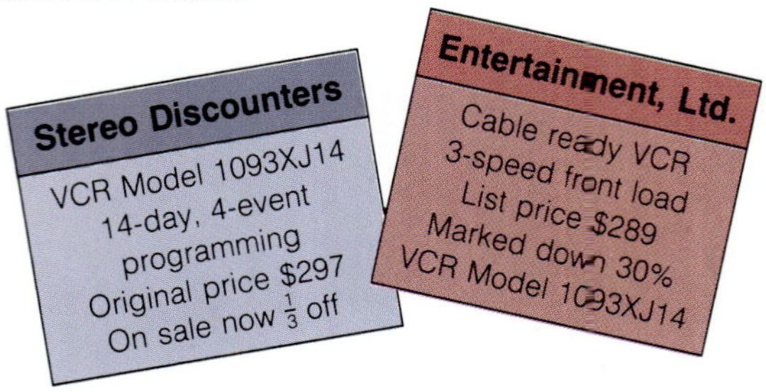

Examples

1. Which is the better buy?

Gail multiplied to find the discount for each VCR.
She then subtracted to find each sale price.

Stereo Discounters

$\frac{1}{3} \times \$297 = \$99$

$\$297 - 99 = \$198$

Entertainment Ltd.

$\$289 \times 0.30 = \$86.70$

$\$289.00 - 86.70 = \$202.30$

Since $198 < $202.30, the better buy is from Stereo Discounters.

2. A video store regularly sells VCR tapes for $5.98. The tapes are now on sale. If you buy 3, you get 1 free Find the sale price.

Buy 3, get 1 free means $\frac{1}{4}$ off the regular price.
If the discount is $\frac{1}{4}$, the sale price is $\frac{3}{4}$ of the regular price.

$\frac{3}{4} \times \$5.98 = \$4.485$, or $4.49 — The sale price of a VCR tape is $4.49.

Try This Find the sale price.

	Item	List Price	Discount	Sale Price
a.	Video cassette	$ 59.95	$\frac{1}{5}$ off	
b.	Television	$670	28% off	
c.	Tape rental	$ 2.50 each	Rent 2, get 1 free	
d.	Audio tapes	$ 1.98	Buy 3, get 1 free	

Exercises

Find the sale price.

	Item	List Price	Discount	Sale Price
1.	Video camera	$875	30% off	
2.	Outside antenna	$198	$\frac{1}{3}$ off	
3.	VCR maintenance	$ 24.95	$\frac{1}{5}$ off	
4.	Compact disc player	$365.90	25% off	
5.	Head cleaner	$ 14.95	2 for 1	
6.	Top 100 45s	$ 1.79	Buy 2, get 1 free	
7.	New movies tapes	$ 3.25	Rent 3, get 1 free	
8.	Compact discs	$ 28.99	Buy 1, get 1 free	

Which is the better buy?

9.

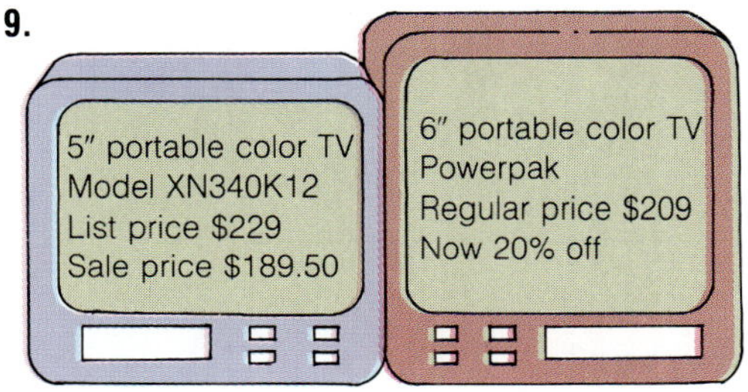

10.

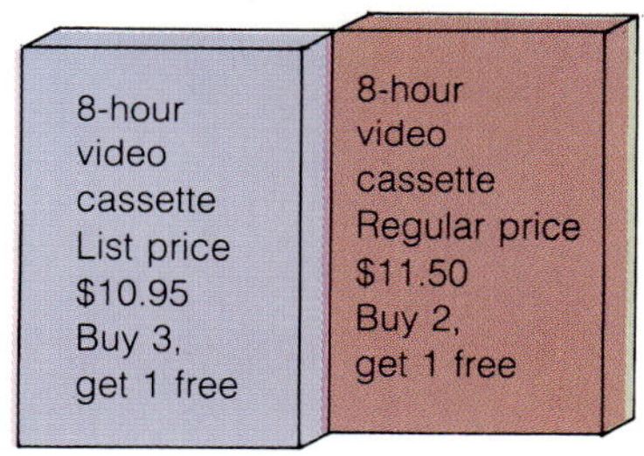

Some stores use a series of planned discounts. The list price of a home entertainment center at the following stores was $1200.

Store	First Discount Jan 1–March 31	Second Discount April 1–Jun 30
Stereo Discounters	20% off list price	15% off sale price
Entertainment, Ltd.	15% off list price	20% off sale price

11. Where would you buy the home entertainment center on January 15?

12. Where would you buy the home entertainment center on April 15?

Mixed Skills Review

1. $1803.6 \div 66.8$ **2.** $296 + $18 + $343 **3.** $34{,}958 \div 154$

4. $12\frac{3}{16} + 2\frac{5}{8}$ **5.** 100.9×32.46 **6.** $4\frac{3}{16} - 3\frac{3}{4}$

7. $15 - 2.078$ **8.** $8\frac{3}{4} \times 1\frac{1}{5}$ **9.** $0.396 + 4.18 + 2.453$

10. $3\frac{1}{11} \div \frac{5}{22}$ **11.** $1493 - 868$ **12.** 1205×67

7-3 Installment Buying

Objective
Find costs of installment plans.

The rock group *Revival* is buying equipment on the installment plan at Martin Ruhl Music. An installment plan helps people to buy expensive items. The customer makes a **down payment** at the time of the purchase and agrees to pay a certain amount of money each month.

Martin Ruhl's Music Installment Plan

Equipment	List Price	Down Payment	Monthly Payment	Term
Keyboard	\$1500	$\frac{1}{3}$	\$352.44	120 days
Synthesizer	\$1250	$\frac{1}{4}$	\$325.15	90 days
Bass guitar w/amp	\$ 670	20%	\$185.90	90 days
Cable	\$ 425	30%	\$101.83	90 days

Understand the Situation

- What part of the price is the down payment?
- How many monthly payments will the buyer have to make?

The **installment price** is the sum of the down payment and the monthly payments. A **carrying charge** is the resulting added cost.

Examples

1. Find the installment price of the synthesizer. *90 days = 3 months*

Find the amount of the down payment. $\frac{1}{4} \times \$1250 = \$312.50$

Find the total of the monthly payments. $3 \times \$325.15 = \$975.45$

Find the sum.
$\$312.50 + \$975.45 = \$1287.95$

The installment price of the synthesizer is \$1287.95.

2. Find the carrying charges.

$\$1287.95 - \$1250 = \$37.95$

The carrying charges on the synthesizer are \$37.95.

Try This

Find the installment price and the carrying charges.

a. Bass guitar **b.** Keyboard **c.** Cable

Exercises

Find the installment price and the carrying charges.

	Equipment	List Price	Down Payment	Monthly Payment	Term
1.	Conga drum	$ 250	$\frac{1}{4}$	$ 97.70	60 days
2.	Microphone w/stand	$ 170	$\frac{1}{5}$	$ 69.82	60 days
3.	Snake cable	$ 450	33%	$104.04	90 days
4.	Rhythm guitar w/amp	$ 360	30%	$ 93.63	90 days
5.	Trap set	$1288	25%	$254.61	120 days
6.	Mixer	$1625	35%	$188.06	180 days
7.	Keyboard amplifier	$ 935	25%	$243.20	90 days
8.	PA cabinet	$ 849	20%	$178.96	120 days
9.	Keyboard	$1750	$900	$153.31	180 days
10.	Bass guitar	$ 499	$150	$121.04	90 days

Calculator Use your calculator to solve the following.

11. The list price on a cello is $950. An orchestra student rents one for $4.75 a month. In how many months will rental fees equal list price?
12. Bess made a down payment of $690 on an organ and 24 monthly payments of $79.95. Find the actual cost of the organ.
13. An insurance company offers policy holders two plans. The customer can make one annual payment of $815.32 or 8 monthly payments of $104.42. How much is saved by paying the full amount?
14. The Salisbury Gas & Electric Company bills customers the same amount each month on their equal payment plan. Every six months the actual cost of gas and electricity is compared to the amount billed. The equal monthly payment is $150. Is there a balance due or a credit? How much is this?

Utility Charges	
January	$180.76
February	$192.37
March	$160.95
April	$145.21
May	$118.09
June	$127.36

Mixed Skills Review

1. 16,382 − 11,548
2. $6\frac{2}{3} \div 1\frac{13}{27}$
3. $220 − $79.38
4. 0.2628 ÷ 1.46
5. 863 × 94
6. 7684 + 38,579 + 585
7. $1\frac{7}{9} \times 1\frac{13}{20}$
8. $7\frac{4}{5} + 6\frac{1}{2}$
9. 56.37 × 6.3
10. 12.6 + 33.51 + 6.7
11. 25,308 ÷ 684
12. $12\frac{2}{3} - 5\frac{7}{9}$

7-4 Credit Cards

Objective
Find cost of using credit cards.

Janice Imwold received the June statement for her credit card. A copy is shown on p. 447 in the Data Bank.

Understand the Situation

- How much does Janice owe on her credit card?
- How many purchases were made?
- How much did she charge during the billing cycle?

When an account is not paid in full by the due date, a finance charge is added. Some credit card companies use the **average daily balance** multiplied by the monthly rate of interest to compute the finance charge.

Example

Janice paid $250 on June 11. She charged $24.50 on June 17. The finance charge is 1.5% per month on the average daily balance.
To find the new balance on July 1, multiply each unpaid balance by the number of days. Add to find the total.

Dates	Purchases and Loans	Payments	Unpaid Balance	Days	Adjusted Balance
June 1–10			$400.64	10	$4006.40
June 11		$250	150.64	1	150.64
June 12–16			150.64	5	753.20
June 17	$24.50		175.14	1	175.14
June 18–30			175.14	13	2276.82
			Total	**30**	**7362.20**

Divide to find the average daily balance. $\$7362.20 \div 30 \approx \$245.41$

Multiply to find the finance charge. $\$245.41 \times 0.015 \approx \$3.68$

Add the unpaid balance and the finance charge. $\$175.14 + \$3.68 = \$178.82$

The new balance is $178.82.

Try This

The finance charge is 1.5% per month on the average daily balance. Find the new balance at the end of the month.

a.

April 1	Unpaid balance	$92
April 16	Payment	$50

b.

May 1	Unpaid balance	$142.90
May 11	Purchase	$ 39.75
May 21	Payment	$125.00

Exercises

The finance charge is 1.5% per month on the average daily balance. Find the new balance at the end of the month.

1. January 1 Unpaid balance \$53
   January 9 Payment \$25
2. February 1 Unpaid balance \$196
   February 14 Payment \$112
3. March 1 Unpaid balance \$86.30
   March 7 Payment \$46.30
   March 10 Purchase \$19.50
4. April 1 Unpaid balance \$20.24
   April 18 Purchase \$35.49
   April 24 Payment \$15.45

The finance charge is 1.8% per month on the average daily balance. Find the new balance for each customer.

5.

Date	Reference Number	Transaction	Amount
06/08	090477602	Standard Plumbing & Heating	229.00
06/16	43550910	Payment	150.72–

Previous Balance	–	Payments & Credits	+	Purchases & Charges	+	Cash Advances	+	Finance Charge	=	New Balance
350.72						.00				

6.

Date	Reference Number	Transaction	Amount
06/12	436901349	Payment	85.40–
06/15	091756002	Harbor Cruises	48.90
06/20	090355142	Pahl's Piano Tuning	125.00

Previous Balance	–	Payments & Credits	+	Purchases & Charges	+	Cash Advances	+	Finance Charge	=	New Balance
155.00						.00				

Mental Math

You have 25 days from each closing date to pay. Give the due date.

7. January 3
8. September 10
9. August 25
10. May 12

Mixed Skills Review

1. $\frac{3}{8} \div 10\frac{1}{2}$
2. 3217×58
3. $8 - 6\frac{2}{11}$
4. $22{,}278 \div 94$
5. $6205.323 + 18.795$
6. $7694.1 \div 92.7$
7. $68{,}239 + 54 + 901$
8. 0.74×10.8
9. $\frac{3}{8} + \frac{1}{4} + 4\frac{1}{2}$
10. $6.025 - 3.864$
11. $54{,}716 - 38{,}498$
12. $9\frac{1}{3} \times 3\frac{3}{4}$

7-5 Buying from a Catalog

Objective
Find costs of buying from a catalog.

Many customers select their purchases from catalogs. In a recent year Americans spent over \$600 billion shopping by mail and by telephone. Shipping and handling charges are added to the cost of catalog orders. The chart on p. 447 in the Data Bank shows the charges for shipments from a company in Virginia.

Understand the Situation

- What are the shipping and handling charges for a 10-lb order sent to South Dakota?
- What are the charges for a 19 lb 13 oz order sent to New Jersey?
- What is the maximum weight that can be mailed?

Example

A customer from Georgia ordered a camping lantern and two sleeping bags. Find the total cost.

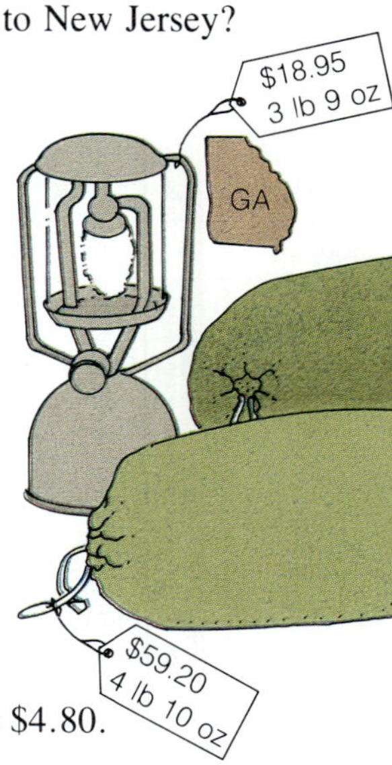

Find the cost of the items.

Lantern		\$18.95
Sleeping bags	2 × \$29.60	+59.20
		\$78.15

Find the weight of the items.

Lantern		3 lb 9 oz
Sleeping bags	2 × 4 lb 10 oz	+9 lb 4 oz
		12 lb 13 oz

Find the shipping and handling charges to Georgia.
The shipping and handling charges for 11–15 lb are \$4.80.

Add the cost and the shipping and handling charges.
\$78.15 + \$4.80 = \$82.95 The total cost is \$82.95.

Try This

Find the total catalog cost.

a. Dome tent \$159.50
Weight 6 lb 12 oz
Backpack \$27.35
Weight 1 lb 15 oz
New Mexico

b. Canoe \$199.98
Weight 14 lb 1 oz
2 paddles \$14.95
Each weighs 2 lb 3 oz
New York

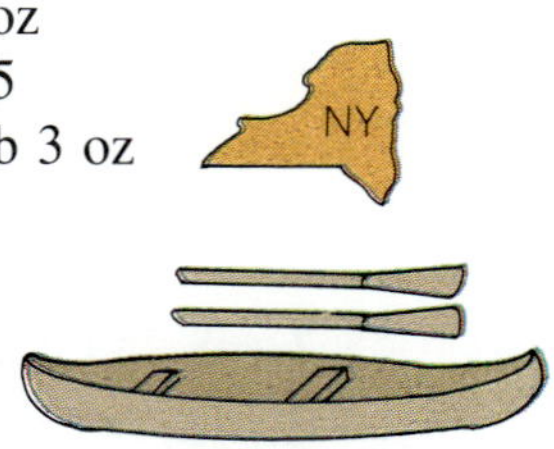

Exercises

Find the total catalog cost.

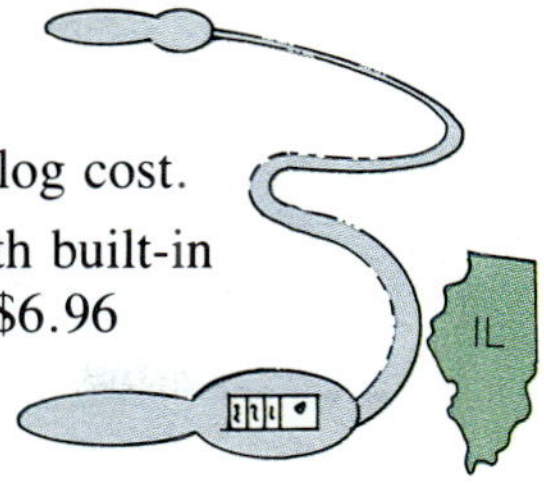

1. Jump rope with built-in digital gauge \$6.96
   1 lb 3 oz
   Illinois

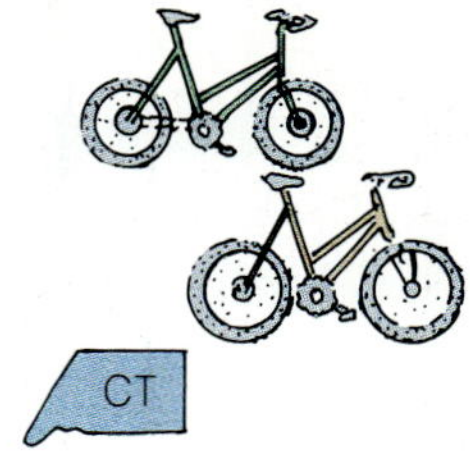

2. 2 26″ ladies' bikes
   \$89.99 each
   7 lb 4 oz each
   Connecticut

3. Badminton and volleyball set
   \$29.56 8 lb 6 oz
   Croquet set \$38.75
   9 lb 10 oz
   Oregon

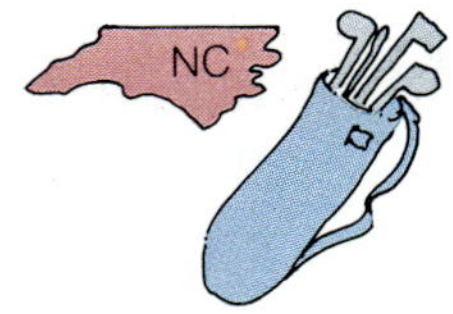

4. Boys' golf clubs
   \$189 7 lb 2 oz
   Lt-weight golf bag
   \$17.50 2 lb 3 oz
   North Carolina

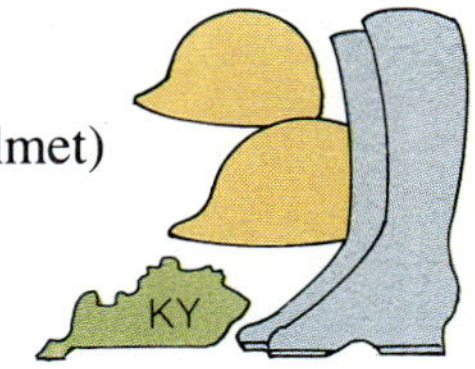

5. Riding boots
   \$84 1 lb 10 oz
   2 riding hats (helmet)
   \$25.64 each
   1 lb 6 oz each
   Kentucky

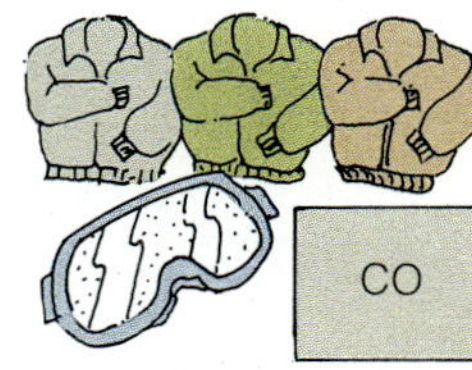

6. 3 ski jackets
   \$64.90 each
   3 lb 12 oz each
   1 pair goggles
   \$12.79 12 oz
   Colorado

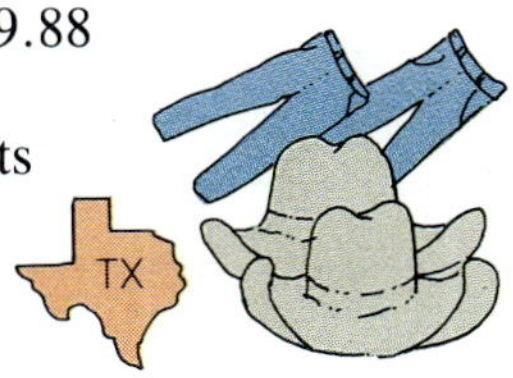

7. 2 pair jeans \$29.88
   1 lb 6 oz each
   2 ten-gallon hats
   \$17.75 each
   12 oz each
   Texas

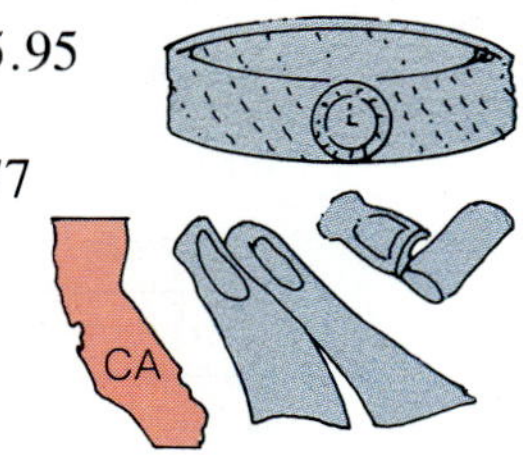

8. 1 scuba watch \$35.95
   11 oz
   1 regulator \$189.77
   2 lb 5 oz
   1 pair fins \$29.21
   3 lb 7 oz
   California

Suppose

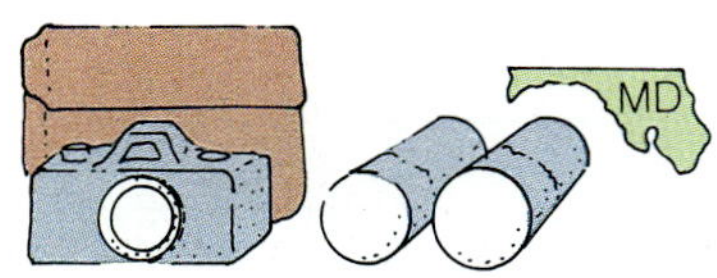

9. Suppose you live in Maryland. A local shop is advertising the camera, case, and two lenses for \$650 plus 5% sales tax. Compare the cost of the store purchase to a catalog sale.

10. Suppose you live in Florida. A local store lists the cycle gym for \$99 plus 6% sales tax. Compare the cost of the store purchase to a catalog sale.

Mixed Skills Review

1. $1\frac{3}{5} - \frac{2}{3}$
2. \$843.64 ÷ 46
3. 0.86 + 0.39 + 0.15
4. $8 \times 16 \times 4$
5. 76 + 48 + 84 + 63
6. $5\frac{1}{7} \times 2\frac{1}{10}$
7. 25,872 ÷ 539
8. 32.157×86
9. \$1200 − \$473
10. $2\frac{3}{5} + 5\frac{1}{3} + 7$
11. 25.3 − 13.383
12. $1\frac{1}{5} \div 4\frac{2}{3}$

7-6 Renting Instead of Buying

Objective
Find the cost of renting items.

Many communities have stores that rent items which are expensive, difficult to store, or only needed for short periods of time. Rental stores display charts showing the rental period and rates.

Anderson Rentals

Item	Hour*	Day	Week	Month
Air drill	$ 3.75	$ 12	—	—
Baby stroller	$ 2.25	$ 9	$ 15	—
Canoe	—	$ 12	$ 30	$100
Chain saw	$ 9.50	$ 38	$160	$260
Concrete mixer	—	$ 22	$120	$320
Forklift	$24.50	$ 95	$380	$950
Horse trailer	—	$ 29	$104	$312
Popcorn machine	$ 3.60	$ 14	—	—
Post-hole digger	—	$ 6	$ 22	$ 38
Punch bowl	$ 3.45	$ 13	—	—
Scooter	$ 6.75	$ 25	$ 98	$295
Skis, snow	—	$ 10	$ 39	—
Snowblower	$12.25	$ 45	—	—
Stump cutter	—	$220	$900	—
Television, color	—	$ 15	$ 55	$170
Tiller	$ 7.75	$ 30	$120	—
Wheelbarrow	$ 2.65	$ 11	$ 42	$121

*Three-hour minimum

Examples Find the rental fee.

1. Stump cutter, 4 days

Multiply the daily rate by the number of days.
$4 \times \$220 = \$880$

The rental fee is $880.

2. Wheelbarrow, 5 hours

Multiply the hourly rate by the number of hours.
$5 \times \$2.65 = \$13.25$
The day rate should be applied.

The rental fee is $11.

Try This Find the rental fee.

a. Snowblower, 2 days
b. Snow skis, 2 days
c. Chain saw, 3 weeks
d. Horse trailer, 2 hours

Exercises

Find the rental fee.

1. Post-hole digger, 3 days
2. Baby stroller, 3 hours
3. Snow skis, 4 days
4. Forklift, 2 months
5. Popcorn machine, 4 hours
6. Air drill, 2 months
7. Scooter, 5 days
8. Television, 3 weeks
9. Tiller, 2 hours
10. Punch bowl, 1 hour
11. Horse trailer, 3 weeks
12. Concrete mixer, 4 hours

Mental Math Find the rental fee. Use mental math.

13. Canoe, 2 weeks
14. Television, 3 days
15. Air drill, 6 hours
16. Punch bowl, 4 days
17. Snow skis, 5 days
18. Horse trailer, 2 days

Estimation A security deposit is required on rentals for expensive equipment. The deposit is about 1.5 times the weekly rate. Estimate the security deposit on the following rentals.

19. Canoe
20. Scooter
21. Chain saw
22. Stump cutter

Talk Math

23. A doctor has prescribed that you ride an exercise bike for 30 days as therapy following surgery. The cost to join a health club is $65 per month with a $100 initiation fee. A rental store charges $32.50 per week for a bike. The cost of an exercise cycle is $166. Which option should you choose? Explain your answer.

Mixed Skills Review

1. $7.98 \div 0.532$
2. $34.008 + 14.39 + 7.054$
3. $6\frac{7}{8} \times \frac{7}{22}$
4. 1078×406
5. $8\frac{5}{6} + 2 + 3\frac{1}{3}$
6. $14.3 - 9.807$
7. $15{,}768 \div 73$
8. $6\frac{3}{8} - 5\frac{3}{4}$
9. $\$658 + \$3918 + \$75$
10. $5274 - 4668$
11. 6.938×75
12. $13 \div 5\frac{1}{5}$

7-7 Local Telephone Service

Objective
Find the cost of local telephone service.

In his job as a service representative at the telephone company, Lyle Keehan explains the rates and arranges service for customers. Lyle uses codes to enter orders for new service into a computer.

Monthly Rates for Local Service

Code	Description of Service	Cost/Month
1FR	Flat-rate service with unlimited local calling	$16.50
1MR	Measured-rate service: 65 outgoing calls/month, each additional call $0.09	$ 9.85
1ER	Economy-rate service: each outgoing local call $0.09	$ 6.60
ABC	Rental charge per telephone A Number B Color C Model Colors W White B Blue E Beige X Black G Green R Red Models D Desk W Wall P Panel	$ 3.75
ADL	Additional directory listing	$ 0.72
NPL	Nonpublished directory listing	$ 1.44
1CW	Call Waiting notifies user of second call on line	$ 3.25
1CF	Call Forwarding transfers calls to another number	$ 3.25
1SD	Speed dialing makes calls using 1- or 2-digit numbers	$ 1.50
	911 Service $0.10 Federal tax 3% Local tax 5%	

Example

Lyle places an order for measured-rate service and one white, desk-model telephone for Clive Simmons. Clive estimates he makes about 90 calls per month. Find the monthly cost for local service.

Simmons, Clive T.
1 MR
1 WD

Total the fixed monthly charges. $\$9.85 + \$3.75 = \$13.60$

Add the cost of the additional calls per month. $90 - 65 = 25$ additional calls
$\$0.09 \times 25 = \$2.25$ $\$13.60 + \$2.25 = \$15.85$

Add the taxes and the 911 service charge. $3\% + 5\% = 8\%$
$0.08 \times \$15.85 \approx \$1.27$ $\$15.85 + \$1.27 + \$0.10 = \$17.22$

The monthly cost for Clive is $17.22

Try This

Find the monthly cost for local telephone service.

a. 1FR 2WD 1CW
85 calls

b. 1MR 1WD ADL
107 calls

c. 1ER NPL
38 calls

Exercises

Find the monthly cost for local telephone service.

1. 1FR 1XD 1GW 95 calls	**2.** 1FR 2ED NPL 87 calls	**3.** 1MR 1XD ADL 75 calls	**4.** 1MR 1BW 1SD 61 calls
5. 1ER 2RD 45 calls	**6.** 1ER 1BP 73 calls	**7.** 1MR 1ED 1CW 42 calls	**8.** 1MR 2XD NPL 101 calls

9. Compare the annual costs of flat-rate service and measured-rate service for a customer who does not exceed the 65-call limit.

During a six-month period a customer kept a record of the calls made.

Month	June	July	August	September	October	November
Total Calls	47	63	82	58	73	51

10. Find the average number of calls made per month.

11. Find the average cost for calls per month for measured-rate service.

12. Find the average cost per month for economy-rate service.

13. Which plan is more economical?

14. What is the annual savings?

Suppose

15. Suppose you purchase a desk-model telephone that costs $89.95. How many months can you rent the telephone for the same price?

16. The monthly rental rate for a portable telephone is $5.50. Suppose you can purchase the same model for $179.50. How many months can you rent the portable for the same price?

Data Hunt

17. Find the rates and types of local service available in your area. Which service would you choose for 1 telephone and 100 calls per month?

Mixed Skills Review

1. $4.306 + 27.35 + 15.8$

2. $13{,}485 \div 465$

3. $1.6 - 0.924$

4. $2\frac{4}{7} \div \frac{28}{35}$

5. $\$12{,}670 - \$8684$

6. 392×56

7. $706.12 \div 27.8$

8. $4\frac{1}{6} \times 1\frac{3}{5}$

9. $4\frac{8}{9} + 3\frac{2}{3}$

10. $6394 + 805 + 4873$

11. $6\frac{7}{12} - 3\frac{2}{3}$

12. 64.07×235.2

7-8 Finding the Cost of Long Distance Calls

Objective
Find the cost of long distance telephone service.

Direct-Dial Interstate Long Distance Rates			
Time	**Holidays* & Sundays**	**Monday through Friday**	**Saturday**
8 A.M. to 5 P.M.	Night rate (N)	Day rate (D)	
5 P.M. to 11 P.M.	Evening rate 60% of day rate (E)		
11 P.M. to 8 A.M.	Night rate 40% of day rate (N)		

*Holidays: New Year's Day, Independence Day, Thanksgiving, Christmas

Example

1. The day rate for a direct-dial call between Baltimore and Columbia is \$0.34 for the first minute and \$0.29 for each additional minute. Find the charge for the call on June 1.

Date	Time	Place Called	Area-Number	Rate	Min	Charge
05 25	1021AM	Columbia, SC	803 555-7957	D	2	
06 01	833PM	Columbia, SC	803 555-7957	E	8	
06 23	1105PM	Columbia, SC	803 555-2204	N	27	

Find the day-rate charge. Then multiply by the evening discount rate.

$\$0.34 + (7 \times \$0.29) = \$2.37 \qquad 0.6 \times \$2.37 \approx \$1.42$

2. The day rate between Waco and Pahrump is \$1.20/\$0.33. Find the cost of an 11-minute, credit-card call dialed at 2:30 p.m. on Tuesday.

The charges to the right are added to the direct-dial rates on operator-assisted calls.

Person to person	\$3.00
Public telephone and collect calls	\$1.50
Customer-dialed credit-card calls	\$0.60

Use the day rate. Add the charge for a credit-card call.

$\$1.20 + (10 \times \$0.33) = \$4.50 \qquad \$4.50 + \$0.60 = \$5.10$

Try This

Find the charge for each call on the bill in Example 1.

a. May 25

b. June 23

Find the cost of each call between Waco and Pahrump.

c. Person to person, 7:10 a.m., Monday, 4 minutes

d. Public telephone call, 1 p.m., New Year's Day, 7 minutes

Exercises

Kelly lives in Salt Lake City and checks her long-distance phone bill. This table lists the rates for calls made from Salt Lake City. Use the table to find the cost of each call on Kelly's bill.

Direct-Dial Day Rates from Salt Lake City		
Place Called	First Minute	Each Additional Minute
Boston	\$1.55	\$0.36
Dallas	\$2.03	\$0.34
Denver	\$0.45	\$0.32
Los Angeles	\$0.47	\$0.33

	Date	Time	Place Called	Area-Number	Rate	Min	Charge
1.	3 1	240PM	Dallas, TX	214 555-0991	D	5	
2.	3 6	355PM	Boston, MA	617 555-8429	D	3	
3.	3 6	630PM	Boston, MA	617 555-8429	E	1	
4.	3 12	1105PM	Denver, CO	303 555-2204	N	4	
5.	3 20	930AM	Los Angeles, CA	213 555-1477	D	12	
6.	3 29	1101PM	Denver, CO	303 555-2204	N	27	

The day rate for a direct-dial call between Brandon and Anderson is \$0.70/\$0.28. Find the cost of each call.

7. Person to person, 10:30 a.m., Friday, 2 minutes
8. Public telephone call, 9:45 p.m., Monday, 9 minutes
9. Credit-card call, 4:00 p.m., Sunday, 14 minutes
10. Collect call, 11:02 a.m., Saturday, 23 minutes
11. Public telephone call, 6:15 p.m., Thanksgiving, 3 minutes

Calculator

12. Jan Stone is stationed in Pensacola, Florida. She budgets \$5 per week for a long-distance call on Sunday to her family in Vermont. The day rate is \$0.60/\$0.32. How many minutes can Jan talk to her family?
13. The day rate for a call between Sugar Loaf and Conte is \$1.35/\$0.35. How many minutes can you talk for \$10 on Christmas?
14. The day rate for a call between Hannibal and Southgate is \$0.85/\$0.30. How many minutes can you talk for \$3.50 on Saturday?

Mixed Skills Review

1. $2\frac{1}{4} \times 3\frac{7}{9}$
2. $8 + 1236 + 447$
3. $17\frac{3}{16} - 11\frac{5}{8}$
4. $6.57 \div 0.045$
5. 176×0.394
6. $70.08 + 4.36 + 16.49$
7. $3\frac{4}{9} + 6\frac{1}{3}$
8. $29 - 16.308$
9. $62{,}356 \div 34$
10. $3 \times 10 \times 27$
11. $2\frac{14}{15} \div 2\frac{2}{5}$
12. $43{,}128 - 37{,}635$

7-9 Enrichment: Add-Ons and Discounts

Objective
Find costs with add-ons and discounts using percent.

Add-ons and **discounts** often affect how much an item really costs.

We calculate add-ons in situations such as:

Sales tax	⟶	Total price = purchase price + sales tax
Food-buying co-ops	⟶	Total price = discount price + percent fee
Pricing items	⟶	Retail price = wholesale price + percent mark-up

Examples

1. An item costing \$312 has 7% sales tax.

\$312 plus 7% of \$312	
7% = .07	Change the percent to a decimal.
.07 × \$312 = \$21.84	Find 7% of \$312.
\$312 + \$21.84 = \$333.84	Add to the purchase price.

The total price of the item is \$333.84.

We calculate discounts in situations such as:

sales ⟶ saleprice = regular price − percent off

2. An item costing \$312 is on sale at a 7% discount.

\$312 − 7% of \$312	
7% = .07	Change the percent to a decimal.
.07 × \$312 = \$21.84	Find 7% of \$312.
\$312 − \$21.84 = \$290.16	Subtract from the regular price.

The sale price of the item is \$290.16.

Exercises

Find the total price. Round to the nearest cent when necessary.

1. \$100 + 6% **2.** \$485 + 7% **3.** \$848 + 25%
4. \$250 + 50% **5.** \$8.98 + 3.5% **6.** \$65.74 + 6.8%
7. \$108.25 + 10.2% **8.** \$784.15 + 7.5% **9.** \$487.15 + 10%
10. \$684 + 25.7% **11.** \$855.43 + 7.9% **12.** \$4862 + 15%

Find the actual price. Round to the nearest cent when necessary.

13. \$300 − 7% **14.** \$928 − 15% **15.** \$46,284 − 6%
16. \$8297 − 25% **17.** \$4.95 − 4.5% **18.** \$64.80 − 10.3%
19. \$674.11 − 15.4% **20.** \$884.19 − 31.4% **21.** \$685 − 6.3%
22. \$47,146 − 6.6% **23.** \$89,148 − 3.9% **24.** \$100,000 − 7.3%

7-10 Computer: Intelligent Cash Registers

Objective
Read a computer-linked cash register receipt.

Cash registers linked to host computers have changed the way people shop. Scanning equipment reads the Universal Product Codes **(UPC)** on items. The receipt shows the item name or store department and other special information, in addition to unit and total cost. An **ATM** bank card may be used to pay for a purchase. The computer deducts the cost directly from your bank account. In addition to the convenience to the shopper, the computer assists the store by keeping track of inventories and total sales.

```
                         Westside Market
                       1856 W. Sand Road
                          Laguna Linda
Taxable item        → xBABY                          2.49
                    → xHSEWARES                      1.00
Department ―――――――――――――――――↑
Special sale price ――――――――→ 2@ 2/ 7.00
                      xHSEWARES                      7.00
                              2@ 2/ .99
                      xHOUSEHLD                      1.98
                      xHOUSEHLD                      1.00
Nontaxable item ――――→ CANDY                          1.99
                              8@  1.00
                      xDOMESTIC                      8.00
                              2@ 2/ 1.00
                       FOOD                          1.00
                              3@  2.00
                       FOOD                          6.00
                              2@  3.99
                      xCAMERA                        7.98
                      TAX                           ____
                      AMT DUE                       ____
                      ATM CARD                      ____
                      CHANGE                          00
                           11/29/87      20.01
Store number ――――――――→ 903034 04 THANKS
Time of purchase (24-h clock) ―― (points to 20.01)
```

0 15300 46002

UPC label

Exercises

1. Find the amount of tax if the tax rate is 6.5%.
2. Find the total amount due.
3. At what time did this transaction occur? Express the time using the 12-hour clock. Is it a.m. or p.m.?
4. How many rolls of film were purchased?

CHAPTER 7 REVIEW

Match each item in column A with the corresponding item in column B.

Column A

1. The unit price of a pound of potatoes when 5 pounds sell for $1.69.
2. The sale price of a CD player that regularly sells for $55.50 during a 20% off sale.
3. The installment price on a TV set that lists at $297. A $\frac{1}{3}$ down payment is required with monthly payments of $68.64 for 90 days.
4. The catalog cost of a lacrosse stick and gloves shipped to Maryland. The stick costs $48.95 and gloves cost $39.60. The total weight is 4 pounds 5 ounces. Use the table on p. 447 in the Data Bank.
5. The carrying charges on a 10-speed bike that lists at $187.60. A 20% down payment is required with monthly payments of $39.40 for 120 days.
6. The monthly costs for local telephone service. Use the rates on p. 162.
   1 FR
   1 WD
   1 CW
   NPL
7. The cost of a 7-minute call from Salt Lake City, Utah, to Dallas, Texas, at the day rate. Use the table on p. 165.
8. The rental cost for a tiller for 6 hours. Use the chart on p. 160.
9. The cost per ounce of the better buy.
   Brand X: 1 lb, $1.96
   Brand Y: 1 lb 15 oz, $2.98
10. The finance charge on an average daily balance of $356.98 at a rate of 1.8% per month.
11. The sale price of batteries, regularly $2.98 per twin-pack, now 3 for the price of 2.

Column B

A. $44.40
B. $27.04
C. $1.99
D. $91.30
E. $0.34
F. $6.43
G. $304.92
H. $30
I. $0.096
J. $4.07
K. $7.52

CHAPTER 7 TEST

Find the better buy.

	Item	Price	Price
1.	Pancake mix	32 oz $1.45	24 oz $1.29

Find the sale price.

	Item	List Price	Discount	Sale Price
2.	Computer	$1299	$\frac{1}{3}$ off	
3.	35 mm film	$4.79/roll	Buy 2, get 1 free	

Find the installment price and the carrying charges.

	Item	List Price	Down Payment	Monthly Payment	Term
4.	Guitar	$599	$\frac{1}{4}$	$86.27	180 days
5.	Piano	$3959	30%	$132.89	2 years

The finance charge is 1.5% per month on the average daily balance. Find the new balance at the end of the month.

6. July 1 — Unpaid balance — $130
July 13 — Payment — $ 40

Find the total catalog cost. Use the chart on p. 447 of the Data Bank.

7. camera: $139.95, 3 lb 4 oz; camera case: $29.49, 2 lb 2 oz; Ohio

Find the rental fee. Use the table on p. 160.

8. canoe, 2 weeks **9.** forklift, 7 hours

Find the monthly cost for local telephone service. Use the chart on p. 162.

10. FR
2 BD
1 ADL
72 calls

11. ER
1 EW, 1 RP
1 SD
104 calls

12. MR
3 XP
1 CF
245 calls

Kelly uses the table on p. 165 to find the cost of calls made from her home in Salt Lake City. Find the cost of each call on her bill.

	Date	Time	Place Called	Area-Number	Rate	Min	Charge
13.	6/23	10:00 A.M.	Los Angeles, CA	213 282-4391	D	12	
14.	6/25	2:00 A.M.	Dallas, TX	214 639-7523	N	23	

CHAPTER 7 CUMULATIVE REVIEW

1. Find the sale price.
television $740,
discount 35%
- A. $259
- B. $479
- C. $471
- D. $481

2. Solve. Use mental math.
At the Hillview High School fair 65% of the booths sold food. What percent were nonfood?
- A. 35%
- B. 46%
- C. 26%
- D. 56%

3. Tell how much money must be repaid.
Use $i = prt$.
principal $13,000,
rate 15.3%, time 9 months
- A. $14,491. 75
- B. $1,491.75
- C. $13,490.75
- D. $14,381.75

4. Find the rental fee. Use mental math.
television, $55 per week for 6 weeks
- A. $300
- B. $360
- C. $600
- D. $330

5. Divide.
$97,638 ÷ 5
- A. $19,527.60
- B. $19,125.60
- C. $11,527.60
- D. $18.027.60

6. Where would you look for data on interest rates on home loans?
- A. car dealer
- B. bank
- C. carpenter
- D. classified ads

7. Find the cost of a 17-minute call. The rate is $0.65 for the first minute and $0.21 for each additional minute.
- A. $3.01
- B. $3.57
- C. $4.01
- D. $11.05

8. Add.

$$\begin{array}{r} \$98.02 \\ 34.52 \\ +55.96 \\ \hline \end{array}$$

- A. $178.40
- B. $188.50
- C. $187.40
- D. $178.50

9. Find the simple interest.
Use $i = prt$.
$72,000 loan, 30 years, 10.6%
- A. $216,000
- B. $228,960
- C. $7,632
- D. $3,456

10. Add.

$3\frac{1}{4} + 2\frac{2}{3} + \frac{1}{6}$

- A. $5\frac{1}{12}$
- B. $6\frac{1}{6}$
- C. $6\frac{1}{12}$
- D. $5\frac{11}{12}$

11. Find the annual salary at the end of the given period.
teacher $17,600, $620 every year for 5 years
- A. $20,500
- B. $20,700
- C. $21,300
- D. $21,700

12. Find the better buy on mayonnaise.
$1.29 for 16 oz
$2.09 for 32 oz
- A. 16 oz is 49¢ less
- B. 32 oz is 39¢ more
- C. 16 oz is 39¢ more
- D. 32 oz is 49¢ less

13. Translate to a number sentence.
320 is what percent of 540?
A. $320 = n\% \times 320$
B. $540 = n\% \times 320$
C. $320 = n\% \times 540$
D. $n\% = 320 \times 540$

14. Find the amount in the account after the given time.
$1155 principal, 12.5% interest, compounded semi-annually for 1 year
Use $I = prt$.
A. $1203.89
B. $1303.89
C. $1302.89
D. $1202.89

15. Find the new balance at the end of the month.
The finance charge is 1.5% per month.

8/1	Unpaid balance	$486
8/17	Payment	$100
8/18	Purchase	$ 76.60

A. $386
B. $391.79
C. $462.60
D. $469.68

16. Find the total catalog cost.
2 skirts @ $66.50,
1 sweater $48.30,
1 blouse $35.70
Total weight 4 lb 10 oz
Shipping costs for 1 lb to 5 lb 5 oz $2.46
A. $219.46
B. $152.96
C. $206.46
D. $209.46

17. Find the annual employee benefit. Ace Laboratories provides 3 lab coats and weekly laundering for each employee. Each coat costs $34.60. Laundry is $5 per week per employee.
A. $108.80
B. $362.40
C. $363.80
D. $353.80

18. Find the installment price for a microwave oven.
list price $176, down payment $\frac{1}{4}$, monthly payment $71.25, term 60 days
A. $142.50
B. $132
C. $44
D. $186.50

19. Lori has a monthly income of $1325. She budgets $199 monthly for food and clothing. What percent of her budget is this?
A. 10%
B. 15%
C. 5%
D. 25%

20. Flat-rate telephone service is $18.15 per month. Measured rate is $10.52. Compare the costs.
A. Measured rate is $7.63 less.
B. Flat rate is $7.63 less.
C. Measured rate is $7.63 more.
D. They are equal.

21. Matt worked three summer jobs. He made a total of $1734. The first job paid one half of jobs two and three combined. Jobs two and three paid the same. What did job one pay?
A. $578
B. $433.50
C. $1156
D. $867

Chapter 8 Overview

Key Ideas

- organizing data
- finding the mean of a set of data
- finding the mode of a set of data
- finding the median of a set of data
- using formulas
- finding sports statistics
- using line graphs
- using bar graphs
- evaluating graphs
- using samples

Key Terms

- data
- frequency table
- interval
- tally
- mean
- mode
- average
- median
- games back
- earned run average (ERA)
- line graph
- horizontal scale
- vertical scale
- comparison
- axis
- misleading
- sample
- prediction
- random

Key Skills

Find the answer.

1. 142 + 339 + 23
2. 56 + 312 + 802
3. 119 + 205 + 29
4. 432 ÷ 9
5. 1622 ÷ 12
6. 3098 ÷ 15
7. 228 ÷ 12
8. 3922 ÷ 16
9. 587 ÷ 8
10. 208 + 119 + 32 + 874
11. 2108 + 173 + 39 + 2295
12. 9907 + 206 + 33.4 + 94.7
13. 403 + 536 + 4.5 + 19.32

Arrange in order from least to greatest.

14. 23, 19, 256, 17, 19, 36, 15, 29
15. 42, 39, 38, 44, 50, 45, 54, 37, 28
16. 118, 134, 129, 107, 156, 137, 143, 129, 145, 129, 136
17. 23.9, 21, 9.23, 4, 29.1, 23, 18.8, 21.8, 19.9, 20.6

Evaluate each expression.

18. $w + 1$ for $w = 19$ and $l = 32$
19. $m - n$ for $m = 108$ and $n = 88$
20. $2m + n$ for $m = 23$ and $n = 56$
21. $3x - y$ for $x = 12$ and $y = 10$

Solve each proportion.

22. $\frac{3}{5} = \frac{n}{15}$
23. $\frac{9}{12} = \frac{12}{n}$
24. $\frac{5}{n} = \frac{2}{3}$
25. $\frac{n}{3} = \frac{4}{12}$
26. $\frac{n}{30} = \frac{2}{3}$
27. $\frac{n}{21} = \frac{21}{49}$

USING CONSUMER INFORMATION 8

8-1 Organizing Data

Objective
Organize data using a frequency table.

Mary Keath is a personnel worker. She made this list of monthly salaries for 35 employees.

Monthly Salaries				
\$465	\$845	\$1100	\$843	\$925
\$817	\$715	\$600	\$1025	\$829
\$900	\$980	\$680	\$1200	\$785
\$680	\$915	\$1250	\$623	\$560
\$490	\$785	\$980	\$590	\$900
\$815	\$755	\$585	\$550	\$650
\$585	\$915	\$725	\$725	\$685

Understand the Situation

- How many salaries are listed?
- What is the lowest salary?
- What is the highest salary?

The data presented in the salary list is not well organized. One way to organize a set of data is to make a **frequency table.** A frequency table shows how many times a certain thing occurs.

A. List the intervals.

B. Use tallies to record each salary within an interval.

C. Record the frequency of tallies in each interval.

A Interval	B Tally	C Frequency
400–499	ǀǀ	2
500–599	卌	5
600–699	卌 ǀ	6
700–799	卌 ǀ	6
800–899	卌	5
900–999	卌 ǀǀ	7
1000–1099	ǀ	1
1100–1199	ǀ	1
1200–1299	ǀǀ	2

The data is now organized. The lowest salaries fall in the \$400–\$499 interval. The highest salaries fall in the \$1200–\$1299 interval. There are 9 intervals and most of the salaries fall in the middle intervals.

Try This

A pushup contest was held on the boardwalk. Here are the 40 scores for the contestants. Make a frequency table using an interval of 3 and starting with 16–18. Make another frequency table using a different interval of your own choice.

Pushup Contest
20, 43, 17, 37, 39, 35, 32, 52
21, 43, 18, 50, 29, 34, 58, 19
21, 18, 24, 47, 30, 18, 29, 22
23, 21, 56, 43, 27, 22, 30, 20
25, 34, 52, 40, 36, 28, 40, 43

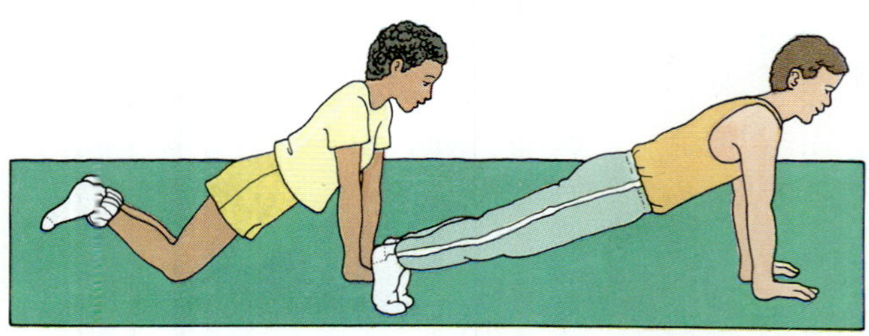

Exercises

1. Nick Sponato works as a quality control manager. To test light bulbs he picked a random sample of 30 bulbs. The table shows the number of hours it took each bulb to burn out. Make a frequency table of this data using an interval of 100 and starting with 800–899.

Hours Till Burnout
1415, 1422, 975, 1023, 1361, 1224, 1026, 1443, 947, 1210, 1320, 1245, 978, 1132, 1519, 1211, 896, 1346, 1496, 1433, 1566, 1112, 1326, 965, 1078, 982, 1322, 998, 1211, 1084

Use the chart on p. 174 for Exercises 2–4.

Estimation

2. Estimate the sum of the highest three incomes.

Mental Math

3. Find the difference between the highest and lowest incomes.
4. Find the sum of the two lowest incomes.

Data Bank

5. The chart on p. 448 shows the lengths, in feet, of 30 of the world's longest suspension bridges. Make a frequency table using 10 intervals.
6. Find the difference between the shortest and longest suspension bridges.
7. Jill Goleke works at the Fitness Center. She recorded the pulse rates of 25 clients before their workouts. Copy and complete this frequency table, which does not use grouped data.

Resting Pulse Rates
72, 70, 68, 71, 69
75, 74, 72, 72, 71
68, 69, 76, 70, 71
68, 67, 73, 74, 72
75, 73, 76, 72, 71

Interval	Tally	Frequency
68		
69		
70		
71		
72		
73		
74		
75		
76		

Mixed Skills Review

1. $6\frac{5}{8} + 2\frac{3}{4} + 5\frac{1}{2}$
2. $6 - 0.724$
3. \$20,000 − \$16,576
4. 18% of 595
5. 46,681 + 53,364
6. What percent is 120 of 500?
7. 186×324
8. $1\frac{5}{8} \div \frac{3}{8}$
9. 12.87 + 16.6 + 13.08
10. $11\frac{3}{4} - 6\frac{4}{5}$
11. $16{,}224 \div 78$
12. 1.2×7.63

8-2 The Mean and Mode

Objective
Find the mean and mode.

Stacey Egerton is planning to buy an all-terrain bike (ATB). In the *Consumer Reports Buyers Guide* she found five ATBs that were rated very good.

Bike	Cost
Bridgestone MB-2	$385
Panasonic All-Terrain	$360
Schwinn Sierra	$300
Univega Alpina Uno	$330
Nishiki Cascade	$349

Source: *Consumer Reports Buyers Guide*, February, 1986

Understand the Situation

- How many bikes were rated very good?
- Which bike cost the most?
- Which bike cost the least?

What single number best represents the set of data? To help answer this question about the bike costs, Stacey found the **mean** of the costs. The mean is the average of a set of numbers.

Example

1. Find the mean cost of the bikes rated very good.

Find the sum of the numbers.	→	Find how many numbers were added.	→	Divide the sum by this number.
$385 + $360 + $300 + $330 + $349 = $1724		5 numbers were added.		$1724 ÷ 5 = $344.80

The mean cost of the bikes rated very good is $344.80.

Try This

a. Six bikes received a rating of poor. Here are the costs of these bikes. Find the mean cost of the bikes rated poor. $130, $110, $80, $100, $110, $100

Another number that sometimes can be used to represent a set of data is the **mode.** The mode of a set of data is the number that occurs most often.

Example

2. Bill's Bike Shop sells caps with a cycle logo. During one week Bill sold these cap sizes: 7, 7, $7\frac{1}{2}$, $6\frac{7}{8}$, 7, $7\frac{1}{8}$, $6\frac{1}{2}$, 7, $6\frac{7}{8}$, 7, $7\frac{1}{8}$, 7. Find the mode. Size 7 occurs more often than any other size. Size 7 is the mode.

Try This

Find the mode.

b. Shoe sizes sold on Wednesday: 6A, $8\frac{1}{2}$B, 7A, 8B, $6\frac{1}{2}$B, 7A, 6B, 7A, 7C, 8D, $8\frac{1}{2}$AA, 6A, 8A, 7A, 8A, 7AA, 8B, 7A

c. Plant heights on the greenhouse floor (in cm): 67, 63, 72, 75, 65, 61, 64, 73, 60, 74

Exercises

1. A magazine rated nine motocross bikes fair. Find the mean cost. \$165, \$125, \$150, \$143, \$100, \$141, \$139, \$160, \$160

2. Find the mode of the costs of the fair-rated bikes.

3. This chart shows a record of pizza sales during a peak time. Find the mode.

Pizza	Tally	Frequency
Cheese	~~IIII~~ II	7
Pepperoni	~~IIII~~ ~~IIII~~ II	12
Sausage	~~IIII~~ ~~IIII~~	10
Onions	III	3

4. Following is a list of the sizes of rings that were sold in a day. Find the mode. 8, 6, 7, 6, 6, 8, 7, 5, 8, 5

Data Hunt

5. Collect a set of data within your school such as the number of students in each homeroom or the height of each member of your math class. Find the mean and mode for the set of data.

Estimation

Estimate the mean cost of these items. Then use your calculator to find the actual mean.

6. 

\$43.99 \$29.00 \$37.50

7. 

\$130 \$89 \$126.99 \$145.55

8. 

\$27 \$45.99

9. 

\$47.50 \$69.99 \$45

Mixed Skills Review

1. \$10 − \$6.49
2. 4.5% of \$30.16
3. 7261 + 5451
4. \$18.72 ÷ 36
5. $1\frac{7}{20} + 2\frac{3}{5}$
6. 508 − 465
7. $1\frac{3}{7} \times 2\frac{4}{5}$
8. 30,272 ÷ 43
9. $8\frac{1}{6} - 6\frac{4}{9}$
10. \$16.23 + \$14.89
11. 16 × 1308
12. $2\frac{4}{7} \div 2\frac{2}{5}$

8-3 The Median

Objective
Find the median.

Monthly Rental Costs	
Bon Air Apartments	$395
Hillcrest Apartments	$435
Kenilworth	$425
Dulaney Towers	$515
Longview Arms	$495
Freedom Heights	$500
Ashburton	$466

Marcy checked the monthly rental costs for one-bedroom apartments in the community near her new job.

Understand the Situation

- How many apartments are shown?
- What is the monthly rental for Dulaney Towers?
- What is the monthly rental for Hillcrest Apartments?

A representative of all the rental costs is the **median** rental cost. The median of a set of data is the middle number when the numbers are arranged in numerical order.

Examples

1. Find the median rental cost.

Arrange the numbers in order.	→	The middle number is the median.

395, 425, 435, **466,**
495, 500, 515

The median is 466.

The median rental cost is $466.

If there are two middle numbers, the median is the mean of those numbers.

2. Find the median number of family members for these families.

Family	No. of Members
A	7
B	4
C	3
D	5
E	2
F	3

Arrange the numbers of family members in order from least to greatest.

2, 3, 3, 4, 5, 7

There are two middle numbers, namely 3 and 4. Find their mean. The mean of 3 and 4 is 3.5.

The median number of family members is 3.5.

Try This

Find the median.

a. Pulse rates: 67, 73, 77, 59, 90, 91, 63, 72, 74, 67

b. Annual incomes: $10,200; $9,800; $14,760; $15,800; $9,675; $16,780; $16,500

Exercises

1. As part of his health maintenance plan, Dave King recorded his blood pressure once a day for a week. He recorded both the systolic and the diastolic pressure. The systolic pressure is the pressure while the heart is beating. The diastolic pressure is the pressure while the heart is relaxing.

   Find the median systolic and median diastolic blood pressure.

Blood Pressure Record		
Day	**Systolic**	**Diastolic**
Sun.	137	87
Mon.	143	83
Tues.	146	79
Wed.	139	81
Thurs.	149	84
Fri.	138	79
Sat.	141	82

2. A real estate agent listed the costs for several houses in a community. Find the median cost. \$98,700; \$101,500; \$67,800; \$88,000; \$79,600; \$114,800; \$98,650
3. An office building held a series of emergency evacuation drills. Here are the evacuation times in minutes. Find the median evacuation time. 3.2, 2.8, 3.0, 3.4, 2.9, 3.1, 3.4, 4.0, 3.8, 2.9
4. Find the median rating for 10 major TV shows. 11.8, 11.1, 10.4, 10.3, 9.7, 9.3, 8.9, 7.8, 7.3, 7.1

Work with a Partner

5. Here is a set of weekly rental rates for vacation condominiums.

   Find the mean and the median rental rate.

   Decide whether the mean or median is the best summary of the rates. Tell why you made this decision.

Rental Rates	
Condominium	**Rate**
Oceanview	\$650
Seabreeze	\$625
The Pyramid	\$995
Surfside	\$590
Sand & Surf	\$700
Tahitian Sun	\$690
Bayside	\$685

Data Bank

6. Find the median number of silver-clad Kennedy half dollars minted from 1965 to 1970. Use the table on p. 448.

Mixed Skills Review

1. $1\frac{2}{3} - \frac{4}{5}$
2. $2\frac{2}{11} \div \frac{9}{22}$
3. $1\frac{1}{4} + \frac{1}{2} + 6\frac{5}{16}$
4. $\frac{3}{4}$ of \$17.20
5. 78% of \$2250
6. 9360 + 284 + 1612
7. 28.3 − 19.54
8. 74,008 − 63,252
9. 40% of what number is 60?
10. 5.5% of \$14.80
11. What percent is 15 of 25?
12. \$16,018.94 + \$7931.47

8-4 Sports Statistics

Objective
Use formulas to calculate statistical information.

Penny Judson's hobby is sports statistics. Penny uses her home computer to keep records on her favorite teams and players.

Understand the Situation

- Which baseball teams are listed?
- How many wins do the Tigers have?
- How many losses do the Orioles have?

The **games back** statistic shows a team's position compared to the leading team. It is the number of games the trailing team must win and the leading team must lose to tie for first place in the standings.

The formula for games back is $G = \frac{dW + dL}{2}$.

G stands for games back. dW stands for the *difference* between wins for the teams. dL stands for the *difference* between losses for the teams.

Examples

1. How many games back of the Blue Jays are the Orioles?

Find the difference in the wins and the difference in the losses.

$dW = 76 - 70 = 6 \qquad dL = 65 - 61 = 4$

Substitute in the formula and find the answer.

$G = \frac{dW + dL}{2} = \frac{6 + 4}{2} = \frac{10}{2} = 5$

The Orioles are 5 games back of the Blue Jays.

Another baseball statistic is Earned Run Average (**ERA**). An earned run is charged to the pitcher. An unearned run is due to an error.

The formula for ERA is $\text{ERA} = 9\frac{R}{I}$.

In the formula R stands for the number of earned runs. I stands for the number of innings pitched. Nine innings make a complete game.

2. Find Fernando Valenzuela's ERA after giving up 63 runs in 218 innings.

$\text{ERA} = 9\frac{R}{I} = \frac{9 \times 63}{218} = 567 \div 218 = 2.60.$

Fernando Valenzuela's ERA was 2.60.

Try This

a. How many games back of the Blue Jays are the Tigers?

b. Find Mike Boddicker's ERA. 81 earned runs, 261 innings pitched

Exercises

Find how many games the second-place team is back of the first-place team.

1.

	Wins	Losses
Angels	28	21
Rangers	24	26

2.

	Wins	Losses
Celtics	33	11
76ers	26	18

3.

	Wins	Losses
Force	16	9
Blast	13	9

4.

	Wins	Losses
Cowboys	8	3
Redskins	6	5

5.

	Wins	Losses
Yankees	54	48
Indians	49	51

6.

	Wins	Losses
Bucks	28	19
Bulls	21	22

Find the ERA for each pitcher.

	Pitcher	Earned Runs	Innings Pitched
7.	D. Steib	108	205
8.	R. Clemens	70	254
9.	M. Davis	62	154
10.	J. Morris	97	267

Data Hunt

11. Bring in the team standings of a major sport that is in season. Check how many games the second-place team is back of the first-place team. How many games is the third-place team back of the second-place team?

Number Sense

12. A lower ERA is considered better than a higher ERA. Here are five pitchers and their ERAs. Arrange them in order from best ERA to worst ERA. Zahn (3.12), Petry (3.24), Niekro (3.09), Tanana (3.25), Steib (2.83).

Mixed Skills Review

1. 9.8% of $122,500 **2.** 3124 − 1825 **3.** $7\frac{5}{6} - 4\frac{4}{15}$

4. 0.27 + 0.8 + 0.363 **5.** 116 × 13 **6.** $8\frac{3}{5} + 12\frac{1}{5}$

7. 1978 ÷ 46 **8.** $\frac{3}{8}$ of 1224 **9.** 6.8 − 5.737

10. $\frac{3}{4} \div 1\frac{5}{7}$ **11.** 7352 + 6248 **12.** 168 ÷ 1.2

8-5 Using Line Graphs

Objective
Construct line graphs from consumer information.

Tina Diehl manages the Microcomputer Market. Tina uses this line graph to show the changes in sales from month to month.

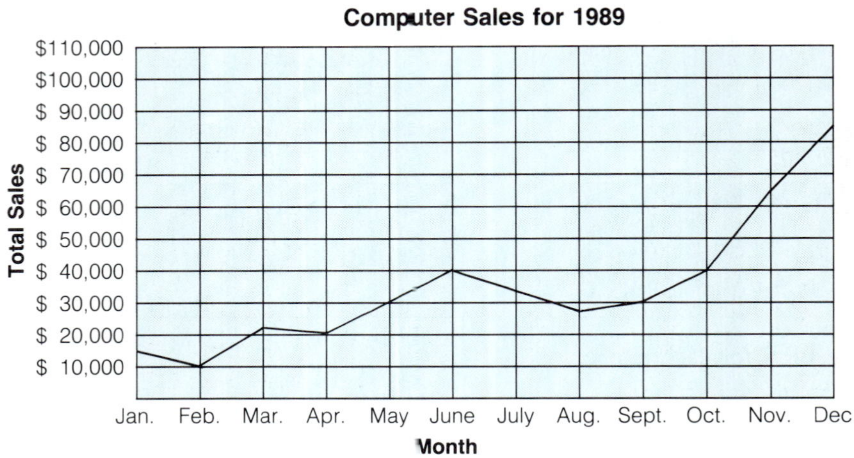

Understand the Situation

- What is the title?
- What is the horizontal scale?
- What is the vertical scale?
- When were sales lowest?
- When were sales greatest?
- What were the total sales in June?

Example Draw a line graph to show the six-month record of computer game sales.

Sales Record	
Jan.	25
Feb.	20
Mar.	36
Apr.	40
May	50
June	47

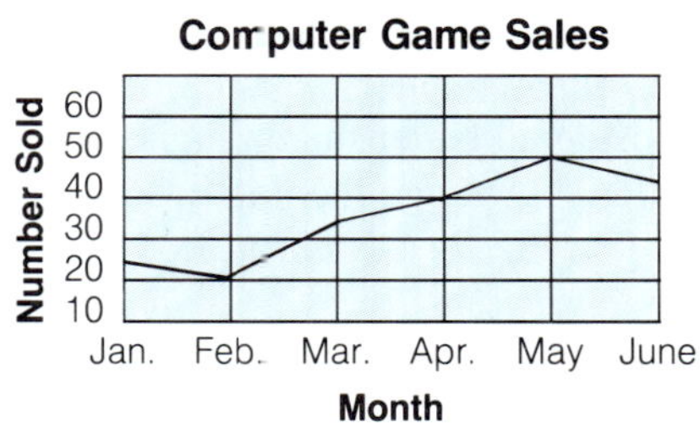

The following steps are helpful when making line graphs:

- Draw a horizontal and a vertical axis with scales that will show your data.
- Label each scale clearly.
- Draw a dot for each data point.
- Connect the dots with line segments.
- Give the graph an appropriate title.

Exercises

Draw a line graph to show the data in each table.

1.

Record of Weight Loss	
Week	**Weight**
1	210
2	200
3	196
4	190
5	185

2.

Value of 1916D Mercury Dime	
Year	**Value**
1950	$400
1960	$650
1970	$800
1980	$1200

3.

Record of Salary	
Year	**Salary**
1960	$9,500
1965	$10,800
1970	$12,000
1975	$13,800
1980	$14,900
1985	$16,800

4.

Value of 1977 Eddie Murray Baseball Card			
Year	**Value**	**Year**	**Value**
1977	$.25	1982	$2.00
1978	$.35	1983	$4.50
1979	$.75	1984	$6.50
1980	$1.00	1985	$10.00
1981	$1.50	1986	$12.50

Use the computer sales graph on the opposite page.

5. Estimate the sales for March.
6. Estimate the total April and May sales.
7. Estimate the difference between the July and August sales.
8. In which month period did the greatest sales increase occur?
9. In which month period did the greatest sales decrease occur?
10. Estimate the total sales for the year.
11. Estimate the mean sales for the year.
12. Estimate the median monthly sales for the year.

Calculator 13. Use the chart on page 182. Find the mean computer games sales.

Mixed Skills Review

1. 18,754 + 26,347
2. $4\frac{1}{2} \div 5\frac{1}{4}$
3. 0.073 − 0.038
4. 8 × 3.27
5. 2259.2 ÷ 3.2
6. $7\frac{3}{8} + 4\frac{5}{8}$
7. 516 × 84
8. 22.7 + 19.5 + 16.6
9. $4\frac{1}{2} \times 3\frac{2}{3}$
10. $8000 − $4261
11. $4\frac{7}{9} - 2\frac{1}{2}$
12. 90% of what number is 45?

8-6 Using Bar Graphs

Objective
Construct bar graphs from consumer information.

Eric Fourhman is a nutritionist. Eric prepared this bar graph to show the comparison of the fat content in some of the popular burgers.

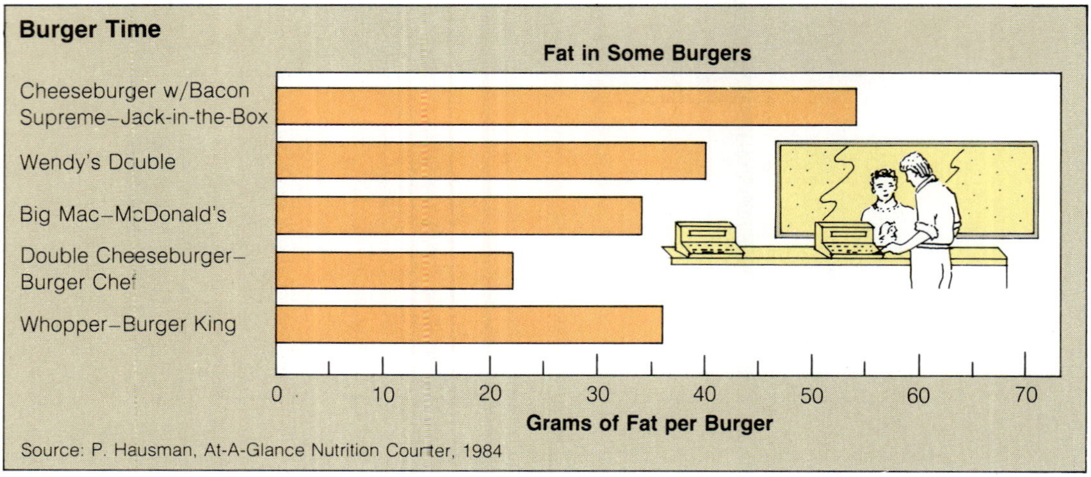

Understand the Situation

- What is the title of the graph?
- What burgers are shown on the graph?
- What does the horizontal scale show?
- Which burger had the least amount of fat?
- Which burger had the greatest amount of fat?

Bar graphs are used to show comparisons among data.

Example Draw a bar graph which compares the amount of protein in the burgers.

Burger	Protein (grams)
Big Mac	26
Bacon Supreme	33
Wendy's Double	26
Double Cheeseburger	23
Whopper	24

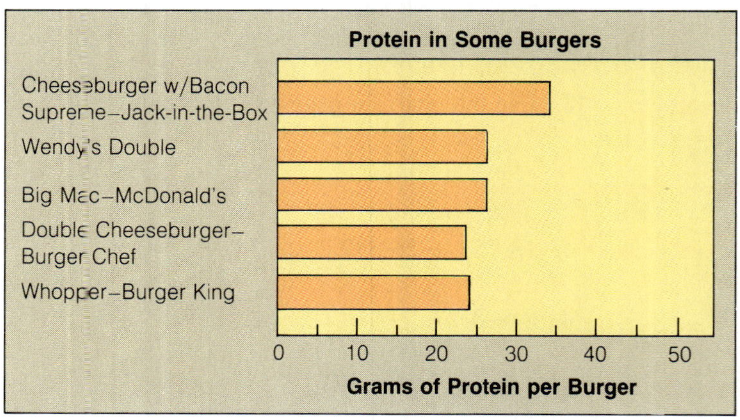

Since the greatest number of grams is 33, the scale is drawn just beyond that. The names of the burgers are spaced evenly. A title is printed at the top.

Exercises

Draw bar graphs to show the data in each table.

1.

U.S. Population Per Square Mile	
Year	**Pop.**
1940	44.2
1950	50.7
1960	50.6
1970	57.4
1980	62.6

2.

Grams of Fat in Some Fast Foods	
Food	**Grams of Fat**
Wendy's chili	8
Hardee's fish	26
Arby's roast beef	15
Shakey's cheese pizza	12
Burger King onion rings	16

3.

Five Busy Airports	
Airport	**Passengers (in millions)**
O'Hare (Chicago)	45.7
Hartsfield (Atlanta)	39.0
Los Angeles Int'l	34.4
Dallas/Ft. Worth	32.3
Kennedy (New York)	29.9

4.

Five Leading Magazines	
Magazine	**Circulation (in millions)**
Reader's Digest	17.9
TV Guide	17.1
National Geographic	10.4
Better Homes and Gardens	8.1
Family Circle	6.9

Estimation Use the graphs on p. 184.

5. Estimate the difference in fat content and in protein content between a Big Mac and a Wendy's Double.

6. Estimate the difference in fat content and in protein content between a Burger King Whopper and a Burger Chef Double Cheeseburger.

7. Estimate the mean fat content and the mean protein content for the burgers mentioned.

8. Find the median fat content and the median protein content for the burgers mentioned.

Mixed Skills Review

1. $6\frac{7}{12} - 3\frac{3}{4}$ **2.** 54% of 76 **3.** $0.16 \div 2.5$

4. $19.7 + 6 + 3.8 + 14.6$ **5.** $263 + 544 + 862$ **6.** $2\frac{2}{3} + 5\frac{1}{6} + 4\frac{1}{2}$

7. 420×51 **8.** $\$2020 - \$1764.32$ **9.** $1\frac{1}{5} \times 7\frac{1}{2}$

10. $1350 \div 48$ **11.** $1001 - 674$ **12.** $3\frac{3}{5} \div 1\frac{1}{20}$

8-7 Misleading Graphs

Objective
Identify misleading data.

Sometimes graphs can be misleading because of the way they are drawn. Here are two different graphs of Manuel's progress in an employee training program.

Graph A
Manuel's Scores

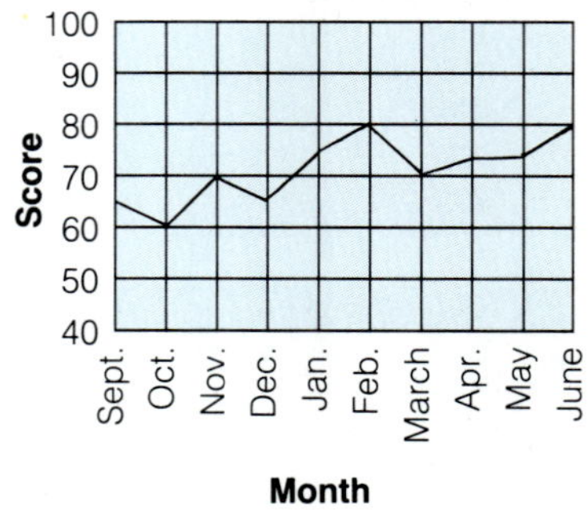

Understand the Situation

- What is the title of each graph?
- What is shown on the vertical scale of each graph?
- What is shown on the horizontal scale of each graph?
- What is the highest possible score?

The way that the scales of a graph are set up can change the impression that a graph gives the reader. Study the two graphs.

Graph B
Manuel's Scores

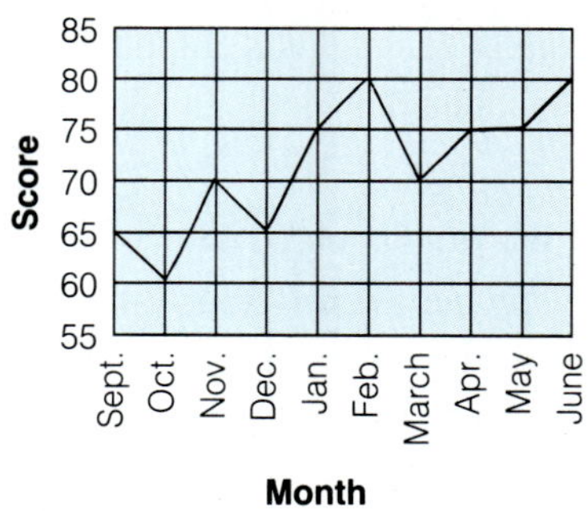

Examples

1. What does Graph A show for Manuel's score in December? The graph crosses the December line between 60 and 70. Manuel's December score is 65.

2. What does Graph B show for Manuel's score in December?
The December score is 65.

Try This

a. Compare Manuel's scores for the other months on both graphs.

b. How does Graph A differ from Graph B?

c. Which graph is more likely to give the impression that Manuel's scores were steady but generally rising? Why?

Hints for evaluating graphs:

- Check for a clear title and labels on the scales.
- Check to see that the numbers on the scales are equally spaced.
- Check to see that the scales have not been arranged to be misleading.
- Check the source of data for the graph.
- What conclusions does the graph help you make?
- Do your conclusions make sense?
- Can you reach the same conclusion without the graph?

Exercises

Study each graph below. Then answer the questions.

Graph A
Marian's Test Scores

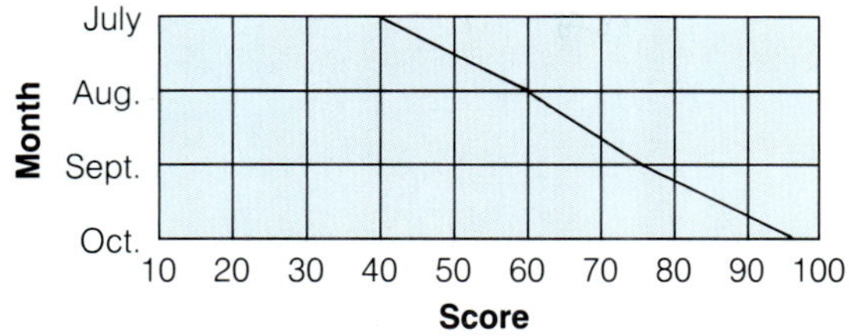

Graph B
Marian's Test Scores

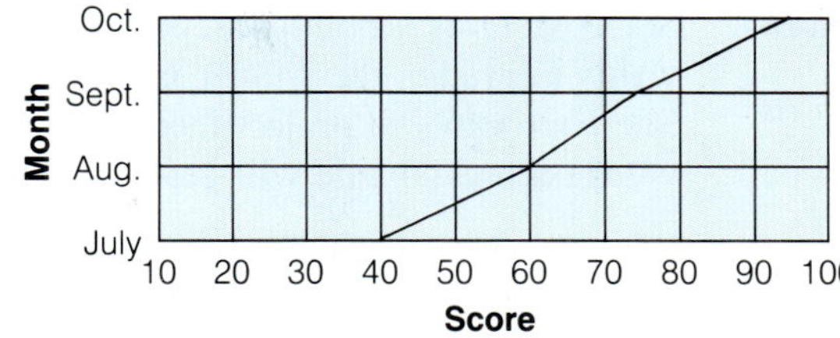

1. How do the scores compare monthly on Graphs A and B above?
2. What does each space on the vertical scale represent on each graph?
3. Which graph is more likely to give a positive impression of Marian's progress? Why?
4. Why is Graph A misleading?
5. This graph has been designed to mislead the reader into thinking that the ACME Manufacturing Company is catching up with the AJAX Manufacturing Company in sales. What has been done to the graph to give this misleading impression?

ACME Overtakes AJAX in Sales

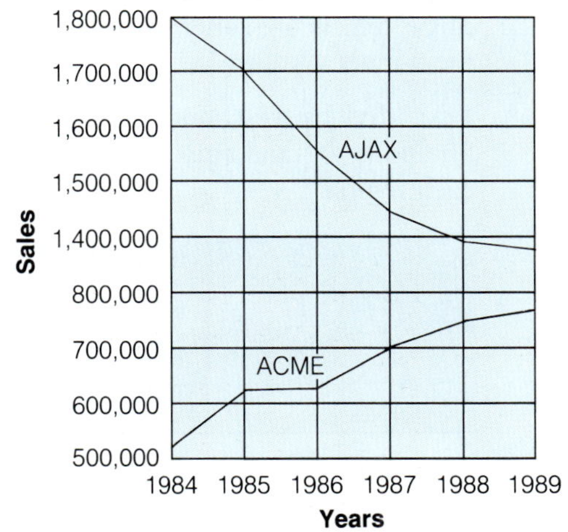

Estimation

6. What are the total number of sales for ACME and AJAX from 1984 to 1989?
7. Would you predict that ACME will catch AJAX in sales? If so, when?

Mixed Skills Review

1. 112% of 220	2. $5\frac{7}{8} + 3\frac{5}{6}$	3. 1720 − 1689
4. 1450 × 34	5. $4\frac{1}{3} \div 2\frac{8}{9}$	6. $\frac{3}{8}$ of 48
7. \$216.07 − \$150.48	8. 16 is what percent of 50?	9. 23.16 + 8.86
10. $10\frac{2}{3} - 4\frac{11}{12}$	11. 14,457 ÷ 61	12. 10,423 + 16,728

8-8 Using Samples

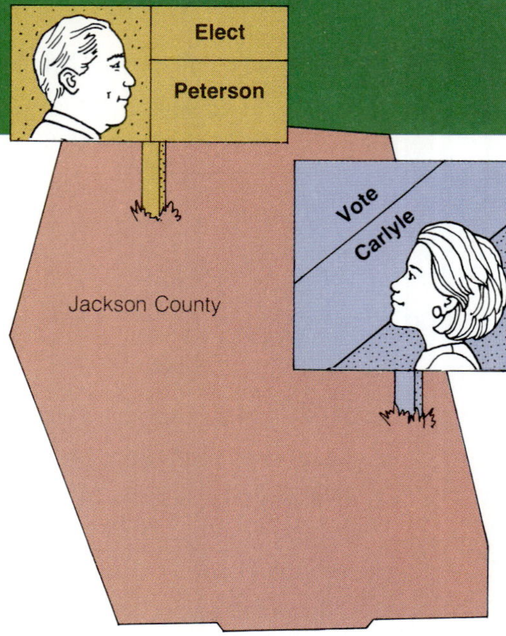

Objective
Make predictions using samples.

Susie Mann is a political reporter. She wants to predict the winner in an election. She also wants to project the number of votes each candidate will get.

Records show that there will be about 20,000 people voting in the election.

Understand the Situation

- What are the names of the candidates?
- In which county are they running?
- How many people are expected to vote?

Susie needs to find the number of people who will vote for each candidate. It is not reasonable to survey all 20,000 voters. Susie selects a smaller **sample** and predicts the winner.

Susie divides the county into four regions. She polls 75 randomly selected voters in each region.

	Regions				
	I	II	III	IV	Total
Carlyle	33	39	30	51	153
Peterson	42	36	45	24	147

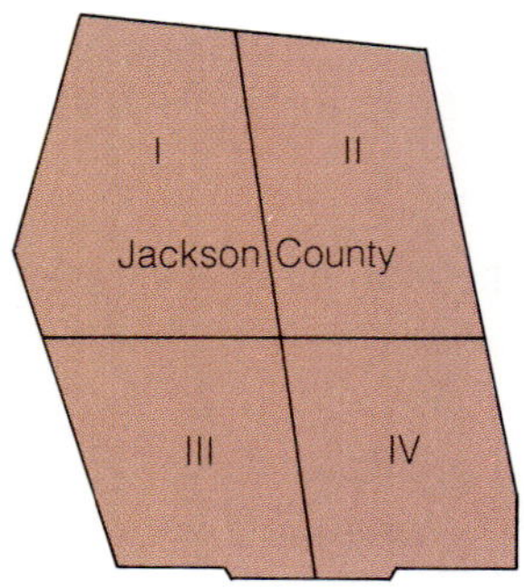

Example Predict the winner in Jackson County.

Write and solve a proportion to project the number of votes Carlyle will get.

$$\frac{153}{300} = \frac{n}{20{,}000}$$

Poll Count / Poll Total = Projected Votes for Carlyle / Total Number of Voters

$$153 \times 20{,}000 = 300 \times n$$

When using a sample to make predictions it is important to select randomly. A sample is random if everyone has the same chance of being selected.

Try This

a. Solve the proportion. You may want to use a calculator.

b. Who do you predict will win the election?

Exercises

1. There are about 12,945,000 registered voters in the state of New York. The choices of a random selection of 50 voters from each of three regions is shown in the table. Predict the winner and the number of votes each candidate will receive.

Senate Race Poll Results	Regions		
	I	II	III
Candidate A	32	28	14
Candidate B	18	22	36

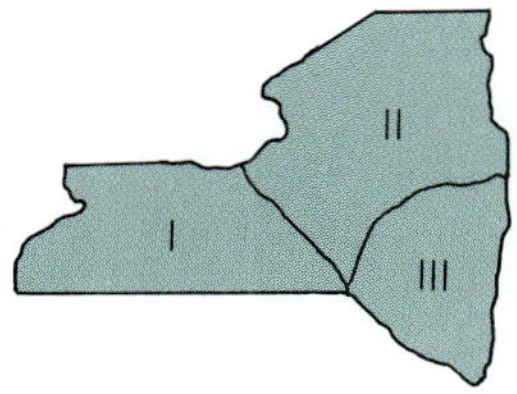

2. There are about 2,114,000 registered voters in Colorado. A poll was taken of 100 randomly selected voters in each of four regions. Predict the winner and the number of votes each candidate will receive.

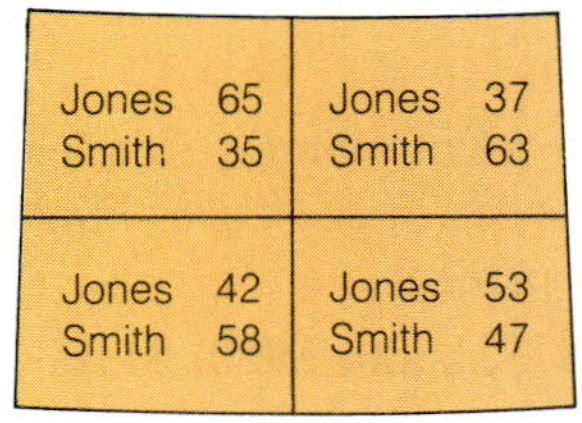

3. In a local election three candidates are running for a seat on the county council. A random sample of 50 voters was selected from each of five regions. There are 12,000 registered voters. Predict the winner and the number of votes each candidate will receive.

County Council Poll Results	Regions				
	I	II	III	IV	V
Candidate Thomas	14	18	12	22	19
Candidate Wynn	20	13	21	18	10
Candidate Selway	16	19	17	10	21

Work with a Group

4. Design a survey of the students in your school on a topic of interest such as favorite singers, tapes, or movies. Predict the results of the survey by selecting random samples from each class or grade.

Mixed Skills Review

1. $2.036 + 0.82 + 9.607$ **2.** $138 + 265 + 541$ **3.** $\frac{9}{18} - 7\frac{3}{16}$

4. $191 - 35.06$ **5.** 14.7×0.24 **6.** $1827 \div 63$

7. $17.64 \div 2.8$ **8.** $1\frac{7}{8} \times 1\frac{3}{5}$ **9.** $\$5385 - \$4776$

10. 70% of 185 **11.** $3\frac{1}{2} + \frac{2}{3} + 1\frac{3}{5}$ **12.** $6\frac{2}{3} \div 1\frac{7}{9}$

8-9 Enrichment: The Trimmed Mean

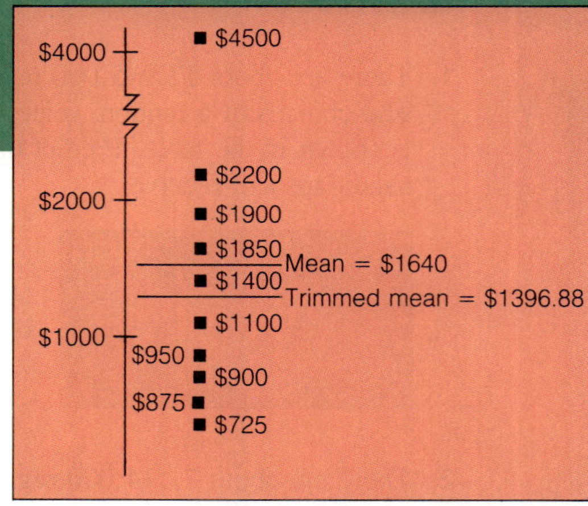

Objective
Find the mean, the trimmed mean, and the trimmed percentage.

Sometimes a mean can be distorted by extreme values, called **outliers.** To avoid this, statisticians order the data from smallest to largest and delete the same number of values from each end. The **trimmed mean** is the mean of the remaining values. The **trimmed percent** is the percent of values deleted from each end.

Here is a set of monthly salaries arranged from smallest to largest for the 10 employees in an office. Find the mean, the trimmed mean, and the trimmed percent.

\$725, \$875, \$900, \$950, \$1100, \$1400, \$1850, \$1900, \$2200, \$4500

$$\text{Mean} = \frac{\$725 + \$875 + \$900 + \$950 - \$1100 + \$1400 + \$1850 + \$1900 + \$2200 + \$4500}{10}$$

$$= \$1640$$

The mean is \$1640. The outlier of \$4500 makes the average unrealistically high for this office. Delete one value from each end (\$725 and \$4500) and find the trimmed mean.

$$\text{Trimmed mean} = \frac{\$875 + \$900 + \$950 + \$1100 + \$1400 + \$1850 + \$1900 + \$2200}{8}$$

$$= \$1396.88$$

The trimmed mean is \$1396.88. Because one salary was deleted from each end, the trimmed percent is 1 out of 10, or 10%.

Exercises

Calculator

Find the mean, the trimmed mean, and the trimmed percent of each set of data.

1. These were the raw scores on a history test.
   39, 55, 60, 61, 63, 67, 68, 70, 71, 73, 75, 76, 78, 80, 96, 98
2. These were the waiting times, in minutes, for patients in a clinic in an afternoon.
   5, 7, 15, 18, 18, 19, 22, 25, 26, 27, 31, 35, 36, 36, 38, 39, 40, 41, 41, 49, 125
3. These were the magazine selling prices on a publisher's special offer.
   \$1.25, \$1.50, \$1.75, \$2, \$2, \$2.25, \$2.25, \$2.50, \$2.50, \$2.50, \$2.75, \$3.50, \$4, \$4.75

8-10 Computer: Finding the Mean

Objective
Use a computer to find a mean and to set up a frequency table.

Find the mean of the following set of numbers. Make a frequency table. Use intervals of 25 and start with 1–25.

54, 23, 96, 72, 79, 45, 34, 95

The following computer program can be useful.

```
10  REM This program finds the sum and mean of a set of up to 50 numbers
20  REM between 1 and 100 and produces a frequency table in intervals of 25
30  DIM N(50): S=0 : A=0 : B=0 : C=0 : D=0
40  INPUT “How many numbers in the set?”; N
50  FOR X = 1 TO N
60  INPUT “Enter a number between 1 and 100”; N(X)
70  IF N(X) >= 1 AND N(X) <= 25 THEN A = A + 1
80  IF N(X) >= 26 AND N(X) <= 50 THEN B = B + 1
90  IF N(X) >= 51 AND N(X) <= 75 THEN C = C + 1
100 IF N(X) >= 76 AND N(X) <= 100 THEN D = D + 1
110 S = S + N(X)
120 NEXT X
130 PRINT “The” N “numbers you entered are:”;
140 FOR X = 1 TO N: PRINT N(X);: NEXT X
150 PRINT: PRINT “The sum of the numbers is” S “and the mean is” S/N “.”
160 PRINT “INTERVAL”, “RANGE”
170 PRINT “ 1–25”, A
180 PRINT “26–50”, B
190 PRINT “51–57”, C
200 PRINT “76–100”, D
210 END
```

The output of the program above is:

```
The 8 numbers in the set are: 54   23   96   72   79   45   34   95.
The sum of the numbers is 498 and the mean is 62.25.
INTERVAL     1–25     26–50     51–75     76–100
RANGE        1        2         2         3
```

Exercises

1. Run the program for the following set of numbers.
   34, 65, 12, 99, 62, 48, 41, 58, 36, 43, 87, 92, 58, 41, 22, 3, 99
2. If you use the program for numbers greater than 100, will it find the correct mean? Will the frequency table be correct?

CHAPTER 8 REVIEW

Choose the one that does not belong in each group.

1. A. mean
   B. frequency table
   C. mode
   D. median

2. A. horizontal scale
   B. title
   C. sample
   D. vertical scale

3. A. tally
   B. interval
   C. mean
   D. frequency

4. A. add all the numbers
   B. find the middle number
   C. count the number of numbers
   D. divide by the number of numbers

5. A. find the middle number
   B. find the most frequent number
   C. find the mean of the two middle numbers
   D. arrange the numbers in order

6. A. most frequent
   B. may not exist
   C. must be calculated
   D. may have more than 1

Tell what the error is in each of these graphs.

7.

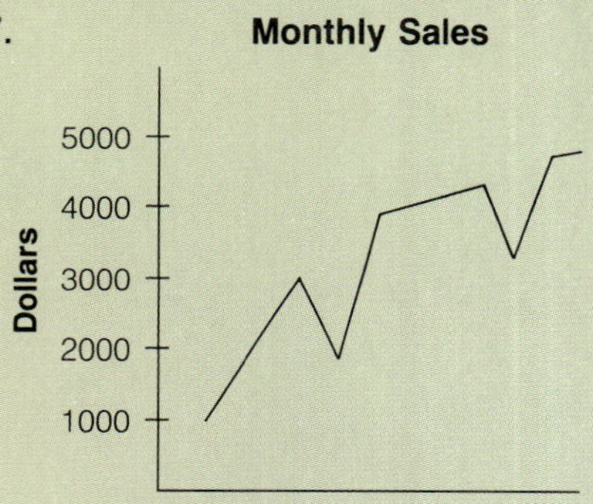

8.

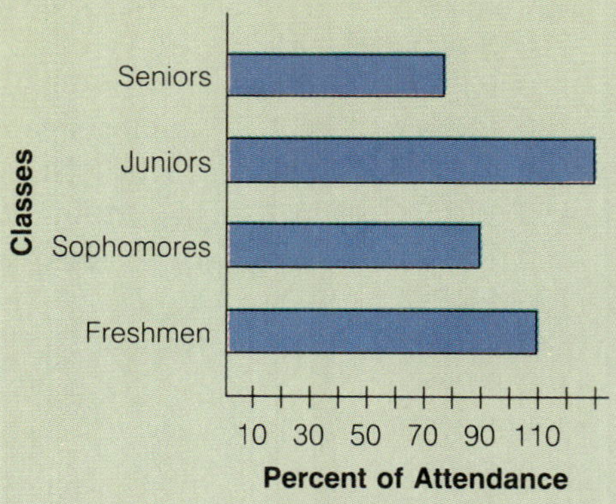

9.

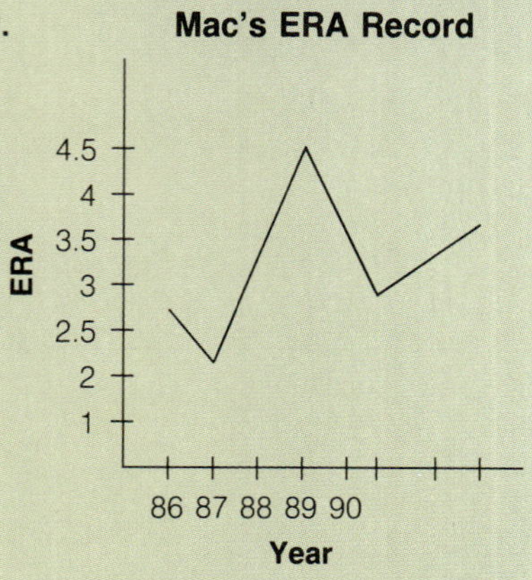

10.

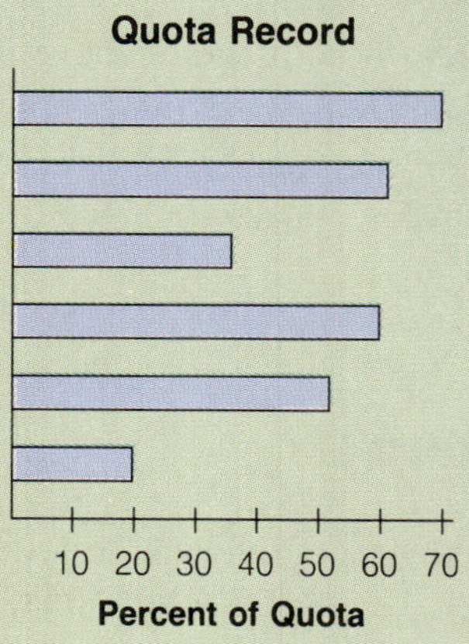

CHAPTER 8 TEST

Miles Jogged						
S	M	T	W	T	F	S
1.1	1.2	1.1	1.5	1.4	1.7	1.4
1.5	1.7	1.9	1.5	1.7	1.6	1.7
1.6	1.8	1.7	1.9	2.1	1.9	2.3
2.6	2.4	2.7	2.4	2.6	2.8	3.1
3.3	3.6	3.8	3.5	3.9	4.1	3.9

1. As part of his fitness program, Mark kept a record of how many miles he jogged. The table shows how many miles he jogged. Make a frequency table of this data. Use an interval of 0.4 and start with 1.0–1.3.
2. Mr. Jackson recorded the number of cars using his car wash every hour. Find the mean and the mode for the set of data.
   14, 16, 14, 17, 14, 18, 21, 25, 22, 24, 22
3. Find the median for the data in Problem 2.
4. Find how many games the second-place team is back of the first-place team.

	Wins	Losses
Pirates	56	16
Mets	45	34

5. Find Johnny Carlton's ERA after giving up 57 runs in 215 innings.
6. Draw a line graph to show the data.
7. Draw a bar graph to show the data.

Raffle Tickets Sold

Week	Number
1	145
2	165
3	254
4	412

ABC Sales Record

Year	Sales in Thousands
1983	9,000
1984	11,500
1985	15,600
1986	21,100

Use the graph to answer items 8–11.

8. What is the title of the graph?
9. What does the vertical scale represent? the horizontal scale?
10. Between which two quarters was there the biggest increase in violations?
11. Why is the graph misleading?

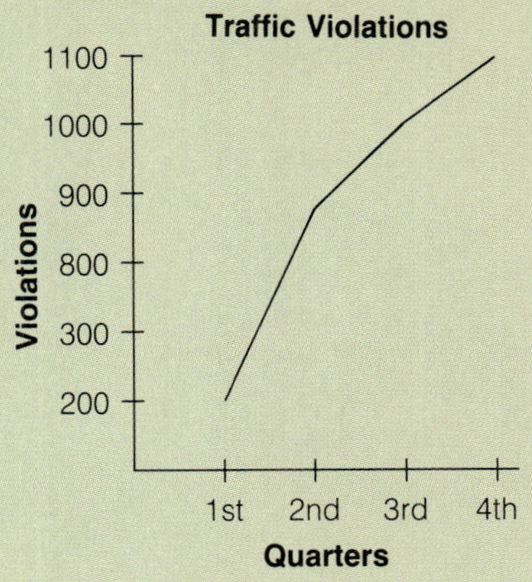

12. Three candidates are running for mayor. Researchers polled 100 randomly selected voters in each of four regions. There are 60,000 registered voters. Predict the number of votes each candidate will receive.

	Regions			
	1	2	3	4
Smith	36	18	26	15
Fong	23	63	39	56
Spera	41	19	35	29

CHAPTER 8 CUMULATIVE REVIEW

1.

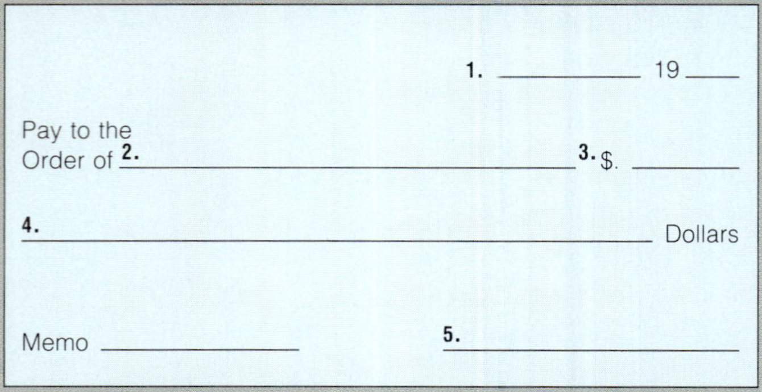

On which line is the amount spelled out?

A. 4 B. 2
C. 5 D. 3

2. Cars passing a corner every hour

56 32 79 53 10 28
37 14 54 29 40 41
75 57 33 42 58 76
14 29 53 27 51 72

What is the frequency for the interval 50–59? Make a table.

A. 5
B. 7
C. 6
D. 9

3. Find the answer.
3.26 + 6.86 − 7.42

A. 17.54
B. 2.6
C. 3.5
D. 2.7

4. Find Carl Schmidt's ERA after giving up 72 runs in 225 innings.

Use ERA $= 9\frac{R}{I}$.

A. 2.60
B. 3.03
C. 3.38
D. 2.88

5. Find 7% of \$59.59.

A. \$8.51
B. \$4.17
C. \$11.75
D. \$4.28

6. Find the sale price of a ski parka.
list price \$150,
discount $\frac{1}{5}$ off

A. \$30
B. \$120
C. \$100
D. \$145

7.

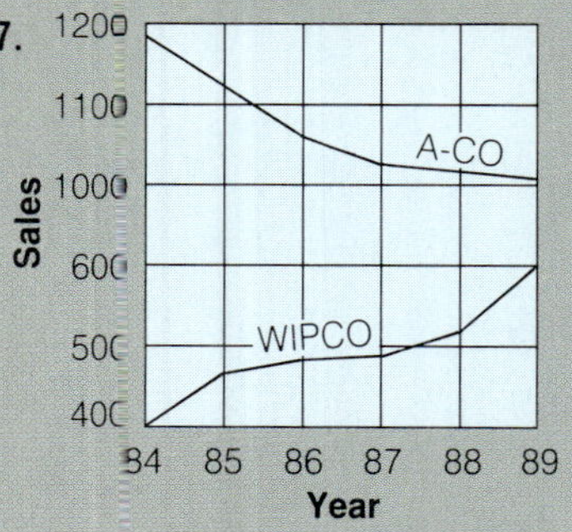

What is the actual difference in 1989 sales?

A. 100
B. 200
C. 300
D. 400

8. Bryan O'Neil had a \$562.12 checking account balance. He deposited \$40.50 and wrote a check for \$7.69. What is his new balance?

A. \$594.93
B. \$554.43
C. \$602.62
D. \$569.81

9. Write an equivalent fraction with a denominator of 48.
$\frac{5}{16}$
A. $\frac{5}{48}$ B. $\frac{3}{48}$
C. $\frac{15}{48}$ D. $\frac{10}{48}$

10. 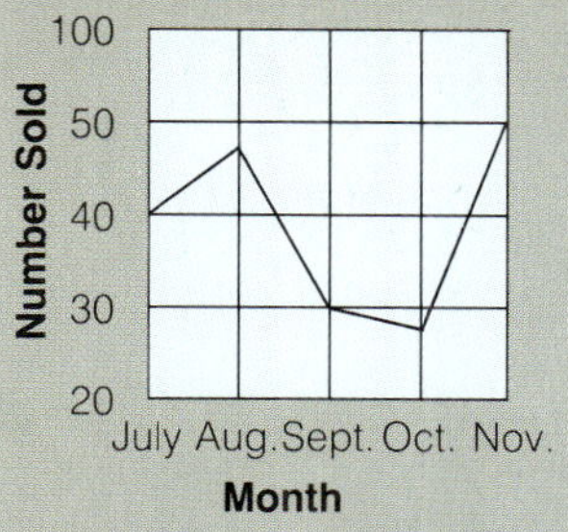

Estimate the difference between October and November sales.
A. 15 B. 25
C. 20 D. 30

11. Find the simple interest.
\$2650 loan, 9 months, 11%
Use the formula $i = prt$.
A. \$218.63
B. \$291.50
C. \$262.35
D. \$211.20

12. The Sandwich Shop uses 1 of 3 kinds of meat in each sandwich. Each sandwich can have any 1 of 3 dressings. How many different sandwiches are possible?
A. 6 B. 5
C. 9 D. 8

13. Following is a list of shoe sizes sold in one day. Find the mode.
5, 9, 6, 8, 7, 5, 9, 6, 5
A. $6\frac{2}{3}$
B. 5
C. 7
D. 8

14. Find the installment price of a TV set.
list price \$325, down payment $\frac{1}{4}$, monthly payment \$63.50, term 120 days
A. \$406.25
B. \$388.50
C. \$335.25
D. \$325.00

15. Find the catalog cost.
Shipping charge for 5 to 10 lb is \$3.25.
croquet set \$21.96, 7 lb 10 oz
A. \$56.25
B. \$18.71
C. \$225
D. \$675

16.

Senate Race Poll Results			
	Regions		
	I	II	III
Candidate A	21	35	11
Candidate B	29	15	39

There are about 78,600 registered voters. The poll covered 50 voters in each region. Predict the number of votes for Candidate A.
A. 89,604 B. 43,492
C. 105,324 D. 35,108

17. These are teachers' ages at Hillview High School.
25, 56, 39, 63, 41, 33, 26, 51, 46, 24, 36, 52, 41, 43, 65
Find the median.
A. 40.5 B. 41
C. 42.7 D. 39

18. Find the better buy for laundry detergent.
32 oz for \$1.49,
128 oz for \$5.59
A. 128 oz
B. 32 oz

19. Floor tiles are sold by the square. Two tiles cover 1 ft^2. How many tiles are needed for a room 8 ft by 6 ft?
Use $A = lw$.
A. 192 tiles B. 96 tiles
C. 48 tiles D. 24 tiles

Chapter 9 Overview

Key Ideas

- finding probabilities
- finding odds
- compound probabilities
- finding expected values
- making predictions
- using percents in equations
- solving counting problems
- making tree diagrams
- the counting principle
- arrangements

Key Terms

- probability
- probability line
- possible outcomes
- favorable outcomes
- odds in favor
- odds against
- independent events
- expectation
- prediction
- tree diagram
- counting principle
- permutation
- factorial
- combinations

Key Skills

Simplify.

1. $\frac{3}{9}$ **2.** $\frac{12}{16}$ **3.** $\frac{14}{16}$

4. $\frac{8}{14}$ **5.** $\frac{4}{10}$ **6.** $\frac{24}{38}$

Find the answer.

7. 0.23×0.52 **8.** 0.19×0.67 **9.** 0.56×0.56

10. 0.67×0.67 **11.** 0.98×0.56 **12.** 0.39×0.43

13. $\frac{1}{6} \times \frac{1}{4}$ **14.** $\frac{2}{3} \times \frac{6}{7}$ **15.** $\frac{4}{5} \times \frac{2}{3}$

16. $\frac{3}{25} \times \frac{8}{25}$ **17.** $\frac{9}{25} \times \frac{6}{23}$ **18.** $\frac{18}{25} \times \frac{2}{17}$

19. $\frac{1}{2} \times 42$ **20.** $\frac{1}{3} \times 52$ **21.** $\frac{1}{10} \times 23$

22. $100\% - 6.7\%$ **23.** $100\% - 5.9\%$ **24.** $100\% - 14.9\%$

25. $17 \times 232 \times 6$ **26.** $9 \times 7 \times 11 \times 3$ **27.** $12 \times 8 \times 9$

28. $24 \times 56 \times 45$ **29.** $23 \times 23 \times 12$ **30.** $26 \times 26 \times 19$

USING PROBABILITY 9

9-1 Problems of Chance

Objective
Find simple probability.

For the grand opening of the Valley Mall, the Sight and Sound shop is giving away a compact disc player. The winner's name will be drawn from a large bowl containing all the entries. There are 1000 entries in the contest and Mark filled out 1 entry. What are his chances of winning?

Drawing one particular card out of a thousand is a chance event. The **probability** of the favorable event can be expressed as a fraction.

$$\text{probability of an event, or P(event)} = \frac{\text{number of favorable events}}{\text{number of possible events}}$$

Examples

1. What is the probability of Mark's winning the CD player?

$$\text{P(Mark's winning)} = \frac{\text{number of entries with Mark's name}}{\text{total number of entries}} = \frac{1}{1000}$$

The probability of Mark's winning is 1 out of 1000, or 0.001, or 0.1%.

The probability of an impossible event is 0. The probability of a sure thing is 1, or 100%. A **probability line** can be drawn to show where probabilities fall between certainty and impossibility.

2. There are 2 white, 3 blue, and 5 green marbles in a jar. Show the probability of picking a particular color out of the jar without looking. Use a probability line.

$\text{P(white marble)} = \frac{2}{10}$ $\quad\text{P(blue marble)} = \frac{3}{10}$ $\quad\text{P(green marble)} = \frac{5}{10}$

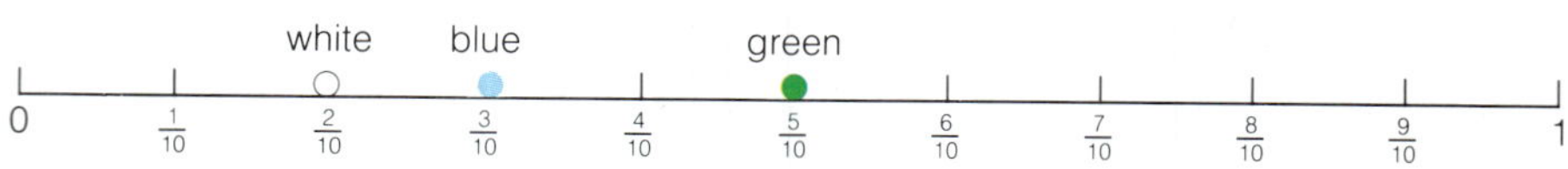

Try This

Add 4 yellow marbles to the jar above. Find the new probabilities of picking each of the colors of marbles.

a. P(white marble) **b.** P(red marble)
c. P(yellow marble) **d.** P(any marble)
e. Draw a probability line showing all the above probabilities of marbles from the jar.

Exercises

1. Some coins are spread on a table. There are 6 pennies, 3 nickels, and 10 quarters. You choose a coin without looking. What is the probability that it is a quarter?
2. Mariana chooses a CD from her collection at random. Five of the CDs are classical music, six CDs are jazz, and eight CDs are rock. What is the probability Mariana will choose jazz?
3. In Tricia's baseball card collection there are 65 cards of the Baltimore Orioles. Eight of these cards are Cal Ripken cards. Tricia selects an Oriole card at random. What is the probability that she will select a Cal Ripken card?
4. The spinner to the right shows the numbers from 1 to 12. Draw a probability line. Draw a dot on the line to show the probability of the spinner's landing on:

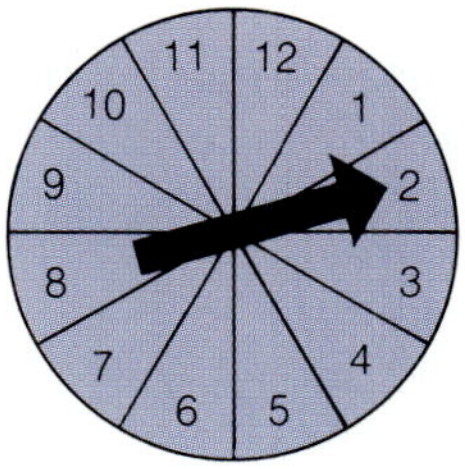

a. a 4	**b.** a multiple of 4
c. a multiple of 5	**d.** a number from 1 to 6
e. a prime number	**f.** 9.25

Data Hunt

5. What is the probability that it will rain in your community tomorrow?

Number Sense

6. This is a list of the probabilities of winning in four different contests.

   Contest A: $\frac{1}{90}$, Contest B: $\frac{1}{50}$, Contest C: $\frac{2}{5}$, Contest D: $\frac{3}{8}$

   Arrange the contests in order from the best chance of winning to the least chance of winning.

Work with a Partner

7. Fold a file card in half as shown below. Toss the folded file card 100 times. Keep a record of the number of times it lands on its side, on its edge, or as a tent. What is the probability of each outcome?

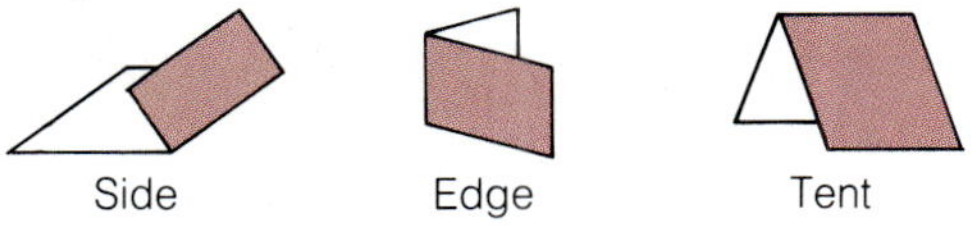

Mixed Skills Review

1. $12\frac{1}{5} - 6\frac{1}{2}$	**2.** $3159 \div 13$	**3.** $\frac{2}{3} + 3\frac{3}{8}$
4. 0.16×98.2	**5.** $1326 + 2844$	**6.** $4\frac{3}{5} \times 15$
7. \$1528 − \$1355	**8.** $8\frac{1}{3} \div \frac{5}{9}$	**9.** 312×83
10. $16.27 + 9.08$	**11.** $12.76 \div 4.4$	**12.** $15.3 - 12.75$

9-2 Probability and Odds

Objective
Find the odds in favor of or against the occurrence of an event.

At a yard sale Eloise bought a box labeled *mixed treasures*. In the box she found an old lock and a key ring containing six keys. If only one key fits the lock, what are the **odds** in favor of Eloise's picking the right key at random?

To find the odds in favor of an event's occurring, find the ratio of the number of ways the event can occur to the number of ways it cannot occur.

Examples

1. Find the odds in favor of Eloise's picking the right key.

   There is 1 way for Eloise to pick the right key.
   There are 5 ways for Eloise to pick the wrong key.

   The odds in favor of picking the right key are 1 to 5.

To find the odds against an event occurring find the ratio of the number of ways the event cannot occur to the number of ways it can occur.

2. In the box Eloise also found a combination lock. The combination was printed on a piece of paper. The last digit was faded and could not be read. Eloise guessed the last digit. What are the odds against her guessing the right digit on the first try?

   There are 9 ways to guess wrong.
   There is 1 way to guess right.

   The odds against guessing the right number are 9 to 1.

Try This

a. What are the odds against Eloise's picking the right key?

b. What are the odds in favor of Eloise's guessing the right number?

c. In the treasure box there was a bag of old coins. The bag contained 8 wheat pennies, 2 buffalo nickels and 1 Mercury dime. Eloise chooses one coin at random. What are the odds in favor of her choosing a buffalo nickel?

Exercises

1. What are the odds against Eloise's choosing a wheat penny from the bag of coins?
2. What are the odds in favor of Eloise's choosing the Mercury dime?
3. At the yard sale the names of all those attending are placed in a hat. The winner is to receive a surprise package gift. There are 139 people at the sale. What are the odds against Eloise's winning the gift?
4. Martin wakes up on his birthday to find on the breakfast table 7 wrapped presents: a pair of shoes, a record, a board game, a jacket, a dress shirt, a tennis racket, and a belt. What are the odds that Martin will first unwrap an article of clothing?

5. The radio weather report says that there is a 60% chance of rain. What are the odds against its raining?
6. The TV weather report says that there is a 90% chance of bright sunshine for the beach. What are the odds in favor of sunshine?
7. Mike says that the odds in favor of his wrestling team's winning the tournament are 5 to 3. What is the probability that Mike's team will win?
8. A political reporter says that the odds in favor of a candidate's winning the Senate race are 5 to 2. What is the probability that the candidate will win?

Write Math

9. Write a formula that tells how to find the odds in favor of an event if you know the probability that the event will occur.
10. The probability that a number cube, when rolled, will come up a 5 is $\frac{1}{6}$. What are the odds in favor of rolling a 6?

Number Sense

11. For each set of odds draw a probability line and place a dot which shows the probability of the event occurring.

    a. 3 to 2 b. 1 to 3 c. 5 to 6 d. 2 to 1

Mixed Skills Review

1. $1\frac{2}{3} \times 2\frac{1}{2}$
2. $7037 + 658$
3. 6.3×0.24
4. $2.4 - 1.061$
5. $4237 \div 19$
6. 402×260
7. $\frac{2}{5} + \frac{2}{3} + \frac{1}{2}$
8. $15\frac{1}{3} \div 23$
9. $10{,}000 - 4{,}622$
10. $5\frac{1}{6} - 2\frac{5}{8}$
11. $75.082 \div 6.2$
12. $60.351 + 72.759$

9-3 Compound Probability

Objective
Find compound probability.

Tom and Bobbi are at the ballpark to see the Angels play a doubleheader. They wonder what the chances are that the Angels will win both games.

Standings	Wins	Losses	Pct.
Angels	53	47	.530
Kansas City	50	49	.505
Rangers	48	49	.495
Twins	45	51	.469
Mariners	45	52	.464
Athletics	46	54	.460
White Sox	39	58	.402

Understand the Situation

- How many games have the Angels won?
- How many games have the Angels lost?
- What percent of the games have the Angels won?

Finding the probability of the Angels' winning both games is a **compound probability** problem. It concerns two events. When one event is not affected by the other event we say that the events are **independent.**

If A and B are independent events, then the probability of A's occurring and B's occurring can be found by multiplying their probabilities.

Examples

1. Find the probability of the Angels' winning both games. You can assume that the two baseball games are independent since the outcome of the first game does not usually affect the second game.

The Angels win about 53% of their games, so the probability of their winning the first game is about 0.53. Likewise, the probability of their winning the second game is about 0.53.

Multiply the probabilities. 0.53×0.53 *Estimate:* $0.5 \times 0.5 = 0.25$

$0.53 \times 0.53 = 0.2809.$

The probability of the Angels' winning both games is about 0.28, or $\frac{28}{100}$.

2. A spinner with 8 spaces is spun and a 6-sided number cube is rolled. What is the probability of spinning a 6 and rolling a 6?

The events are independent. Spinning the spinner does not affect the rolling of the cube.

P(spinning a 6) $= \frac{1}{8}$ P(rolling a 6) $= \frac{1}{6}$

P(spinning a 6 and rolling a 6) $= \frac{1}{8} \times \frac{1}{6} = \frac{1}{48}$

Try This

a. Find P(rolling a 4 on a number cube and tossing a head with a coin).

b. Find P(spinning a 2 and tossing a tail).

c. Find P(the Athletics' winning a doubleheader).

Exercises

1. Find P(the White Sox' winning a double header).
2. Find P(spinning a 3 and rolling a 5 on a number cube).
3. Find P(spinning greater than 4 and rolling less than 4).
4. Find P(tossing a head and spinning a number which is a multiple of 2).

There are 4 yellow marbles, 5 green marbles, 9 blue marbles, and 7 red marbles in a bag. Another bag contains 4 white blocks, 5 red blocks, and 2 blue blocks. A marble and a block are drawn from the bags without looking.

5. Find P(green marble and red block).
6. Find P(blue marble and blue block).
7. Find P(red marble and white block).
8. Find P(red marble and red block).
9. Find P(red marble and white block and tossing a head on a coin).
10. Find P(spinning a number less than 5 and green marble and red block).

Suppose Suppose that for the bag of marbles described above a first marble is drawn and then put back in the bag. A second marble is then drawn.

11. Find P(red marble then blue marble).
12. Find P(blue marble then yellow marble).
13. Find P(green marble then red marble).
14. Find P(red marble then red marble).
15. Suppose that the second marble is put back in the bag and then a third marble is drawn. Find P(red marble then blue marble then green marble).
16. Larry Bird of the Boston Celtics makes about 86% of his free throws. In a key game Larry must make two free throws to win the game. What is the probability that the Celtics will win the game?

Write Your Own Problem

17. Make up a compound probability problem involving the spinner, coins, bag of marbles, and bag of blocks described above.

Mixed Skills Review

1. $2\frac{2}{9} \div \frac{5}{12}$
2. $8\frac{1}{2} - 4\frac{2}{3}$
3. What percent is 15 of 200?
4. $1253 - 1106$
5. $3\frac{5}{8} + 2\frac{3}{4}$
6. $6.3 - 5.94$
7. $\$1.32 + \$0.89 + \$0.94$
8. $1\frac{2}{5} \times 1\frac{1}{9}$
9. 17.1×2.06
10. $338.1 \div 2.3$
11. 4218×12
12. $539 + 276 + 118$

9-4 Expectation

Objective
Find the expectation of an event.

The probability of precipitation on any spring day in Washington, D.C. is about $\frac{1}{2}$. How many rainy days should the White House gardener expect during the month of April?

Understand the Situation

- What is the probability of rain on any April day in Washington, D.C.?
- How many days are there in the month of April?

This is a problem in **expectation.** When we know the probability of an event and the number of possibilities for the event we can find the number of ways we can expect the event to occur by multiplying.

Examples

1. Find the number of rainy days that can be expected in April in Washington, D.C.

There are 30 days in April. The probability of rain is $\frac{1}{2}$.

$\frac{1}{2} \times 30 = 15$

Rain can be expected on about 15 days.

2. The probability of Keith Hernandez' getting a hit in any time at bat is about 31%. If Hernandez comes to bat about 600 times in a season, about how many hits can he expect to get?

Multiply the probability times the number of possibilities.
$0.31 \times 600 = 186$

Keith Hernandez can expect to get about 186 hits in 600 times at bat.

Try This

a. The probability of rain on any day in July in the Dallas-Ft. Worth area is about $\frac{1}{10}$. How many rainy days can be expected in July?

b. The school social is giving away 10 prizes in a raffle. Each ticket has a 10% chance of winning a prize. If Jenny buys 20 tickets, how many prizes can she expect to win?

c. Jericho is going door to door trying to get donations for a charitable project that his class is sponsoring. Jericho knows that the chances of getting a donation at any door are 1 in 4. Jericho visits 60 houses. About how many donations can he expect?

Exercises

1. The probability that it will rain on any day in October in Bakersfield, California is about $\frac{1}{15}$. About how many days can rain be expected in Bakersfield in October?
2. The probability that it will rain on any day in July in Miami, Florida is about $\frac{2}{3}$. About how many days can rain be expected in Miami in July?
3. The probability of James Worthy's making a field goal in basketball is about 0.57. If Worthy takes about 1100 shots in a season about how many can he expect to make?
4. An advertising agency indicated that the probability of an ad being heard on local radio is $\frac{2}{5}$, on local TV is $\frac{1}{5}$, and on both $\frac{1}{20}$. A corporation arranges for an ad to be placed on both radio and TV in a community of about 200,000 people. About how many people can the corporation expect to hear the ad on radio, on TV, and on both?
5. The probability that a home fire is due to a faulty heating system is about 0.22. The statistics of a local fire department showed 56 home fires. About how many of these fires could have been expected to be from faulty heating systems?

Work with a Partner

6. Suppose that the weather forecast for 10 cities calls for about an 80% chance of snow. To estimate how many of the cities can expect snow, complete the following experiment. Roll a number cube once for each city. For 1 to 5 we agree that it will snow. For a 6 we agree that it will not snow. For instance, if the outcomes are

   3, 4, 6, 2, 4, 5, 3, 1, 2, 5

   we conclude that 9 of the 10 cities got snow. Repeat the experiment at least 25 times and then find the mean number of cities that received snow. Is the result what you expected?

Number Sense

7. Arrange these probabilities in order of decreasing expectation.
   $\frac{1}{5}$, 30%, $\frac{3}{4}$, 0.56, $\frac{5}{6}$, 65%

Mixed Skills Review

1. 23,765 + 38,157	2. 6003 − 584	3. 1320 × 64
4. 1760 ÷ 0.23	5. $16\frac{1}{2} - 5\frac{5}{9}$	6. $\frac{3}{8} \times \frac{4}{9} \times \frac{2}{3}$
7. 1 − 0.444	8. $2\frac{1}{2} \div \frac{3}{4}$	9. 11,584 ÷ 64
10. \$4023.16 + \$385.89	11. $3\frac{5}{12} + 2\frac{2}{3}$	12. 72% of 900

9-5 Making Predictions

Objective
Make predictions based on probability.

At a local carnival a merchant donated a video cassette recorder for the charity raffle. The organizers sold 250 chances at \$2 each. Vicki bought 12 chances. Predict whether Vicki will win or not.

Understand the Situation

- How many chances were sold?
- What was the cost of a chance?
- How many chances did Vicki buy?
- How much did Vicki pay?

When you know the probability of an event's occurring, you may make reasonable predictions about the event. If the probability of the event's occurring is greater than the probability of the event's not occurring, it is reasonable to predict that the event will occur.

Examples

1. Either Vicki will win the VCR or someone else will. Predict whether Vicki will win the VCR. Vicki bought 12 of the 250 chances sold. The probability of Vicki's winning is $\frac{12}{250}$, or 4.8%.
The probability that someone else will win is $100 - 4.8\% = 95.2\%$.

Someone else will probably win the VCR. Vicki spent \$24 on a good cause but a bad risk.

2. A coaches' poll shows that the odds in favor of the Wildcats' winning the basketball championship are 5 to 2. Predict if the Wildcats will win. If the coaches are right, the probability of the Wildcats' winning is $\frac{5}{7}$. This means the probability of their not winning is $\frac{2}{7}$.

The Wildcats will probably win the championship.

Try This

a. Josh bought 15 chances on the VCR. Predict whether he will win.

b. How much would you have to spend to predict your winning the VCR?

c. Raul must kick a field goal for his team to win the football game. Thus far he has made 28 out of 40. Predict whether he will make the field goal.

d. A coin is tossed. Predict whether it will be heads or tails.

Exercises

1. Jamie Selway must make a free throw in order for his team to win a basketball game. Thus far during the season Jamie has made 55 out of 85 free throws. Predict whether he will make the shot.
2. The weather forecast for a World Series game calls for a 60% chance of rain. Predict whether the game will be played or not.
3. At the carnival there is a booth for the bottle toss. People try to toss a coin into a milk bottle. The odds against a person's making a good toss are 49 to 1. Marcie tries the toss. Predict whether she will be successful.
4. The students at Hereford High are hoping that schools will be closed because of a snowstorm. The weather forecast calls for a 50% chance of snow. Predict whether schools will be closed.

Work with a Partner

5. A booth at the carnival is to be run by the high school sports boosters club. Juan, Meta, Donna, Chuck, and Kirsten volunteer to work. Only three students are needed. To select the students, each of the ten possible selections of three students is written on a slip of paper and placed in a box. One slip is selected. The students whose names are on the slip will work. Predict whether Donna will get to work at the carnival.

Number Sense

Following is a list of probabilities of events. Predict if the event will occur. Then rate your prediction on the scale from confident (C) to not confident (NC). Use probability lines.

6. The probability that the Giants will win is $\frac{4}{7}$.
7. The probability that it will rain is 55%.
8. The odds in favor of winning a contest are 7 to 5.
9. The odds against being selected as a member of the team are 3 to 8.
10. The probability that Jason will hit a home run is $\frac{1}{40}$.
11. The probability that your name will be drawn as a winner of a contest is 3%.

Talk Math

12. Explain how a prediction can be made directly from knowing the odds in favor of an event without first finding the probability.

Mixed Skills Review

1. 426×113
2. 63% of 890
3. $16.43 + $21.08 + $0.33
4. $\frac{9}{16} - \frac{3}{8}$
5. $5681 + 1989$
6. $1\frac{1}{8} + \frac{3}{16} + 2\frac{3}{4}$
7. $1\frac{7}{8} \div 2\frac{1}{12}$
8. $17{,}600 \div 55$
9. 85 is what percent of 400?
10. $1042 - 963$
11. 18.2×1.9
12. $16 − $7.29

9-6 Counting Problems—Tree Diagrams

Objective
List all possible outcomes using tree diagrams.

Tom Vasold manages the pizza stand at the beach. The different choices that Tom can prepare are posted on a sign.

Understand the Situation

- What are the choices for crust?
- What are the choices for toppings?
- What are the sizes that Tom sells?

Example

What are the different combinations of pizza that can be ordered with only one topping? To find all the different combinations, we use a **tree diagram.**

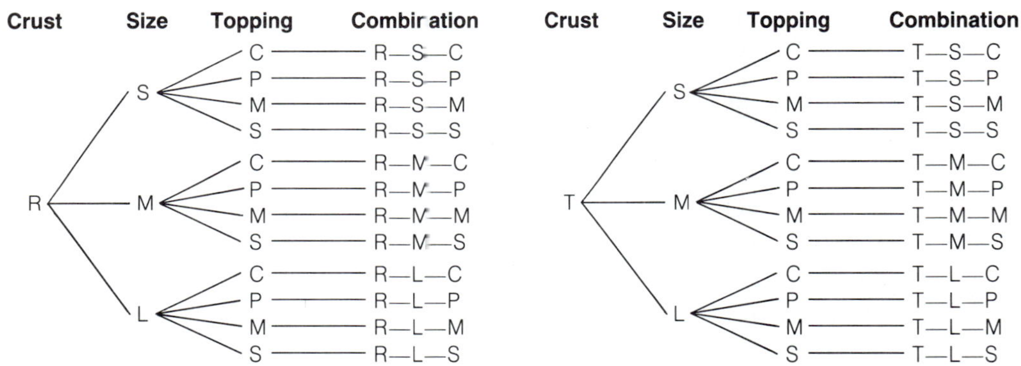

You can read every possible combination of Tom's pizza in the tree diagram. For instance, T-L-P means a large, thick-crust pizza with pepperoni topping.

Try This

Sidena manages the Texas Chili stand at the beach. She offers the following choices. Find all the different combinations of chili that can be ordered with 2 toppings. Use a tree diagram.

Size	Chili	Toppings
Cup	Regular	Cheese
Bowl	Hot	Sour cream
		Onions
		Jalapeño peppers
		Black olives

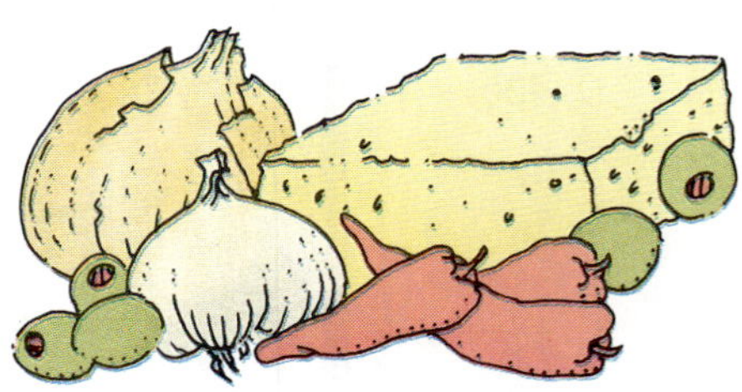

Exercises

Use tree diagrams to solve.

1. When visiting Thresher's Fries at the boardwalk, what are all the possible choices that you have for fries?
2. Herb Santucci runs the hoagie shop on the boardwalk. What are all the possible choices for hoagies that Herb sells?
3. When packing for a trip to the beach, Erin includes her blue bathing suit, her pink bathing suit, and her yellow bathing suit. She packs her beach sandals and her sneakers along with her sun visor and her straw beach hat. What are all the different beach outfits that Erin can use?
4. While at the beach Erin visits the Sun and Surf Shop. She finds that there are cover-ups available in sizes small, medium, and large. The cover-ups come in terry cloth or mesh and in white, yellow, or orange. What are all the different possible combinations for cover-ups?
5. The Goldstein family lives in Pikesville. To get to the beach they have the choice of several routes as shown on the map. What are all the possible ways for them to get to the beach?

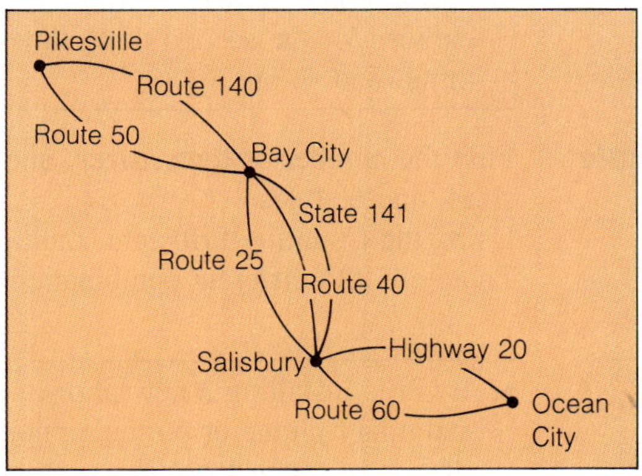

Suppose

6. How many possible round trips are there if the Goldsteins never go home exactly the same way they took to get to the beach?

Mixed Skills Review

1. $1.526 - 1.387$	2. 68×87	3. $1638 \div 39$
4. $0.32 + 8.697$	5. $\frac{13}{15} \times \frac{3}{4}$	6. $89{,}276 + 42{,}054$
7. $7210 - 5633$	8. $0.978 \div 0.06$	9. $5\frac{3}{7} + 2\frac{2}{3}$
10. 3.06×12.4	11. $6\frac{3}{10} - 4\frac{3}{8}$	12. 28 is what percent of 50?

9-7 Counting Problems—The Counting Principle

Objective
List all possible outcomes using the counting principle.

Allen uses a word processor in his work. His word processor gives him the choices of 6 different fonts, 9 font sizes, and 6 styles of print. How many different combinations of fonts, sizes, and styles can Allen use?

Understand the Situation

- How many different fonts are there?
- How many different font sizes are there?
- How many different styles of print are there?

To find all the different combinations, you could make a tree diagram. In the tree diagram there would be 6 different fonts. For each font there would be 9 sizes and for each size there would be 6 styles of print. This would be a very large tree diagram. Since we only want the number of combinations, we can multiply using **the Counting Principle.**

> To find the total number of combinations available in a series of choices, multiply the number of choices for the first thing times the number of choices for the second thing times the number of choices for the third thing, and so on.

Example

Find the number of fonts, sizes, and styles that Allen can use. 6 fonts, 9 sizes, and 6 styles
Using the Counting Principle, multiply. $6 \times 9 \times 6 = 324$
There are 324 different combinations of fonts, sizes, and styles.

Try This

a. Irwin is installing a new phone system in his office. He can choose from a stationary phone or portable phone and for each of these there is pulse dialing or touch-tone. For each type of dialing he can also choose redial, call waiting, or three-way calling. How many different choices for a phone system are there?

b. Kira has the following matching skirts, jackets, and blouses. How many different outfits can be formed?

Skirts	Jackets	Blouses
Tan	Blue	White
Gray	Striped	Pink
Blue		Yellow

Exercises

Use the Counting Principle to solve the following.

1. On a trip from Baltimore to Dallas to San Francisco, there are 3 possible flights from Baltimore to Dallas and 5 possible flights from Dallas to San Francisco. How many possible combinations of flights are there from Baltimore to San Francisco via Dallas?
2. Fred has 4 different sport coats, 3 pairs of trousers, and 5 shirts. How many days can Fred go without wearing the same outfit twice?
3. On a dinner menu there are 5 choices for the main dish, 4 choices for vegetables, 3 salad dressings, and 5 desserts. How many different dinner choices are possible?
4. At an ice cream shop there are 24 flavors of ice cream and 12 toppings. A sundae can be made with two scoops or three scoops. How many different sundae combinations are there?
5. At an Italian restaurant the pasta choices are spaghetti, linguini, fettuccine, and capelli d'angelo. The sauces available are marinara, meat sauce, clam sauce, and Alfredo sauce. These can be topped with parmesan cheese, locatelli cheese, or hot peppers. How many different combinations of pasta dinners are possible?
6. The automobile license plates in a certain state are made with three letters followed by three digits. How many different license plates are possible?
7. How many seven-digit telephone numbers can be made if the first and second digits cannot be 0 or 1?

Write a Problem

8. Write your own problem that can be solved using the Counting Principle.

Data Hunt

9. Check on the method for labeling license plates in your state. Find the total number of license plates that are possible.
10. If you used the letters rather than the digits of your telephone number, how many different combinations would there be? Find out from the telephone company how they handle special requests for numbers that spell out words or names.

Mixed Skills Review

1. 19% of 45	2. $7.2 - 3.758$	3. What percent is 75 of 50?
4. $\frac{19}{20} + \frac{4}{5} + \frac{1}{4}$	5. \$9040 + \$1674	6. $8 - \frac{7}{45}$
7. $14{,}402 \div 19$	8. $947 - 648$	9. $4\frac{1}{3} \times \frac{3}{5}$
10. $6.234 + 8.646$	11. 184×41	12. $80.8 \div 2.02$

9-8 Arrangements

Objective
Find all possible arrangements using permutations.

On a ski vacation Erin made three friends: Chris, Carley, and Mike. She decided to take a picture of her new friends. In how many ways can Chris, Carley, and Mike arrange themselves for the picture?

Understand the Situation

- How many new friends did Erin make?
- Where was Erin vacationing?
- Name one way that three friends can arrange themselves.

Example

1. Find the number of ways that Erin's friends can arrange themselves.
There are 3 choices for the first person in the picture. Then there are 2 choices for the second person. Finally, there is only 1 person remaining for the third space.

Using the Counting Principle, we have $3 \times 2 \times 1 = 6$.

There are 6 ways that Chris, Carley, and Mike can arrange themselves.

Try This

a. Erin, Chris, Carley, and Mike buy tickets for the ski lift. In how many different ways can they line up to buy tickets?

An arrangement of all the objects in a set is called a **permutation** of the objects. Multiply to find the total number of permutations of a set of n objects.
$n \times (n - 1) \times (n - 2) \times \ldots \times 1$

We use the symbol **n!** to represent $n \times (n - 1) \times (n - 2) \times \ldots \times 1$.
We read this symbol **n factorial.**

Example

2. There are four pairs of ski boots on a shelf in the ski shop. How many permutations of the four pairs are there?
There are 4! permutations of the four pairs.
$4! = 4 \times 3 \times 2 \times 1 = 24$
There are 24 permutations of the four pairs of boots.

Try This

b. There are six pairs of skis on a rack. How many permutations of the pairs of skis are there?

Exercises

1. Five people rented snowmobiles. In how many ways can the snowmobiles be arranged to travel one behind the other?
2. How many ways can 7 snowmobiles be parked in a row?
3. The ski lodge snack bar sells 8 kinds of hot drinks. In how many ways can the names of the 8 drinks be arranged in a column on a sign?
4. In the evening at the ski lodge a trivia game is organized. In this game players try to earn wedges by answering questions in 6 different subjects. There are 6 different colors of wedges. In how many different orders can the wedges be won by a single player?
5. Erin, Chris, Carley, and Mike ride the ski lift. Two people can sit in a seat on the lift. In how many ways can 2 seats be filled from the four friends? (Hint: How many choices for the first seat?)
6. How many different 3-digit numbers can be named by using the five digits 1, 2, 3, 4, and 5? Each digit may be used only once.

Talk Math

7. Ashlie says that 5! is the same as $5 \times 4!$. Is she correct? Explain.
8. In how many ways can 7 message flags be arranged on a ship's flagpole?

Suppose

9. The 7 flags in Exercise 8 are used to make 3-flag messages. How many different 3-flag messages are possible?
10. In baseball, how many different ways can the 9 players in the starting lineup be arranged for a batting order?
11. Coach Foster has a problem with the girls softball team. She has 5 excellent outfielders. All are good fielders and hit very well. In how many ways can Coach Foster select 3 outfielders?
12. What is the greatest factorial that you can find using your calculator?

Calculator

13. How many different ways can you line up all of the students in your math class? How can you use your answer to find the probability that everyone in the class will arrive in the classroom in the same order two days in a row?

Mixed Skills Review

1. $6\frac{1}{4} - 2\frac{5}{6}$	2. $16.008 - 13.635$	3. 93×1108
4. $2\frac{6}{7} + 2\frac{1}{5}$	5. $8459 + 2637$	6. 2.41×0.8
7. $1083 - 96$	8. $14.22 \div 4.5$	9. $3186 \div 54$
10. $1.365 + 7.377$	11. $\frac{14}{15} \div \frac{9}{10}$	12. $8 \times 1\frac{3}{5}$

9-9 Enrichment: Number Poles and Palindrome Numbers

Objective
Find the pole of a number. Create palindrome numbers.

You find **number poles** by repeated subtraction. Choose any 3-digit number such that at least 2 of the numbers are different. — 3-digit number: 318

Write the largest and smallest numbers possible with these digits. — largest: 831 smallest: 138

Subtract the smallest number from the largest number. — 831 − 138 = 693

Repeat the procedure. — largest: 963 smallest: 369; 963 − 369 = 594

Repeat the procedure again. — largest: 954 smallest: 459; 954 − 459 = 495

Since the digits in 495 are the same as those in 594, if we repeat the procedure again we will get 495. It is the number pole.

Palindrome numbers are the same if you read them forward or backward, like the number 1551. You find palindromes by repeated addition, as in the following example.

Choose a number with at least 3 digits. — 628

Reverse the digits. Add the numbers. — 628 + 826 = 1454

Is the number a palindrome? If not, reverse the digits and add. — 1454 + 4541 = 5995

5995 is a palindrome number.

Exercises

Use the procedure on the following 3-digit numbers. If the difference is a number with less than 3 digits, like 32 or 5, place zeroes in the missing places.

1. 824 **2.** 906 **3.** 751 **4.** 373 **5.** 625 **6.** 119

7. Write a statement about the pole of all 3-digit numbers in which at least 2 digits are different.

8. Do 2-digit numbers have a pole? Do 4-digit numbers have a pole?

Follow the procedure to make palindrome numbers.

9. 235 **10.** 1452 **11.** 7835

12. 3084 **13.** 888 **14.** 6492

9-10 Computer: Simulated Number Cube

Objective
Use a computer to simulate rolling a number cube.

We know that in rolling a 6-sided cube, the chance of rolling a "1" or a "3," for example, is 1 in 6, or 1/6. If we roll the cube 60 times, we will roll a "1" **approximately** 10 times, and a "6" **approximately** 10 times. The following computer program simulates rolling a 6-sided number cube.

```
10  REM Simulation of rolling a 6-sided number cube
20  INPUT "How many times do you want to roll"; N
30  LET A=0:B=0:C=0:D=0:E=0:F=0
40  PRINT "THE" N "VALUES ROLLED WERE:";
50  FOR X = 1 TO N
60  LET Z = INT(RND(1)*6)+1
70  IF Z = 1 THEN A = A + 1
80  IF Z = 2 THEN B = B + 1
90  IF Z = 3 THEN C = C + 1
100 IF Z = 4 THEN D = D + 1
110 IF Z = 5 THEN E = E + 1
120 IF Z = 6 THEN F = F + 1
130 PRINT Z " ";
140 NEXT X
150 PRINT:PRINT:PRINT TAB(7) "TOSS = ";TAB(15)"1";TAB(22)"2";
TAB(29)"3";TAB(36)"4";TAB(43)"5";TAB(50)"6"
160 PRINT " NO. TIMES =";TAB(14)A;TAB(21)B;TAB(28)C;TAB(35)D;TAB(41)E;
  TAB(48)F
170 END
```

The following runs show results for rolling the number cube.

```
1st run    THE 12 VALUES ROLLED WERE: 6 5 3 5 2 2 1 4 2 5 3 5
           TOSS      = 1 2 3 4 5 6
           NO. TIMES = 1 3 2 1 4 1

2nd run    THE 12 VALUES ROLLED WERE: 6 2 1 5 2 5 5 5 6 5 5 6
           TOSS      = 1 2 3 4 5 6
           NO. TIMES = 1 2 0 0 6 3
```

Exercises

1. Why are the results for the two 12-roll trials different?
2. Roll a number cube 12 times and compare your results to those above.
3. How can you change line 60 in the program to simulate the toss of a coin?
4. If it takes a computer 1 min 16 sec to "roll" a number cube 6000 times, how long will it take to "roll" it 120,000 times? 1,200,000 times?

CHAPTER 9 REVIEW

Match the following.

1.	probability	**A.**	compound probability
2.	4!	**B.**	$5 \times 4 \times 3 \times 2 \times 1$
3.	0.5	**C.**	probability of 0
4.	1 to 5	**D.**	24
5.	P(A then B)	**E.**	P(certain event)
6.	permutation	**F.**	odds when $P = \frac{1}{6}$
7.	an impossible event	**G.**	P(tossing a head)
8.	independent events	**H.**	$\frac{\text{favorable outcomes}}{\text{possible outcomes}}$
9.	1	**I.**	arrangement
10.	factorial	**J.**	rolling a die and spinning a spinner

Fill in the blanks for numbers 11–15. Choose from: probability, expectation, prediction, 0, 5 to 1, number, compound, independent.

11. Tree diagram is to a list of all possible combinations as counting principle is to the ______ of all combinations.

12. Factorial is to permutation as ratio is to ______.

13. A certain event is to 1 as an impossible event is to ______.

14. The odds in favor of an event are to 1 to 5 as the odds against the event are to ______.

15. $n \times (n - 1) \times (n - 2) \times \ldots \times 1$ is to factorial as probability $\times$ number of possibilities is to ______.

There are 5 red marbles, 3 blue marbles, 6 green marbles, and 2 black marbles in a jar.

16. If one marble of each color is selected from the jar, in how many ways can the marbles be arranged in a line?

17. Arrange these in order from least to greatest: P(green), 0.5, P(red), $\frac{3}{4}$, 3!.

18. What is the probability of drawing a black marble and then drawing a blue marble without replacing the black one?

19. Draw 8 marbles from the jar. How many green ones can you expect?

CHAPTER 9 TEST

1. Each of the 3 suits (gold, silver, and ruby) in a 36-card deck contains a 2, 3, 4, 5, 6, 7, 8, sphere, cube, cone, tetrahedron, and hyperlozenge. If you pick one card from this deck, what is the probability that you will pick a tetrahedron? a ruby? a card greater than 7? the silver four? a gold hyperlozenge or any cube?
2. In a dart game, Tim hit the bull's-eye 3 times in the last 30 tries. What are the odds against his hitting the bull's-eye?
3. Jar A contains 6 red marbles and 4 green marbles. Jar B contains 3 red marbles, 6 green marbles, and 6 blue marbles. What is the probability of picking a red marble from both jars? a green marble from both jars? a red marble from Jar A together with a blue marble from Jar B?
4. Suppose that the probability of rain in the desert is $\frac{1}{100}$. About how many rainy days can be expected from June through September?
5. Arnold won 12 out of 18 wrestling matches this season. Predict if he will win his next match or not.
6. The carnival had a basketball toss game. Of the last 52 people who played, 13 made baskets. Predict if the next person will make a basket.
7. Joe wants to buy a new bowling ball. They are made of rubber, urethane, or plastic. The colors are green, red, and black. The textures are speckled, swirls, and solid. How many choices does Joe have?
8. Ted's flower shop has the following selection: daisies, roses, and tulips. The daisies come in white and yellow. The roses and the tulips each come in both red and yellow. How many different kinds of flowers does Ted have for sale?
9. Margaret is setting tables for the school picnic. The plastic silverware is blue, red, and yellow. The napkins are red, blue, green, and white. The plates are green, yellow, purple, and orange. How many different place settings can Margaret set up?
10. Mr. Taylor was shopping for a new convertible. They come in red, white, blue, green, and tan. The interiors come in black, white, and tan. The convertible tops come in white, tan, and black. How many different convertible color combinations are there?
11. The students at Arnold High built six floats for the county parade. In how many ways can the floats travel one behind the other?

CHAPTER 9 CUMULATIVE REVIEW

1. The weather report says there is a 20% chance of rain. What are the odds against rain?
   A. 4 to 1
   B. 1 to 2
   C. 5 to 1
   D. 1 to 5

2. Find the repayment. principal $8500, rate 15.75%, time 1.5 years Use *Repayment* = $p + prt$.
   A. $10,635.63
   B. $9,923.75
   C. $2135.63
   D. $1423.75

3. Find the rental fee for a floor sander. $1.35 per hour for 7 hours
   A. $7.45
   B. $7.15
   C. $9.15
   D. $9.45

4. In the sophomore class raffle each ticket has a 12.5% chance of winning. If Mai-Ling buys 16 tickets, how many prizes can she expect to win?
   A. 6.5
   B. 8
   C. 2
   D. 1

5. Eileen bought 3 tubes of paint for $2.75 each, 2 sheets of paper for $1.80 each, and 4 brushes for $3.95 each. She gave the clerk $40. What was her change?
   A. $12.35
   B. $1.55
   C. $2.45
   D. $3.15

6. Find the new balance. finance charge 1.9% of the average daily balance, unpaid balance 12/1: $269.50, payment 12/16: $150
   A. $227.05
   B. $123.15
   C. $121.77
   D. $119.50

7. Four friends are playing tennis. How many possible combinations of partners are there?
   A. 2
   B. 12
   C. 3
   D. 5

8. Find the amount after 2 years. $600 principal, 8% interest, compounded semiannually Use $i = prt$.
   A. $648.96
   B. $674.96
   C. $624
   D. $701.92

9. Add. $3\frac{3}{8} + \frac{1}{4} + 2\frac{1}{2}$
   A. 6
   B. $5\frac{5}{8}$
   C. $6\frac{1}{8}$
   D. 5

10. Write as a percent. 0.137
    A. 1.37%
    B. $1\frac{37}{100}\%$
    C. 13.7%
    D. 137%

11. What is the median score on the math test? 82, 67, 75, 90, 98, 86, 63, 76, 67
    A. 67
    B. 78
    C. 76
    D. 90

12. Subtract. $7\frac{3}{8} - 2\frac{5}{6}$
    A. $4\frac{1}{3}$
    B. $4\frac{13}{24}$
    C. $5\frac{13}{24}$
    D. $5\frac{17}{24}$

13. Mr. Samson has 3 red ties, 7 green ties, and 1 white tie. What is the probability that he will select a blue tie out of his dark closet?
   A. $\frac{5}{16}$
   B. $\frac{5}{11}$
   C. $\frac{1}{3}$
   D. 0

14. For his new job at CompuLab, Tom bought 1 pair each of tan, blue, and black slacks; 1 each plaid, white, blue, striped, and tan shirt; 1 pair each brown and black shoes. How many possible outfits does he have?
   A. 45
   B. 10
   C. 30
   D. 15

15. Doris takes one piece of fruit out of a bowl containing 6 limes and 4 oranges. She also picks one nut out of a bowl containing 7 cashews and 8 walnuts. What is the probability that she will pick a lime and a cashew?
   A. $\frac{14}{25}$
   B. $\frac{7}{25}$
   C. $\frac{6}{15}$
   D. $\frac{13}{25}$

16. Find the median of these math book prices.
   \$20, \$18, \$21, \$15, \$17, \$22, \$12, \$22
   A. \$12
   B. \$22
   C. \$18
   D. \$19

17. Ms. Sanchez says there is a 45% chance of a test on Friday. Predict if there will be a test.
   A. yes
   B. no

18. Michelle has better grades in math and science than in English and history. Which job might she prefer?
   A. gossip columnist
   B. lab technician
   C. proofreader
   D. ad writer

19. An automobile seat comes in 3 styles, 5 colors, and 4 choices of fabric. How many combinations are possible?
   A. 60
   B. 20
   C. 15
   D. 45

20. The Panthers have won 12 games and lost 5. The Gorillas have won 11 and lost 8. The Gorillas are how many games back of the Panthers?
   Use $G = \frac{dW + dL}{2}$.
   A. 5
   B. 2
   C. 3
   D. 1

21. Find the quotient. Use mental math.
   $1605 \div 5$
   A. 313
   B. 33
   C. 301
   D. 321

Chapter 10 Overview

Key Ideas

- finding calorie intake
- comparing calorie counts
- finding calories used
- using a meal planning guide
- finding catering fees
- comparing dining costs
- comparing food production figures

Key Terms

- calorie
- ideal body weight
- calorie intake
- serving size
- trace
- carbohydrates
- protein
- vitamin
- RDA
- nutritionist
- energy allowance
- catering
- service factor
- acre

Key Skills

Find the answer.

1. 160 + 110 + 85
2. 190 + 135 + 80 + 125
3. 65 + 125 + 80
4. \$25.98 − \$18.67
5. \$45.19 − \$27.63
6. \$62.19 − \$46.28
7. 9.2×200
8. 19.3×200
9. 8.7×500
10. $\frac{3}{4} \times 426$
11. $\frac{5}{12} \times 642$
12. $\frac{2}{3} \times 644$
13. $2\frac{1}{2} \times 24$
14. $3\frac{1}{2} \times 36$
15. $5\frac{1}{4} \times 28$
16. $232 \div 22$
17. $195 \div 22$
18. $198 \div 22$
19. 3 × \$1.89
20. 4 × \$2.27
21. 18 × \$1.38

Solve.

22. 2% of 90 = n
23. n = 4% of 80
24. 25% of 120 = n
25. What is 10% of 95?
26. What is 100% of 87?
27. What is 2% of 60?

Simplify.

28. $\frac{12}{60}$
29. $\frac{35}{60}$
30. $\frac{27}{60}$
31. $\frac{105}{300}$
32. $\frac{115}{23}$
33. $\frac{1024}{256}$

FOOD AND DIET

10

10-1 Calculating Calorie Intake

Objective *Calculate daily calorie intake.*

Good health requires proper diet, exercise, and rest. One way to insure a proper balance is to measure the energy we store in our bodies when we eat and the energy we burn when we exercise. A **calorie** is a unit of heat used to measure the energy values of food and exercise.

Dr. Suh Kang Kim has developed a weight program for teenagers. She uses this chart to calculate the number of calories her patients need each day to maintain their ideal body weight.

First 22 pounds of patient's weight	1000 calories
Next 22 pounds	500 calories
Each additional 22 pounds	200 calories
Active athlete	500 calories

Examples

1. Bill Gordon plays football at 225 pounds. Find the number of calories Bill needs to maintain his weight.

Calculate how many multiples of 22 pounds are in Bill's body weight. Add the number of calories needed per day for each 22 pounds of weight.

$225 \div 22$ pounds $= 10.2$		*round to the nearest 0.1*
	1000 calories	*first 22 pounds*
	500 calories	*next 22 pounds*
$10.2 - 2 = 8.2 \times 200 =$	1640 calories	*remaining 8.2×22 pounds*
	500 calories	*active athlete*
Total	3640 calories per day to maintain Bill's body weight	

Dr. Kim advises that weight should be gained or lost gradually. She adds 500 calories per day for weight gain and subtracts 500 calories per day for weight loss.

2. Bill wants to lose some weight for wrestling. How many calories does Bill need each day to begin losing weight? Deduct 500 calories from the amount needed to maintain the current weight.
$3640 - 500 = 3140$ calories

Try This Find the recommended daily calorie intake.

		Current Weight	Weight Goal	Sport
a.	Doug Tyree	185 lb	185 lb	Baseball
b.	Betty Couper	140 lb	125 lb	Tennis
c.	Jack Sprat	98 lb	140 lb	None

Exercises

Find the recommended daily calorie intake.

		Current Weight	Weight Goal	Sport
1.	Patty Robinson	101 lb	101 lb	Cheerleading
2.	Pablo Martinez	135 lb	135 lb	Track
3.	Ryland Vest	215 lb	198 lb	Wrestling
4.	Francine Roberts	97 lb	109 lb	Soccer
5.	Joan Switzer	98 lb	96 lb	Jockey
6.	Mark Smith	145 lb	148 lb	Scuba diving
7.	Cindy Hoyt	165 lb	130 lb	None
8.	Mike Ryan	107 lb	128 lb	None

Strategy Practice

9. Margarita Ross is a ballerina and watches her diet carefully. She eats 2750 calories per day. How much weight will this calorie intake maintain?

Data Hunt

10. Talk to your physician about your weight.

a. Find the desired weight for a teenager of your age and height.

b. What is your recommended daily calorie intake? Learn how your doctor computes this number.

Mixed Skills Review

1. $14.2 - 8.026$
2. $8635 + 9461$
3. 1218×43
4. $7481 - 5937$
5. 1.18×4304
6. $6\frac{1}{3} - 3\frac{7}{8}$
7. $6.835 + 3.547$
8. $1840 \div 24$
9. $\frac{5}{9} + 4\frac{1}{4}$
10. $2751 \div 3.5$
11. $\frac{8}{15} \times \frac{3}{16}$
12. $9 \div 1\frac{1}{2}$

10-2 Counting Calories

Objective
Calculate and compare calorie and nutrient counts in foods.

We count the calories in the food we eat to determine our daily total. To maintain good health, we look for foods that contain the proper amounts of nutrients. The chart on p. 449 in the Data Bank lists many familiar foods and their nutrient contents.

Calories Per Serving		
1 pancake	60 calories	$3 \times 60 = 180$ calories
1 tablespoon syrup	60 calories	$2 \times 60 = 120$ calories
2 slices bacon	100 calories	$\frac{3}{2} \times 100 = 150$ calories
1 cup milk, nonfat	90 calories	= 90 calories
		Total = 540 calories

Example

1. Find the total number of calories.
3 pancakes, 4″ in diameter   3 slices bacon   2 tbsp syrup   1 cup milk, nonfat
Use the chart to find the number of calories per serving. Multiply the number of calories by the serving size. Add to find the total.

This meal contains 540 calories.

Try This

Find the number of calories in each meal.

a. roast beef, 3 oz
$\frac{1}{2}$ c mashed potatoes
$\frac{1}{2}$ c green beans
1 c milk, whole
1 medium apple

b. 2 lamb chops, 4.8 oz each
$\frac{3}{4}$ c rice
$\frac{2}{3}$ c Brussels sprouts
1 navel orange
1 c tea

Example

2. Compare the calories and fat in 1 cup of macaroni and cheese to the calories and fat in 1 cup of spaghetti.

Macaroni and cheese	470 calories	24 grams of fat
Spaghetti	−260 calories	− 9 grams of fat
	210 calories	15 grams

One cup of macaroni and cheese contains 210 more calories and 15 more grams of fat than 1 cup of spaghetti.

Try This

Solve.

c. Compare the number of calories and the amounts of carbohydrates in 1 cup of rice and 1 baked potato.

Exercises

Find the total number of calories in each meal.

1. 1 c orange juice
1 oz corn flakes
1 c whole milk
2 slices whole wheat bread
3 tbsp jam
1 tbsp peanut butter

2. 2 oz boiled ham
2 slices white bread
2 large leaves lettuce
$1\frac{2}{3}$ oz processed cheese
1 c raisins
1 c milk, nonfat

Solve.

3. Compare the number of calories and the amounts of carbohydrates in 1 cup of raisins and 1 raw apple.

4. Compare the number of calories and the amounts of carbohydrates in 1 pat of margarine and 1 tablespoon of jam.

5. Compare the number of calories and the amounts of protein in $\frac{1}{2}$ fried chicken breast and 1 pork chop.

6. Compare the number of calories and the amounts of protein in 1 oz. processed cheese and 1 tbsp. peanut butter.

7. Compare the number of calories and the amounts of fat in 1 cup mashed potatoes and 1 baked potato.

8. Compare the number of calories and the amounts of fat in 1 cup whole milk and 1 cup nonfat milk.

Mental Math

9. Researchers asked 500 people what they look for when they compare foods. Estimate the numbers of people who responded *yes* in each category.

[ ] Less sugar 20%
[ ] Less salt 19%
[ ] Fewer chemical additives 18%
[ ] High vitamin/mineral content 17%
[ ] Nutritional value 14%
[ ] No preservatives 13%
[ ] Low fat content 13%
[ ] Cost per unit of measure 87%

Mixed Skills Review

1. 24,418 + 63,594
2. $0.67 × 24
3. 8.75 is what percent of 35?
4. 4400 ÷ 16
5. $6\frac{2}{5} + 2\frac{7}{8}$
6. $12.75 + $13.80 + $6.32
7. $20 - 12\frac{3}{7}$
8. $\frac{16}{25} \times 1\frac{3}{4}$
9. 7291 − 436
10. 6% of 82.5
11. $35 − $19.22
12. $3\frac{9}{10} \div 3\frac{1}{4}$

10-3 Using Calories

Objective *Calculate the number of calories used in physical activities.*

Exercise uses energy from the foods we eat. Strenuous activities require more energy and therefore burn more calories. The chart below lists the approximate number of calories used per hour for different activities. The actual number of calories burned will vary with the weight of the individual.

Activity	Calories Used (per hour)	Activity	Calories Used (per hour)
Badminton	374	Rowing	672
Basketball	420	Skating	690
Bicycling	332	Skiing, downhill	576
Calisthenics	300	water	438
Golfing	300	Swimming	660
Handball	612	Tennis	426
Horseback riding	198	Volleyball	374
Jogging	648	Walking, outdoors	366
Paddleball	540	Wrestling	774
Racquetball	632	Running, 6 mph	726

Example Find the total number of calories burned.

Walter Noell ran 6 mph for 45 minutes.

Convert the exercise time to hours.

$\frac{45 \text{ minutes}}{60 \text{ minutes}} = \frac{3}{4}$ hour

Find the number of calories used per hour in the chart. Multiply by the number of hours of activity.

Running at 6 mph burns 726 calories per hour.

$\frac{3}{4} \times 726 = 544\frac{1}{2} \approx 545$ calories

Walter burns 545 calories running at a rate of 6 mph for 45 minutes.

Try This Find the total number of calories burned.

a. Corinne Strouse practiced ice skating 50 minutes.

b. Al Korman walks outdoors at a rate of 3 mph. Tuesday he walked 55 minutes in the morning and 1 hour 15 minutes in the evening.

c. Clay Basket rides her Appaloosa for 2 hours in the evening.

Exercises

Find the total number of calories burned.

1. Sarah Rogers played tennis for 35 minutes.
2. Donna Rochin practiced badminton for 52 minutes.
3. A marathon participant ran at a rate of 6 mph for 1 hour 50 minutes before stopping due to a leg cramp.
4. A cyclist rode 4 hours 15 minutes preparing for an upcoming race.
5. In a match David played golf for 45 minutes before it rained. Later he played an additional 2 hours 25 minutes.
6. In a volleyball tournament Naoku played in matches of 1 hour 38 minutes and 1 hour 12 minutes.
7. A member of the York Manor Swim Team practiced 1 hour 25 minutes in the morning and 55 minutes in the afternoon.
8. Greg waterskied for 1 hour 27 minutes and again later for 1 hour 13 minutes.
9. Ceil went horseback riding for 1 hour 45 minutes and then walked outdoors for 30 minutes.

Calculator Solve. Use a calculator.

10. Walking at a moderate pace burns 1.5 calories per hour per pound of weight. How long must a 150-pound man walk to burn 900 calories?
11. Doing homework burns 0.8 calories per hour per pound of weight. How many calories does a 135-pound student burn in $2\frac{1}{2}$ hours of study?
12. Strenuous exercise burns 2.5 calories per hour per pound of weight. How many calories does a 170-pound weightlifter burn in 40 minutes of practice?
13. Eight hours of sleep burns 432 calories for a 108-pound sleeper. How many calories per hour does sleeping burn for each pound of weight?
14. Make a chart of the five most strenuous sports listed. Calculate the calories used per minute for each sport.

Mixed Skills Review

1. $8\frac{3}{14} + 9\frac{1}{2}$	2. \$0.37 + \$0.18 + \$0.41	3. $2 - 0.76$
4. 49×626	5. $4845 \div 51$	6. 3.5% of what number is 1.75?
7. $16{,}235 + 7644$	8. $6\frac{2}{3} \times 1\frac{3}{5}$	9. \$8807 − \$7638
10. $3\frac{3}{7} \div \frac{8}{21}$	11. 26% of 482	12. $1\frac{1}{8} - \frac{15}{16}$

10-4 Reading Labels

Nutrition Information per Serving	
Serving Size	$7\frac{3}{8}$ oz (209 g)
Servings per container	$3\frac{1}{2}$
Calories	220
Protein (grams)	9
Carbohydrates (grams)	27

Objective
Find nutrient count using product labels.

This label shows the nutrients in a can of spaghetti. The label shows the number of calories and the amount of protein and carbohydrates per serving.

Example

1. Find the total number of grams of carbohydrates in a can of spaghetti.

Multiply carbohydrates in grams per serving by the number of servings.

$3\frac{1}{2} \times 27 = \frac{7}{2} \times 27 = \frac{189}{2} = 94\frac{1}{2}$ g

Try This

Use the label to find the total amount of the following nutrients in the can.

a. calories **b.** protein

The chart on p. 449 in the Data Bank shows U.S. **Recommended Daily Allowances (RDA)** for teenagers aged 15–18.

Example

2. Use the RDA chart and this label from a box of raisin bran. Find the amount of protein in one serving of cereal with whole milk.

The RDA of protein is 85 grams. One serving of raisin bran with whole milk contains 15% of the RDA of protein.

Multiply the RDA of protein by the percent found in one serving.

0.15×85 grams $= 12.75$ grams

One serving contains 12.75 grams of protein.

Percentages of U.S. Recommended Daily Allowances (RDA)

	Cereal With Raisins	With Skim Milk	With Whole Milk
Protein	4	15	15
Calcium	*	15	15
Iron	100	100	100
Vitamin A	25	30	30
Niacin	25	25	25
Vitamin C	*	2	2
Vitamin D	10	25	25

*contains less than 2%

Try This

Find the amount of each nutrient in one serving of green beans.

c. Iron

d. Vitamin B_1

e. Protein

Green Beans

Protein	*	Vitamin B_2	2
Vitamin A	4	Niacin	2
Vitamin C	4	Calcium	2
Vitamin B_1	2	Iron	4

Exercises

Use the label to find the total amount of nutrients in each package of pancake mix and vegetable soup.

1. calories
2. protein
3. carbohydrates
4. fat
5. sodium

Vegetable Soup	
Serving Size: 8 oz (226 grams)	
Servings per Container	$2\frac{1}{2}$
Calories	50
Protein (g)	0
Carbohydrates (g)	9
Fat (g)	0
Sodium (mg)	630

Find the amount of each nutrient in one serving of pancake mix.

6. protein
7. Vitamin A
8. Vitamin C
9. Vitamin B_1
10. Vitamin B_2
11. niacin

Pancake Mix	
Serving Size: three 4-inch pancakes	
Servings per Container	26
Mix plus egg, oil, and milk	
Calories	210
Protein (g)	6
Carbohydrates (g)	30
Fat (g)	7
Sodium (mg)	490

Mental Math Your doctor recommends a maximum daily intake of 1500 milligrams of sodium. What part of the recommended total is found in one serving of the following foods? Use compatible numbers.

12. spaghetti
13. vegetable soup
14. pancakes

Percentages of U.S. Recommended Daily Allowances (U.S. RDA)

Pancake Mix	
Protein (g)	10
Vitamin A	4
Vitamin C	*
Vitamin B_1	15
Vitamin B_2	10
Niacin	6
Sodium (mg)	490

*contains less than 2%

Data Hunt 15. Bring a label to class and answer 1–11 above.

Mixed Skills Review

1. 87×102
2. $5\frac{2}{3} - 4\frac{5}{9}$
3. $406 + 873 + 638$
4. $1349 \div 71$
5. $2193 - 808$
6. $12.753 + 6.068$
7. $\frac{6}{7} \times 2\frac{1}{3}$
8. $11.3 - 4.23$
9. 9.6 is what percent of 60?
10. 0.15×8.3
11. $9\frac{1}{2} + 6\frac{5}{6}$
12. $1\frac{7}{9} \div \frac{2}{3}$

10-5 Diet Planning

Objective
Use a meal planning guide.

Nutritionists use meal planning guides to prepare daily diets.

Meal Planning Guide

Serving Size Food Group	Provision	12–19 yrs M	12–19 yrs F	20–50 yrs M	20–50 yrs F	over 50 yr M	over 50 yr F
Meats, fish or seafood, eggs, beans, peas, or nuts	200–300 cal 12–20 g protein	2	2	2	2	2	2
Milk products	150–200 cal 250–300 mg calcium 8 g protein	4	4	3	3	3	3
Breads and cereals, whole grains or enriched	70 cal 2 g protein	6	6	4	4	4	4
Dark green and deep yellow fruits and vegetables	100–200 cal 3000 IU Vitamin A	1	1	1	1	1	1
Citrus fruits	100 cal 30 mg Vitamin C	1	1	1	1	1	1
Other fruits and vegetables	100–200 cal carbohydrates, fiber, minerals, and vitamins	2	2	2	2	3	3
Essential fats	40–70 cal 1 g fatty acid	3	$2\frac{1}{2}$	3	2	2	2
Supplemental sources of energy *as needed for total	150–300 cal	*	*	*	*	*	*
Energy allowances (cal)		3000	2250	2700	2000	2400	1800

Example Sam Broadbelt is 25 years old. What is his total energy allowance in the milk group? Use mental math.

A male, age 25, should plan 3 servings per day from the milk group.
Each serving should contain 150–200 calories.
3 servings × 150 calories = 450 calories. 3 servings × 200 calories = 600 calories
The total energy allowance is 450–600 calories.

Try This Find the total energy allowance and nutrients from the given food group.

	Person	Age	Food Group	Allowance	Nutrients
a.	Diane Diven	18	Citrus Fruit		
b.	Eric Tucker	13	Meat and Fish		
c.	Wayne Miller	54	Bread		

Exercises

Name the food group for each food.

1. potatoes **2.** bran flakes **3.** spinach **4.** yogurt

5. oranges **6.** perch **7.** green beans **8.** peanuts

Find the total energy allowance and nutrients from the given food group.

	Person	Age	Food Group	Allowance	Nutrients
9.	Laura Baker	15	Other fruits, vegetables		
10.	Aki Adams	12	Bread		
11.	Randy Dodd	28	Citrus fruit		
12.	Carol Hayes	47	Milk		
13.	Tom Gill	32	Meat		
14.	Isabel Cohen	55	Fats		
15.	John Baxter	62	Bread		
16.	Sue Small	21	Dark fruits, vegetables		

17. Write a daily meal plan for yourself. List each food group and the number of servings. Name one food in each group.

The average American ate 337 pounds of vegetables in 1985.

Vegetables (lb/person)	1983	1984	1985
Fresh (excluding potatoes)	75	80	81
Canned (excluding potatoes)	38	47	43
Frozen (excluding potatoes)	8	8	9
Potatoes	120	119	121
Sweet potatoes	5	5	5

18. How many pounds of vegetables did the average person eat in 1983?

19. What was the percent of increase from 1983 to 1985?

20. Estimate what part of the vegetables eaten in 1985 were fresh.

21. What part of the vegetables eaten in 1985 were frozen?

Mixed Skills Review

1. $901 \div 53$ **2.** $3\frac{1}{2} \times 2\frac{1}{3}$ **3.** $3972 + 48 + 73$

4. $1\frac{11}{16} - 1\frac{1}{2}$ **5.** 0.036×105 **6.** $1800 - 1256$

7. 88×64 **8.** $2\frac{7}{10} \div 1\frac{4}{5}$ **9.** $163.4 + 216.9 + 0.4$

10. $0.78 \div 0.065$ **11.** $\frac{2}{3} + \frac{5}{8} + 1\frac{1}{4}$ **12.** $3 - 1.407$

10-6 Food Catering

Objective *Calculate fees and recipe amounts from tables.*

Julee Williams owns and operates Complete Caterers, Inc. Julee and her staff help customers plan the menus, decorations, table service and linen, and even the number of tables and chairs for parties and special events.

Each time Julee caters an event, she begins with a plan for 24 guests. Then she makes changes in the plan based on the number of invited guests. Julee uses a worksheet like this to calculate the costs.

Number of guests ____			Service factor = Number of guests/24			**Country Buffet Menu**
Food and Beverage Service			**Table Service**			Country ham and biscuits
Country Buffet	$216	____	Tables w/chairs	$34	____	Smoked turkey
Deluxe Service	310	____	Centerpieces	31	____	Cold rice salad
Gourmet Gala	474	____	Candles w/sticks	7	____	Broccoli salad vinaigrette
			Linen	18	____	Scalloped pineapple
Menu cost		____	China and silver	36	____	Assorted breads
Rental fees		____	Breakage insurance	11	____	Banana nut cake
Total	____		Rental fees	____		Beverages
Total × factor		____				

Examples

1. One of Julee's clients is planning a reception for 18 people. Find the rental fee for centerpieces.

Multiply the rate by the service factor. $\frac{18}{24} \times \$31 = \$23.25$

The rental fee for a centerpiece for a party of 18 is $23.25.

2. Find the total cost to cater Deluxe Service for 54 with linen and candles.

Add the cost for each item. Then multiply by the service factor.

$\$310 + \$18 + \$7 = \$335$

Julee uses a calculator to compute costs. $\frac{54}{24} \times \$335$

The cost of catering for 54 is $753.75.

Try This

Find the cost to cater the following.

a. Country Buffet for 20
China and silver
Insurance

b. Gourmet Gala for 40
Tables
Centerpieces

Exercises

Find the rental fee.

1. Linen for 32
2. Candles for 96
3. Insurance for 80
4. Tables for 8
5. China and silver for 36
6. Centerpieces for 56
7. Linen for 150
8. Tables for 180
9. China and silver for 200

Find the menu cost.

10. Country Buffet for 9
11. Deluxe Service for 50
12. Gourmet Gala for 20
13. Country Buffet for 33
14. Gourmet Gala for 15
15. Deluxe Service for 100
16. Country Buffet for 150
17. Gourmet Gala for 300

Find the cost to cater the following.

18. Country Buffet for 18
    Tables and chairs
    Linen
19. Deluxe Service for 40
    China and silver
    Insurance
20. Deluxe Service for 84
    Tables
    Linen
    Candles
21. Gourmet Gala for 8
    Linen
    Centerpieces
    Candles
22. Country Buffet for 60
    Tables and chairs
    China and silver
    Linen
    Insurance
23. Gourmet Gala for 44
    Linen
    Centerpieces
    China and silver
    Insurance

Strategy Practice

Find the maximum number of guests possible on the given budget.

24. Budget $400
    Country Buffet
    Tables
    Linen
25. Budget $800
    Gourmet Gala
    Centerpieces
    Candles
26. Budget $650
    Deluxe Service
    China and silver
    Insurance

Mixed Skills Review

1. $\frac{8}{11} \times 1\frac{2}{3}$
2. $902 - 18$
3. $1.56 \div 0.312$
4. $6\frac{5}{8} - 6\frac{3}{7}$
5. $3.8 + 0.26 + 4.6$
6. 15.6×9.4
7. $\$29.95 - \$18.39$
8. $29 + 8 + 15 + 64$
9. $11{,}745 \div 29$
10. $2\frac{2}{3} + \frac{1}{2} + 1\frac{1}{3}$
11. 1743×16
12. $7 \div 1\frac{3}{4}$

10-7 Comparing Costs

Objective *Compare costs of dining out and at home.*

Bill and Karen Anderson enjoy both dining out in restaurants and preparing meals at home. They save on their food budget by raising vegetables in a backyard garden, buying some items in large quantities, and comparing prices in different markets.

Example Bill and Karen like shrimp scampi. Compare the cost for each meal.

Restaurant Comparison

Restaurant		Prepared at Home	
Dinner	$9.95 each	$\frac{3}{4}$ lb shrimp	$8.50/lb
Beverage	$1.20 each	Seasonings	$1.65
Sales tax	5%	Salad ingredients	$0.75
Tip	15%	Beverage	$0.45

Find the total restaurant cost.

Dinner	$2 \times \$9.95 =$	$19.90
Beverage	$2 \times \$1.20 =$	$2.40
		$22.30
Sales Tax	$0.05 \times \$22.30 \approx$	$1.12
Tip	$0.15 \times \$22.30 \approx$	$3.35
		$26.77

Find the cost of preparing the meal at home.

Shrimp $\frac{3}{4} \times \$8.50 \approx$	$6.38
Seasonings	$1.65
Salad	$0.75
Beverage	$0.45
	$9.23

Find the difference. $\$26.77 - \$9.23 = \$17.54$

The Andersons save $17.54 by preparing the meal at home.

Try This Compare the costs.

a.

Gardening Expenses	
3 packets seeds	$0.79/pkt.
Tiller rental	$6.60
Yield	
9 dozen ears corn	
Supermarket Costs	$2.25/dozen

b.

Pizza Carry-Out	$8.25
Beverage for 4	$2.20
Sales tax	6%
Pizza Prepared at Home	
Crust	$1.29
Sauce	$0.89
Cheese	$0.75
Beverage for 4	$0.80

Exercises

Compare the costs.

1.

Cucumbers, Summer Squash			
Gardening Expenses		**Yield**	**Supermarket Costs**
7 market packs	\$1.69/pk.	96 cucumbers	\$0.35 each
Fertilizer	\$1.05	84 summer squash	\$0.49 each

2.

Peas, Lettuce			
Gardening Expenses		**Yield**	**Supermarket Costs**
4 packets seeds	\$0.89/pkt.	45 lb peas	\$0.95/lb
Water	\$3.72	18 hd. lettuce	\$0.69/hd.

3.

Carrots, Broccoli, Potatoes			
Gardening Expenses		**Yield**	**Supermarket Costs**
3 packets seeds	\$0.59/pkt.	15 lb carrots	\$0.35/lb
2 market packs	\$1.90/pk.	24 hd. broccoli	\$1.49/hd.
2 lb seed potatoes	\$1.05	34 lb potatoes	\$0.75/lb

4.

Chicken Carry-out		Chicken Prepared at Home	
Bucket of 9 pieces	\$5.96	$1\frac{1}{4}$ lb chicken	\$1.19/lb
Coleslaw for 3	\$1.29	$\frac{1}{3}$ hd. cabbage	\$0.87/hd.
Beverage for 3	\$1.74	Seasonings	\$0.38
Sales tax	4%	Beverage for 3	\$0.75

Estimation

5. An estimated 18 million families grow fruits and vegetables in backyard gardens. 64% grow cucumbers. How many gardening households grow cucumbers?

6. In a survey of 800 people, 23% of those polled chose Chinese food as their favorite ethnic food. How many people picked Chinese food?

Mixed Skills Review

1. $2116 \div 92$
2. $26.91 \div 0.23$
3. $2.08 - 0.936$
4. $11\frac{1}{3} \times \frac{9}{17}$
5. $20{,}003 - 11{,}497$
6. $0.3 + 0.8 + 0.6 + 0.5$
7. 42×0.89
8. $1\frac{6}{7} \div \frac{8}{21}$
9. $9\frac{2}{3} + 3\frac{7}{10}$
10. 1161×47
11. $16\frac{1}{2} - 4\frac{5}{7}$
12. $23{,}814 + 62{,}518$

10-8 Food Production

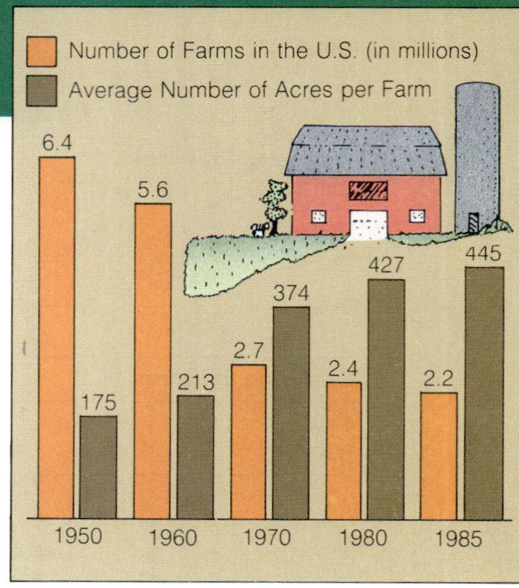

Objective
Compare food production figures.

Two hundred years ago, almost 90% of the people in the United States worked on a farm. Today Americans who live or work on a farm account for only 2.4% of the population.

Understand the Situation

- How many farms were there in 1950? in 1980?
- What was the average number of acres per farm in 1950? in 1980?

Example

1. Compare the total acres farmed for the years 1980 and 1985. Multiply the number of farms by the average number of acres per farm for each year. Compare the results.

1980 → 427 acres/farm × 2.4 million farms = 1024.8 million acres
1985 → 445 acres/farm × 2.2 million farms = 979.0 million acres
1024.8 − 979.0 = 45.8

From 1980 to 1985, there was a decrease of 45.8 million acres.

Try This

a. Compare total acres farmed for the years 1970 and 1980.

The chart on p. 450 in the Data Bank shows yield rates of various produce.

Example

2. Find the yield from one lug of cherries. Round to the nearest quart.

Divide the pounds per lug by the high and low amounts needed for one quart.

22 lb ÷ 2 lb/qt (low) = 11 qt
22 lb ÷ $2\frac{1}{2}$ lb/qt (high) = $8\frac{4}{5}$ rounded to 9 qt

One lug of raw cherries is 9 to 11 quarts of prepared cherries.

Try This

Find the yield to the nearest quart.

b. 1 bushel of peas **c.** 1 box of pears **d.** $\frac{1}{2}$ lug of peaches

Exercises

Solve. Use the chart on p. 236.

1. Find the decrease in the number of farms between 1960 and 1970.
2. Find the decrease in the number of farms between 1950 and 1985.
3. Find the increase in the average number of acres per farm between 1970 and 1980.
4. Find the total number of acres farmed in each year. Make a chart.
   a. 1950 b. 1960 c. 1970 d. 1980 e. 1985

Compare the number of acres used for farming for the given years.

5. 1950 and 1960
6. 1960 and 1970
7. 1970 and 1980
8. 1950 and 1970
9. 1950 and 1980
10. 1950 and 1985

Find the yield to the nearest quart.

11. 1 bu carrots
12. 1 lug plums
13. 1 box apricots
14. 1 lug tomatoes
15. $\frac{1}{2}$ lug peaches
16. $\frac{1}{3}$ bu apples
17. $\frac{3}{4}$ lug plums
18. 3 boxes pears
19. $2\frac{1}{2}$ boxes apricots

Estimation The average farmworker produces enough food for 76 people. Estimate the number of farmers needed to feed the people in the following states.

20. Georgia 5,463,000
21. Texas 14,229,000
22. California 23,668,000
23. Wyoming 470,000
24. Nevada 801,000
25. West Virginia 1,950,000

Data Bank Use the World Production of Food Crops chart on p. 450 of the Data Bank.

26. Cereal grains are the seeds or fruits produced from cereal plants. How many million metric tons of cereal are produced each year?
27. The United States produces 20% of the world's cereal grains. How many million metric tons are grown by American farmers?

Mixed Skills Review

1. $32 \div \frac{8}{13}$
2. $341.15 − $18.74
3. $2\frac{3}{8} + \frac{1}{6} + \frac{1}{4}$
4. $115.2 \div 14.4$
5. $4644 \div 43$
6. $12\frac{3}{8} \times \frac{4}{11}$
7. $1812 - 1776$
8. $160 + 429 + 212$
9. $8\frac{5}{12} - 4\frac{7}{16}$
10. 748×60
11. $1.60 + $0.19 + $8.53
12. $6\frac{3}{4}\%$ of $60

10-9 Enrichment: World Food Production

Objective
Find and compare world food production percent figures.

In a recent year Bangladesh and the United Kingdom each produced about 1000 metric tons of corn. What percent of the world's corn did each country produce?

The table on p. 450 in the Data Bank lists world grain production figures. The total corn production was 449,255 thousand metric tons. This is about 450 million metric tons of corn. Choose a calculation method.

$$\text{Percent of world production} = \frac{1 \text{ thousand t}}{449255 \text{ thousand t}}$$
$$= 1 \div 449{,}255$$
$$= 0.000002, \text{ or about } 0.0002\%$$

Bangladesh and the United Kingdom each produced about 0.0002%, or 2 ten-thousandths of 1 percent of the world's corn.

What percent of the world's rice did Canada, the United States, and Mexico produce?

Add the figures for the three countries' rice production. Assume that Canada's rice production was negligible.

1755 + 6216 = 7971, or 7971 thousand metric tons

$$\text{Percent of world production} = \frac{7971 \text{ thousand t}}{469959 \text{ thousand t}}$$
$$= 0.016961, \text{ or about } 2\%$$

The three North American countries produced about 2% of the world's rice.

Exercises

Solve. Choose a calculation method. Use the table on p. 450 in the Data Bank.

1. What percent of the world's rice did Panama produce?
2. What percent of the world's wheat did Czechoslovakia produce?
3. What percent of the world's corn did the South American countries of Argentina, Brazil, Chile, Ecuador, Peru, Uruguay, and Venezuela produce?
4. What percent of the world's wheat did the European Economic Community countries produce?
5. Which is greater, Nepal's percent of world wheat production or its percent of world rice production?
6. Which is greater, Hungary's percent of world wheat production or its percent of world corn production?

10-10 Calculator: Exponents

Objective
Read and use exponents with a calculator.

Scientists and mathematicians often use **exponents** to describe very large or very small numbers. In the expression 4^2 the ^{2} is the exponent, and 4 is the **base.** The exponent tells how many times to use the base as a factor.

1 googol = 10^{100} = 10,000,000,000,000,000,000,000,000,000,000,000,000,000,000,000,000,000,000,000,000,000,000,000,000,000,000,000,000,000,000,000,000,000,000,000,000

$4^2 = 4 \times 4 = 16$ is read "4 to the second power" or "4 squared equals 16."
$5^3 = 5 \times 5 \times 5 = 125$ is read "5 to the third power" or "5 cubed equals 125."
$8^6 = 8 \times 8 \times 8 \times 8 \times 8 \times 8 = 262{,}144$ is read "8 to the sixth power equals 262,144."

Finding 8^6 on a calculator could take 12 key punches. Here is a faster method, requiring only 7 key punches, which works on many models of calculators.

8^6 = 8 [×] [=] [=] [=] [=] [=]

The display shows each intermediate power of 8. Notice that you hit [=] 5 times, not 6. Consider hitting the [×] as 8^1 or 8 to the first power. The [=] shows 8^2, the next shows 8^3, and so on through 8^6.

Use your calculator to solve. $2^4 \times 3^2$

Think $2^4 \times 3^2 = 2 \times 2 \times 2 \times 2 \times 3 \times 3$. Use the [M+] and [MR] to help solve the problem.

$2^4 \times 3^2$ = 2 [×] [=] [=] [=] [M+] 3 [×] [=] [×] [MR] [=] 144
$2^4 \times 3^2 = 144$

Exercises

Write as a numeral. Find the value of each expression. Choose a calculation method.

1. 2 to the second power
2. 7 squared
3. 12 to the third power
4. 9 cubed
5. 8 to the seventh power
6. 16 to the sixth power
7. 25 to the fourth power
8. 53 to the third power
9. 142 to the third power
10. 8924 squared

Use your calculator to find the value of each expression.

11. $8^3 \times 2^5$
12. $6^3 \times 4^3$
13. $10^3 \times 3^4$
14. $12^3 \times 5^3$

CHAPTER 10 REVIEW

Tell what the items in each group have in common. Match 1–4 with A–D.

1. serving size calorie protein carbohydrates	**2.** protein carbohydrates vitamins fat	**3.** vegetables milk products bread & cereal meat & fish	**4.** A C B_1 niacin
A. nutrients	**B.** label information	**C.** vitamins	**D.** food groups

Match the recommended daily calorie intake for 5–10 with E–J. Use the chart on p. 222.

		Current Weight	Weight Goal	Sport	Daily calorie Intake
5.	Lew	190	190	Tennis	**E.** 3060
6.	Lila	140	130	None	**F.** 2700
7.	Glenn	160	160	Hockey	**G.** 3640
8.	Norman	170	190	Football	**H.** 1880
9.	Kelly	120	120	Lacrosse	**I.** 3320
10.	Ilene	115	120	Softball	**J.** 3140

11. Arrange these three meals in order from least calories to greatest calories. Use the calorie count table on p. 438.

Meal A	Meal B	Meal C
3 oz steak	2 pork chops	6 oz fish sticks
$\frac{1}{2}$ c peas	$\frac{1}{2}$ c rice	$\frac{1}{2}$ c canned corn
1 c mashed potatoes	$\frac{1}{2}$ c green beans	1 c asparagus
2 slices of white bread	1 slice rye bread	1 slice whole wheat bread
1 c fruit cocktail	2 watermelon wedges	1 c sherbet
2 c skim milk	1 c apple juice	1 c lemonade

Use >, <, or =. Use the charts on pp. 226 and 230.

12. The number of calories burned jogging for 1 hour ☐ the number of calories recommended daily from the citrus fruit group for a 13-year-old girl.

13. The number of calories burned playing basketball for 1 hour ☐ the number of calories recommended daily in the bread group for teenagers.

14. The number of calories recommended in the meal-planning guide for a 16-year-old girl ☐ the number of calories burned by doing calisthenics for 7 hours per week.

CHAPTER 10 TEST

Find the recommended daily calorie intake.

		Current Weight	Weight Goal	Sport	Daily Calorie Intake
1.	Tom Summers	173	173	Swimming	
2.	Betty Abelson	152	130	None	
3.	Jack Roberts	136	155	Wrestling	

Find the total number of calories. Use the calorie count table on p. 449.

4. 2 slices white bread $1\frac{1}{2}$ c whole milk $\frac{1}{2}$ c raisins
6 oz corn flakes   two 4″ × 8″ wedges of watermelon

Find the total number of calories burned.

5. Donna ice skated for 2 hours 35 minutes and then jogged for 28 minutes.

Use the label to find the total amount of nutrients in the package.

6. calories

7. protein

8. carbohydrates

Rapid Rice	
Serving size:	11 oz
Servings per package:	24
Calories	120
Protein	3 g
Carbohydrates	27 g

Find the total energy allowance and nutrients for each food group.

		Age	Food Group	Energy Allowance	Nutrients
9.	Howard Peters	57	Meats, fish		
10.	Kathy Walker	17	Essential fats		
11.	Molly Gruber	33	Bread		

Find the cost to cater the following.

12. Deluxe Service for 16, tables w/chairs, candles, linens, china

Compare the costs.

	Gardening Expenses	Yield	Supermarket Costs
13.	4 packets seeds @ $0.79	12 lb carrots	$0.49/lb

Solve. Use the chart on p. 236.

14. What was the decrease in the number of farms from 1970 to 1985?

15. What was the increase in the average number of acres per farm from 1950 to 1985?

Find the yield. Use the chart in the Data Bank on p. 450.

16. 2 boxes pears   **17.** $2\frac{1}{2}$ lugs tomatoes   **18.** 1.6 bushels corn

CHAPTER 10 CUMULATIVE REVIEW

1. Bobby Rigo wins 47% of his tennis sets. What are the chances he will win 2 sets in a row?

A. $\frac{47}{100}$ B. $\frac{22}{100}$

C. $\frac{25}{100}$ D. $\frac{94}{100}$

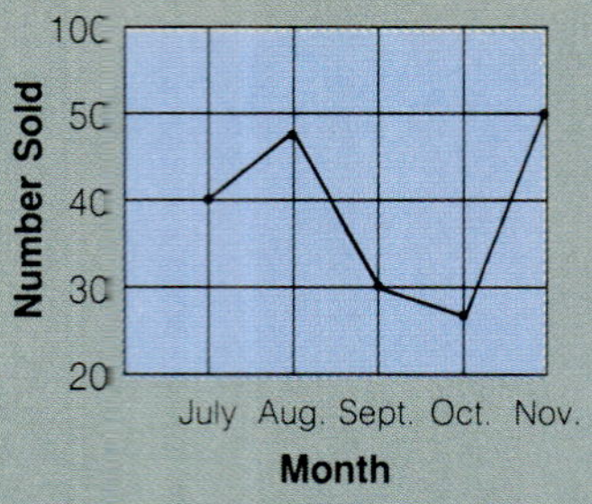

2. Estimate the total sales for August.

A. 45
B. 30
C. 40
D. 50

3. Write as a whole number.
100% = □

A. 100
B. 10
C. 1
D. 0.1

4. The average farmworker produces enough food for 76 people. Estimate the number of farmers needed to feed 6,462,000 people.

A. 1,000,000
B. 850
C. 8,500
D. 85,000

5. Find the rental fee.
Car luggage rack: $5 per week for 28 days

A. $15
B. $20
C. $35
D. $25

6. Round to the nearest ten.
567

A. 600
B. 570
C. 568
D. 667

7. Ken Chan budgeted 6.7% of his monthly income of $1050 for savings. How much was he planning to save?

A. $7.35
B. $70.35
C. $71.45
D. $6.25

8. Mike entered the bicycle giveaway 6 times. There were 1200 entries. What are his chances of winning?

A. 0.5%
B. 5%
C. 0.05%
D. 0.2%

9. There are 78,600 registered voters. A poll of 50 voters per region is shown. Predict the number of votes for Candidate B.

A. 35,108
B. 89,604
C. 43,492
D. 105,324

Senate Race Poll Results

	Regions		
	I	II	III
Candidate A	21	35	11
Candidate B	29	15	39

10. Rental fees:
Service factor = 12
Linen = $15
China and silver = $33
Breakage insurance = $10
Find the cost for a party of 30.

A. $215
B. $58
C. $116
D. $145

11. Find the sale price for a refrigerator.
list price \$549,
discount 25% off
A. \$411.75
B. \$137.25
C. \$414.25
D. \$437.75

12. Bonnie Marsh is a swimmer and weighs 121 pounds. Find the number of calories she needs to maintain her weight.
A. 2700 calories
B. 2600 calories
C. 2200 calories
D. 3100 calories

Use:
First 22 pounds = 1000 calories
Next 22 pounds = 500 calories
Each additional 22 pounds = 200 calories
Active athlete = 500 calories
Lose/gain weight: ±500 calories

13. The bill for four at a restaurant is \$21.65 plus 6.5 % tax and 15% tip. The cost to make the same food at home is \$1.96 + \$0.89 + \$0.33 + \$3.03 + \$2.50 + \$2.40. How much is saved by cooking at home?
A. \$11.10
B. \$26.52
C. \$15.19
D. \$17.14

14. A word processing workstation can have 3 styles of chairs, 2 styles of desktops, and 5 styles of storage bins. How many different combinations are there?
A. 6
B. 15
C. 30
D. 10

15. Golfers use 300 calories per hour. Glenda McDonald played golf for 2.5 hours. How many calories did she use?
A. 600 calories
B. 700 calories
C. 650 calories
D. 750 calories

16. Multiply. Use mental math.
$6 \times 2\frac{1}{5}$
A. $13\frac{1}{5}$
B. 30
C. $12\frac{1}{5}$
D. $13\frac{2}{5}$

17. One serving of bread provides 70 calories and 2 grams protein. How much energy and protein would an adolescent receive from a daily requirement of 6 servings?
A. 140 calories, 14g protein
B. 360 calories, 12g protein
C. 490 calories, 14g protein
D. 420 calories, 12g protein

18. Pencils are 20¢ each. Packages of 10 cost \$1.50. What is the least expensive way to buy 15 pencils?
A. 1 package and 5 singles
B. 2 packages
C. 15 singles
D. Either A or C

Chapter 11 Overview

Key Ideas

- health insurance costs
- company health plans
- using formulas
- health statistics
- using a fitness center
- fitness programs
- measuring fitness
- types of life insurance

Key Terms

- deductible
- premium
- annual cost
- benefits
- copayment
- anatoglyph
- blood plasma
- hemoglobin
- oxygen pressure
- pulse rate
- respiration rate
- vital capacity
- toning program
- maximum weight
- lean body weight
- ideal body weight
- term insurance
- whole life insurance
- limited pay life
- endowment
- universal life

Key Skills

Find the answer.

1. \$82.56 + \$7.29 + \$563.80

2. \$293.60 + \$45.98 + \$22.86

3. \$533.88 + \$34.82 + \$247.08 + \$15.78

4. \$41.92 + \$139.40 + \$63.05

5. \$34.82 + \$2.98 + \$16.04

6. \$5.07 − \$2.46

7. \$18.30 − \$11.57

8. \$730.00 − \$57.76

9. \$408.19 − \$273.99

10. \$110.08 − \$82.19

11. \$200 − \$19.39

12. 0.80 × \$39.88

13. 0.76 × \$14.93

14. 1.3 × 54.41

15. 3.2 × 76.56

16. 2.03 × 98.65

17. 0.78 × \$23.50

18. 115.64 ÷ 0.98

19. 123 ÷ 0.97

20. 163.35 ÷ 0.99

Solve.

21. What is 60% of 82?

22. What is 45% of 200?

23. What is 85% of 30?

24. What is 19% of 654?

25. What is 28% of \$45,882?

26. 18 is what percent of 30?

27. What percent of 1000 is 230?

28. 2 is what percent of $\frac{1}{2}$?

29. What percent of 90 is 18?

30. 93 is what percent of 465?

31. Five years ago the population of Mannheim was 12,464. Today the population is 15,200. Find the percent of increase.

HEALTH AND FITNESS

11

11-1 Health Insurance

Objective
Find health insurance costs.

Many corporations now provide nutrition and fitness programs for their employees. These programs help to reduce the number of absences from work. Employers also provide medical insurance at reasonable costs for their employees.

Health insurance policies often require you to pay a **deductible,** or fixed amount of your medical costs. The insurance company then pays a percent of the remaining costs.

Example Find the amounts paid by the insurance company and by the insured.

Covered Medical Expense	Deductible	Rate Paid by Insurance Company	Amount Paid by Insurance Company	Amount Paid by Insured
\$1245	\$100	80%		

Find the company payment.

Subtract the deductible.
$\$1245 - \$100 = \$1145$
Multiply by the rate.
$0.80 \times \$1145 = \$916.00$

The insurance company payment is \$916.00.

Find the insured member payment.
$\$1245 - \$916 = \$329$

The insured member's payment is \$329.

Try This Find the amount paid by the company and the insured.

	Covered Medical Expense	Deductible	Rate Paid by Insurance Company	Amount Paid by Insurance Company	Amount Paid by Insured
a.	\$975	\$100	75%		
b.	\$2760	\$50	80%		
c.	\$1200	\$75	90%		
d.	\$197	\$50	80%		

Exercises

Find the amount paid by the company and the insured.

	Covered Medical Expense	Deductible	Rate Paid by Insurance Company	Amount Paid by Insurance Company	Amount Paid by Insured
1.	$575	$100	90%		
2.	$891	$50	75%		
3.	$1243	$200	80%		
4.	$1767	$100	70%		
5.	$2085	$50	100%		
6.	$4533	$200	85%		
7.	$6729	none	80%		
8.	$9733	none	75%		

9. Billy broke his ankle sliding into first base. The doctor charged $335 to mend it. Billy's insurance company will pay 85% of the medical expense after Billy's $150 deductible. How much must Billy pay? How much must his insurance company pay?

10. How much would Billy pay if his deductible were $200 and the rate paid by his insurance company were 70%? How much would the company pay?

11. Mr. Whittaker works for Dawn Industries. His family is covered by one health insurance policy. He pays via a biweekly deduction of $44.50 from his paycheck. Dawn Industries contributes $15.75 to his insurance cost every two weeks. How much does Mr. Whittaker pay per year for his family's health coverage? How much does Dawn Industries pay?

Data Hunt

12. Find the cost for the following services at a local hospital. Make a chart comparing the benefits under various medical plans.

a. Chest X-ray
b. Semi-private room
c. Emergency room visit
d. Setting a broken leg
e. An injury requiring stitches
f. Complete physical exam

Mixed Skills Review

1. $\frac{7}{8} \div 1\frac{1}{2}$
2. $236 + 439$
3. 11.3×0.17

4. $16.3 + 12.1 + 18.6$
5. $3\frac{7}{8} - 2\frac{2}{3}$
6. $2\frac{3}{4} \times 4\frac{1}{2}$

7. $506 - 427$
8. $704 \div 16$
9. $93.68 - 14.59$

10. $32.76 \div 91$
11. 157×26
12. $6\frac{2}{3} + 10\frac{1}{6}$

11-2 Comparing Health Plans

Objective *Compare health plan costs and benefits.*

Bonnie Dyer's employer offers its employees the two health insurance plans on p. 451 of the Data Bank. Before making a decision, Bonnie considers the benefits and services available under each plan.

Examples

1. A prescription costs \$21.95. Compare the benefits under each plan after the annual deductible has been paid

Find the amount paid by the World Health Plan.
The World Health Plan pays 80% of prescription costs.
$0.80 \times \$21.95 = \$17.56$

Find the amount paid by the Capital Care Medical Plan.
The insured makes a \$2 copayment for each prescription.
$\$21.95 - \$2.00 = \$19.95$

Find the difference in benefits.
$\$19.95 - \$17.56 = \$2.39$

The Capital Care Medical Plan pays \$2.39 more.

2. The bill for the emergency room care following a skateboarding accident was \$345.30. Compare the benefits under each plan.

Find the amount paid by the World Health Plan.
The World Health Plan pays the first \$300 plus 80% of the amount over \$300.

$\$345.30 - \$300 = \$45.30$	Find excess over \$300.
$0.80 \times \$45.30 = \$36.24$	Multiply excess by rate.
$\$300 + \$36.24 = \$336.24$	Add the benefits.

The World Health Plan pays \$336.24.

Find the amount paid by the Capital Care Medical Plan.
The Capital Care Medical Plan pays \$10 for each emergency room visit.

Find the difference in benefits. $\$336.24 - \$10 = \$326.24$

The World Health Plan pays \$325.24 more.

Try This

Compare the benefits under each plan. Assume the annual deductible has been paid.

a. Surgery, cost \$2365.90

b. Routine physical, cost \$45

c. Doctor's office visits (2), cost \$75 each

d. X-rays, cost \$125.40

Exercises

Compare the benefits of the World Health Plan and Capital Care Medical Plan for these services.

1. Debbie Tomkins is insured under the World Health Plan. Find the benefits for the following services. Subtract $50 from the total benefits for her annual deductible.

Hospital room and board	$11,345.60
X-rays and lab tests	$619.45
Surgery	$9,942.95
Doctor's hospital visits	$1,270.00

2. Carl Ratcliffe is insured under the Capital Care Medical Plan. Find the benefits for the following services.

Emergency room care	$481.62
X-rays and lab tests	$237.45
Doctor's office visits (3)	$190.00
Prescriptions (2)	$56.81

3. Routine physical, cost $60
4. Prescription, cost $37.72
5. Doctor's home visit, cost $48
6. Surgery, cost $1049.90
7. Eye examination, cost $82
8. X-rays, cost $143.94
9. Doctor's office visits (3), cost $78
10. Prescription, cost $12.50
11. Emergency room care, cost $89.50
12. Routine physical, cost $62.50
13. Lab tests, cost $220.25
14. Hospital room and board, cost $4590.20

A magazine poll of 200 people found that 60% of those surveyed postponed going to the dentist. Use mental math to tell how many of those who postponed their visits gave each of the following reasons.

15. dislike going 82%
16. fear 61%
17. time inconvenience 79%
18. no reason 5%
19. money 70%
20. forget 8%

Mixed Skills Review

1. $6\frac{2}{3} - 4\frac{7}{9}$
2. 347×12
3. $4.2 - 3.826$
4. $65,365 \div 85$
5. $4926 + 8654$
6. $8.254 + 6.074$
7. $2\frac{3}{8} + 4\frac{3}{4}$
8. 7.7×6.78
9. $4\frac{3}{7} \div \frac{3}{14}$
10. $742.9 \div 1.7$
11. $84,630 - 18,739$
12. $8 \times 4\frac{5}{16}$

11-3 Working in the Hospital

Objective
Use formulas.

A test tube contains a 100-mL sample of blood. Julie Troy, a hospital lab technician, uses the formula $\boldsymbol{W = 0.003\,P}$ to determine the number of milliliters of oxygen dissolved in the sample.

W stands for the number of mL of oxygen dissolved in 100 mL of **whole blood plasma.** P stands for the machine reading in millimeters for the oxygen pressure.

Examples

1. Find the amount of oxygen dissolved in the 100 mL of whole blood plasma with a P reading of 91 mm.

$W = 0.003P$ $\quad\quad$ $0.003\ (91) = 0.273$ mL

There is 0.273 mL of oxygen dissolved in the blood plasma.

To find the number of milliliters of oxygen in the **hemoglobin** of red blood cells, formula $\boldsymbol{H = 1.36\,gs}$ is used.

H stands for the milliliters of oxygen in the hemoglobin per 100 mL of whole blood.
g stands for the number of grams of hemoglobin in 100 mL of whole blood.
s stands for the percent of oxygen in the hemoglobin.

2. The sample contains 13 g of hemoglobin with 88% oxygen. Find the amount of oxygen in the hemoglobin.

$H = 1.36gs$ $\quad\quad$ $1.36\ (13)(0.88) = 15.5584$ mL

There are about 15.56 mL of oxygen dissolved in the hemoglobin.

To find the total amount of oxygen in the blood, Julie adds the amount of oxygen in the plasma, W, to the amount in the hemoglobin, H.

3. Find the total amount of oxygen in the 100-mL blood sample.

0.273 mL + 15.56 mL = 15.833 mL

The total amount of oxygen in the sample is 15.833 mL.

Try This

Find the total amount of oxygen for each 100-mL blood sample.

	Oxygen Pressure in Plasma P	Oxygen in Plasma W (mL)	Hemoglobin g	Percent of Oxygen s	Oxygen in Hemoglobin H (mL)	Total Oxygen in Sample $W + H$ (mL)
a.	88 mm		14 g	93%		
b.	95 mm		16 g	89%		
c.	79 mm		12 g	90%		

Exercises

Use the formula $W = 0.003P$ to find the amount of oxygen dissolved in 100 mL of whole blood plasma with the following P readings.

1. 85 mm **2.** 91 mm **3.** 87 mm

4. 95 mm **5.** 89 mm **6.** 93 mm

Use the formula $H = 1.36\ gs$ to find the amount of oxygen in a 100-mL blood sample containing the following amounts of hemoglobin and oxygen.

7. $g = 13$ g, $s = 82\%$

8. $g = 15$ g, $s = 91\%$

9. $g = 12$ g, $s = 93\%$

10. $g = 11$ g, $s = 86\%$

11. $g = 14$ g, $s = 87\%$

12. $g = 15$ g, $s = 89\%$

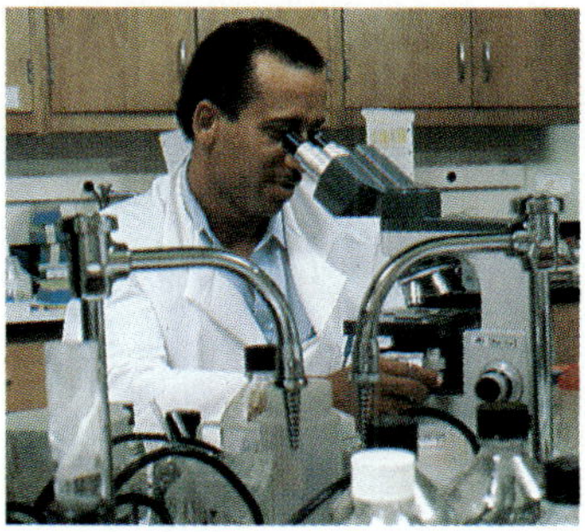

Find the total amount of oxygen for each 100-mL blood sample.

	Oxygen Pressure in Plasma P	Oxygen in Plasma W (mL)	Hemoglobin g	Percent of Oxygen s	Oxygen in Hemoglobin H (mL)	Total Oxygen in Sample $W + H$ (mL)
13.	84 mm		11 g	91%		
14.	93 mm		15 g	88%		
15.	78 mm		13 g	92%		
16.	90 mm		12 g	90%		
17.	96 mm		15 g	96%		
18.	85 mm		10 g	87%		
19.	76 mm		9 g	84%		
20.	88 mm		14 g	95%		

Mixed Skills Review

1. 163×0.007 **2.** $21{,}482 + 68{,}394$ **3.** $7 - 2\frac{3}{4}$

4. $6\frac{1}{2} \times 2\frac{1}{5}$ **5.** $6 - 2.732$ **6.** $1525 - 878$

7. $48.236 + 68.991$ **8.** $\frac{5}{9} \div 1\frac{1}{2}$ **9.** 4.2 is 12% of what number?

10. $30{,}786 \div 60$ **11.** 422×384 **12.** $14\frac{3}{8} + 2\frac{4}{9}$

11-4 Health and Statistics

Objective *Find statistical data.*

The life expectancy for Americans has risen to 74.6 years. More than 50% of the population is over 30 years of age. This increase means a greater number of people are needed to work in the health care field.

Example

1. There are approximately 500,000 physicians practicing in the United States. By the year 2000, that number should increase to 700,000. Find the percent of increase.

$700{,}000 - 500{,}000 = 200{,}000 \qquad \frac{200{,}000}{500{,}000} = \frac{2}{5}$, or 40%

There will be a 40% increase in the number of physicians by the year 2000.

Try This

Solve.

a. The average American life expectancy rose from 49.2 years in 1900 to 73.7 years in 1980. Find the percent of increase.

b. In a recent year Americans spent an average of $1580 per person on health care. How much would a family of four have spent?

Example

2. A certain hospital's records show patients requiring X-rays over a 30-day period. Find the average number of patients per day.

Number of patients	20	21	22	23	24	25	26
Number of days	1	0	3	7	5	8	6

Multiply the number of patients per day by the number of days. Add.

$(1 \times 20) + (0 \times 21) + (3 \times 22) + (7 \times 23) + (5 \times 24) + (8 \times 25) + (6 \times 26)$

$20 + 0 + 66 + 161 + 120 + 200 + 156 = 723$

Divide by the number of days in the period. $723 \div 30 = 24.1\ldots \approx 24$

The hospital averaged 24 patients per day for X-rays.

Try This

Solve.

c. A hospital recorded the number of births each month over a two-year period. Find the average number of births per month.

Number of births	48	49	50	51	52	53	54
Number of months	5	3	6	4	3	1	2

Exercises

Solve.

1. The number of live births in the United States decreased from 4,257,850 in 1960 to 3,612,258 in 1980. Find the percent of decrease.
2. Four million people work in 7000 hospitals in the U.S. What is the average number of employees per hospital?
3. There are about 1,400,000 beds in hospitals in the U.S. What is the average number of beds per hospital?
4. The occupancy rate for hospital beds is 77.4%. How many hospital beds is this?
5. The number of people working in health care services increased from 4.2 million in 1970 to 7.9 million in 1984. Find the percent of increase.

In 1984 Americans spent $387.4 billion on health care services. Find the amount spent for each type of service.

6. hospital
7. physicians
8. dentists
9. nursing home care

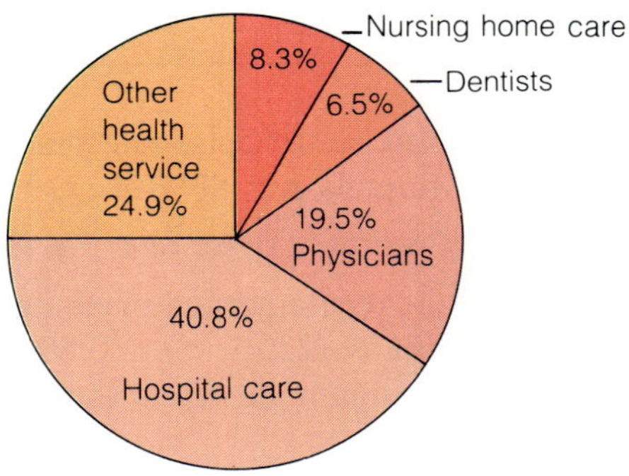

An outpatient clinic recorded the number of patient visits over a four-week period.

	M	T	W	Th	F	S
Week 1	42	36	45	32	50	18
Week 2	32	30	36	47	29	20
Week 3	27	25	31	41	37	17
Week 4	35	29	38	36	43	21

10. During which week did the clinic treat the greatest number of patients?
11. On which day of the week did the most patients visit the clinic?
12. What was the total number of patients treated in 4 weeks?
13. What percent of the patients visited the clinic in the second week?
14. What percent of the patients visited the clinic on Wednesdays?

Mixed Skills Review

1. 39,246 + 27,863
2. $3\frac{2}{3} + 6\frac{1}{8}$
3. 10.03 − 8.6
4. 16.891 + 12.46
5. 2716 − 385
6. $4\frac{3}{4} - 1\frac{7}{8}$
7. 1521 × 14
8. 13.212 ÷ 0.06
9. 51,480 ÷ 858
10. $2\frac{3}{16} \times \frac{1}{2}$
11. 21.3 × 0.8
12. $4\frac{1}{2} \div \frac{5}{12}$

11-5 The Fitness Center

Objective *Find health data using formulas.*

Lynne Brick teaches classes in aerobics.

During each class students check their pulse rate several times. The **pulse rate** tells how many times the heart beats per minute.

Example

1. Lynne counted 11 beats in 10 seconds. Find the pulse rate.
Think: 60 seconds ÷ 10 seconds = 6   6 × 11 beats = 66 beats
Lynne's pulse rate is 66 beats per minute.

Try This Work with a partner to find your pulse rate.

a. before exercise **b.** after exercise **c.** Find the percent change from **a** to **b.**

During exercise the pulse rate should not exceed a limit based on age. Lynne uses the formula $\boldsymbol{p = 80\% \times (220 - a)}$ to find the maximum rate.

p is the maximum pulse rate. a is the person's age.

Example

2. Find the recommended maximum pulse rate for an 18 year old.
$p = 80\% \times (220 - a) = 0.80 \times (220 - 18) = 0.80 \times 202 = 161.60 \approx 162$
The maximum pulse rate should be 162 beats per minute.

Try This Find the recommended maximum pulse rate during exercise.

d. Tom, age 49 **e.** Jana, age 13 **f.** Suh Kang, age 21

Respiration rate is the number of breaths taken in one minute.
Vital capacity is the amount of air exchanged per breath.
Lynne uses the formula $\boldsymbol{V = 0.041\,h - 0.018\,a - 2.69}$ to estimate the vital capacity for women.

V is vital capacity in liters. h is height in centimeters. a is age in years.

Example

3. Find the vital capacity for Anne Elliott, 143 cm tall and 29 years old.
$V = 0.041\,h - 0.018\,a - 2.69 = 0.041(143) - 0.018(29) - 2.69 = 2.651 \approx 2.7$
Anne's vital capacity is about 2.7 liters.

Try This Find the vital capacity.

g. Charlotte Keith, 160 cm, 19 years **h.** Beverly, 149 cm, 30 years

Exercises

Mental Math

Find the pulse rate. Use mental math.

1. 10 seconds
9 beats

2. 15 seconds
24 beats

3. 20 seconds
13 beats

4. 12 seconds
19 beats

5. 30 seconds
78 beats

6. 6 seconds
13 beats

Find the recommended maximum pulse rate during exercise.

7. Anne, age 14

8. Charles, age 17

9. Felipe, age 29

10. Ashley, age 34

11. Megan, age 45

12. Sam, age 58

Find the vital capacity.

	Name	Height, h (cm)	Age, a (yr)	Vital Capacity, V
13.	Barbara Day	170	38	
14.	Melba Williams	153	41	
15.	Beverly Snyder	146	26	
16.	Grace Patton	161	18	
17.	Eileen O'Reilly	165	35	
18.	Mardi Sicca	157	16	

19. Typical respiration rates are between 16 and 20 breaths per minute. The number of breaths is counted for 15 seconds. How many breaths should be counted to produce normal rates?

20. Lynne has a respiration rate of 20 breaths per minute. How many seconds does one complete breath take?

21. Following strenuous exercise, Lynne counted 7 breaths in 15 seconds. How much has the rate increased?

22. About 500 mL of air per breath are exchanged during quiet breathing. How many milliliters of air are exchanged during 20 breaths?

23. How many liters of air are exchanged in one hour of quiet breathing?

Mixed Skills Review

1. 656,000 − 358,012

2. 823×106

3. What percent is 210 of 250?

4. \$12 − \$7.63

5. $\frac{1}{2} + \frac{1}{3} + \frac{1}{4} + \frac{1}{8}$

6. 436,275 + 276,849

7. 30% of 235.5

8. $6\frac{3}{8} \times 2\frac{2}{3}$

9. 3.8 + 12 + 16.07 + 0.365

10. $7\frac{1}{2} - 2\frac{5}{16}$

11. 14,688 ÷ 544

12. $13\frac{1}{3} \div 4\frac{1}{6}$

11-6 Fitness Programs

Objective
Find fitness measures using formulas.

Instructors at physical fitness centers and health clubs plan weight lifting programs. A toning program is designed to maintain fitness. A training program prepares athletes for competition.

Student: John Carlos **Instructor: C. Cavolo**

Exercise	Maximum Weight	Toning 50%	Training 80%	Number of Sets	Repetitions per Set
Tricep curl	45 lb	—	36	3	10
Bicep curl	40 lb	—	32	4	15
Bench press	82 lb	—	66	2	12

Understand the Situation

- What is the maximum weight John can bench press?
- How much weight should John bench press in his training program?
- How many sets of bicep curls should John do in each workout?
- How many times should John do bicep curls in each set?

The amount of weight to be lifted is found using these formulas.

Toning program: $w = 50\%$ of m
Training program: $w = 80\%$ of m

w is the weight a person should use.
m is the maximum weight a person can lift.

Example

Steve is in a training program. The maximum weight he can bench press is 170 lb. How much weight should he lift in training?

$w = 80\%$ of 170 lb
$0.80 \times 170 = 136$ lb

Steve should bench press 136 pounds.

Try This

Find the weight to be used.

a. Toning program: leg lifts. The maximum weight Kay can lift is 128 pounds.

b. Training program: shoulder press. The maximum weight Greg can lift is 190 pounds.

Exercises

Find the weight to be used.

	Program	Event	Maximum Weight	Weight to Be Used
1.	Toning	bicep curl	70 lb	
2.	Training	leg lift	58 lb	
3.	Training	bench press	189 lb	
4.	Toning	standing calf press	97 lb	
5.	Toning	military press	105 lb	
6.	Training	tricep curl	84 lb	

To increase muscle size and strength, you should exercise with at least 75% of the maximum weight you can lift. Find the minimum weight to be lifted when exercising to increase strength.

		Maximum Weight	Weight to Be Used
7.	Judy Scheper	98 lb	
8.	Jim Shank	175 lb	
9.	Tony Cho	201 lb	
10.	Chanda Walker	83 lb	

People who exercise regularly want to improve their body proportions. They divide their waist measure by their hip measure and compare the answer to the average. Use percent to compare the body proportions to the average.

		Waist	Hip
11.	Laura	22	30
12.	Jim	42	50
13.	Helen	24	35
14.	Stuart	40	45
15.	Kamin	34	40

Average Body Proportions	
Women: 0.7	Men: 0.85

Mixed Skills Review

1. $\frac{3}{8}$ of 37
2. $94{,}026 + 127{,}638$
3. $7\frac{5}{9} - 3\frac{1}{12}$

4. $1.006 - 0.935$
5. 54% of 1218
6. $4100 - 3651$

7. $16.234 + 8.659$
8. $52{,}200 \div 120$
9. $37.72 \div 2.3$

10. $6.3 \times 2.4 \times 3$
11. $4\frac{3}{5} + 3\frac{1}{2}$
12. $3\frac{3}{5} \div \frac{3}{20}$

11-7 Measuring Fitness

Objective
Find fitness measures using formulas.

Exercise physiologists often use instruments to measure various parts of the body. Formulas are then used to calculate fitness. One measure of fitness is percent of body fat. This table groups people by the percent of body fat.

Percent of Body Weight	Group
Below 10%	Very lean
10.0–12.9	Trim
13.0–16.9	Average
17.0–19.9	Plump
20.0–24.9	Fat
25% or more	Obese

Lean body weight is the difference between the actual weight and the body fat weight. Exercise physiologists use the formula $\boldsymbol{L = W(1 - F)}$.

L is lean body weight in pounds. W is actual body weight in pounds. F is percent of body fat.

Example

1. Tom weighs 180 pounds with 13.7% body fat. Find his lean body weight.

$L = W(1 - F) = 180(1 - 0.137) = 180(0.863) = 155.34 \approx 155$ lb

Tom's lean body weight is 155 lb.

Try This

Find the lean body weight to the nearest pound.

a. Craig weighs 225 lb. His percent of body fat is 23.9%.

Physiologists use the formula $I = \dfrac{L}{1 - d}$ to find the **ideal body weight.**

I is ideal body weight in pounds. L is lean body weight in pounds. d is desired percent of body fat.

Example

2. Laura's lean body weight is 105 lb. She wants her percent of body fat to be 15%. Find her ideal body weight to the nearest pound.

$I = \dfrac{L}{1 - d} = \dfrac{105}{1 - 0.15} = \dfrac{105}{0.85} \approx 124$ lb

Laura's ideal body weight is about 114 lb.

Try This

Find the ideal body weight to the nearest pound.

b. Leigh is a ballerina. Her lean body weight is 85 pounds. Leigh wants her percent of body fat to be 10%.

Exercises

Use the formula $L = W(1 - F)$ to find the lean body weight to the nearest pound.

1. $W = 105, F = 13\%$ **2.** $W = 188, F = 18\%$

3. $W = 230, F = 20.1\%$ **4.** $W = 76, F = 11.7\%$

Use the formula $I = \frac{L}{(1 - d)}$ to find the ideal weight to the nearest pound.

5. $L = 132, d = 14\%$ **6.** $L = 145, d = 9\%$

7. $L = 170, d = 8\%$ **8.** $L = 108, d = 9\%$

Find the lean body weight.

		Weight, *W*	Percent Body Fat, *F*	Lean Body Weight, *L*
9.	Katie Welch	175	24%	
10.	Brigid Weiss	124	13%	
11.	John Foley	165	11.6%	
12.	Pat Conroy	198	15.3%	

Find the ideal body weight.

		Lean Body Weight, *L*	Desired Percent Body Fat, *d*	Ideal Body Weight, *I*
13.	Pam Christian	105	9%	
14.	Tom Hoffman	152	11%	
15.	Bob Scannell	138	13%	
16.	Brenda Lee	126	8%	

Data Hunt

17. Call a fitness center, your doctor, or other health professional. Ask what methods are used to find the percent of body fat.

Mixed Skills Review

1. $0.7 - 0.257$ **2.** 32.5×0.034 **3.** $1{,}218{,}436 + 6{,}251{,}934$

4. $130 \div 3\frac{1}{3}$ **5.** $1\frac{3}{4} + 2\frac{7}{10}$ **6.** $22{,}792 \div 407$

7. $2.06 + 5 + 0.834$ **8.** 416×208 **9.** $\frac{2}{3} \times \frac{5}{6} \times \frac{9}{10}$

10. $2\frac{16}{21} - 1\frac{5}{7}$ **11.** $449.36 \div 16.4$ **12.** $1820 - 655$

11-8 Life Insurance

Objective
Find life insurance rates from a table.

Many people purchase life insurance. When the insured person dies, the insurance company pays money to the person chosen as the **beneficiary.** Family wage earners buy life insurance so their dependents will have some money for support. Some policies also pay money when a person retires.

Type of Insurance	Coverage
Term	Provides temporary protection at a low cost
Straight or whole life	Provides lifetime coverage; premiums are paid as long as you keep the policy
Limited-pay life	Provides lifetime coverage; premiums are paid for a certain amount of time, usually 20 or 30 years
Endowment	Provides coverage for certain amount of time, usually 20 or 30 years; the beneficiary receives the amount of the policy
Universal life	Provides lifetime coverage or until you cancel the policy; premiums may decrease through investment

The premiums for life insurance vary depending on the sex, age, type of policy, and the amount of coverage. Premiums are calculated from tables.

Example Find the annual life insurance premium. Use the table on p. 452 in the Data Bank.

Sex	Age Purchased	Type	Amount
M	32	10-yr term	$15,000

Use the table to find the annual rate per $1000.
Since the insured is a male less than 35 years of age, use the annual rate for a 30-year-old male.

The annual rate per $1000 is $3.30.

Multiply the rate by the number of thousands. $15,000 = 15 thousands
15 × $3.30 = $49.50 The annual premium is $49.50.

Try This Solve.

a. Sam Herman's annual premium for life insurance is $192.40. Find the amount deducted from his weekly paycheck.

b. Kayla's employer pays 40% of her annual life insurance premium of $378. Find the amount deducted from her monthly paycheck.

Exercises

Find the annual life insurance premium. Use the table on p. 452.

	Sex	Age Purchased	Type	Amount	Premium
1.	M	26	5-yr term	$40,000	
2.	F	17	Straight life	$50,000	
3.	F	35	30-yr limited pay	$65,000	
4.	M	15	20-yr endowment	$25,000	
5.	M	22	Universal life	$90,000	
6.	F	27	10-yr term	$60,000	
7.	F	37	30-yr endowment	$125,000	
8.	M	20	20-yr limited pay	$100,000	

Solve.

9. Saul Cohen has group life insurance. His annual premium is $154.96. Find the amount deducted from his weekly paycheck.

10. Eileen O'Reilly pays an annual premium of $146.64 for group insurance. Find the amount deducted from her biweekly paycheck.

11. Lou Ferrari buys group life insurance. His employer pays 80% of the annual premium of $256. Find the amount deducted from Lou's weekly paycheck.

12. An employer contributes 45% of an annual premium of $164. Find the amount deducted from the employee's monthly paycheck.

13. The annual group life premium for Amy Pierson is $382. Her employer contributes 60% of this amount. Find Amy's bimonthly deduction.

14. Hector Lugo's employer contributes $33\frac{1}{3}$% of his annual group life insurance premium of $312. Find the amount deducted from Hector's weekly paycheck.

15. A company pays 75% of its employees' annual life insurance premiums. How much is deducted from one employee's wages each quarter if her annual premium is $426?

Mixed Skills Review

1. $3\frac{9}{10} \times 1\frac{3}{7}$

2. $3\frac{2}{3} + 6\frac{1}{8} + 4\frac{3}{4}$

3. What percent is 20 of 32?

4. $7168 - 6939$

5. $8\frac{5}{6} - 2\frac{7}{36}$

6. $2154 + 6378 + 426$

7. $6\frac{3}{4} \div 1\frac{11}{16}$

8. $9.1 - 6.011$

9. 26.5% of what number is 116.6?

10. 65% of 3010

11. 645×219

12. $16.25 + $13.56 + 0.18

11-9 Enrichment: Anatoglyphs

Objective *Construct anatoglyphs.*

In some hospitals pictures called **anatoglyphs** are used to show the condition of patients. The vital organs are shaded according to injury or damage. An unshaded glyph is 100% functional. A fully shaded glyph is 0% functional.

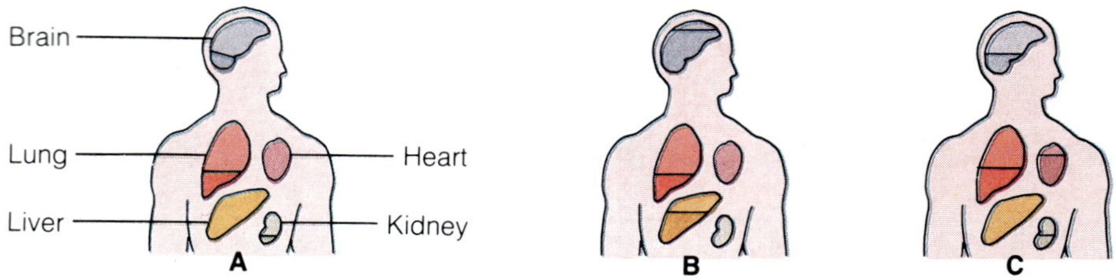

Exercises

Use the anatoglyphs above.

1. Which patient appears to have had only minor injuries?
2. Which patient may have suffered severe brain damage?
3. Which patient may have had a heart attack?
4. Which patients may not survive?

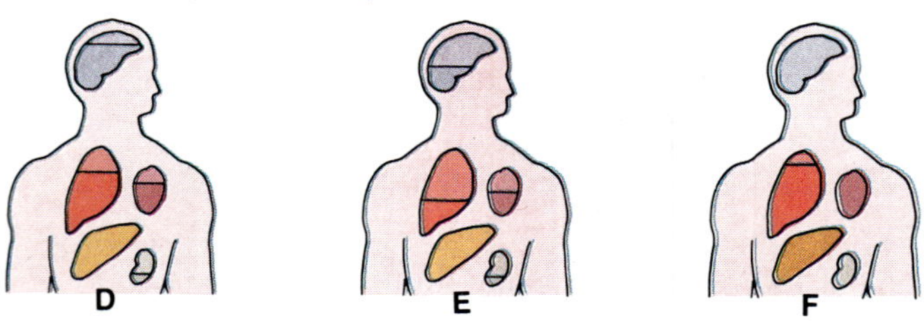

5. Which patients are having severe respiratory problems?
6. Which patient may suffer kidney failure?
7. Which patient appears to have suffered liver damage?
8. Which patients may not survive?

Draw anatoglyphs for the following patients.

9. a patient experiencing cardiac arrest
10. a patient with 50% pulmonary function
11. a patient with 25% renal function
12. a patient with hepatitis
13. a patient with meningitis

11-10 Computer: Health Statistics

Objective
Use the computer to generate health statistics.

The following computer program can be used to give lean body weight (no fat), desired body weight, maximum recommended pulse rate during exercise, and vital capacity for women.

```
10  REM Calculating health statistics
20  INPUT ''Are you male or female (M/F)'';S$
30  INPUT ''How much do you weigh'';W
40  INPUT ''What is your percent body fat'';F
50  INPUT ''What is your desired body fat'';D
60  INPUT ''How old are you'';A
70  INPUT ''What is your height in feet and inches (separate with
    comma)'';X,Y
80  H = (X * 12 + Y) * 2.54 : L = W * (1-F/100) : I = L/(1−D/100)
90  P = .8 * (220 − A) : V = .041 * H − .018 * A − 2.69
100 L = INT(L + .5) : I = INT(I + .5) : P = INT(P + .5)
110 IF S$ = ''M'' OR S$ = ''m'' THEN V = 0
115 V = INT(V * 10 + .5)/10
120 PRINT ''Lean Body'';TAB(20) ''Desired Body'';TAB(43)
    ''Maximum'';TAB(61) ''Vital Capacity''
130 PRINT TAB(2) ''Weight(lb)'';TAB(23) ''Weight(lb)'';TAB(42)
    ''Pulse Rate'';TAB(65) ''(Women)(L)''
140 PRINT TAB(2) L;TAB(23) I;TAB(44) P; TAB(66) V
150 INPUT ''Do you want to do another'';A$
160 IF A$ = ''y'' OR A$ = ''Y'' THEN 20
170 END
```

The example shows the calculation of some health statistics for the data given.

Weight: 135   Body fat: 11.7%   Desired body fat: 8.5%
Age: 15   Height: 5 ft 6 in.

Lean Body Weight (lb)	Desired Body Weight (lb)	Maximum Pulse Rate	Vital Capacity (Women) (L)
119	130	164	3.9

Exercises

1. Find the health statistics for a 32-year-old man, 187 lb, 6 ft 1 in., 17% body fat, 9% desired body fat.
2. Find the health statistics for a 145-lb woman, 28 years old, 13% body fat, 8% desired body fat, 5 ft 7 in.

CHAPTER 11 REVIEW

Match the item in column A with the correct amount in column B.

Column A

1. the cost of a $16.95 prescription through World Health Plan (Use the chart on p. 451.)
2. the amount paid by Capital Care Plan for a routine physical costing $110.00 (Use the chart on p. 451.)
3. the amount of oxygen in 15 g of hemoglobin with 86% oxygen (Use the formula on p. 250.)
4. the amount of oxygen dissolved in 100 mL of whole blood plasma with a P reading of 94 mm (Use the formula on p. 250.)
5. the recommended maximum pulse rate for Tammy, who is 22 years old (Use the formula on p. 254.)
6. the vital capacity for Robin Wertz, who is 141 cm tall and 26 years old (Use the formula on p. 254.)
7. the weight to be lifted in a training program by Mary Jacque, who can lift a maximum of 80 lb (Use the table on p. 257.)
8. the minimum weight to be lifted by Rico, who can lift a maximum of 180 lb, to increase strength (Use the formula on p. 256.)
9. the lean body weight of Tyler, who weighs 190 lb with 14.1% body fat (Use the table on p. 258.)
10. the annual life insurance premium for a $90,000, 10-year term policy on a 35-year-old male (Use the table on p. 452 in the **Data Bank.**)

Column B

A. 135 lb
B. $13.56
C. $106
D. 17.544 mL
E. $405
F. 0.282 mL
G. 158
H. 2.6 L
I. 64 lb
J. 163 lb

CHAPTER 11 TEST

Find the amount paid by the company and the insured.

	Covered Medical Expense	Deductible	Rate Paid by Insurance Company
1.	$ 637	$ 75	85%
2.	$8635	$200	90%

Find the annual cost for the insured and for the employer.

	Pay Period	Coverage	Deduction per Pay	Employer's Contribution
3.	Biweekly	Family	$23.65	$15.75
4.	Monthly	Individual	$ 5.44	$ 8.37

Compare the benefits under the Capital and World medical plans.

	Service	Cost
5.	Routine physical	$ 74
6.	Doctor's office visits (5)	$ 190
7.	Hospital room and board	$6392

Find the total amount of oxygen for a 100-mL blood sample.

	Oxygen Pressure in Plasma P	Oxygen in Plasma W (mL)	Hemoglobin g	Percent of Oxygen s	Oxygen in Hemoglobin H (mL)	Total Oxygen in Sample $W + H$ (mL)
8.	88 mm		13 g	89%		

9. The hospital laboratory analyzed 287 blood samples over a 30-day period. Find the average number of samples analyzed per day.

10. Find the maximum pulse rate during exercise for a 37-year-old male.

11. Find the vital capacity for Elaine, height 150 cm, age 28.

12. Debbie can lift a maximum of 92 lb. What is the minimum weight she should lift to increase her strength?

13. Use the formula $L = W(1 - F)$ to find the lean body weight to the nearest pound. $W = 153$ lb, $F = 15.6\%$

Find the annual life insurance premium.

	Sex	Age Purchased	Type	Amount
14.	M	33	Universal life	$ 80,000
15.	F	26	30-year endowment	$130,000

CHAPTER 11 CUMULATIVE REVIEW

1. Find the amounts paid by the insurer and by the insured.
covered medical expense $1572, deductible $100, rate paid by insurer 85%

A. insurer, $718.28; insured, $853.76
B. insurer, $1336.20; insured, $235.80
C. insurer, $1251.20; insured, $320.80
D. insurer, $1472.00; insured, $100.00

2. Mavis Washington weighs 101 pounds. Her weight goal is 115 pounds. She does not play any sports. Find the number of calories per day she needs to reach her goal.

A. 2057 calories
B. 2101 calories
C. 2114 calories
D. 2520 calories

Use:
First 22 pounds = 1000 calories
Next 22 pounds = 500 calories
Each additional 22 pounds = 200 calories
Active athlete = 500 calories
Lose/gain weight: ±500 calories

3. Subtract.
$2 - \frac{5}{6}$

A. $1\frac{1}{5}$ B. $2\frac{5}{6}$
C. $1\frac{1}{6}$ D. $2\frac{1}{6}$

4. The odds against the Panthers winning the pennant are 4 to 11. What is the probability the Panthers will win?

A. $\frac{7}{11}$ B. $\frac{11}{15}$
C. $\frac{4}{11}$ D. $\frac{4}{15}$

5. Estimate. Use rounding.
$$\begin{array}{r} 4012 \\ -2204 \\ \hline \end{array}$$

A. 1000
B. 2000
C. 200
D. 6000

6. Multiply.
$$\begin{array}{r} 5079 \\ \times \ .36 \\ \hline \end{array}$$

A. 1871.10
B. 1558.44
C. 1828.44
D. 2171.40

7. Find Dean Jones' ERA after giving up 21 runs in 90 innings. Use the formula $ERA = 9\frac{R}{I}$.

A. 2.6   B. 2.1
C. 4.3   D. 2.3

8. Use the formula $L = W(1 - F)$ to find the lean body weight to the nearest pound. $W = 265$ lb, $F = 21.1\%$

A. 211 lb   B. 208 lb
C. 210 lb   D. 209 lb

9. Compare.
$\frac{5}{12} \square 0.41\frac{2}{6}$

A. $<$ B. $=$
C. $>$ D. $\approx$

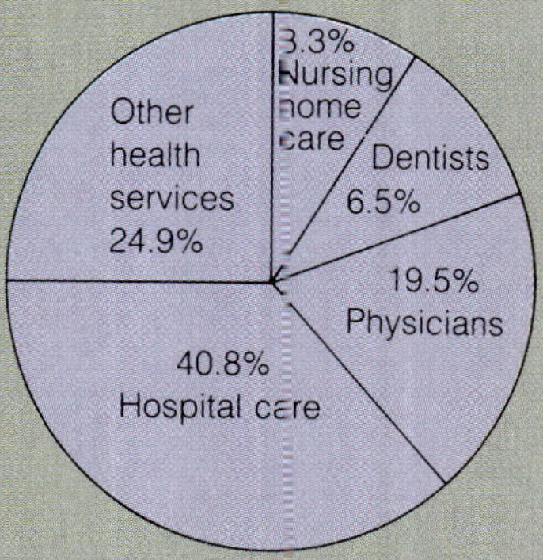

10. How much more was spent on hospital care than on physicians?

A. 21.3%
B. 39.3%
C. 29.3%
D. 20.3%

11. Sal's Sundaes offers 2 sizes, 3 flavors of ice cream, and 4 different toppings. There is one choice each of ice cream and topping. How many different sundaes are available?
A. 24 B. 9
C. 12 D. 36

12. Find the net deposit.
3 \$100 bills, 2 \$50 bills, 6 \$20 bills, a check for \$62.26
A. \$582.26
B. \$562.26
C. \$482.26
D. \$662.26

13. Find the pulse rate. Use mental math. 20 seconds, 24 beats
Use $\frac{60}{S} \times b$.
A. 82
B. 62
C. 92
D. 72

14. Find the carrying charges for a motor scooter.
list price \$1560, down payment 30%, monthly payment \$190, term 180 days
A. \$1140
B. \$48
C. \$38
D. \$108

15. A prescription costs \$15.42. Green Health Plan pays 85%. White Health Plan requires a copayment of \$3. Find the difference in benefits.
A. Green pays 69¢ more.
B. White pays 69¢ more.
C. Green pays \$13.11 more.
D. White pays \$12.42 more.

16. Find the weight to be used. Use $w = 0.5m$.
toning leg lift max. = 38 lb
A. 29 lb
B. 57 lb
C. 16 lb
D. 19 lb

17. Compare the cost of raising 34 pounds of potatoes from 2 pounds of seed costing \$1.05 per pound with the market price of 45¢ per pound for potatoes.
A. save \$13.20
B. save \$14.21
C. save \$21.76
D. save \$15.30

18. Find the annual employee benefit. Amerink Company runs a free shuttle van between the train station and the plant. A monthly bus pass for the same route is \$15.
A. \$150
B. \$450
C. \$180
D. \$280

19. Find the annual life insurance premium. Use *Premium* = 0.001 (*Amount* × *rate*).
For \$50,000 straight life for a 27-year-old female the rate is \$11.40.
A. \$750
B. \$308
C. \$550
D. \$570

Chapter 12 Overview

Key Ideas

- finding sales tax
- finding withholding tax
- finding FICA deductions
- finding federal income tax
- finding state and local income taxes
- finding property taxes
- finding gift, estate, and excise taxes
- comparing receipts and expenses

Key Terms

- sales tax
- sales tax schedule
- withholding tax
- exemptions
- FICA deductions
- gross pay
- net pay
- adjustable gross income
- taxable income
- filing status
- W-2 form
- market value
- assessed value
- rate of assessment
- estate value
- federal excise tax

Key Skills

Find the answer.

1. \$26.78 + \$1.89 + \$132.06

2. \$21.32 + \$319.43 + \$2.99 + \$217.60

3. \$48.87 + \$2.09 + \$302.61

4. \$638.99 + \$21.82 + \$10.65 + \$7.41

5. \$23.62 − \$18.94

6. \$21,438 − \$7,000

7. \$52,981 − \$9,000

8. \$54,987 − \$11,000

9. \$15.98 − \$4.09

10. \$23,083 − \$13,000

Solve.

11. What is 20% of \$34,218?

12. What is 4% of \$28,872?

13. What is 5% of \$42,986?

14. What is 2% of \$18,994?

15. What is 28% of \$45,882?

16. What is 15% of \$38,889?

17. What is $4\frac{1}{2}$% of \$89.16?

18. What is $5\frac{1}{2}$% of \$48.03?

19. What is $3\frac{1}{2}$% of \$50.04?

20. What is $6\frac{1}{4}$% of \$28.63?

21. What is 7.51% of \$38,459?

22. What is 8.49% of \$12,560?

23. What is 4.1% of \$12,765?

24. What is 1.45% of \$98,564?

25. 289.6 is what percent of 783.2?

26. 342.4 is what percent of 863.9?

27. 317 is what percent of 986.8?

28. 568.4 is what percent of 1089.5?

TAXES 12

12-1 Sales Tax

Objective
Find sales tax from a table. Find sales tax using percent.

Most purchases we make are subject to a sales tax. The rate of tax is normally a percent of the purchase price and varies among states. Some cities and counties add their own tax. Sales clerks frequently use a table to find the amount of tax.

Example

1. Find the sales tax on $18.53. Use the 6% Sales Tax Schedule on p. 452.
$18.53 is between $18.42 and $18.58. The tax is $1.11.

Try This

Find the sales tax. Use the 6% Sales Tax Schedule.

a. $5.85 **b.** $0.64 **c.** $20.06 **d.** $17.53

Example

2. Anna Chin bought some tennis gear. Find the total bill with sales tax.

Totally Tennis

Quantity	Item	Unit Price		Price
1	Racket	$65.98	1 × 65.98	$65 | 98
3	Cans of balls	1.29	3 × 1.29	3 | 87
2	Tennis skirts	14.79	2 × 14.79	29 | 58
		Subtotal	Add	99 | 43
		$4\frac{1}{2}$% tax	0.045 × 99.43	4 | 47
		Total		$103 | 90

Try This

e. Find the total bill including sales tax.

Watson's Garden Center

Quantity	Item	Unit Price	Price
3 bags	Fertilizer	$ 8.79	
2	Pear trees	27.50	
		Subtotal	
		7% tax	
		Total	

Exercises

Find the sales tax. Use the tax table.

1. \$19.88 **2.** \$16.71 **3.** \$7.43 **4.** \$5

5. \$0.87 **6.** \$0.91 **7.** \$2.04 **8.** \$9.56

Find the total bill including sales tax.

9.

Lucy's Linen Press

Quantity	Item	Unit Price	Price
4	Bath towels	\$12.49	
4	Hand towels	7.99	
6	Face cloths	1.75	
		Subtotal	
		6% tax	
		Total	

10.

Clancy's Hardware Store

Quantity	Item	Unit Price	Price
4 cans	Motor oil	\$ 1.25	
3 qt	Interior paint	12.95	
2	Brushes	2.49	
		Subtotal	
		$5\frac{1}{2}$% tax	
		Total	

Strategy Practice

11. Make a list of the currency used to make change for a \$20 bill for the following amounts. Use the smallest number of bills and coins possible.

A. \$15.75 B. \$13.39 C. \$9.51 D. \$7.02 E. \$0.16

Mixed Skills Review

1. $13.2 - 12.728$ **2.** $3\frac{3}{5} \times 2\frac{2}{9}$ **3.** $12.3 + 16.07 + 19.21 + 9.4$

4. 1.34×12.6 **5.** $16{,}903 + 18{,}267$ **6.** 318×236

7. $1226 - 762$ **8.** $1426.28 \div 39.4$ **9.** $2\frac{1}{3} + 8\frac{1}{2}$

10. $245{,}912 \div 521$ **11.** $4\frac{2}{9} - 3\frac{2}{3}$ **12.** $1\frac{5}{8} \div 1\frac{5}{7}$

12-2 Withholding Tax

Objective
Find withholding tax from a table.

A portion of an employee's gross pay is withheld for federal income taxes during each pay period. The amount deducted depends upon the employee's gross pay, marital status, and the number of **exemptions** claimed.

A tax table is used by an employer to determine the amount withheld.

Single Persons—Weekly Payroll Period

And the wages are–		And the number of withholding allowances claimed is–			
At least	But less than	0	1	2	3
		The amount of income tax to be withheld shall be–			
110	115	14	8	3	0
115	120	14	9	4	0
120	125	15	10	4	0
125	130	16	10	5	1
270	280	38	33	27	22
280	290	40	34	29	23
290	300	41	36	30	25

Examples

1. Joan Hewitt is single and claims one exemption. Her weekly gross pay is $286. Find the amount withheld for federal income tax.

Use the table to find the wage interval:
$286 is at least $280 but less than $290.

Read across the table to find the amount withheld for 1 exemption: $34.

2. John Kahnert is married and claims 4 exemptions. He works 36 hours per week and earns $11.65 per hour. Use the withholding tax table for married persons on pages 456–458 to find the amount withheld.

Find the weekly gross pay. $36 \times \$11.65 = \$419.40$

$419.40 is at least $410 but less than $420.

Read across the table to find the amount withheld for 4 exemptions: $33.

Try This Find the amount withheld for federal income tax. Use the tax tables on p. 453–458.

	Job	Marital Status	Exemptions	Hours per Week	Hourly Rate
a.	Baker	single	2	36	$6.75
b.	Cashier	married	3	21.5	7.40

Exercises

Find the amount withheld for federal income tax. Use the tables on p. 453–458.

	Job	Marital Status	Exemptions	Hours per Week	Hourly Rate
1.	Painter	single	1	24	\$11.40
2.	Stock clerk	married	3	40	8.45
3.	Custodian	single	2	36	10.64
4.	Plumber's helper	married	4	32	7.53
5.	Seamstress	married	6	40	9.60
6.	Library aide	single	0	32	6.90
7.	Floral designer	single	1	20	8.25
8.	Delivery person	single	0	12	8.25
9.	Operator	married	5	37.5	10.60
10.	Typist	single	1	16.5	7.30

11. Carlton Hawpe worked as a stock clerk for five weeks during the holiday season. His weekly gross pay was \$98. He is single and claimed one exemption. Find the total amount withheld for federal income tax.

12. Susie Culbertson works thirteen weeks as a summer lifeguard. Her weekly gross pay is \$145. She is single and claims no exemptions. Find the total amount withheld from Susie's paychecks.

13. Deaf Smith runs fishing and hunting trips near Big Bend. His base pay from West Inc. is \$300 per week. Tips and bonuses are worth another \$100 per week. Deaf is married and claims his entire family of 10 as dependents. What amount must be withheld from his weekly pay for federal income tax?

14. Jerry Cohen is single and claims one exemption. His employer deducts \$22 from his weekly paycheck for federal income tax. What is Jerry's weekly gross pay?

15. Elaine Geisz is a single parent with two children. Each week her employer deducts \$117 from her paycheck for federal income tax. What is Elaine's weekly gross pay?

Mixed Skills Review

1. $2\frac{7}{15} + 3\frac{3}{5} + \frac{1}{3}$

2. $199{,}248 \div 48$

3. $1.18 + 3.06 + 6.23$

4. $0.814 \div 0.044$

5. $13{,}030 - 9736$

6. $\frac{2}{3} - \frac{5}{12}$

7. $8 - 0.632$

8. $16 + 34 + 19 + 67 + 49$

9. $\$12.78 \times 4.5$

10. $1\frac{7}{8} \times 2\frac{4}{5}$

11. 1641×13

12. $\frac{5}{9} \div \frac{5}{6}$

12-3 Social Security Taxes

Objective *Calculate the FICA deduction and net pay.*

Each employer must also deduct **FICA** taxes from an employee's earnings. Of the first $42,000 of the employee's gross income, 7.15% is withheld each year.

Example

1. Find the FICA deduction for the gross weekly pay of $487.50.

Multiply gross pay by 7.15%. $0.0715 \times \$487.50 \approx \$34.856$

FICA deduction is $34.86.

Try This

Find the FICA deduction.

a. Gross pay $385.40 **b.** Gross pay $822.75

An earnings statement is usually attached to a paycheck. It shows the gross pay, deductions, and net pay. The net pay is the gross pay minus deductions.

Example

2. Find the FICA deduction and the net pay.

Hours	Amount	Deductions	
40.0 Regular	$160.00	Fed. withholding tax	$20.90
5.5 Overtime	33.00	State and local tax	11.58
Gross pay	$193.00	FICA	
		Insurance	6.33
Net pay			

Find the FICA deduction. $0.0715 \times \$193 = \$13.80.$

Add the deductions. $\$20.90 + \$11.58 + \$13.80 + \$6.33 = \$52.61$

Subtract the total from the gross pay $\$193 - \$52.61 = \$140.39$

Net pay is $140.39.

Try This

Find the FICA deduction and the net pay.

c.		**d.**	
Gross pay	$241.67	Gross pay	$175.42
Fed. withholding tax	26.40	Fed. withholding tax	15.10
State and local taxes	14.56	State and local taxes	11.53
FICA		FICA	
Credit Union	7.50	United Fund	5.00
Net pay		Net pay	

Exercises

Find the FICA deduction.

1. Gross pay $165.42 **2.** Gross pay $427.63

3. Gross pay $198.91 **4.** Gross pay $98.36

5. Gross pay $273.05 **6.** Gross pay $380.16

Find the FICA deduction and the net pay.

7.

Gross pay	$154.99
Deductions	
Fed. withholding tax	$ 13.70
State and local taxes	$ 9.77
FICA	_______
Credit Union	$ 25.00
Net pay	

8.

Gross pay	$375.02
Deductions	
Fed. withholding tax	$ 58.80
State and local taxes	$ 15.88
FICA	_______
Credit Union	$ 8.33
Net pay	

9.

Gross pay	$231.88
Deductions	
Fed. withholding tax	$ 21.20
State and local taxes	$ 12.79
FICA	_______
Credit Union	$ 13.41
Net pay	

10.

Gross pay	$409.00
Deductions	
Fed. withholding tax	$ 62.30
State and local taxes	$ 24.54
FICA	_______
Credit Union	$ 5.00
Net pay	

Calculator Find the annual FICA deduction. Use a calculator.

11. Annual gross pay $29,432 **12.** Monthly gross pay $1025.44

13. Annual gross pay $42,000 **14.** Monthly gross pay $1377.62

15. Annual gross pay $57,751 **16.** Weekly gross pay $821.80

17. If in 1990 the FICA deduction increases to 7.65% of the annual gross pay, what would be the maximum annual deduction?

Strategy Practice

18. Jason Kampes' employer deducted $15.02 for FICA taxes from his paycheck. What was Jason's gross pay? (Hint: Translate to a number sentence.)

Mixed Skills Review

1. 680,820 − 46,756 **2.** $4\frac{2}{3} - \frac{11}{12}$ **3.** 0.82 + 0.16 + 0.7

4. 11,952 ÷ 83 **5.** 276 × 364 **6.** 11.47 ÷ 0.062

7. 42,653 + 18,934 **8.** $1\frac{5}{8} + 3\frac{3}{4}$ **9.** $7 \times 5\frac{3}{8}$

10. 12.22 × 126 **11.** 12.2 − 10.64 **12.** $6\frac{2}{3} \div 6$

12-4 Federal Income Tax

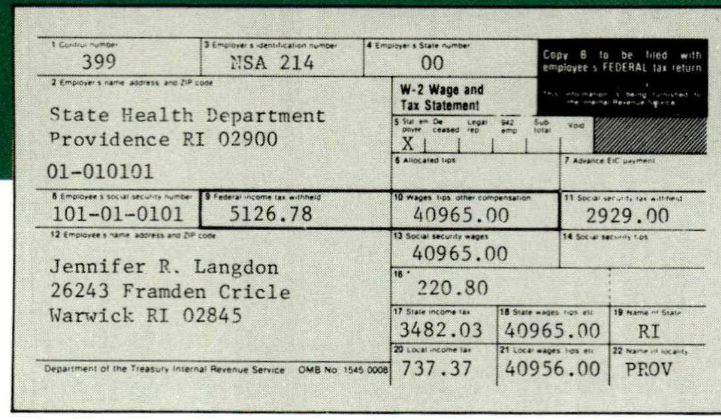

1 Control number: 399
3 Employer's identification number: NSA 214
4 Employer's State number: 00
Copy B to be filed with employee's FEDERAL tax return
W-2 Wage and Tax Statement
2 Employer's name, address, and ZIP code
State Health Department
Providence RI 02900
01-010101
5 Statutory employee: X
6 Allocated tips
7 Advance EIC payment
8 Employee's social security number: 101-01-0101
9 Federal income tax withheld: 5126.78
10 Wages, tips, other compensation: 40965.00
11 Social security tax withheld: 2929.00
12 Employee's name, address, and ZIP code
Jennifer R. Langdon
26243 Framden Cricle
Warwick RI 02845
13 Social security wages: 40965.00
14 Social security tips
16: 220.80
17 State income tax: 3482.03
18 State wages, tips, etc.: 40965.00
19 Name of State: RI
20 Local income tax: 737.37
21 Local wages, tips, etc.: 40956.00
22 Name of locality: PROV
Department of the Treasury Internal Revenue Service OMB No. 1545-0008

Objective
Find federal income tax.

Generally, every U.S. resident under 65 with a gross income of $3560 or more must file an income tax return. Children claimed on their parents' return must file if they have an income of $1000 or more.

The **W-2 form** is a record of earnings prepared by an employer. It shows the gross income and the amount withheld for federal, state and local taxes, and FICA. Each taxpayer must report the total wages, salaries, and tips earned during the tax year. The sum of earnings and income from all sources is the **adjusted gross income.** The W-2 is attached to the tax return when filing.

Taxpayers are allowed certain deductions from the adjusted gross income. The amount deducted depends on the taxpayer's **filing status** and the number of exemptions claimed.

Standard Deduction	Personal Exemptions
$3000 for single taxpayers $5000 for married couples $4400 for heads of households	$2000 for each exemption claimed

The taxable income is the difference between the adjusted gross income and the total amount deducted. The amount of tax is based on the taxable income.

Example

Jennifer and Jim Langdon are married and claim 3 exemptions. Their adjusted gross income is $40,965. Find their taxable income and the amount of tax.

Total the deductions for a married couple with 3 exemptions.
$\$5000 + (3 \times \$2000) = \$11,000$

Subtract the total from the adjusted gross income.
$\$40,965 - \$11,000 = \$29,965$ The taxable income is $29,965.

The tax rate is 15% up to $29,750 plus 28% of any amount over $29,750.

$0.15 \times \$29,750 = \$4462.50$
$0.28 \times (\$29,965 - \$29,750) = 0.28 \times \$215 = \$60.20$
$\$4462.50 + \$60.20 = \$4522.70$ Their tax is $4522.70.

Try This

Find the taxable income and the amount of tax.

	Adjusted Gross Income	Filing Status	Exemptions
a.	$14,956	single	1
b.	$54,067	married	4
c.	$27,804	head of household	3

Exercises

Use the tax rates given on p. 276. Find the taxable income and the amount of tax.

	Adjusted Gross Income	Filing Status	Exemptions
1.	$24,044	single	1
2.	$37,225	married	3
3.	$19,106	head of household	2
4.	$44,067	married	4
5.	$32,956	single	1
6.	$38,470	married	5
7.	$17,229	single	1
8.	$56,804	head of household	3

9. Barbara Kirk's W-2 form showed that $7399.24 had been withheld for federal income tax. The actual tax on her income was $6947.30. Find the amount of Barbara's tax refund.

10. The W-2 form showed $5308.58 was withheld for federal income tax from Paul Krysiak's earnings. The actual tax on his income was $5733.40. Find the balance due.

11. Patrick McGee makes $41,000 a year. He is single and claims 2 exemptions. The federal government withholds $12,000 of his adjusted gross income. Will Patrick have a balance due or will he receive a refund from the government? How much will he pay or receive?

Given the taxable income, find the amount of tax.

12. $28,000 **13.** $18,600 **14.** $20,400 **15.** $14,800

Estimation Given the taxable income, estimate the amount of tax.

16. $32,000 **17.** $45,000 **18.** $54,600 **19.** $49,500

20. John Paul made $38,597 last year. He is married and claims 5 exemptions. Estimate the amount of tax he must pay.

Mixed Skills Review

1. $2.7 \times 6.3 \times 3$ **2.** $2\frac{2}{5} + 4 + \frac{2}{3}$ **3.** $1\frac{1}{5} \div \frac{8}{15}$

4. $152{,}460 \div 330$ **5.** $43.005 \div 23.5$ **6.** $13.285 - 1.636$

7. $127 + 38 + 9 + 1764$ **8.** $2765 - 1837$ **9.** $3\frac{5}{8} \times 6\frac{4}{5}$

10. $1\frac{1}{2} - \frac{7}{8}$ **11.** 911×1600 **12.** $16.085 - 16.036$

12-5 State and Local Income Tax

Objective
Find state and local income tax.

David Rause lives in Dearborn, Michigan. He works part-time at a local newspaper. His parents helped him fill out his first tax forms. They explained that many states and some cities have an income tax.

State Income Tax Rates	
Texas	None
Nebraska	18% of federal income tax paid
Michigan	5.35% of adjusted gross income
Virginia	2% of first \$3000 of adjusted gross income 3% of next \$2000 of adjusted gross income 5% of next \$7000 of adjusted gross income $5\frac{3}{4}$% of adjusted gross income over \$12,000

City Income Tax Rates	
Baltimore, MD	50% of state income tax paid
Dayton, OH	$2\frac{1}{4}$% of adjusted gross income
Philadelphia, PA	4.96% of adjusted gross income

Example

1. David had an adjusted gross income of \$5610. Find his Michigan tax.

Multiply the adjusted gross income by the rate.

$0.0535 \times \$5610 = \$300.135$ David's state tax is \$300.14.

Try This

Find the state income tax.

a. Caroline lives in Nebraska. She paid federal income tax of \$1587.

b. Reggie lives in Virginia. His adjusted gross income was \$19,046.

Example

2. Bill Welsh had an adjusted gross income of \$23,644. He lives and works in Dayton, Ohio. Find Bill's city income tax.

Multiply the adjusted gross income by the city tax rate.

$0.0225 \times \$23{,}644 = \$531.99$ Bill's city income tax is \$531.99.

Try This

Find the city income tax.

c. Dustin lives in Baltimore. He paid state income tax of \$659.

d. Shirley lives in Philadelphia. Her adjusted gross income was \$8296.

Exercises

Find the state income tax.

1.	Michigan	Adjusted gross income \$15,000
2.	Texas	Adjusted gross income \$37,495
3.	Virginia	Adjusted gross income \$20,950
4.	Nebraska	Federal income tax paid \$1875
5.	Nebraska	Federal income tax paid \$2435
6.	Michigan	Adjusted gross income \$9078
7.	Texas	Adjusted gross income \$12,899
8.	Virginia	Adjusted gross income \$23,790

Find the city income tax.

9.	Dayton, Ohio	Adjusted gross income \$6950
10.	Baltimore, Maryland	State income tax paid \$1689
11.	Philadelphia, Pennsylvania	Adjusted gross income \$8460
12.	Philadelphia, Pennsylvania	Adjusted gross income \$12,940
13.	Baltimore, Maryland	State income tax paid \$2077

States use different formulas to determine the amount of tax to be paid by state residents. One state uses this formula.

Tax Rate Schedule

If the taxable income is: Over	But not over	The amount of tax is:
\$0	\$1000	2% of the taxable income
\$1000	\$2000	\$20 plus 3% of excess over \$1000
\$2000	\$3000	\$50 plus 4% of excess over \$2000
\$3000	—	\$90 plus 5% of excess over \$3000

Calculate the state tax on the following amounts.

14. \$756 **15.** \$1230 **16.** \$2150 **17.** \$9675

Mixed Skills Review

1. \$470 − \$324.73 **2.** $3\frac{1}{2} + 2\frac{5}{16} + \frac{1}{4}$ **3.** 6.82×25.4

4. $2\frac{1}{9} \div 6\frac{1}{3}$ **5.** $32{,}430 \div 705$ **6.** \$23 + \$16.07 + \$31.84 + \$0.67

7. $12{,}038 - 6335$ **8.** $1\frac{9}{10} - 1\frac{5}{8}$ **9.** $20{,}433 \div 49$

10. $2\frac{1}{3} \times 2\frac{5}{7}$ **11.** 245×459 **12.** $17.09 + 0.635 + 21.8$

12-6 Property Tax

Objective
Find property tax.

Taxes from real estate help pay for many local services, such as schools and fire and police protection. Property taxes are based on the following.

The **market value** is the selling price of the property on the open market. The **assessed value** is the value used for tax purposes. It is usually some percent of the market value.

Examples

1. Find the assessed value and the property tax in Wyoming.

Market Value	Rate of Assessment	Assessed Value	Tax Rate	Property Tax
$43,500	15%		$8 per $100	

Multiply the market value by the rate of assessment to find the assessed value.
$0.15 \times \$43,500 = \$6525$

Multiply the assessed value by the tax rate to find the property tax.

$$\$6525 \times \frac{\$8}{\$100} = \frac{52,200}{100} = \$522.00$$

Tax rates are often given as a percent of the assessed value.

2. Find the assessed value and the property tax in Georgia.

Market Value	Rate of Assessment	Assessed Value	Tax Rate	Property Tax
$65,950	40%		2.47%	

Find the assessed value.
$0.40 \times \$65,950 = \$26,380$

Find the property tax.
$0.0247 \times \$26,380 = \$651.59$

Try This

Find the assessed value and the property tax.

	Market Value	Rate of Assessment	Assessed Value	Tax Rate	Property Tax
a.	$39,000	25%		$3 per $100	
b.	$60,500	100%		$0.75 per $100	

Exercises

Find the assessed value and the property tax.

	Market Value	Rate of Assessment	Assessed Value	Tax Rate	Property Tax
1.	\$40,400	25%		\$5 per \$100	
2.	\$81,300	40%		\$3 per \$100	
3.	\$39,500	30%		\$4 per \$100	
4.	\$67,400	65%		\$2 per \$100	
5.	\$56,000	35%		\$4.25 per \$100	
6.	\$70,000	50%		\$2.15 per \$100	
7.	\$29,900	30%		3.7%	
8.	\$67,200	45%		2.6%	
9.	\$44,700	25%		4.1%	
10.	\$51,900	31%		5.5%	
11.	\$83,000	15%		8.49%	
12.	\$95,000	100%		1.54%	

13. Suppose you were appointed tax assessor. What factors would you consider in assigning value to a piece of property?

14. Look at these pieces of property. Estimate the market value of each in your area.

Data Hunt

15. Call the assessor's office in your area.

A. Find the rate of assessment, the tax rate, and the filing deadline.

B. Ask if there is a discount when you pay property taxes before the annual filing deadline.

C. Calculate the property tax on an \$80,000 home in your area.

Mixed Skills Review

1. $14 - \frac{13}{15}$

2. $946 + 1210 + 1578$

3. $0.6 \times 1.34 \times 5$

4. $0.38 + 1.218 + 1.054$

5. $\frac{1}{2} \div 6\frac{3}{8}$

6. $16{,}397 \div 863$

7. $1849.252 \div 0.554$

8. 1764×34

9. $1\frac{7}{10} + 2\frac{3}{4}$

10. $936 - 548$

11. $9\frac{1}{4} \times 1\frac{1}{2}$

12. $124.58 - 47.493$

12-7 Gift, Estate, and Excise Taxes

Objective
Find gift, estate, and excise taxes.

An inheritance or a gift is subject to tax by the federal government.

If the net amount is: From	To	The tax is: Tax +	%	On excess over
$ 0	$10,000	$ 0	(18)	$ 0
10,000	20,000	1,800	20	10,000
20,000	40,000	3,800	22	20,000
40,000	60,000	8,200	24	40,000
60,000	80,000	13,000	26	60,000
80,000	100,000	18,200	28	80,000
100,000	150,000	23,800	30	100,000
150,000	250,000	38,800	32	150,000
250,000	500,000	70,800	34	250,000

Example

1. Find the amount of estate and gift tax due on an inheritance in the amount of $75,600.

$75,600 is between $60,000 and $80,000.
The amount of tax is $13,000 plus 26% of the excess over $60,000.
$\$75{,}600 - \$60{,}000 = \$15{,}600$ $\quad$ $\$13{,}000 + \$4056 = \$17{,}056$
$0.26 \times \$15{,}600 = \$4056$ $\quad$ The tax is $17,056.

Try This

Find the federal estate and gift tax.

a. $23,400 **b.** $189,000 **c.** $309,500

Certain goods and services are subject to federal excise tax.

Item	Federal Excise Tax Rate	Item	Federal Excise Tax Rate
Telephone service	3%	Electrical outboard motors	3%
Airline transportation	8%	Bows and arrows	11%
Shipment by air	5%	Gasoline	9%
Sport fishing equipment	10%	Gasohol	3%

Example

2. Find the amount of federal excise tax on a bow for $209.95.

Multiply the price by the rate of tax. $0.11 \times \$209.95 \approx \$23.09$ The tax is $23.09.

Try This

Find the amount of federal excise tax.

d. Local telephone service, $18.54 **e.** Outboard motor, $459.80

Exercises

Find the federal estate and gift tax.

1. \$4500 **2.** \$39,200 **3.** \$21,700 **4.** \$60,300

5. \$49,900 **6.** \$87,100 **7.** \$250,400 **8.** \$432,300

Find the amount of federal excise tax.

9. Airline ticket, \$431

10. Gasoline, \$9

11. Rod and reel, \$159.80

12. Arrows, \$45.60

13. Telephone service, \$84.25

14. Gasohol, \$65.72

Each year the Internal Revenue Service examines a portion of all the returns filed for accuracy and proper payment. If the returns checked in each category on the graph below are chosen at random, what is the probability in each category that a return is examined?

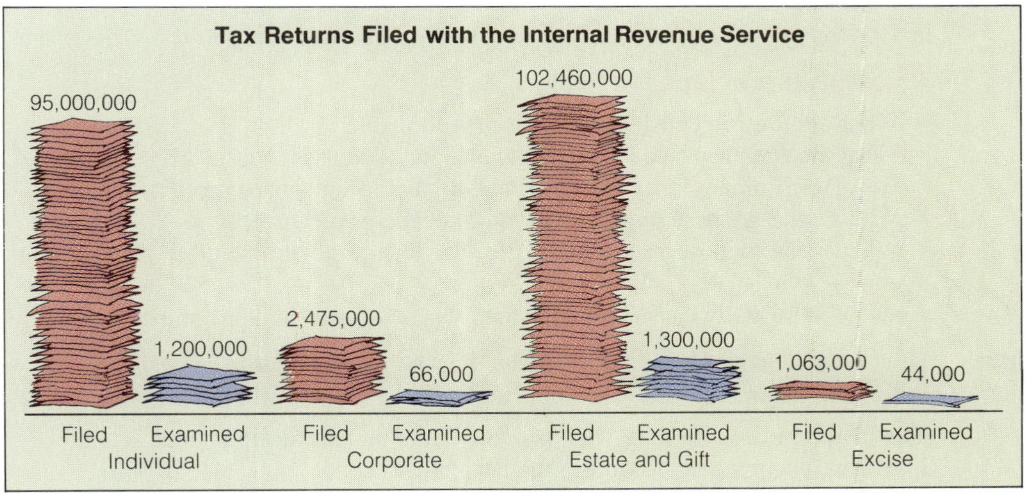

15. Individual Returns

16. Corporate Returns

17. Estate and Gift Returns

18. Excise Returns

Mixed Skills Review

1. 805×2209 **2.** $13.908 \div 2.44$ **3.** $1990 - 1896$

4. $4\frac{3}{10} + 5\frac{2}{3}$ **5.** $102,654 + 267,634$ **6.** $1\frac{5}{7} \div \frac{6}{7}$

7. 206.3×0.041 **8.** $6\frac{1}{8} - 4\frac{5}{9}$ **9.** $9.2 + 3.4 + 6.3 + 2.4$

10. $114 - 3.226$ **11.** $9728 \div 128$ **12.** $8\frac{1}{2} \times 1\frac{1}{3}$

12-8 How Your Tax Dollars Are Spent

Objective
Compare receipts and expenses from the federal budget.

The annual budget for the federal government is about \$1,000,000,000,000. These graphs project receipts and expenditures for one tax year.

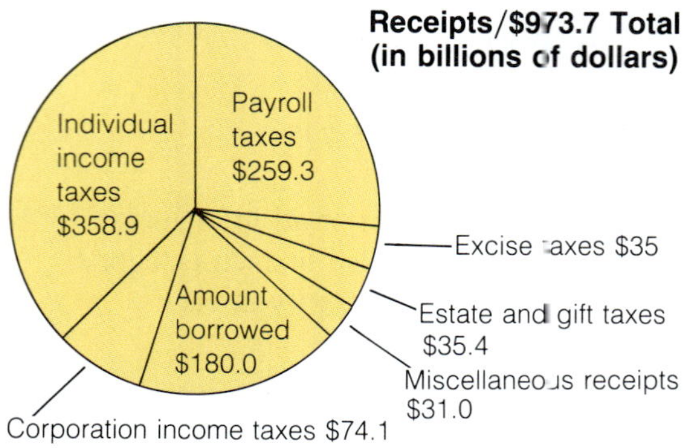

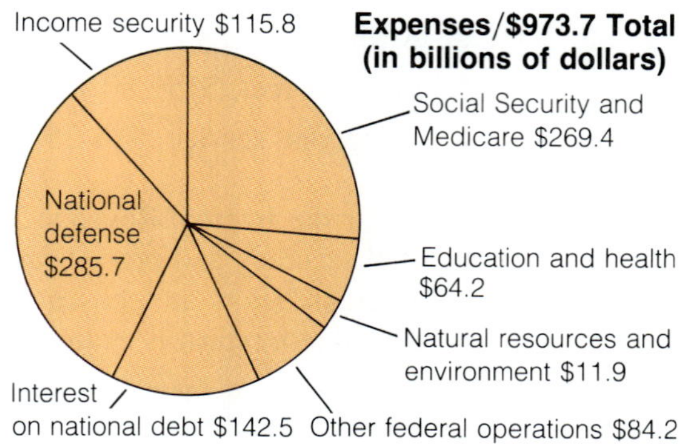

Understand the Situation

- What are the projected receipts for payroll taxes?
- What are the projected expenses for national defense?
- How much money does the government have to borrow to meet its budget?
- What is the greatest source of revenue for the government?
- What is the total operating budget for the federal government?

Budgets are usually easier to understand when we use smaller numbers.

Example How many cents of each dollar received in the federal budget come from individual income taxes?

This is the same as finding what percent of the budget receipts come from individual income tax. Round to the nearest percent.

$$\frac{\text{individual income tax receipts}}{\text{total budget receipts}} = \frac{\$358.9 \text{ billion}}{\$973.7 \text{ billion}} \approx 37\%$$

A reasonable estimate is $\frac{\$400 \text{ billion}}{\$1000 \text{ billion}} = 40\%$

which is close to 37%. 37% of \$1 is \$0.37.

About 37¢ of each federal dollar received comes from individual income taxes.

Try This Find the number of cents per dollar allocated for each budget item.

a. Receipts from corporation taxes

b. Expenses for health and education

c. Expenses for other federal operations

Exercises

Use the federal budget to answer the following questions.

1. What are the projected receipts from corporation taxes?
2. What is the difference between the receipts for individual income taxes and payroll taxes?
3. What is the projected total for excise and estate and gift taxes?
4. Which item is the greatest expense for the government?
5. How much more money is spent on education and health than on natural resources and environment?
6. The amount spent on interest is what percent of the amount spent on national defense?
7. Find the number of cents per dollar allocated for each budget item in receipts. The total should be $1. Add to check.
8. Find the number of cents per dollar allocated for each budget item in expenditures. Add to check.

Number Sense

Use the given information to find about how many hours per week each person works to pay federal income taxes.

9. A. John Cranston works 40 hours per week.
   B. He earns $24,960 per year.
   C. He paid $2394 in federal income taxes.
10. A. Faith Hermann works 36 hours per week.
    B. She earns $2901.60 per month.
    C. She paid $8349.38 in federal income taxes.

Estimation

11. What percent of federal receipts are from individual income taxes?
12. What percent of federal receipts are from corporate income taxes?
13. What percent of federal expenses are used for national defense?
14. What percent of federal expenses are used for education and health?

Mixed Skills Review

1. $5618 + 12 + 264$
2. $2\frac{3}{8} - 1\frac{2}{3}$
3. 6135×73
4. $24{,}768 \div 516$
5. $4\frac{1}{2} \times 5\frac{1}{2}$
6. $8\frac{2}{5} + 1\frac{1}{6}$
7. $3.38 - 2.385$
8. $0.54 + 0.27 + 0.36$
9. $2958 \div 435$
10. 19.6×2.35
11. $1430 - 847$
12. $2\frac{5}{6} \div \frac{1}{4}$

12-9 Enrichment: Checking a Tax Form

Objective
Find errors in a completed 1040EZ Income Tax Return.

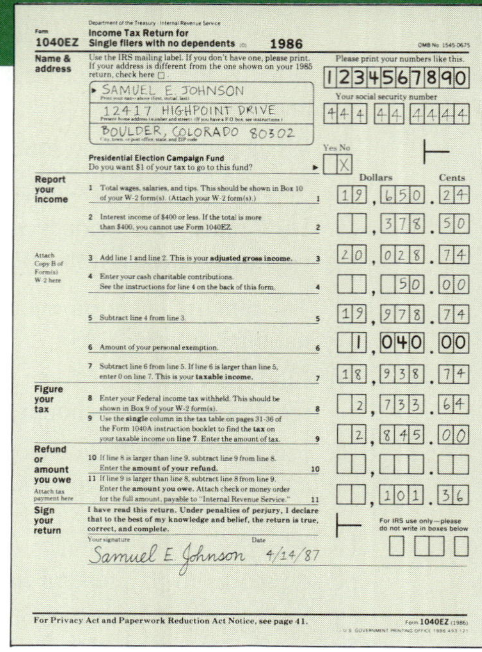

Sam Johnson checked the figures on his 1040EZ form. He did the following calculations.

19,650.24 + 378.50 − 50 − 1040 = 18,938.74

This checks with the amount on line 7.

Sam checked the amount of tax on line 11.

He subtracted line 8 from line 9.

2,845 − 2,733.64 = 111.36

This does not check with line 11. Sam must correct the amount entered.

When checking a tax form, you must carry forward corrections throughout the calculations. For instance, if Sam had found an error on line 7, he would have had to recalculate the amount of tax on line 9. This would require a correction of the amount of taxes owed on line 11.

Correct the errors. You may wish to use a calculator. Use the table on p. 453 of the Data Bank if appropriate.

1.

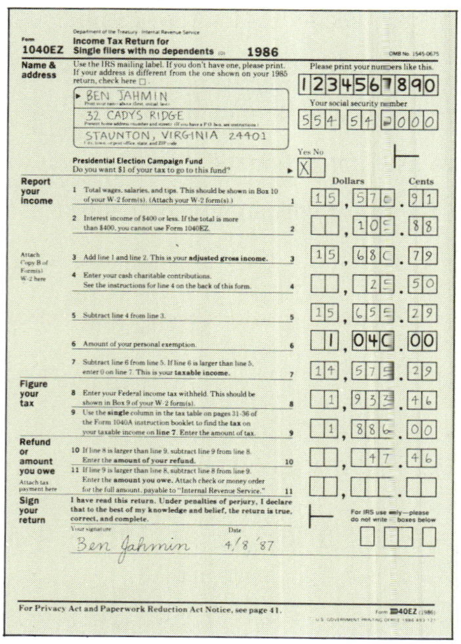

2.

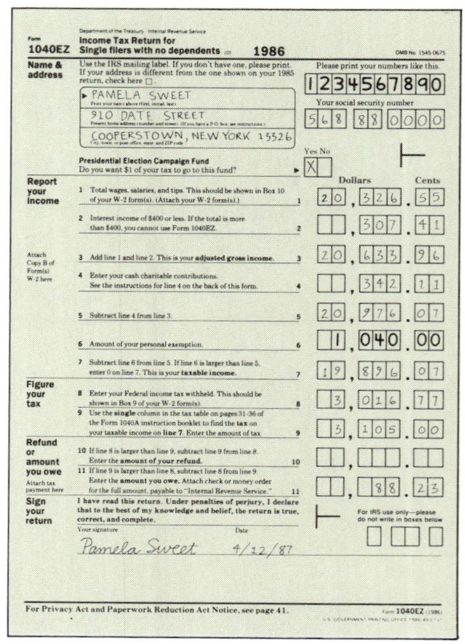

12-10 Calculator: Square Roots

N	N^2	$\sqrt{N}$
1	1	1
2	4	1.414
3	9	1.732
4	16	2
5	25	2.236
6	36	2.449
7	49	2.646
8	64	2.828
9	81	3
10	100	3.162
11	121	3.317
12	144	3.464
13	169	3.606
14	196	3.742
15	225	3.873
16	256	4
17	289	4.123
18	324	4.243
19	361	4.359
20	400	4.472

Objective
Find square roots.

On p. 238 you learned that 4^2 means "4 squared", or 4×4, which equals 16. The **square root** of a number is the opposite of squaring a number. The square of the square root of a number is the number. The radical symbol ($\sqrt{\ }$) directs you to find the square root of a number.

$4^2 = 16 \qquad \sqrt{16} = 4$
$7^2 = 49 \qquad \sqrt{49} = 7$
$10^2 = 100 \qquad \sqrt{100} = 10$
$20^2 = 400 \qquad \sqrt{400} = 20$

Numbers like 16, 49, 100, and 400 are **perfect squares** because each has an exact whole number square root.

Many calculators have a $\boxed{\sqrt{\ }}$ key. This function is especially useful when a number is not a perfect square. Numbers like 2 and 5 are not perfect squares. There is no whole number which, when multiplied by itself, equals 2 or 5. Tables like the one above give approximate square roots.

Find $\sqrt{2}$ to the nearest thousandth.

Using the table, $\sqrt{2} \approx 1.414$ rounded to the nearest thousandth.

Using a calculator, press 2 $\boxed{\sqrt{\ }}$. The display shows 1.4142136. Rounded to thousandths, it is 1.414.

Exercises

Find the square root. Use mental math.

1. $\sqrt{36}$ **2.** $\sqrt{81}$ **3.** $\sqrt{9}$ **4.** $\sqrt{4}$ **5.** $\sqrt{64}$
6. $\sqrt{121}$ **7.** $\sqrt{289}$ **8.** $\sqrt{169}$ **9.** $\sqrt{900}$ **10.** $\sqrt{10{,}000}$

Find the square root. Use a calculator.

11. $\sqrt{289}$ **12.** $\sqrt{529}$ **13.** $\sqrt{1936}$ **14.** $\sqrt{11{,}449}$ **15.** $\sqrt{253{,}009}$
16. $\sqrt{0.09}$ **17.** $\sqrt{1.44}$ **18.** $\sqrt{0.0009}$ **19.** $\sqrt{0.0169}$ **20.** $\sqrt{12.25}$

Find the square root. Use a calculator. Round to thousandths.

21. $\sqrt{30}$ **22.** $\sqrt{46}$ **23.** $\sqrt{75}$ **24.** $\sqrt{99}$ **25.** $\sqrt{120}$
26. $\sqrt{250}$ **27.** $\sqrt{347}$ **28.** $\sqrt{361}$ **29.** $\sqrt{489}$ **30.** $\sqrt{683}$

CHAPTER 12 REVIEW

Choose the one that does not belong in each group.

1. A. exemption
   B. deduction
   C. W-2 form
   D. assessed value

2. A. FICA
   B. market value
   C. rate of assessment
   D. tax rate

3. A. excise tax
   B. gift tax
   C. FICA
   D. property tax

4. A. national debt interest
   B. national defense
   C. Medicare
   D. payroll taxes

5. A. federal withholding
   B. property tax
   C. state & local tax
   D. FICA

6. A. excise
   B. property
   C. sales
   D. deduction

7. Solve. Arrange the letters in order from least to greatest answer.
   A. FICA deduction on a gross weekly pay of $565.90
   (Use the rate on p. 274.)
   B. federal tax on an adjusted gross income of $38,520, married, with 4 exemptions
   (Use the table on p. 276.)
   C. city income tax on adjusted gross income of $28,682 for a resident of Philadelphia
   (Use the table on p. 278.)
   D. Michigan state income tax for an adjusted gross income of $34,800
   (Use the table on p. 278.)
   E. federal income tax withheld weekly for Jason Fisher, who is single, claims 3 exemptions, and has a weekly gross pay of $312.90
   (Use the table on pages 453–455 of the Data Bank.)
   F. sales tax of 6% on sales of $219.85

CHAPTER 12 TEST

Find the total bill including sales tax.

	Quantity	Item	Unit Price	Price
1.	3 gallons	Paint	$9.99	
	4	Paintbrushes	$2.79	
			Subtotal	
			6% Tax	
			Total	

Find the amount withheld for federal income tax. Use the tables in the Data Bank.

	Job	Marital Status	Exemptions	Hours per Week	Hourly Rate
2.	Store clerk	Single	1	28	$ 4.25
3.	Electrician	Married	4	40	$14.75

Find the FICA deductions and the net pay.

4. Gross pay $375.45
   Deductions
   Fed. withholding tax $ 33.47
   State & local taxes $ 21.55
   FICA
   Credit Union $ 30.00

5. Gross pay $250.50
   Deductions
   Fed. withholding tax $ 23.56
   State & local taxes $ 11.89
   FICA
   Credit Union $ 12.46

Find the taxable income and the amount of tax.

	Adjusted Gross Income	Status	Exemptions
6.	$35,051	Married	2
7.	$18,539	Single	1

8. Find the state income tax in Virginia for an adjusted gross income of $10,678.

Find the assessed value and the property tax.

	Market Value	Rate of Assessment	Assessed Value	Tax Rate	Property Tax
9.	$63,400	30%		$4.50 per $100	
10.	$39,200	75%		2.2%	

11. Find the amount of federal excise tax on an airline ticket costing $599.

Use the Federal Budget on p. 284 to answer the following question.

12. How much more is spent for national defense than for Social Security and Medicare?

CHAPTER 12 CUMULATIVE REVIEW

1. Find the amount withheld weekly for federal income tax. Use the table on p. 453. data entry clerk, single, one exemption, 40 hours per week, \$7.30 per hour

A. \$30 **B.** \$34
C. \$41 **D.** \$36

2. Estimate.
$4945 \div 62$

A. 80
B. 800
C. 60
D. 90

3. Find the taxable income. standard deduction \$3000; each exemption \$2000; adjusted gross income \$19,983; single, 1 exemption

A. \$14,983 **B.** \$16,983
C. \$17,983 **D.** \$15,983

4. Find the volume of a packing carton,
$3 \text{ ft} \times 1\frac{3}{4} \text{ ft} \times 1\frac{1}{2} \text{ ft}.$
Use $V = l \times w \times h$.

A. $5\frac{1}{2} \text{ ft}^3$ **B.** $7\frac{7}{8} \text{ ft}^3$
C. $6\frac{1}{4} \text{ ft}^3$ **D.** 60 in.^3

5. The Hometown Kids win about 63% of their games. What are the chances they will win two games in a row?
Use $P = p \times p$.

A. about 0.12
B. about 0.4
C. about 0.3
D. about 0.5

6. Find the assessed value market value \$79,500, rate of assessment 22%

A. \$17,380
B. \$16,390
C. \$16,280
D. \$17,490

7. Divide.
$4\frac{4}{9} \div 7\frac{5}{9}$

A. $\frac{7}{9}$ **B.** $\frac{3}{4}$
C. $\frac{10}{17}$ **D.** $\frac{3}{5}$

8. Find the federal excise tax.
tax rate = 3%
telephone bill = \$25

A. \$0.75 **B.** \$0.55
C. \$0.65 **D.** \$0.70

9. Estimate total sales in July, August, and September.

A. \$90,000 **B.** \$200,000
C. \$120,000 **D.** \$85,000

Computer Sales for 1989

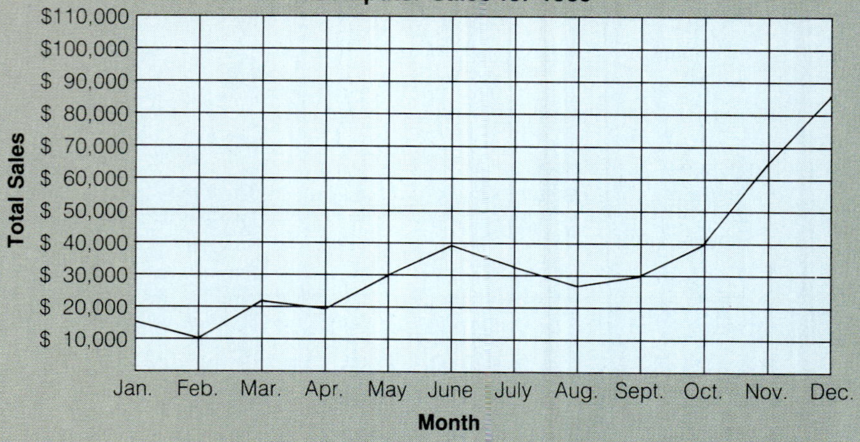

10. Find the sales tax.
The tax rate is 3.5%.
2 shirts @ $17.95 each, 1 hat @ $5.68 each
A. $1.46
B. $2.06
C. $1.25
D. $1.56

11. Use the formula, $H = 1.36gs$ to find the amount of oxygen in a 100-mL blood sample.
$g = 15\text{g}$, $s = 88\%$
A. 14.5 mL
B. 12.2 mL
C. 13.0 mL
D. 18.0 mL

12. Marlis earns $0.56 for each electronic component assembled. Find the total earned for 195 components.
A. $104.90
B. $106.20
C. $109.20
D. $105.20

13. Federal government expenses are about $973.7 billion. $64.2 billion is spent for education and health. About what percent of the total is this?
A. about 7%
B. about 15%
C. about 10%
D. about 2%

14. Give the percent. Use a decimal.
125% of 125
A. 125.25
B. 156.25
C. 100.25
D. 155.05

15. Find the state income tax. Rate is 5.35% of adjusted gross income. Adjusted gross income is $2253.
A. $118.54
B. $120.54
C. $119.24
D. $121.24

16. Find the amount after 2 years.
Use $i = prt$.
$600 principal, 8.5% interest, compounded annually
A. $705.35
B. $651
C. $702
D. $706.34

17. Find the annual cost for family insurance coverage.
$56.40 paid biweekly
A. $1466.40
B. $676.80
C. $2932.80
D. $1256.40

18. Find the catalog cost.
15 packs of beads $1.25 each, 5 oz per pack, $2.35 per lb shipping
A. $18
B. $18.75
C. $29.77
D. $19.10

19. Find the FICA deduction.
7.15%, gross pay $250
A. $19.78
B. $8.75
C. $16.87
D. $17.88

20. Find the number of calories a 124-lb swim team member needs to maintain her weight. Use the chart on p. 222.
A. 3100
B. 2600
C. 2720
D. 1750

21. Find the amount of protein in one serving of the spaghetti on p. 228.
A. 9 g
B. $66\frac{3}{8}$ g
C. $31\frac{1}{2}$ g
D. 1980 g

Chapter 13 Overview

Key Ideas

- buying a new car
- comparing fuel costs
- buying a used car
- financing a car purchase
- finding insurance costs
- collision and comprehensive insurance
- finding the cost of a rental car
- finding maintenance costs

Key Terms

- base price
- sticker price
- selling costs
- annual fuel costs
- trade-in allowance
- driver rating factor
- deferred payment price
- down payment
- amount financed
- liability insurance
- basic rate
- rental cost
- annual premium
- collision insurance
- comprehensive insurance
- deductible
- discount
- labor charge

Key Skills

Find the answer.

1. \$3400 + \$899 + \$187 + \$273

2. \$8712 + \$434.18 + \$113

3. \$24 + \$7624.99 + \$562 + \$212

4. \$8762 + \$220.08 + \$45

5. \$9076 + \$231 + \$25 + \$614

6. \$10,872 + \$453 + \$208

Solve.

7. What is 56% of 284?

8. What is 15% of 8.1?

9. What is 32% of 59?

10. What is 62.5% of 72?

11. 60% of 650 is what number?

12. 40% of 55 is what number?

13. 74 is what percent of 296?

14. 56 is what percent of 800?

15. 33.18 is what percent of 158?

16. 6 is what percent of 40?

17. 29.34 is what percent of 2934?

18. 48.16 is what percent of 56?

USING A CAR

13

13-1 Buying a New Car

Objective
Find new car selling prices.

The **base price** of a new car is set by the manufacturer. It includes a profit for the dealer. The **sticker price** is the base price plus the cost of any optional equipment.

Sales tax, destination charge, local dealer preparation charges, and the license fee must be added to the sticker price. The total is the **selling price** of the car.

This chart shows some typical base prices and costs for optional equipment.

	Compact	Sedan	Sport Coupe	Luxury
Base price	$5570	$7725	$9345	$11,270
Body side molding	55	60	standard	standard
Pin striping	39	45	52	60
Air conditioning	750	790	790	800
Tinted glass	115	115	125	135
AM-FM stereo radio	162	162	standard	standard
with cassette player	284	284	130	130
Power brakes	129	145	standard	standard
Power steering	275	295	325	standard
Power windows	unavailable	195	270	295
Power door locks	unavailable	130	180	225
Automatic transmission	175	190	245	315
Rear window defogger	145	155	155	165
Tilt steering wheel	unavailable	115	145	standard

Example Find the selling price. You may wish to use a calculator.

Sedan	Sales tax	5%
Air conditioning	Destination charge	$185
Automatic transmission	Dealer preparation	$112
AM-FM stereo radio with cassette player	License fee	$ 25

Base price	$7725.00	Sales tax (0.05 × $8989)	449.45
Air conditioning	790.00	Destination charge	185.00
Automatic transmission	190.00	Dealer preparation	112.00
AM-FM radio with tape	284.00	License fee	25.00
Sticker price	$8989.00	Selling price	$9760.45

Try This Find the selling price.

Sport coupe	Power steering	Sales tax 6%	Dealer prep. $145
	Tilt steering wheel	Dest. charge $178	License fee $25

Exercises

Find the selling price.

1. Luxury
   Power windows
   Power locks
   Automatic transmission

Sales tax	7%
Destination charge	$295
Dealer preparation	$361
License fee	$32

2. Sport coupe
   Air conditioning
   Tilt steering wheel
   Rear window defogger

Sales tax	5%
Destination charge	$139
Dealer preparation	$79
License fee	$25

3. Compact
   Pin striping
   Tinted glass
   Body side molding
   Rear window defogger

Sales tax	4%
Destination charge	$85
Dealer preparation	$66
License fee	$20

4. Sedan
   AM-FM stereo radio with cassette player
   Power brakes
   Power windows
   Tilt steering wheel

Sales tax	6%
Destination charge	$129
Dealer preparation	$110
License fee	$30

5. Here is a sticker from a showroom automobile. Most new cars in the dealer's showroom display the optional equipment available. Some dealers give a special added discount when a customer purchases a certain package of options. The special added discount on this car is $600. Use the sticker to find the total vehicle and options price, the total before discount, and the selling price.

X-GS 2-DOOR, SPECIAL PAINT	11421.00
REGATTA BLUE CLOTH TWIN COMFORT SEATS,	
993 3.8L V6 EFI ENGINE	NC
172 INTERVAL WINDSHIELD WIPERS	50.00
440 AUTOMATIC TRANSMISSION	NC
T7E P215/70R14 WSW TIRES	NC
520 TILT STEERING WHEEL	115.00
524 LEATHER-WRAPPED STEERING WHEEL	59.00
525 SPEED CONTROL	176.00
534 2-WAY POWER DRIVER'S SEAT	238.00
570 REAR WINDOW DEFROSTER	145.00
572 AIR CONDITIONER	762.00
58G ELECTRONIC AM/FM STEREO/CA	127.00
628 DUAL POWER REMOTE MIRRORS	96.00
64R POWER SIDE WINDOWS	207.00
65N LOCKING WIRE-STYLE WHEEL COVERS	178.00
677 DUAL ILLUMINATED VISOR MIRROR	106.00
923 TINTED GLASS	115.00
943 LIGHT GROUP	35.00
963 POWER LOCK GROUP	220.00
SPECIAL ORDER 177501	
SPECIAL TUTONE PACKAGE	315.30
PT CODE M5909 GRAY	
PT CODE M6008 BLUE	
TOTAL VEHICLE & OPTIONS	
DESTINATION & DELIVERY	432.00
TOTAL BEFORE DISCOUNTS	
##SPECIAL ADDED DISCOUNT#	600.00

Mixed Skills Review

1. 19,064 − 16,556
2. $\frac{3}{5} \times \frac{4}{3} \times \frac{15}{16}$
3. 12.2 − 6.186
4. 27,720 ÷ 42
5. 3.06 + 24.24
6. $5\frac{2}{3} - 4\frac{7}{9}$
7. 2318 + 4864
8. 1826 × 37
9. 2.16 × 0.9
10. $4\frac{1}{2} \div \frac{3}{4}$
11. $1\frac{3}{4} + 6\frac{7}{8}$
12. 18.72 ÷ 0.24

13-2 Comparing Fuel Costs

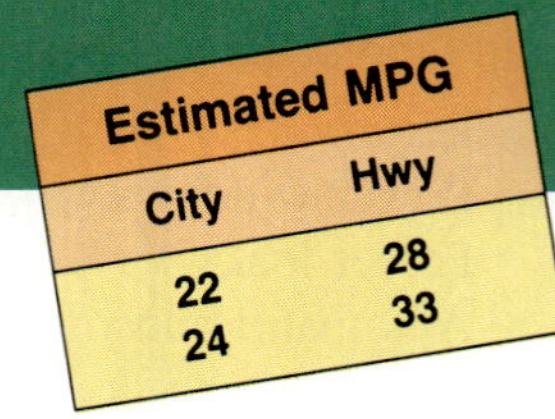

Objective
Compare fuel costs.

New cars have labels that show the average number of miles per gallon for city and highway driving. Since driving conditions vary, these numbers are estimates.

Example

1. Chip Downs had a full tank on July 1. He filled the tank again with 8.7 gallons on July 12. Find the actual fuel economy rate of his car.

3 7 9 4 5 2

3 8 2 2 0 2

Subtract to find the number of miles driven. $38{,}220.2 - 37{,}945.2 = 275.0$ mi

Divide the number of miles driven by the number of gallons of gasoline used.
$275.0 \text{ mi} \div 8.7 \text{ gal} \approx 31.6 \text{ mpg}$

Try This

Find the actual fuel economy rate.

a.
0 4 1 7 1 5
0 4 3 1 4 1

Gas purchased: 7.4 gallons

b.
5 3 9 0 1 4
5 4 0 3 0 9

Gas purchased: 4.9 gallons

Example

2. Stan Neumann is purchasing a new station wagon. He estimates 20% of his driving is in the city. Calculate his annual fuel cost.

Miles Driven	City mpg	Highway mpg	Price per Gallon
15,000	16	25	$1.25

Use the chart on p. 459 in the Data Bank to find the annual fuel costs for the estimated mpg in the city and on the highway. Multiply by the appropriate percent to find city and highway costs.

City	16 mpg	$1.25	$1172 × 20% = $234.40
Highway	25 mpg	$1.25	$ 750 × 80% = $600.00

Add to find the total annual fuel cost. $234.40 + $600 = $834.40

Stan's annual fuel cost should be $834.40.

Try This

Calculate the annual fuel cost based on 15,000 miles driven.

	Highway Driving	City mpg	Highway mpg	Price per Gallon	Annual Fuel Cost
c.	60%	31	35	$1.45	
d.	25%	26	32	$1.05	

Exercises

Find the actual fuel economy rate for each car.

1. 1 8 9 0 5 7
   1 9 0 5 9 0
   Gas purchased: 5.1 gallons

2. 0 7 7 9 1 6
   0 8 0 0 0 9
   Gas purchased: 6.2 gallons

3. 4 5 0 1 2 3
   4 5 2 5 2 1
   Gas purchased: 9.8 gallons

4. 2 4 7 6 3 5
   2 4 9 0 9 0
   Gas purchased: 7.6 gallons

Calculate the annual fuel cost based on 15,000 miles driven.

	Highway Driving	City mpg	Highway mpg	Price per Gallon	Annual Fuel Cost
5.	30%	18	21	\$1.45	
6.	65%	24	29	\$1.15	
7.	75%	21	27	\$1.25	
8.	45%	32	36	\$1.05	

Estimate the amount of fuel left in each gasoline tank.

9.
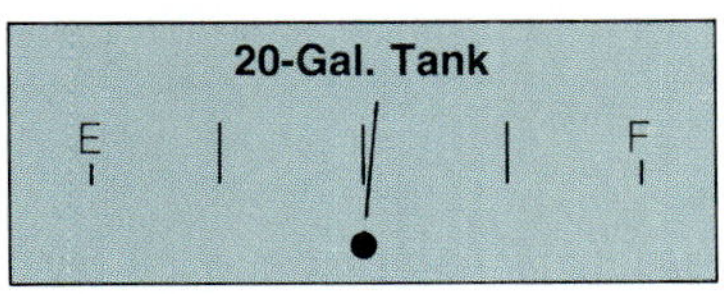

10.
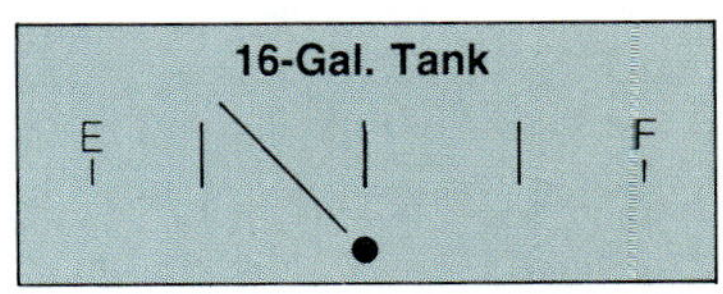

11.
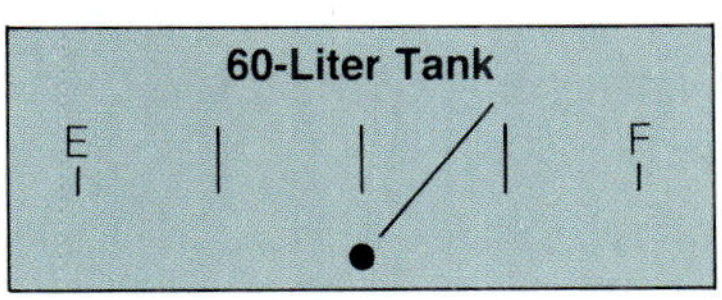

12.
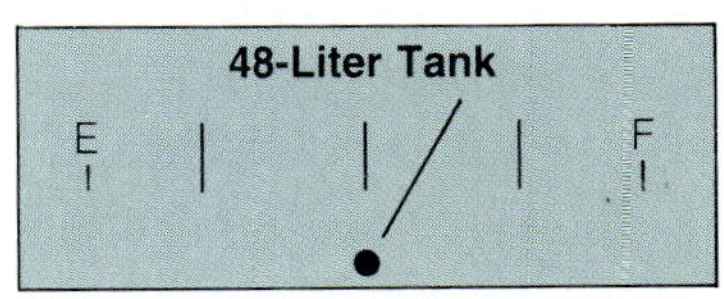

Mixed Skills Review

1. $4\frac{3}{8} + 6\frac{3}{4}$ **2.** 3120×56 **3.** $2.081 + 3.267$

4. $10{,}317 \div 181$ **5.** $\$16 - \$4.36$ **6.** $\frac{5}{8} \div \frac{5}{16}$

7. $2\frac{5}{12} - 1\frac{1}{2}$ **8.** $34.51 \div 2.03$ **9.** $800 - 572$

10. $43{,}615 + 29{,}028$ **11.** 16.43×2.8 **12.** $1\frac{2}{3} \times 1\frac{4}{5}$

13-3 Buying a Used Car

Objective
Find used car selling prices and dealer profit.

For many people, a new car is an unnecessary or unwanted expense. Used cars frequently offer good value at a lower price.

A car dealer uses a guidebook that lists the average retail prices on used models. The average retail price is adjusted based on mileage.

Example

1. The average retail price on a 1982 two-door coupe is \$3625. The car has been driven 35,000 miles. Find the selling price.

Use the mileage chart on p. 460 of the Data Bank to find the amount to add to or deduct from the average retail price.

Add \$240 to the retail price.
\$3625 + \$240 = \$3865 The selling price is \$3865.

Try This Find the selling price.

	Model	Retail Price	Miles Driven	Selling Price
a.	'82 station wagon	\$5210	62,000	
b.	'86 sedan	\$9270	4,900	

Example

2. A dealer allowed \$4150 on a 1985 station wagon as a trade-in allowance on a new car. The dealer made repairs costing \$350 and sold the car for \$5950. Find the dealer's percent of profit.

Add the trade-in allowance and the cost of repairs to find the dealer's cost.
\$4150 + \$350 = \$4500

Subtract the dealer's cost from the selling price.
\$5950 − \$4500 = \$1450 This is the dealer's profit.

Divide the amount of profit by the dealer's cost.
\$1450 ÷ \$4500 ≈ 0.32 The dealer's percent of profit is about 32%.

Try This Find the dealer's percent of profit.

	Model	Trade-in Allowance	Cost of Repairs	Selling Price	Percent of Profit
c.	'80 compact	\$1590	\$620	\$2500	
d.	'86 luxury	\$5400	\$280	\$7200	

Exercises

Find the selling price. Use the mileage chart on p. 460 in the Data Bank.

	Model	Retail Price	Miles Driven	Selling Price
1.	'82 compact	$3130	50,000	
2.	'86 luxury	$9500	25,000	
3.	'87 mid-size	$7525	15,600	
4.	'83 pickup	$4550	48,500	
5.	'84 subcompact	$2190	38,000	
6.	'80 sport van	$8400	32,300	

Find the dealer's percent of profit.

	Model	Trade-in Allowance	Cost of Repairs	Selling Price	Percent of Profit
7.	'82 4-D sedan	$4135	$295	$5250	
8.	'81 station wagon	$2525	$415	$3630	
9.	'79 luxury	$6400	$580	$9000	
10.	'84 hatchback	$4515	$105	$5930	
11.	'78 motorcycle	$ 500	—	$1750	
12.	'80 4WD pickup	$2312	$385	$3490	

A dealer accepted a compact as a trade-in. After a service tune-up, the fuel economy rate of the car increased from 20.1 to 23.9 miles per gallon. Estimate each of the following answers.

13. What was the percent of increase in the car's fuel economy rate?

14. The capacity of the gas tank on the compact is 12.5 gallons. How far can the car travel on a full tank of gasoline?

15. How many gallons of gasoline will be needed for a trip of 1400 miles?

Mixed Skills Review

1. $216.305 + 56.97$ **2.** $2\frac{5}{9} \times \frac{9}{11}$ **3.** 9.008×46

4. $2\frac{17}{20} - 1\frac{4}{5}$ **5.** $28{,}764 \div 705$ **6.** $57{,}845 + 30{,}937$

7. 4624×33 **8.** $3\frac{7}{16} + 2\frac{1}{2}$ **9.** $6.003 - 4.724$

10. $18.752 \div 0.64$ **11.** $20{,}065 - 13{,}274$ **12.** $\frac{15}{16} \div 2\frac{1}{2}$

13-4 Financing Your Purchase

Objective
Find financed amounts, finance charges, and deferred payment prices.

Many customers purchase cars, trucks, and motorcycles on a **deferred payment plan.** A **down payment** is made in the form of a trade-in, cash payment, or some combination of these two. The difference between the selling price and the down payment is the **amount financed.**

Example

1. A down payment of 15% is required on a sport coupe that costs $12,355. Find the amount financed.

Multiply to find the down payment.
0.15 × $12,355 = $1853.25

Subtract to find the amount financed.
$12,355.00 − $1853.25 = $10,501.75

Try This

Find the amount financed.

a. A down payment of 20% is made on a motorcycle that costs $2965.

b. A down payment of one third is made on a sedan that costs $11,400.

A loan is usually repaid in equal monthly payments. The interest paid on a car loan is called the finance charge. It is based on the **annual percentage rate (APR)** charged by the lender and is the difference between the total of the monthly payments and the amount financed. The **deferred payment price** is the sum of the monthly payments and the down payment. This is the actual cost of the car.

Example

2. Juanita Angeles made a down payment of 25% on a new car that cost $5100 and financed the difference. She repaid the loan in 3 years in equal monthly payments of $127.09. Find the deferred payment price.

Find the down payment.	$5100 × 0.25 = $1275
Find the total of the monthly payments.	$127.09 × 36 = $4575.24
Find the sum.	$1275 + $4575.24 = $5850.24

The deferred payment price is $5850.24.

Try This

Find the deferred payment price.

c. Maryann Murdzak bought a racing motorcycle for $2760. She made a down payment of $1000 and financed the difference. She repaid the loan in 2 years with monthly payments of $81.97.

Exercises

Complete the chart.

	Model	Selling Price	Down Payment	Amount Financed	Term	Monthly	Deferred Payment Price
1.	Coupe	$ 9000	$2000		4 yr	$177.54	
2.	Pickup	$12,000	$2200		5 yr	$208.23	
3.	Street bike	$ 2400	25%		18 mo	$111.74	
4.	Sedan	$ 9650	30%		42 mo	$187.19	
5.	Compact	$ 5500	20%		2 yr	$219.64	
6.	Motorcycle	$ 2790	$\frac{1}{4}$		36 mo	$ 79.87	
7.	Moped	$ 675	$\frac{1}{5}$		18 mo	$ 38.47	
8.	Van	$19,800	$\frac{1}{3}$		5 yr	$333.69	

9. John Foley financed $6500 on the purchase of a new car. He repaid the loan in 4 years in equal monthly payments of $164.85. Find the finance charge.

10. Mariquita Lopez financed $8600 on the purchase of a pickup truck. She repaid the loan in 54 months in equal monthly payments of $197.03. Find the finance charge.

Salespeople use a **compound interest table** to compute monthly payments for their customers. Use a calculator and the compound interest table on p. 446 in the Data Bank to find the monthly payment.

	Amount Financed	APR	Term in Months	Monthly Payment
11.	$3000	12.50%	30	
12.	$7000	17.25%	24	
13.	$5800	13.00%	36	
14.	$1900	19.25%	48	

Mixed Skills Review

1. $37{,}268 - 28{,}557$
2. $76{,}687 \div 347$
3. $8\frac{2}{3} \div 4$
4. $10\frac{1}{8} - 2\frac{5}{6}$
5. $15{,}236 + 26{,}867$
6. $3\frac{3}{5} \times 2\frac{2}{9}$
7. 305×1654
8. $146.2 - 15.841$
9. $2\frac{6}{25} + 1\frac{3}{5}$
10. $46.03 + 6 + 9.8$
11. 65.8×1.26
12. $1.07738 \div 5.23$

13-5 Automobile Liability Insurance

Objective
Find auto liability insurance costs.

Liability insurance pays for any injury and damage for which you are liable (at fault). Some states require this insurance for licensing and registration.

Liability insurance is classified by maximum amounts payable. With 25/50/25 insurance the company will pay \$25,000 to each person injured, \$50,000 to all persons injured in one accident, and \$25,000 for property damaged.

The annual **liability premium** is figured on a **basic rate** and the **driver rating factor** based on age, sex, driving record, and residence.

Example

1. Find the annual liability premium for 20/40/15 coverage for a driver with a 1.7 rating.

Find the basic rate in the table.
Multiply the rate by the driver rating factor.
$1.7 \times \$196 = \$333.20$

The annual liability premium is \$333.20.

Liability Insurance

Coverage	Basic Rate
15/30/15	\$180
20/40/15	\$196
25/50/25	\$207
50/100/50	\$230
100/300/50	\$246

Try This

Find the annual liability premium. Use the Liability Insurance table.

a. 25/50/25 coverage, 2.5 driver rating factor

b. 50/100/50 coverage, 3.8 driver rating factor

c. 15/30/15 coverage, 0.8 driver rating factor

The driver rating factor is often higher for people who are single, live in large cities, or drive their car more than ten miles to work.

Example

2. Jan moved to the city. Her driver rating factor changed from 1.1 to 1.9. Find the increase in her annual premium for 20/40/15 coverage.

Multiply the rate in the chart by the difference in the driver rating factors.
$1.9 - 1.1 = 0.8$ $\quad$ $0.8 \times \$196 = \$156.80$

Jan's annual liability premium increased \$156.80.

Try This

d. Stan Butler got married. His driver rating factor changed from 2.3 to 2.1. Find the decrease in his annual premium for 50/100/50 coverage.

Exercises

Find the annual liability premium. Use the Liability Insurance table on p. 302.

	Coverage	Driver rating factor
1.	15/30/15 coverage	3.7 driver rating factor
2.	20/40/15 coverage	0.9 driver rating factor
3.	50/100/50 coverage	2.1 driver rating factor
4.	15/30/15 coverage	1.9 driver rating factor
5.	100/300/50 coverage	3.4 driver rating factor
6.	25/50/25 coverage	2.8 driver rating factor
7.	20/40/15 coverage	4.15 driver rating factor
8.	50/100/50 coverage	3.2 driver rating factor

Solve.

9. Jim Madden had two accidents within six months. His driver rating factor changed from 1.3 to 2.8. Find the increase in the annual premium for 100/300/50 coverage.

10. Tara Reardon moved and now takes the bus to work. Her driver rating factor changed from 2.7 to 1.8. Find the decrease in the annual premium for 15/30/15 coverage.

11. Sam Tupper changed his liability coverage from 25/50/25 to 50/100/50. His driver rating factor is 3.1. Find the increase in the annual premium.

12. Carol Parks has a driver rating factor of 0.8. She changed her coverage from 25/50/25 to 20/40/15. Find the decrease in the annual premium.

13. The driver rating factor for a new driver is 2.5. Fred reduced this number 15% by taking driver training in school. He reduced the new factor 20% by earning a B average in school. Find the new driver training factor.

14. Find the amount Fred saves on his annual liability premium for 25/50/25 coverage.

Mixed Skills Review

1. $1\frac{1}{16} \div 1\frac{1}{2}$ **2.** $9036 - 877$ **3.** 132.4×28.7

4. $129{,}426 \div 53$ **5.** $0.1464 + 3.375$ **6.** $31.978 \div 5.42$

7. $\frac{5}{3} + \frac{7}{5}$ **8.** 5211×901 **9.** $4\frac{3}{10} - 3\frac{1}{2}$

10. $\$103 - \$97.26$ **11.** $2\frac{6}{7} \times \frac{14}{15}$ **12.** $31{,}685 + 8624$

13-6 Collision and Comprehensive Insurance

Objective *Calculate collision and comprehensive insurance costs.*

Collision insurance pays for damage to your car if you are in an accident. **Comprehensive insurance** pays for damage from fire, theft, vandalism, or natural causes such as floods or falling trees.

The basic rate for collision and comprehensive coverage depends on the car class rating and the deductible. Cars with higher resale values are in higher-rated classes. Increasing the **deductible,** the amount you may pay on each claim before the insurance coverage begins, lowers the basic starting rate.

Basic Rate

Car Class	Collision Deductible			Comprehensive Deductible		
Rating	$50	$100	$200	$0	$50	$100
1–10	$100	$87	$74	$28	$25	$22
11–20	$163	$150	$137	$70	$62	$55
21–30	$227	$207	$187	$89	$81	$73
31–40	$255	$238	$221	$133	$122	$111

Example Find the total annual collision and comprehensive premium.

Driver rating factor 2.9 Collision deductible $100
Car class 13 Comprehensive deductible $50

Use the Basic Rate table. Car class 13 is between 11 and 20. Move across to the $100 deductible column under Collision. The rate for collision insurance is $150.

Move across to the $50 column under Comprehensive deductible. The rate for comprehensive insurance is $62.

Multiply the two rates by the driver rating factor to find each premium.

Collision $\$150 \times 2.9 = \$435$
Comprehensive $\$62 \times 2.9 = \$179.80$

Add to find the annual premium. $\$435 + \$179.80 = \$614.80$

The total annual premium for collision and comprehensive is $614.80.

Try This Find the total annual collision and comprehensive premium. Use the Basic Rate table.

	Driver Rating Factor	Car Class Rating	Collision Deductible	Collision Premium	Comp. Deductible	Comp. Premium	Total Premium
a.	1.8	22	$50		$0		
b.	3.3	35	$200		$100		

Exercises

Find the total annual premium. Use the table on page 304.

	Driver Rating Factor	Car Class Rating	Collision Deductible	Collision Premium	Comp. Deductible	Comp. Premium	Total Premium
1.	1.3	32	$50		$0		
2.	2.7	9	$200		$100		
3.	3.4	15	$100		$50		
4.	1.8	27	$50		$0		
5.	2.2	10	$200		$100		
6.	4.15	38	$200		$100		
7.	3.9	7	$50		$0		
8.	0.8	21	$100		$50		

Solve.

9. Zane Thomas owns a class-9 car, with a $100 collision deductible and a $50 comprehensive deductible. His driver rating factor is 2.6. How much more will his total collision and comprehensive premium be if he purchases a class-15 car and keeps the same deductibles?

10. Paula Drake owns a class-23 car, with a $100 collision deductible and a $100 comprehensive deductible. Her driver rating factor is 1.8. How will her total premium change if she increases her collision deductible to $200?

11. Oma Edwards owns a class-26 car and has $100 deductibles for collision and comprehensive insurance. Her driver rating factor is 2.5. When she gave her car stereo to her sister, she dropped the comprehensive deductible to $0. By how much did her total insurance premium change?

12. Ginnie Carmelli owns a class-36 car and has a $200 deductible for collision and $100 for comprehensive insurance. Her driver rating factor is 0.9. Scott Montefusco drives a class-5 car, with a $50 deductible for collision and a $0 deductible for comprehensive insurance. His driver rating factor is 3.9. Who pays the higher premiums? How much more?

Mixed Skills Review

1. $21 + 8.4 + 19.37$

2. 18.42×55

3. $90{,}623 + 426{,}395$

4. $2\frac{2}{3} - \frac{8}{15}$

5. $39.552 \div 12.8$

6. $1\frac{17}{25} \div 2\frac{4}{5}$

7. 485×76

8. $\frac{5}{8} + \frac{27}{40}$

9. $4.2821 - 2.6354$

10. $15{,}624 \div 68$

11. $12{,}000 - 645$

12. $14\frac{5}{32} - 6\frac{1}{2}$

13-7 Renting a Car

Objective
Calculate the cost of renting a car.

Many people rent cars when they are away from home. Stu and Jennie Tucker spent their vacation in Massachusetts. They rented a car to drive from Boston to Cape Cod.

Example

1. A mid-size car rents for \$42.50 per day and \$0.41 per mile. The Tuckers kept the car for three days and drove 260 miles. Find the rental cost.

Multiply to find the cost for 3 days.	$3 \times \$42.50 = \$127.50$
Multiply to find the mileage cost.	$260 \times \$0.41 = \$106.60$
Add to find the rental cost.	$\$127.50 + \$106.60 = \$234.10$

Try This

Find the rental cost.

	Cost per Day	Number of Days	Cost per Mile	Number of Miles	Rental Cost
a.	\$49.25	4	\$0.48	320	
b.	\$36.75	7	\$0.35	440	

Example

2. The Barney Corporation rents cars for company visitors. The cost to rent a station wagon is usually \$47.85 per day and \$0.32 per mile. The rental agency gave the company a 25% discount on a four-day rental that was driven 127 miles. Find the rental cost.

Find the total daily rental.	Find the total mileage charge.
$4 \times \$47.85 = \$191.40$	$127 \times \$0.32 = \$40.64$

Apply the 25% discount to the sum of the daily rental and mileage.

$\$191.40 + \$40.64 = \$232.04$ $\quad 0.75 \times \$232.04 = \$174.03$

Try This

Find the rental cost.

	Cost per Day	Number of Days	Cost per Mile	Number of Miles	Discount	Rental Cost
c.	\$37.60	3	\$0.35	384	100 free miles	
d.	\$45.75	6	\$0.40	260	15%	

Exercises

Find the rental cost.

	Cost per Day	Number of Days	Cost per Mile	Number of Miles	Discount	Rental Cost
1.	$39.50	5	$0.33	520	None	
2.	$44.20	6	$0.47	210	None	
3.	$49.75	3	$0.45	192	None	
4.	$41.00	10	$0.41	511	None	
5.	$33.40	4	$0.39	388	100 free miles	
6.	$48.30	2	$0.51	65	40 free miles	
7.	$38.75	9	$0.44	673	75 free miles	
8.	$43.00	12	$0.43	928	15%	
9.	$32.25	3	$0.49	256	25%	
10.	$53.15	7	$0.48	445	30%	

An airport rental agency has the inventory shown here.

Model	Total	Blue	Green	Gray	White
Subcompact	8	2	1	2	3
Compact	20	5	2	8	5
Mid-size	15	10	1	1	3
Luxury	5	1	1	0	3

A customer does not make a reservation. The agent chooses a car at random. Find the probability of each.

11. a white luxury **12.** a blue mid-size **13.** a gray compact

14. a mid-size **15.** a compact **16.** a subcompact

17. blue **18.** red **19.** white

Strategy Practice

20. The rental cost for a mid-size was $184. The cost of the car was three times as much as the mileage cost. What was the cost to rent the car? What was the mileage cost?

Mixed Skills Review

1. $8.37 + 6.025 + 3.8$ **2.** $138.024 \div 56.8$ **3.** $16\frac{2}{3} \div 2\frac{1}{12}$

4. 426×1007 **5.** $6\frac{2}{3} + 3\frac{8}{9}$ **6.** 0.841×1.4

7. $1\frac{3}{13} \times \frac{15}{16}$ **8.** $0.83 - $0.58 **9.** $5 + 8 + 3 + 6 + 5 + 7 + 2 + 6$

10. $10{,}380 - 9604$ **11.** $4716 \div 72$ **12.** $9\frac{1}{12} - 2\frac{3}{4}$

13-8 Maintenance and Repairs

Objective
Calculate car maintenance and repair costs.

Bailey's Auto Repairs provides major repairs and basic maintenance service. A work order form is completed for each job. The cost of parts, gas, oil, and grease are subject to sales tax. The charge for labor is based on a fixed rate per hour. Labor charges are not taxable.

Example This is the total repair bill. Use 6% tax.

A. Add charges for parts and oil to find taxable charges.

B. Multiply to find the tax.

C. Enter labor charges.

D. Add the subtotal, tax, and labor costs.

The total bill is \$115.15.

Bailey's Auto Repairs Name Joan Seller Job Number 71507
Date Oct. 15 Address Rt. 4, Johnstown Telephone 377-2154

Quantity	Part Number	Name of Part	Price
1	V-297	PCV Valve	3.89
2	N-016	Wipers	3.45
1	78-A	Filter	4.85
1	J-667	Alternator	39.90
Total Parts			52.09

Year 85 Model FORD LTD License BA2-274
Mileage 29997 Promised 4:00 p.m.

Description of Work	Amount
30,000 Mile Service	54.00
Repair/replace alternator	
Total Labor	54.00

Gas, Oil, Grease	Amount
Gasoline	
Oil 4 quarts	5.60
Grease	
Brake fluid	
Other	
Total	5.60

Oil Changes (√): Engine, Transmission, Differential, Wash, Wax, Other

Totals		
Parts	52.09	
Gas, Oil, Grease	5.60	
Subtotal	57.69	A
Tax	3.46	B
Labor	54.00	C
Total	\$115.15	D

Try This Find the total repair bill. Use 4% tax.

Bailey's Auto Repairs Name Cole Greer Job Number 34-275
Date June 6 Address 28 Seminary St. Telephone 321-4960

Quantity	Part Number	Name of Part	Price
1	643-A	Oil Filter	7.85
1	374-H	Hub Cap	12.16
1	J-2198	Ring	17.54
Total Parts			

Year 86 Model CHEV License JLE-098
Mileage 10,204 Promised NOON

Description of Work	Amount
10,000 Mile SVC	27.00
REPLACE HUB CAP	NC
Total Labor	

Gas, Oil, Grease	Amount
Gasoline	
Oil 4 quarts	5.40
Grease	3.00
Brake fluid	
Other	
Total	

Oil Changes (√)	
Engine	√
Transmission	√
Differential	√
Wash	
Wax	
Other	

Totals	
Parts	
Gas, Oil, Grease	
Subtotal	
Tax	
Labor	
Total	

Exercises

Find the total repair bill. Use 6% tax.

1.

Bailey's Auto Repairs Name Pam Cherundolo Job Number A-2351
Date August 30 Address 135 Greenridge Telephone 252-6321

Quantity	Part Number	Name of Part	Price
4	35R120	Tires	246 50
1	G-635	Gasket	1 98
1	F-364	Belt	8 65
Total Parts			

Year 86 Model Plym License 101-KP2
Mileage 8561 Promised 11:00

Description of Work	Amount
Service Tune-up	27 00
Replace tires, repair spare	15 00
Total Labor	

Gas, Oil, Grease	Amount	Oil Changes (√)		Totals	
Gasoline		Engine		Parts	
Oil 4 quarts	3 56	Transmission	√	Gas, Oil, Grease	
Grease		Differential		Subtotal	
Brake fluid		Wash		Tax	
Other		Wax		Labor	
Total		Other			
				Total	

2.

Bailey's Auto Repairs Name Gus Forrest Job Number 3960
Date December 10 Address 1042 Woods St. Telephone 942-8596

Quantity	Part Number	Name of Part	Price
1	A-2965	Air Filter	9 42
1	B-6371	48 Mo. Batt.	59 97
1	P-4980	Points	3 75
Total Parts			

Year 86 Model Regal License JFE-108
Mileage 1465 Promised 5:00

Description of Work	Amount
Check engine knock	42 00
Won't start on cold a.m.s	
Total Labor	

Gas, Oil, Grease	Amount	Oil Changes (√)		Totals	
Gasoline	15 40	Engine	√	Parts	
Oil ___ quarts		Transmission		Gas, Oil, Grease	
Grease		Differential		Subtotal	
Brake fluid		Wash		Tax	
Other		Wax		Labor	
Total		Other			
				Total	

Mixed Skills Review

1. $146 - 0.084$ **2.** $41.86 \div 16.1$ **3.** $7\frac{1}{2} \times \frac{3}{5}$

4. 924×18 **5.** $950{,}016 + 237{,}652$ **6.** $18.061 + 4.994$

7. $9594 \div 117$ **8.** $1064 - 158$ **9.** $6 \div 3\frac{3}{5}$

10. $4\frac{2}{5} + 1\frac{3}{4}$ **11.** 6.301×464 **12.** $6\frac{2}{3} - 5\frac{11}{12}$

13-9 Enrichment: Annual Expenses

Objective
Find annual automobile expenses.

Bob Chapman keeps track of his car expenses and the number of miles he drives each year. The sum of loan payments and the charges for insurance, gasoline, and repairs and maintenance are Bob's annual operating expenses.

Annual Expenses	
Loan payments	\$2628
Insurance	\$ 940
Gasoline	\$ 907.50
Repairs and maintenance	\$ 119.82
Miles driven:	13,795

To find the mean or average cost per mile of owning and operating a car during the past year, he follows these steps.

Add the expenses to find the annual operating costs.
\$2628 + \$940 + \$907.50 + \$119.82 = \$4595.32

Divide the sum by the number of miles driven to find the average cost per mile.
\$4595.32 ÷ 13,795 ≈ \$0.33   Round to the nearest cent.

The average cost per mile to own and operate Bob's car is about \$0.33.

Exercises

Find the average cost per mile to own and operate each car.

1.

Annual Expenses	
Loan payments	\$1988
Insurance	\$ 944.50
Gasoline	\$1189.16
Repairs and maintenance	\$ 410.79

Miles driven: 19,026.6

2.

Annual Expenses	
Loan payments	\$1862
Insurance	\$ 566.16
Gasoline	\$ 185.53
Repairs and maintenance	\$ 24.98

Miles driven: 4055.8

3.

Annual Expenses	
Loan payments	\$2977.64
Insurance	\$ 759.68
Gasoline	\$1150.88
Repairs and maintenance	\$1274.50

Miles driven: 26,879.4

13-10 Computer: Deferred Payment Price

Objective
Use a computer to find the amount financed and the deferred payment price of a car loan.

A compact car lists for $5995. You agree to make a down payment of $875 and pay $158.32 a month for three years. How much do you really pay for the car?

The following computer program is helpful in doing the calculations.

```
10  REM Program for Calculating Cost of a Car Loan
20  CLS:INPUT "What is the model";M$
30  INPUT "What is the selling price"; P
40  INPUT "How much is the down payment"; D
50  INPUT "How many years are the payments"; Y
60  INPUT "How much are the monthly payments"; M
70  PRINT TAB(65) "Deferred"
80  PRINT TAB(15) "Selling"; TAB(25) "Down"; TAB(35) "Amount"; TAB(55)
    "Monthly"; TAB(65) "Payment"
90  PRINT "Model"; TAB(15) "Price"; TAB(25) "Payment"; TAB(35) "Financed";
    TAB(47) "Term"; TAB(55) "Payment"; TAB(65) "Price"
100 PRINT "---------------------------------------------------------------------------
    ------------------------------"
110 PRINT M$; TAB(15) "$"P; TAB(25) "$"D; TAB(35) "$"P-D; TAB(45) Y
    "years"; TAB(55) "$"M; TAB(65) "$"D+Y*12*M
120 END
```

Here is the computer output for this example.

Model	Selling Price	Down Payment	Amount Financed	Term	Monthly Payment	Deferred Payment Price
Compact	$5995	$875	$5120	3 years	$158.32	$6574.52

Exercises

Use the program to complete the chart.

	Model	Selling Price	Down Payment	Amount Financed	Term	Monthly Payment	Deferred Payment Price
1.	Compact	$6287	$1350		2 years	$224.22	
2.	Van	$16,739	$6500		4 years	$231.44	
3.	Pickup	$10,493	$4300		3 years	$194.39	
4.	Coupe	$13,067	$6000		3 years	$219.86	

CHAPTER 13 REVIEW

Match the car and the selling price. Use the chart on p. 294.

Car	Selling Price
1. Compact, air conditioning, power brakes, tinted glass, AM-FM radio, 5% sales tax, $125 dealer preparation fee, $175 destination charge, and $25 license fee	**A.** $7387.30
2. Sedan, power brakes and steering, AM-FM radio, body side molding, automatic transmission, tilt steering wheel, 6% sales tax, $115 dealer preparation fee, $165 destination charge, and $20 license fee	**B.** $14,539.58
3. Sport Coupe, air conditioning, AM-FM radio with cassette player, power brakes, steering, window and door locks, tinted glass, rear window defogger, 6% sales tax, $156 dealer preparation fee, $185 destination charge, and $25 license fee D	**C.** $9513.52
4. Luxury, pin striping, air conditioning, tinted glass, power windows, automatic transmission, rear window defogger, power door locks, 5.5% sales tax, $320 dealer preparation fee, $200 destination charge, and $25 license fee	**D.** $12,365.20
	E. $8176.57
	F. $16,972.89

5. Arrange the fuel consumption for A–D in order from best to worst.

A. 476295 / 478361 — 8.4 gallons

B. 748126 / 752939 — 14.8 gallons

C. 092686 / 096098 — 10.6 gallons

D. 365819 / 369845 — 20.3 gallons

6. Which tank contains the greatest number of liters of fuel?

A.

44-liter tank

B.
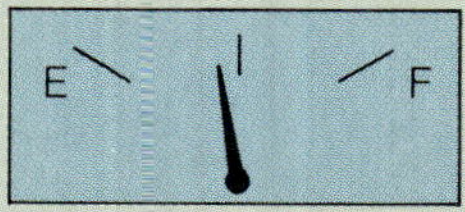

60-liter tank

C.

80-liter tank

D.

50-liter tank

CHAPTER 13 TEST

1. Find the selling price.
   Sedan: body side molding, pin striping, air conditioning, AM-FM stereo radio with cassette player, power brakes, power steering, rear window defogger; sales tax 6%, destination charge \$342, dealer preparation \$287, license fee \$30

Calculate the annual fuel cost based on 35,000 miles driven.

	Highway Driving	City mpg	Highway mpg	Price per Gallon
2.	80%	17	23	\$1.10
3.	40%	19	26	\$1.25

Find the dealer's percent of profit.

	Model	Trade-in Allowance	Cost of Repairs	Selling Price
4.	'83 luxury	\$9300	\$622	\$11,100
5.	'80 4-D sedan	\$2500	\$289	\$3900

Complete the chart.

	Model	Selling Price	Down Payment	Amount Financed	Term	Monthly Payment	Deferred Payment Price
6.	Sedan	\$9550	\$3000		3 years	\$207.42	
7.	Compact	\$6320	\$1500		4 years	\$115.48	

Find the annual liability premium. Use the table on p. 302.

8. 25/50/25 coverage, 2.8 driver rating factor
9. 100/300/50 coverage, 3.65 driver rating factor

Find the total annual collision and comprehensive premium. Use the table on p. 304.

10. Driver rating factor 3.2, car class 28, collision deductible \$200, comprehensive deductible \$100

Find the rental cost.

	Cost per Day	Number of Days	Cost per Mile	Number of Miles	Discount
11.	\$37.25	5	\$0.22	803	150 free miles
12.	\$42.50	10	\$0.30	2593	10%

13. The labor charge for Lisa's car repair was \$64.50. The parts included 6 spark plugs at \$2.75 each, 1 fan belt at \$14.75, and 1 oil filter at \$8.10. Find the total repair bill. Use 6% tax.

CHAPTER 13 CUMULATIVE REVIEW

1. Find the equivalent fraction. Use mental math.
If $13\frac{1}{3}\% = \frac{2}{15}$, then $40\% = \square$.
   A. $\frac{4}{15}$ B. $\frac{1}{3}$
   C. $\frac{6}{15}$ D. $\frac{1}{5}$

2. Find the selling price.
base price $7725, air conditioning $790, tinted glass $115, power steering $295
   A. $8925   B. $8825
   C. $9015   D. $8905

3. Find the gross pay.
regular hours 37, regular rate $6.40, overtime hours 8, overtime rate $9.60
   A. $303.60   B. $76.80
   C. $236.80   D. $313.60

4. Calculate the annual fuel cost based on 12,000 miles driven.
highway driving 45%, city mpg 24, highway mpg 29, price per gallon $0.98
   A. $209.43   B. $451.98
   C. $287.75   D. $450.88

Car Class Rating	Collision deductible $50	$100	$200
21–30	$227	$207	$187

A 29 rated car,
Driver rating factor 1.5,
$200 deductible.

5. Find the total annual collision premium.
   A. $277.50
   B. $340.50
   C. $310.50
   D. $280.50

6. The odds in favor of winning a prize are 2 to 7. Predict if Arlene will win one.
   A. yes
   B. no

7. Find the rental fee.
concrete mixer, $22 per day for 3 days
   A. $22   B. $44
   C. $66   D. $7.50

8. Find the total number of calories burned.
Jogging uses 648 calories per hour for $1\frac{1}{2}$ hours.
   A. 862   B. 972
   C. 1072   D. 970

9. Subtract.
$$\begin{array}{r} 8.02 \\ -5.228 \\ \hline \end{array}$$
   A. 3.208
   B. 2.802
   C. 2.792
   D. 3.792

10. What is the actual rise in WIPCO sales between 1984 and 1989?
   A. 100
   B. 250
   C. 200
   D. 300

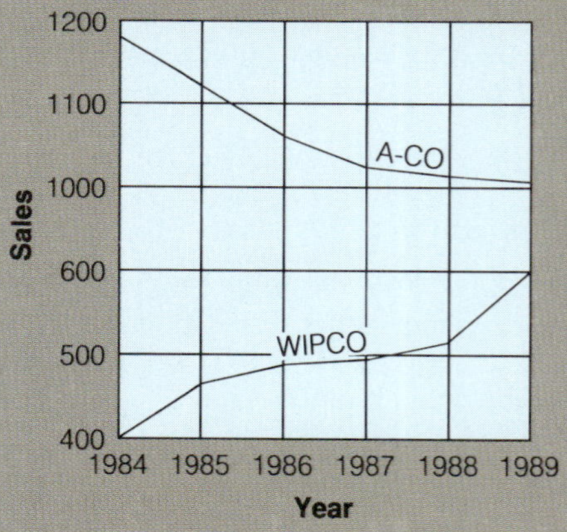

11. Find the annual liability premium.
15/30/15 coverage, basic rate $180, 2.8 driver rating factor
A. $384
B. $504
C. $404
D. $484

12. Find the balance carried forward.

Balance	$105.26
Check	4.09
Total	
Other	
Withdrawal	20.00
Balance	

A. $101.17
B. $129.35
C. $81.17
D. $89.35

13. Find the finance charge.
new subcompact $6700, monthly payments for 4 years $169.75
A. $1448
B. $8148
C. $679
D. $203.70

14. Find the property tax.
assessed value $87,000; tax rate $2.15 per $100
A. $1870.50
B. $187.50
C. $2070.50
D. $1769.50

15. Jack and 14 friends went backpacking. The average weight of their packs was 42 lb. Is 600 lb a reasonable estimate of the total weight they carried?
A. unreasonable 400 lb
B. unreasonable 800 lb
C. unreasonable 200 lb
D. reasonable

16. Find the rental cost for a car.
$25.20 per day, $0.39 per mile, 10% discount, for 3 days, 210 miles
A. $68.04
B. $81.90
C. $149.94
D. $141.75

17. In a recent year, Americans spent 6.5% of health care money on dentists and 19.5% on physicians. How much more was spent on physicians than dentists?
A. 13% B. 27%
C. 7.5% D. 6%

18. Find the total repair bill. Use 6.5% tax on parts.

spark plugs	$10
fuel filter	$14
rotor	$ 9.28
anti-freeze	$17
labor	$20

A. $70.28
B. $73.55
C. $53.55
D. $74.85

19. Find the total cost with tax. Use 3.5%.
2 bottles aspirin @ $2.50 each, 3 boxes bandages @ $1.67 each, 1 tube skin cream @ $3.75 each
A. $14.24
B. $14.25
C. $8.20
D. $13.76

Chapter 14 Overview

Key Ideas

- using a bus schedule
- traveling by taxicab
- comparing transportation costs
- using a mileage chart
- giving directions
- using a train schedule
- using an airline schedule
- finding airfare

Key Terms

- timetable
- taxi fare
- commuter
- mileage
- legend
- schedule
- host computer

Key Skills

Find the answer.

1. $\begin{array}{r} \$17.25 \\ 3.27 \\ \underline{+54.18} \end{array}$ **2.** $\begin{array}{r} \$\ 2.96 \\ 12.14 \\ \underline{+20.19} \end{array}$ **3.** $\begin{array}{r} \$19.86 \\ 8.04 \\ \underline{+17.21} \end{array}$ **4.** $\begin{array}{r} \$109.19 \\ 21.00 \\ \underline{+\quad 4.26} \end{array}$

5. $21.19 − $16.84 **6.** $24.38 − $12.99 **7.** $13.70 − $9.85

8. $\frac{7}{10} \div \frac{1}{4}$ **9.** $2\frac{4}{5} \div \frac{1}{4}$ **10.** $2\frac{7}{10} \div \frac{1}{4}$

11. 936 ÷ 21.5 **12.** 832 ÷ 23.5 **13.** 568 ÷ 18.2

14. 456 ÷ 16.9 **15.** 716 ÷ 21.9 **16.** 924 ÷ 32.5

17. $\begin{array}{r} 2\text{ h }15\text{ min} \\ \underline{+1\text{ h }37\text{ min}} \end{array}$ **18.** $\begin{array}{r} 3\text{ h }\ 2\text{ min} \\ \underline{+1\text{ h }59\text{ min}} \end{array}$ **19.** $\begin{array}{r} 1\text{ h }12\text{ min} \\ \underline{+4\text{ h }49\text{ min}} \end{array}$

Solve.

20. n = 19% of 54 **21.** 80% of 600 = n **22.** 75% of $85 = n

23. 65% of $320 = n **24.** 60% of $240 = n **25.** n = 15% of 85

26. How much time has elapsed between 7:45 a.m. and 8:22 a.m.?

27. How much time has elapsed between 8:10 a.m. and 9:08 a.m.?

28. How much time has elapsed between 11:47 a.m. and 12:17 a.m.?

TRANSPORTATION

14

14-1 Riding the Bus

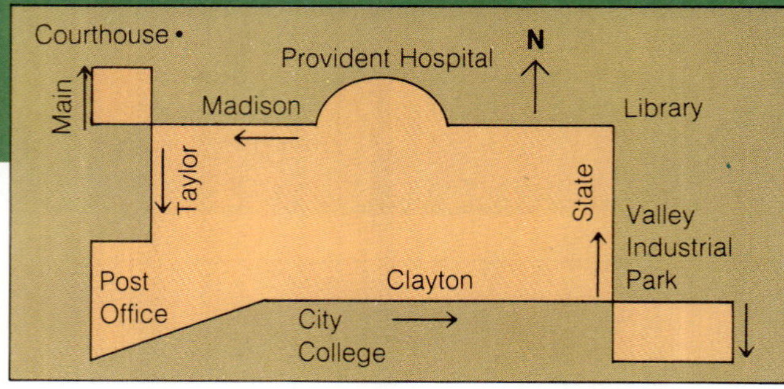

Objective
Use a bus schedule.

A bus schedule is also called a **timetable.** Below is a timetable for the Clayton bus during the early morning rush hours. The map shows the bus route.

Bus Schedule

Post Office	City College	Clayton & State	Valley Park	State & Madison	Provident Hospital	Madison & Main	Main & Taylor	Post Office
6:00	6:03	6:08	6:15	6:23	6:27	6:31	6:35	6:42
6:20	6:23	6:28	6:35	6:43	6:47	6:51	6:55	7:02
6:45	6:49	6:54	7:02	7:10	7:15	7:20	7:24	7:32
7:05	7:09	7:14	7:22	7:30	7:36	7:42	7:47	7:57
7:15	7:19	7:24	7:32	7:41	7:48	7:55	8:01	8:11
7:35	7:39	7:45	7:53	8:02	8:09	8:16	8:22	8:32
8:00	8:04	8:10	8:18	8:27	8:34	8:41	8:47	8:57
8:15	8:19	8:25	8:33	8:42	8:49	8:56	9:02	9:10
8:35	8:39	8:45	8:53	9:02	9:08	9:14	9:19	9:27
9:00	9:04	9:10	9:18	9:24	9:29	9:34	9:39	9:47
9:30	9:34	9:40	9:48	9:54	9:59	10:04	10:09	10:17

Examples

1. Cecilia Kopp works the 8:30–4:30 day shift at Provident Hospital. When should she catch the bus at Clayton and State streets?

Look under the destination to find the time of arrival.

Cecilia needs to be at work at 8:30. She will get to work on time if she takes the bus that arrives at 8:09.

Look to the left to see what time to board the bus.

The bus leaves Clayton and State streets at 7:45.

2. How long is Cecilia's bus ride?

Think: From 7:45 to 8:00 is 15 minutes. From 8:00 to 8:09 is 9 minutes. The bus ride lasts 24 minutes.

Try This

Find the answer.

a. Jud catches the bus at the Post Office. He has to be at work in Valley Park by 8:00. What time should he board?

b. A bus leaves State and Madison at 7:41. When does it arrive at Main and Taylor?

Exercises

Use the timetable to complete the chart.

	Destination	Boarding Point	Arrival Time	Boarding Time	Travel Time
1.	Valley Park	City College	7:22		
2.	Madison & Main	State & Madison	8:16		
3.	Courthouse	Clayton & State	by 8		
4.	Library	City College	by 9		
5.	Post Office	Provident Hospital		7:48	
6.	Madison & Main	Post Office		8:35	
7.	Main & Taylor	State & Madison		first bus after 8	
8.	Provident Hospital	Main & Taylor		first bus after 9:15	

9. Bill Crocker overslept and took the 8:19 from City College to Main and Taylor. He has a ten-minute walk after he gets off the bus. What time will Bill arrive for a 9 o'clock appointment?

10. Rhoda Wallace boards the bus at Main and Taylor at 6:35. The bus leaves the Post Office at 6:45. What is the total travel time to City College?

Estimation

11. The Clayton bus travels about 12 miles per hour during the early morning rush hour. Estimate the length of the route.

12. Bus driver Suzy Parker makes six complete trips each weekday and fills her 30-gallon tank twice a week. Estimate the miles per gallon of gasoline.

Suppose

13. Suppose the 7:05 from the Post Office loses 1 minute between stops due to engine trouble. At what stop will the 7:15 from the Post Office catch up with the 7:05?

Mixed Skills Review

1. $0.68 × 243 **2.** 1401 − 1293 **3.** $10\frac{1}{8} + \frac{2}{3}$

4. 13,664 + 209,165 **5.** $100 \div 1\frac{12}{13}$ **6.** $10.08 − $7.25

7. $5\frac{3}{4} - 1\frac{7}{8}$ **8.** 3.08 + 16.7 + 0.468 **9.** 1509 × 84

10. 394.2 × 14.6 **11.** $\frac{5}{6} \times \frac{3}{8} \times \frac{4}{5}$ **12.** 20,655 ÷ 243

14-2 Traveling by Taxi

Objective
Find costs of using a taxi. Confirm fares by finding distances.

Carrie Small took a taxi from her hotel in Brinkley to a business meeting at the Convention Center, a distance of 1.5 miles. The taxi fares in Brinkley are set by the local government.

Taxi Fares
$1.25 first $\frac{1}{4}$ mile
$0.50 each add'l $\frac{1}{4}$ mile
$30/h Sightseeing Tour
$24/h Wait time

Examples

1. How much was Carrie's fare?

Think:
$1.25 + (5 × $0.50) = $3.75

The fare was $3.75.

1.5 miles

$\frac{1}{4}$	$\frac{1}{4}$	$\frac{1}{4}$	$\frac{1}{4}$	$\frac{1}{4}$	$\frac{1}{4}$
$1.25	$0.50	$0.50	$0.50	$0.50	$0.50

2. The distance between City Hall and the train station is 2.3 miles. Find the fare.

Think: How many $\frac{1}{4}$-mile units are in $2\frac{3}{10}$?

$2\frac{3}{10} \div \frac{1}{4} = \frac{23}{10} \times \frac{4}{1} = \frac{92}{10} = \frac{46}{5} = 9\frac{1}{5}$

$1.25 + (9 × $0.50) = $5.75

The taxi fare from City Hall to the train station is $5.75.

Try This

Find the taxi fare for each distance.

a. $\frac{3}{4}$ mi **b.** $3\frac{1}{2}$ mi **c.** 0.9 mi **d.** $2\frac{3}{5}$ mi

3. John Cavanaugh spent $14.75 for a taxi ride. The driver had to wait 20 minutes for John. About how far did he travel?

Subtract the waiting time cost.
20 min = $\frac{1}{3}$ h

$\frac{1}{3} \times \$24 = \$8$

$14.75 − $8 = $6.75

Subtract the fare for the first $\frac{1}{4}$-mile unit.
$6.75 − $1.25 = $5.50

For each additional unit, divide the balance by the cost.
$5.50 ÷ $0.50 = 11 additional units

Jim rode about 12 quarter-mile units, or 3 miles.

Try This Find the distance traveled by taxi.

e. fare \$3.25 **f.** fare \$9.75 **g.** fare \$16.25, 30 minutes waiting time

Exercises

The Jefferson Hotel provides local travel information for hotel guests and visitors to the city of Brinkley. Find the taxicab fare for each distance.

	Distance in Miles	From the Hotel to . . .
1.	$\frac{3}{4}$	Hudson Park
2.	$1\frac{1}{4}$	City Hall
3.	$1\frac{3}{8}$	Union Station
4.	$2\frac{3}{5}$	County Stadium
5.	3	Walters Gallery
6.	$1\frac{9}{10}$	Davis Planetarium
7.	$3\frac{7}{10}$	Liberty Amusement Park
8.	$5\frac{1}{8}$	Holiday Museum
9.	$9\frac{1}{2}$	Cliburn Gardens
10.	11	Airport

Find the distance traveled by taxicab.

11. fare \$2.25 **12.** fare \$4.75 **13.** fare \$9.75 **14.** fare \$12.25

15. fare \$9.75, 15-min wait **16.** fare \$10.25, 20-min wait **17.** fare \$27.75, 45-min wait **18.** fare \$30.25, 40-min wait

In Waterford Township the rate for a taxicab is \$1.15 for the first $\frac{1}{5}$ mile and \$0.65 for each additional $\frac{1}{5}$ mile. Find the fare for the following distances.

19. 1 mile **20.** $1\frac{4}{5}$ miles **21.** $2\frac{3}{5}$ miles **22.** $3\frac{2}{5}$ miles

23. $1\frac{1}{2}$ miles **24.** $1\frac{5}{8}$ miles **25.** $7\frac{3}{4}$ miles **26.** $6\frac{7}{10}$ miles

Mental Math Many people give taxicab drivers a 15% tip. Find the tip for the following fares using mental math.

27. \$3 **28.** \$7 **29.** \$4.50 **30.** \$8.60

Mixed Skills Review

1. $6\frac{3}{5} - 2\frac{17}{20}$ **2.** \$236 + \$1594 + \$33 **3.** 18 − 11.0984

4. 11,692 ÷ 74 **5.** $1\frac{1}{5} \times \frac{3}{8}$ **6.** 9152 × 43

7. 0.8 + 2.34 + 12 + 2.9 **8.** 406,080 − 17,956 **9.** 89.408 ÷ 5.08

10. $\frac{7}{9} \div \frac{1}{6}$ **11.** 63% of 507 **12.** $3\frac{2}{9} + 2\frac{5}{6}$

14-3 Comparing Transportation Costs

Objective *Compare costs of different modes of transportation.*

Clark Larson has been hired as a teller at the downtown branch of a local bank. He lives 8.3 miles from the bank and estimates that the cost to drive would be $0.19 per mile. Clark wants to find the most economical means of transportation for commuting to and from his job five days per week.

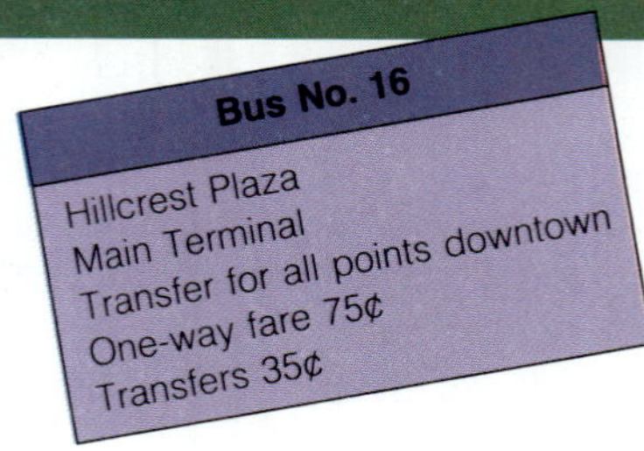

Understand the Situation

- How many miles would Clark drive each day?
- How many trips would Clark take if he rode the bus to and from work?

Examples

1. Parking costs $8.50 per week downtown. Compare the weekly cost of driving to riding the bus if Clark must make one transfer.

Find the travel cost per week.

car: 8.3 miles × 2 trips × 5 days × $0.19 = $15.77
bus: ($0.75 + $0.35) × 2 trips × 5 days = $11.00

Add the driving and parking costs.
$15.77 + $8.50 = $24.27

Subtract to compare.
$24.27 − $11.00 = $13.27

Driving costs $13.27 more per week than riding the bus.

2. Tom lives 7 miles from his work. He estimates that the cost to drive his car would be $0.18 per mile. A commuter van service in Tom's area costs $34.60 per month. How much will Tom save per week by using the van service? Assume a 5-day work week and a 4-week month.

Find the travel cost per week.

car: 7 miles × 2 trips × 5 days × $0.18 = $12.60
van: $34.60 ÷ 4 weeks = $8.65

$12.60 − $8.65 = $3.95 Subtract to compare.

It costs $3.95 less per week to use the van service.

Try This

Compare the weekly transportation costs.

	Distance to Work	Driving Cost/Mile	Parking Fees	Driving Cost/Week	Cost of Public Transportation	Cost per Week	Weekly Difference
a.	1.9 mi	$0.18	$3.25/day		Bus 65¢/tkt.		
b.	29.4 mi	$0.21	$36/mo		Train $90/mo		

Exercises

Compare the weekly transportation costs. Assume a 4-week month.

	Distance to Work	Driving Cost/Mile	Parking Fees	Driving Cost/Week	Cost of Public Transportation	Cost per Week	Weekly Difference
1.	2.5 mi	\$0.22	\$5.50/wk		Bus 95¢/tkt.		
2.	14.3 mi	\$0.18	\$60/mo		Subway 85¢/tkt.		
3.	29.4 mi	\$0.21	\$1.65/day		Train \$185/mo		
4.	5.7 mi	\$0.19	\$9/wk		Bus 95¢ and trolley 35¢		
5.	13.5 mi	\$0.12	\$0		Tokens \$42/mo		
6.	9.8 mi	\$0.23	\$35/mo		Van \$47.60/mo		
7.	7.4 mi	\$0.23	\$1.90/day		Bus \$1.25/tkt.		
8.	33.0 mi	\$0.19	\$45/mo		Taxi \$7.50/day		

Mental Math

Answer the following questions about the business traveler's guide.

City	Miles, Airport to City	Taxi	Airport Limo	Mass Transit
Atlanta	8	\$13.50	\$6.00	\$0.60
Baltimore	10	12.00	5.00	1.00
Chicago	18	19.50	9.00	1.00
Dallas	7	9.00		0.50
Los Angeles	15	27.50	6.30	0.85
Miami	7	11.00	6.75	0.75
Minneapolis	13	16.50	6.50	0.90
Phoenix	5	8.50	5.00	0.75
San Francisco	14	23.25	6.00	1.25

9. How much more expensive is a taxi than a limo in Atlanta?

10. How much will you save by taking mass transit in Dallas?

11. What is the cost per mile for an airport limo in Minneapolis?

12. What is the round-trip cost for an airport taxi in San Francisco?

13. Six passengers took a limo in Miami. What was the total fare?

Mixed Skills Review

1. 5680×31

2. $188.42 + 36.07 + 0.56$

3. $6\frac{3}{4} \times 1\frac{1}{9}$

4. $1\frac{2}{5} - 1\frac{2}{15}$

5. $2915 - 886$

6. 1532.3×0.6

7. $36{,}756 \div 18$

8. $1\frac{7}{9} \div \frac{2}{3}$

9. $27 + 6 + 53 + 68 + 5$

10. 9.061×450

11. $11.2 - 9.48$

12. $60.858 \div 6.21$

14-4 Using a Mileage Chart

Objective
Find distances using a mileage chart.

More than 1 out of every 2 Americans will travel over one million miles in a lifetime. Most of this distance is traveled by automobile. A mileage chart is useful when planning long trips.

Road Mileage Between U.S. Cities

	Boston	Buffalo	Detroit	Memphis	Miami	Philadelphia	Phoenix	St. Louis
Birmingham	1194	947	754	249	777	894	1680	508
Boston	—	457	702	1345	1539	304	2664	1179
Cleveland	639	192	175	773	1325	432	2052	567
El Paso	2358	1928	1696	1095	1957	2065	402	1179
Indianapolis	940	510	290	466	1225	647	1713	239
Los Angeles	3036	2606	2369	1847	2757	2743	398	1864
Memphis	1345	965	756	—	1025	1045	1464	295
Minneapolis	1402	955	698	854	1801	1177	1644	559
New Orleans	1541	1294	1101	401	892	1241	1523	696
New York	213	436	671	1134	1328	92	2442	968
Washington	440	386	526	902	1101	140	2278	836

Examples

1. Find the distance between Los Angeles and Detroit.

Look across the Los Angeles row to the number in the Detroit column.

The distance between the cities is 2369 miles.

2. The O'Reilly family is planning a trip from Memphis through El Paso to Phoenix. They estimate they will drive about 275 miles per day. About how many days will the trip take?

Find the total distance.

from Memphis to El Paso	1095
from El Paso to Phoenix	+ 402
	1497 miles

Divide to find the number of days. Use compatible numbers.

$\frac{1497}{275} \approx \frac{1500}{300}$, or 5

The trip will take about 5 days.

3. On a trip from Washington, D.C. to St. Louis, a car averaged 23.5 miles per gallon. How many gallons of gasoline (to the nearest tenth) were used?

The distance from Washington, D.C. to St. Louis is 836 miles.

Divide the distance by the number of miles per gallon **(mpg).**

$836 \div 23.5 \approx 35.6$ gallons

The car will use 35.6 gallons.

Try This

a. Find the distance between Cleveland and Miami.

b. Jim Crenshaw drove from Birmingham to Buffalo in 19 hours. What was his average speed?

c. Sophie Trestle drove from Philadelphia through New York to Boston. She used 9.9 gallons of gasoline. About how many miles per gallon did her car get?

Exercises

Find the distance in miles.

1. from Memphis to Miami
2. from Boston to Detroit
3. from Indianapolis to St. Louis
4. from Cleveland to Phoenix
5. from Birmingham through Memphis to Minneapolis
6. from Boston through Philadelphia to New Orleans

Solve.

7. Wilbur and Helen Su are driving from Buffalo to Indianapolis. Their car will get about 21 mpg. To the nearest tenth, how many gallons of gasoline will they need?
8. On a trip from Minneapolis to Miami a car uses 76 gallons of gasoline. To the nearest tenth, find the number of mpg.
9. Clay Timberlane drove his horse van from St. Louis to El Paso in 24 hours. About how many miles per hour did he average?
10. The Tisdales are traveling cross-country from Boston to Los Angeles. At 325 miles per day, about how many days will the trip take?

Calculator

11. The cost of gasoline on a trip from Philadelphia to Memphis was $38.54. Find the cost per mile.
12. The total cost of tolls on an interstate between New York and Philadelphia was $2.60. Find the toll cost per mile.

Work with a Partner

13. San Francisco is 2540 miles from Cleveland. Estimate the hours of driving time from San Francisco through Cleveland to Boston at 55 mph.
14. Baltimore is 90 miles from Philadelphia. Estimate the hours of driving time from Baltimore through Philadelphia to Indianapolis at 55 mph.

Show Math

Draw a map. Use a mileage chart and a line segment for each 200 miles. Show the direction and distance from Kansas City.

15. ENE New York
16. NE Chicago
17. NW Seattle
18. W Denver
19. SSW Dallas
20. E Louisville

Mixed Skills Review

1. $4.8 \div 0.75$
2. $1\frac{3}{5} \times \frac{10}{21}$
3. $6.5356 + 12.0836$
4. $8\frac{9}{10} + 3\frac{2}{5}$
5. $114{,}264 \div 54$
6. 473×653
7. $71{,}827 - 45{,}608$
8. $3.264 - 0.768$
9. $5930 - 1764$
10. 6.091×450
11. $90{,}615 + 28{,}586$
12. $\frac{11}{16} - \frac{3}{8}$

14-5 Giving Directions

Objective
Practice giving directions.

Our highway system uses different symbols to identify interstate highways, U.S. routes, state roads, and county roads. A legend on each map helps the traveler.

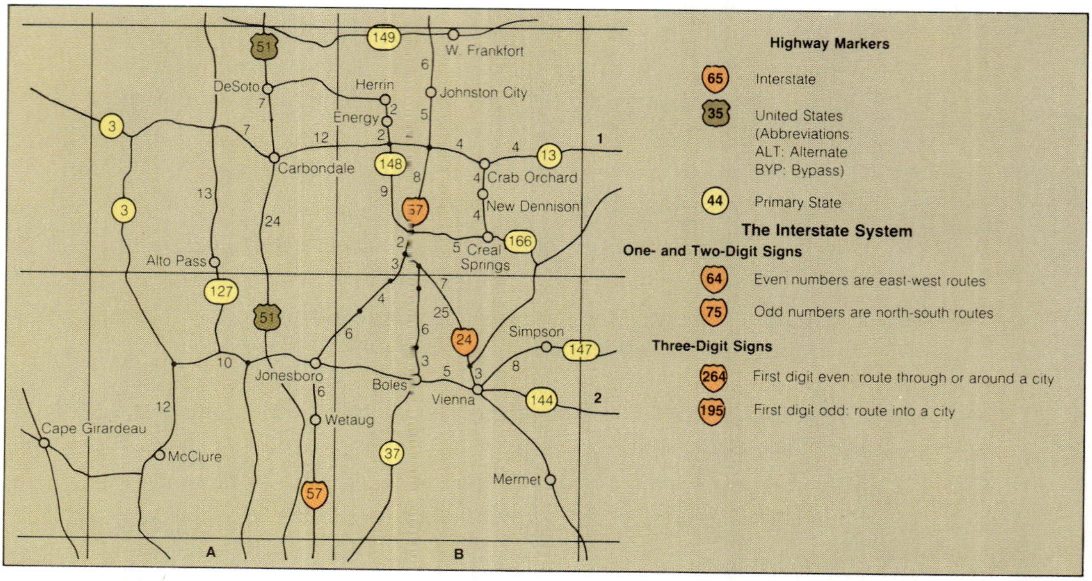

Understand the Situation

- What legend symbol is used for primary state roads?
- On a three-digit interstate highway, what does the first digit indicate?
- Cape Girardeau is the most westerly city on the map. What is the northernmost city?

Example Give directions for a trip from Carbondale (A, 1) to W. Frankfort (B, 1).

Road	Direction	Distance
State Route 13	East	16 miles
Interstate 57	North	11 miles
State Route 149	East	About 1 mile

Try This Give directions for the following trips.

a. from McClure (A, 2) to Jonesboro (A, 2)

b. from Energy (B, 1) to Vienna (B, 2)

Exercises

Give directions for the following trips. Use the map on p. 326.

1. from Creal Springs (B, 1) to Crab Orchard (B, 1)
2. from DeSoto (A, 1) to Alto Pass (A, 1)
3. from Johnston City (B, 1) to Simpson (B, 2)
4. from Boles (B, 2) to New Dennison (B, 1)
5. from Herrin (B, 1) to Mermet (B, 2)
6. from W. Frankfort (B, 1) to Wetaug (A, 2)

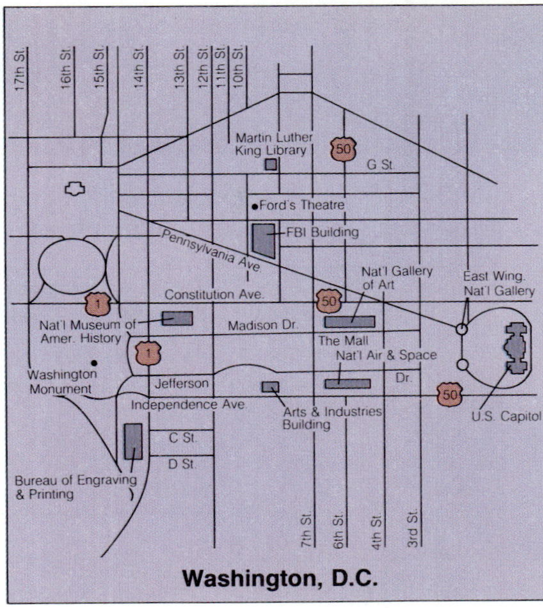

Think Math Many visitors to Washington, D.C. ask for directions to historic places of interest. Follow the directions and the map to identify each attraction in the nation's capital.

7. From the Washington Monument, go north on 14th Street two blocks to Constitution Avenue; turn right and go east two blocks; turn left and go north two blocks.
8. From the Arts and Industries Building, go east on Independence Avenue to 7th Street; turn left and go north one block; turn right.
9. From Ford's Theater go south on 10th Street to Pennsylvania Avenue; turn left and go two blocks past the East Wing of the National Gallery.
10. From the National Gallery of Art, go west on Madison Drive past the National Museum of American History; go south across Independence Drive to the building facing C and D streets.

Work with a Partner

11. Draw a map of your hometown. Include important points of interest. Write directions from your school to your house.

Mixed Skills Review

1. $0.7 - 0.163$
2. $2\frac{11}{20} + 3\frac{1}{4}$
3. 1003×103
4. $0.221 \div 0.034$
5. $1568 + 7 + 265$
6. $6\frac{5}{6} + 2\frac{7}{12}$
7. $0.17 + 0.26 + 0.43$
8. $1\frac{7}{15} \times 1\frac{1}{2}$
9. $5930 - 1764$
10. $6\frac{3}{8} - 4\frac{5}{6}$
11. 1.2×183.6
12. $351{,}549 \div 583$

14-6 Riding the Train

Objective
Use schedules and find train fare.

Jim Binko is planning to take the train from McKinney to visit his grandparents in Shreveport. He picked up a schedule from the train station.

		Eastbound		Westbound	
	Frequency	**Silver Streak**	**Rio Grande**	**Silver Streak**	**Rio Grande**
Wichita Falls	Daily	dp 8 00 a	dp 2 57 p	arr 10 31 p	arr 12 41 p
Denton	Daily	dp 9 38 a	dp 4 40 p	dp 8 53 p	dp 11 03 a
McKinney	Ex Sa Su	dp 10 17 a	dp 5 21 p	dp 8 14 p	dp 10 22 a
Greenville	Su only	dp 10 49 a	dp 5 59 p	dp 7 42 p	dp 9 44 a
Gilmer	Sa only	dp 11 25 a	dp 6 45 p	dp 7 06 p	dp 8 58 a
Shreveport	Daily	arr 1 01 p ↓	arr 8 13 p	dp 5 30 p ↑	dp 7 30 a

Understand the Situation

- How many trains travel between McKinney and Shreveport every week?
- On what days does a train stop in McKinney?

Examples

1. What is the travel time on the Silver Streak between McKinney and Shreveport?

Find the departure and arrival times on the schedule.

The Silver Streak leaves McKinney at 10:17 a.m. and arrives in Shreveport at 1:01 p.m.

Count forward and add to find the travel time.

10:17 a.m. → 11:00 a.m.	=	43 min
11:00 a.m. → 1:01 p.m.	=	2 h 1 min
		2 h 44 min

The travel time is 2 hours 44 minutes.

2. The one-way fare between McKinney and Shreveport is \$37.50. The round-trip discount is 25%. Find the cost of a round-trip ticket.

Find the cost of 2 one-way tickets. $\$37.50 \times 2 = \$75$

Find the discount fare. $0.75 \times \$75 = \$56.25$

The round-trip fare is \$56.25.

Try This

Find the answer.

a. The westbound Rio Grande from Gilmer to Greenville left 20 minutes late. What time did the train depart?

b. A one-way ticket from Wichita Falls to Denton costs \$25. The discount on a round-trip fare is 15%. Find the cost of a round-trip ticket.

Exercises

Use the train schedule on p. 328 to answer the following questions.

1. What time does the eastbound Silver Streak leave Denton?
2. What time does the westbound Rio Grande leave Denton?
3. On what days does the train stop in Greenville?
4. The Rio Grande from Shreveport arrives in Gilmer at 8:51 a.m. How long do passengers have to board?
5. A passenger arrived at the train station in McKinney at 10:30 a.m. How long does the passenger have to wait for the next train to Wichita Falls?
6. A rainstorm slowed all eastbound trains 5 minutes between stops. At what time will the 5:21 from McKinney arrive in Gilmer?

Mental Math

Find the travel time. There is a 2-minute wait at each stop.

	Silver Streak	Rio Grande
7. Denton to Greenville		
8. McKinney to Gilmer		
9. Wichita Falls to McKinney		
10. Shreveport to Greenville		
11. Gilmer to Wichita Falls		
12. Greenville to McKinney		

Estimation

Estimate the cost per mile for each of the following train rides.

13. \$25 for 93 miles
14. \$17 for 56 miles
15. \$37.50 for 149 miles
16. \$189.95 for 970 miles

Write Math

17. The Coastliner travels at an average speed of 70 mph. Write a train schedule for a train from Sudbrook to Owings Mills.

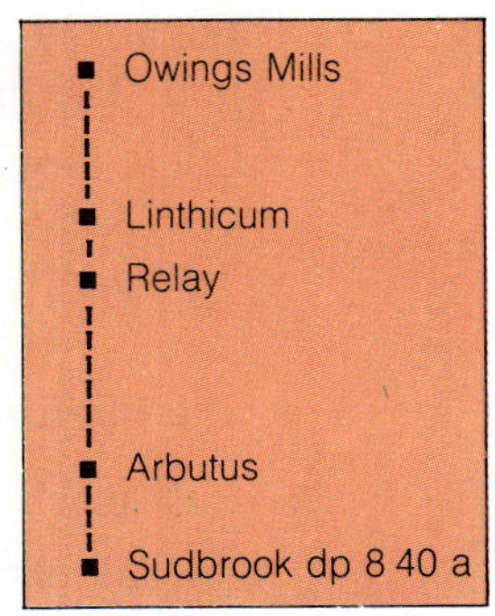

Scale $\frac{1}{2}$ inch: 100 miles

Mixed Skills Review

1. 12.76×90.5
2. $396{,}153 + 16{,}298$
3. $19.406 - 7.84$
4. $121{,}012 - 27{,}572$
5. $4.62 + 8.365 + 16.2$
6. $6\frac{2}{5} \div \frac{16}{35}$
7. $588{,}303 \div 807$
8. $1\frac{2}{3} - \frac{5}{7}$
9. 862×434
10. $2\frac{4}{7} \times 4\frac{2}{3}$
11. $1069.5 \div 150$
12. $\frac{7}{8} + \frac{1}{4}$

14-7 Air Travel

ENTER DEPARTURE CITY NAME OR CODE:
Baltimore, MD
ENTER DESTINATION CITY NAME OR CODE:
Memphis, TN
ENTER DEPARTURE DATE:
09 Aug

Objective
Use airline flight schedules.

Travel agents and some people at home use computers to make airline reservations. First, the user enters initial data, in this case a trip from Baltimore to Memphis on 8/9. The computer displays a schedule menu.

The customer sees that breakfast Flight 107 leaves from Baltimore at 9:00 a.m. and makes one stop before arriving in Memphis at 10:57 a.m.

DIRECT FLIGHTS — THUR–09 AUG
FROM BALTIMORE, MD, USA — 72S $149; 80% for M–Th travel
TO MEMPHIS, TN, USA — 095 $175; first-class, $209

	Depart	Port	Arrive	Port	Line	No.	Rate	M	S
1.	900A	BWI	1057A	MEM	PI	107	72S	B	1
2.	1015A	BWI	1139A	MEM	AA	173	72S	L	0
3.	135P	BWI	341P	MEM	PI	159	72S	S	1
4.	300P	BWI	816P	MEM	EA	141	095	D	3

Examples Use the schedule menu to answer the following questions.

1. What is the travel time on Flight 159?

Flight 159 arrives in Memphis at 3:41 p.m. The time in Baltimore is 4:41 p.m.

Count forward and add to find the travel time.

1:35 p.m. → 2:00 p.m. = 25 min
2:00 p.m. → 4:41 p.m. = 2 h 41 min
2 h 66 min = 3 h 6 min

The travel time is 3 hours 6 minutes.

Baltimore
1:35 p.m.

Memphis
12:35 p.m.

2. How much more will an August 9 reservation on Flight 141 cost?

The discount rate for 72S is in effect on Thursday.

Find the 72S discount rate. $0.80 \times \$149 = \$119.20$

Subtract the 72S rate from the 095 rate. $\$175 - \$119.20 = \$55.80$

Try This Use the schedule menu to answer the following questions.

a. When does the first afternoon flight depart?

b. How many stops does Flight 141 make en route to Memphis?

c. What is the travel time on the 9:00A flight?

Exercises

Use this schedule menu to answer the following questions.

NONDIRECT FLIGHTS — TUES–12 MAY
FROM ALPENA, MI, USA
TO CHICAGO, IL, USA

	Depart	Port	Arrive	Port	Line	No.	Rate	M	S
1.	932A	ALP	1021A	DET	NW	063	85P	S	0
	1059A	DET	1148A	CHI	RE	057	S	0	
2.	215P	ALP	307P	FLT	NW	039	067	S	0
	435P	FLT	539P	CHI	EA	159	S	1	

85P $101; 90% rate for F–Su travel
067 $125; 66 2/3% rate for 30-day pre-reservations

1. One nondirect flight from Alpena goes through Flint. What is the stopover city for the other flight?
2. What is the layover time for a passenger leaving Alpena on Flight 039?
3. How much is a one-way ticket on the 9:32A flight on May 12?
4. What rate is in effect for Saturday travel on the 9:32A flight?
5. What is the price of a one-way ticket on the 2:15P flight if the ticket is purchased on April 1?
6. When it is 12 noon in Chicago, the time is 1:00 p.m. in Alpena. What is the travel time between Alpena and Chicago on Flights 039 and 159?

Mental Math

Use the time zone map on p. 461 in the Data Bank to find the arrival time for each flight.

	Flight	Departs	Flying Time
7.	Helena, MT/Sweetwater, TX	1040A	6 h 07 min
8.	Columbus, OH/Minneapolis, MN	238P	5 h 12 min
9.	Boulder, CO/Wheeling, WV	627P	7 h 43 min
10.	New York, NY/Los Angeles, CA	815A	10 h 32 min
11.	Seattle, WA/Albuquerque, NM	947P	3 h 20 min

Mixed Skills Review

1. $4\frac{4}{9} \div 16\frac{2}{3}$
2. $2.35 + 8.061 + 15.82$
3. $2346 + 854 + 5523$
4. $3\frac{7}{12} + 5\frac{3}{4}$
5. $78{,}512 \div 28$
6. $71.8 - 53.337$
7. $\frac{9}{16} \times 2\frac{2}{5}$
8. $6400 - 1888$
9. $12\frac{2}{5} - 4\frac{5}{6}$
10. 684×32
11. $\$6.58 \times 126$
12. $37.04 \div 46.3$

14-8 Airfare

Objective *Find costs of air travel.*

Lisa Bollinger is a travel agent. When Lisa writes an airline ticket, she must show both the base fare and the tax. Use the chart on p. 462 in the Data Bank.

Example

1. Lisa writes a ticket totaling \$388. Find the base fare and the tax.

Find \$388 in the Total column.
Read across to find the base fare and the tax.

The base fare is \$359.26. The tax is \$28.74.

Try This Given the cost, find the base fare and the tax.

a. \$403 **b.** \$396 **c.** \$423 **d.** \$375

Examples Travel agents usually receive a commission on the base fare. The rate of commission varies with the distance traveled, season, and the travel package sold. Some airlines offer economy packages on certain flights.

2. Craig Laferty is flying from Minneapolis to Denver. The full fare is \$391. Craig has a discount rate of $66\frac{2}{3}\%$. Find the cost of his ticket.

Look in the \$391 row and the $66\frac{2}{3}\%$ column.
The base fare is \$241.37. The tax is \$19.31.
\$241.37 + \$19.31 = \$260.68.

The total fare is \$260.68.

3. Lisa earned a commission of 8% on Craig's ticket. Find the amount.

Multiply the base fare by the rate of commission.
$0.08 \times \$241.37 \approx \$19.31$

Lisa's commission is \$19.31.

Try This Find the cost of each ticket and the amount of commission.

	Regular Fare	Discount Rate	Base Fare	Tax	Total Cost	Rate of Commission	Amount of Commission
e.	\$375	—				8%	
f.	\$403	75%				6%	
g.	\$429	50%				7%	

Exercises

Given the cost, find the base fare and the tax. Use the chart on p. 462 in the Data Bank.

1. \$393 **2.** \$371 **3.** \$426 **4.** \$419 **5.** \$405 **6.** \$378

Find the cost of each ticket and the amount of commission.

	Regular Fare	Discount Rate	Base Fare	Tax	Total Cost	Rate of Commission	Amount of Commission
7.	\$372	75%				8%	
8.	\$394	75%				6%	
9.	\$417	50%				7%	
10.	\$380	$66\frac{2}{3}\%$				8%	
11.	\$423	$66\frac{2}{3}\%$				6%	
12.	\$414	50%				5%	

Flight Coupon 1 Date of Issue 3 MAR 89
Passenger Name LAFERTY/C MR

Itinerary	Flight	Class	Date	Time	Status	Fare Basis
LV MINNEAPOLIS	0129	M	7APR	0730A	OK	MXP66
AR CHICAGO				0835A		
LV CHICAGO	0247	M	7APR	0912A	OK	MXP66
AR DENVER				1038A		

Fare	241.37	Fare Calculation
Tax	19.31	7APR ORD EWR 130.34 MXP66 130.34 MXP66 260.68
Total	260.68	Form of Payment Credit Card

13. On what date was Craig Laferty's ticket issued?

14. On what date did Craig fly?

15. At what time did his plane leave Minneapolis?

16. Craig changed planes in Chicago. How long was his layover?

17. At what time was Flight 247 scheduled to arrive?

18. How did Craig pay for his ticket?

Mixed Skills Review

1. $68{,}544 \div 136$ **2.** $2\frac{8}{9} - \frac{2}{3}$ **3.** $43{,}863 + 5446$

4. 15.37×20.4 **5.** $\frac{3}{8} + \frac{1}{2} + 1\frac{1}{4}$ **6.** $4\frac{8}{9} \div 3\frac{2}{3}$

7. $0.8 + 3 + 6.34 + 0.17$ **8.** $539.328 \div 48$ **9.** $211.4 - 82.462$

10. $5\frac{1}{4} \times 18$ **11.** $10{,}305 - 4536$ **12.** 742×358

14-9 Enrichment: Road Rally

Objective
Calculate distance, rate, and time problems.

To prepare for a road rally the navigator calculates how fast the driver should travel between checkpoints. Teams get penalty points for arriving before or after a preset time. The team with the fewest penalty points, rather than the fastest time, wins the rally.

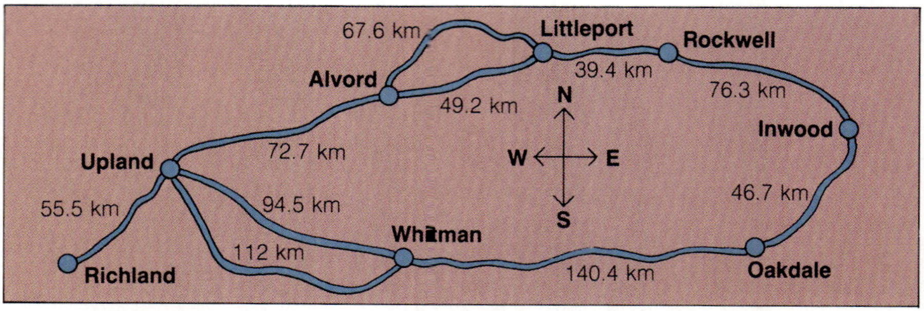

The navigator uses this formula to determine the rate of speed.

$r = \frac{d}{t}$

r = rate in km/h
d = distance in km
t = time in hours

How fast should you drive from Richland to Upland to arrive in 45 minutes? Express time in hours.

$$45 \text{ min} = \frac{45}{60} \text{ h} = 0.75 \text{ h}$$

$$r = \frac{d}{t} = \frac{55.5 \text{ km}}{0.75 \text{ h}} = 74 \text{ km/h}$$

Exercises

Find how fast you should drive between checkpoints to arrive in the given times. Round to the nearest kilometer per hour.

1. from Upland to Whitman, northern route, in 1 h 12 min
2. from Whitman to Oakdale in 1 h 48 min
3. from Oakdale to Rockwell in 1 h 36 min
4. from Rockwell to Alvord, northern route, in 1 h 20 min
5. from Rockwell to Alvord, southern route, in 66 min
6. from Alvord to Richland in 1 h 42 min

14-10 Calculator: Bicycle Gear Ratios

Objective
Calculate bicycle gear ratios and speeds.

The speed of a bicycle depends on the ratio between the front and rear gears and the speed at which the bicyclist can pedal. A ten-speed bicycle has two front sprocket gears, one with 42 teeth, the other with 52 teeth. On the rear wheel are five gears with 12, 16, 19, 23, and 26 teeth, respectively.

To find the ratio for first gear divide the 42 teeth on the front gear by the 26 teeth on the rear gear. Use a calculator. Round to the nearest hundredth.

$42 \div 26 \approx 1.62$

Use a calculator to find each gear ratio. Round to the nearest hundredth.

	Gear Number	Sprocket/ Wheel	Gear Ratio
1.	1	42/26	
3.	3	52/26	
5.	5	52/23	
7.	7	52/19	
9.	9	42/12	

	Gear Number	Sprocket/ Wheel	Gear Ratio
2.	2	42/23	
4.	4	42/19	
6.	6	42/16	
8.	8	52/16	
10.	10	52/12	

To find the speed of a bicycle in miles per hour use $r = 60dng$.
r = the rate in miles per hour
d = distance traveled on one wheel revolution
n = number of wheel revolutions per minute
g = gear ratio

Find the speed of a bicycle in miles per hour traveling in first gear. The 27″ wheels revolve at a rate of 70 revolutions per minute. Round to the nearest hundredth.

A bicycle with a wheel diameter of 27 inches travels 0.0013 mile on each revolution. Use the formula to calculate the speed.

$r = 60dng = 60 \times 0.0013 \times 70 \times 1.62 \approx 8.85$ mph

The ten-speed is traveling at about 8.85 mph in first gear.

Use the gear ratios found in Exercises 1–10 to find the speed in miles per hour to the nearest hundredth. You may wish to use a calculator.

	Gear Number	Revolutions per Minute	Speed
11.	5	60	
13.	1	90	

	Gear Number	Revolutions per Minute	Speed
12.	3	85	
14.	7	50	

CHAPTER 14 REVIEW

Choose the one that does not belong in each group.

1. A. timetable
   B. destination
   C. arrival
   D. legend

2. A. fare
   B. rate
   C. port
   D. taxi

3. A. Silver Streak
   B. bus
   C. parking
   D. commute

4. A. mileage
   B. timetable
   C. distance
   D. miles per gallon

5. A. route
   B. legend
   C. interstate
   D. direct

6. A. highway
   B. round-trip
   C. arrival
   D. departure

7. A. flight
   B. reservation
   C. port
   D. track

8. A. base fare
   B. travel agent
   C. ticket
   D. mileage

Match the item in column A with the correct amount in column B.

Column A	Column B
9. a 15% tip for a \$24.75 taxicab fare	A. 23
10. taxicab fare: 4.5 miles at a rate of \$1.75 for the first $\frac{1}{4}$ mile, 50¢ for every additional $\frac{1}{4}$ mile	B. \$3.71
11. Jennie drove 32 miles and used 1.4 gallons of gasoline. About how many miles per gallon does her car get?	C. \$55.92
12. A one-way train ticket from Portland to Eugene costs \$34.95. The round-trip discount is 20%. What is the cost of a round-trip ticket?	D. \$10.25
13. Frank's car gets 24 miles to the gallon of gas. He spent \$18 to fill his car with gas that cost \$1.25 per gallon. How far can he travel on a full tank?	E. 345.6 mi

CHAPTER 14 TEST

Use the timetable on p. 318 to complete the chart.

	Destination	Boarding Point	Arrival Time	Boarding Time	Travel Time
1.	Main & Taylor	Valley Park	9:39		
2.	State & Madison	Post Office		7:05	
3.	Provident Hospital	City College	By 7:45		

4. Veteran's Cab charges \$1.50 for the first $\frac{1}{2}$ mile and \$0.20 for each $\frac{1}{5}$ mile thereafter. Rodney Karr paid \$5.70. How far did he travel?

Compare the weekly transportation costs.

	Distance to Work	Driving Cost/Mile	Parking Fees	Driving Cost/Week	Cost of Public Transportation	Cost per Week	Weekly Difference
5.	9.7 mi	\$0.20	\$1.85/day		Bus \$1.75 one way		
6.	23.6 mi	\$0.18	\$35/mo		Train \$175/mo		

Use the mileage chart on p. 324. Find the distance in miles.

7. from Indianapolis to Miami

8. from Boston to El Paso

Use the map on p. 326. Give directions for the following trip.

9. from Simpson (B,2) to McClure (A,2)

Use the train schedule on p. 328 to answer the following question.

10. The eastbound Silver Streak is delayed 10 minutes leaving McKinney and 15 minutes leaving Gilmer. What time does it arrive in Shreveport?

Use the schedule on p. 330 to answer the following questions.

11. How much more would it cost to fly from Baltimore to Memphis first-class than for rate 72S on Wednesday?

Find the cost of each ticket and the amount of commission.

	Regular Fare	Discount Rate	Base Fare	Tax	Total Cost	Rate of Commission	Amount of Commission
12.	\$389	75%				5%	
13.	\$378	$66\frac{2}{3}$%				8%	

CHAPTER 14 CUMULATIVE REVIEW

1. Taxi fare is $1.25 for the first $\frac{1}{4}$ mile and $0.50 for each additional $\frac{1}{4}$ mile. Find the distance traveled for a fare of $8.75.
   A. 8 miles
   B. 3 miles
   C. 6 miles
   D. 4 miles

2. Find the fuel economy rate.
   present odometer reading 60,321, previous odometer reading 60,001, gas purchased 12.8 gallons
   A. 25 mpg
   B. 24.5 mpg
   C. 26 mpg
   D. 30.2 mpg

3. Joe Chu's W-2 form showed $6288.13 federal income tax withheld. His actual tax was $5935.20. Find the amount of Joe's refund.
   A. $252.93
   B. $351.95
   C. $353.13
   D. $352.93

4. Sue and Helen plan to take 5 days to drive the 1864 miles from Los Angeles to St. Louis. How many miles per day will they average?
   A. 372.8
   B. 293.2
   C. 79.5
   D. 452.8

5. Use the formula
   $I = \frac{L}{(1 - d)}$
   to find the ideal weight to the nearest pound.
   $L = 165$, $d = 9\%$
   A. 183
   B. 181
   C. 182
   D. 180

6. Estimate the amount of tip. Use mental math.
   dinner bill $15.86, tip 15%
   A. $1.40
   B. $3.00
   C. $2.40
   D. $1.00

7. Find the mode of math test scores.
   91, 75, 52, 75, 90, 51, 83, 98, 75, 52, 90, 89
   A. 98
   B. 75
   C. 83
   D. 76.8

8. Find the rental fee.
   tables and chairs $110, service factor of 24 at $34
   A. $150
   B. $48
   C. $96
   D. $120

9. Compare. Use <, >, or =.
   $\frac{5}{9} \square 0.5$
   A. <
   B. >
   C. =

10. In her shopping cart Maria has items with these prices.
    $1.69, $3.89, $0.56, $0.99, $2.29
    Will her $10 cover these purchases? Use estimation.
    A. yes
    B. no

11. How many ways can 5 bicycles be parked in a row? Use $n \times (n - 1) \times (n - 2) \times \ldots \times 1$.
    A. 25
    B. 120
    C. 240
    D. 15

12. Find the better buy on tomato sauce.
    two 4-oz cans for 44¢
    three 6-oz cans for 62¢
    A. two 4-oz cans
    B. three 6-oz cans

13. Subtract.
$\frac{8}{21} - \frac{1}{3}$
A. $\frac{1}{21}$
B. $\frac{1}{3}$
C. $\frac{1}{7}$
D. $\frac{1}{8}$

14. Timetable:

Willow & Main	Maple & 2nd	Oak & 5th
7:55	8:09	8:18
8:10	8:24	8:33

Tom lives near Willow and Main. His job at Oak and 5th starts at 8:30. What time should he board the bus? How long is the ride?
A. 7:55, 23 min
B. 8:10, 33 min
C. 8:18, 19 min
D. 8:33, 21 min

15. The total cost of an airline ticket is $410. The tax is $30.32. The travel agent earns an 8% commission on the base fare. Find the agent's commission.
A. $32.80
B. $30.32
C. $30.37
D. $31.27

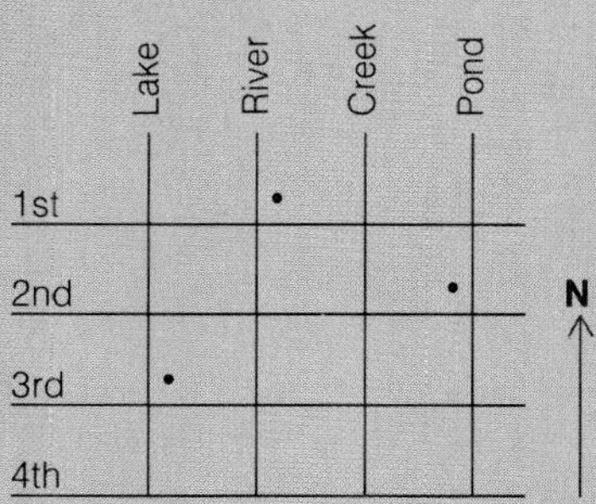

16. Give directions from the bank at 3rd and Lake to the market at 1st and River to Sal's at 2nd and Pond.
A. S to River, E to 1st, W to Pond, N to 2nd
B. N to 1st, W to River, S to Pond, E to 2nd
C. N to 1st, east to River, S to 2nd, E to Pond
D. S to 3rd, E to River, N to Pond, W to 2nd

17. The train leaves Denton at 9:38 a.m. and arrives in Gilmer at 11:13 a.m. What is the travel time?
A. 2 h 25 min
B. 2 h 35 min
C. 1 h 35 min
D. 1 h 25 min

18. Divide.
369 ÷ 24
A. 15.275
B. 14.375
C. 15.265
D. 15.375

19. The Monroe family has a monthly income of $1843. They budget 26% for housing. What amount is budgeted for housing?
A. $479.18
B. $497.18
C. $480.08
D. $708.88

20. Compare the weekly cost of driving to riding the bus.
car $11.50 per day, parking $5 per day, bus $78 weekly pass
A. bus, $4.50 more
B. driving, $4.50 more
C. driving, $15.50 less
D. bus, $15.50 less

Chapter 15 Overview

Key Ideas

- budgeting for housing costs
- finding rental costs
- finding the down payment
- finding mortgage payments
- finding housing costs
- calculating utility costs
- finding homeowner insurance costs
- finding energy savings

Key Terms

- monthly housing costs
- weekly gross income
- rule of thumb
- utility costs
- deposits
- advanced rental payment
- mortgage loan
- down payment
- terminating debts
- principal
- closing costs
- loan fee
- escrow fee
- kilowatt hour
- premium
- replacement cost
- personal property

Key Skills

Find the answer.

1. $340 + 278 + 45 + 240$

2. $430 + 129 + 78 + 490$

3. $130 + 23 + 46$

4. $1285 − $210

5. $3263 − $1586

6. $2061 − $879

7. $67{,}431 - 66{,}907$

8. $76{,}330 - 74{,}998$

9. $56{,}037 - 55{,}972$

10. $28,800 × 2.5

11. $43,900 × 2.5

12. $39,800 × 2.5

13. $665 × 12

14. $598 × 12

15. $755 × 12

16. $8.25 × 40 × 50

17. $12.80 × 40 × 50

18. $11.25 × 40 × 50

19. 0.062×1472

20. 0.071×1176

21. 2308×0.083

22. $1385 \div 4.3$

23. $945 \div 15$

24. $1765 \div 4.3$

Solve.

25. n = 10% of $76,900

26. 20% of $78,900 = n

27. 18% of $76,300 = n

28. n = 60% of $120,000

29. 90% of $98,000 = n

30. 3% of $26,800 = n

31. 10% of $78,900 = n

32. n = 5% of $54,600

33. n = 2% of $103,000

HOUSING

15

15-1 Housing Costs

Objective
Budget housing costs.

Ruxton Towers	
Apartments for Rent	
1 Bedroom	$510
2 Bedroom	$625

A good rule of thumb is to spend no more than one week's gross income for monthly housing costs. Such costs include rent or mortgage payment, property taxes, and insurance.

Understand the Situation

- What rule of thumb should be followed when finding monthly housing costs?
- What are the costs for 1- and 2-bedroom apartments at Ruxton Towers?
- What is the minimum weekly gross income that should be earned to rent a 2-bedroom apartment at Ruxton Towers?

Examples

1. Rich Salla works 40 hours per week and earns $7.65 per hour. His wife, Angela, averages $1295 a month in salary and commissions. Can they afford a 2-bedroom apartment at Ruxton Towers?

Find Angela's gross weekly pay. Divide by 4.3 weeks per month.
$1295 ÷ 4.3 = $301.16

Find Rich's gross weekly pay.
$7.65 × 40 = $306

Find their total gross weekly pay.
$306 + $301.16 = $607.16

Rich and Angela cannot afford the 2-bedroom apartment.

Another rule of thumb is that you should not pay more than about 2.5 times your annual gross pay when you buy a house.

2. Dick Steigelman works 40 hours per week for $9.75 per hour, 50 weeks a year. His wife, Emmy, earns $565 a month, 12 months a year. Estimate the maximum amount they should pay for a house.

Estimate Dick's gross annual income.
$9.75 × 40 × 50
Estimate: $10 × 2000 = $20,000

Estimate Emmy's gross annual income.
$565 × 12
Estimate: $600 × 12 = $7200

Find their approximate total annual income. $20,000 + $7200 = $27,200

Multiply by 2.5 to find the suggested maximum.
$27,200 × 2.5 Estimate: $27,000 × 2.5 = $54,000 + $13,500 = $67,500

Dick and Emmy should pay no more than about $67,500 for a house.

Try This

Jan earns $8.25 per hour and works 40 hours per week. How much should be budgeted for monthly housing costs?

Exercises

How much should be budgeted for monthly housing costs?

1. Gary earns $11.50 per hour, 40 hours per week.
2. Rita Buxton is under contract for a year and earns $29,800.
3. Larry Goldstein earns $529 per week. His wife, Marsha, earns $6.79 per hour, 40 hours per week.
4. Fred earns $1120 biweekly. His wife, Jessica, earns $7.75 per hour and works 30 hours per week.
5. Nicolas Bailey earns $425 per week. His roommate, Jose, earns $5.65 per hour and works 45 hours per week.

Estimation Estimate the maximum amount to pay for a house.

6. Jack Mitchell earns $2175 per month.
7. Marv and Ruby Browning have an annual income of $39,800.
8. Jason Lee has average earnings of $1950 per month in commissions. Jennifer Lee earns $9.25 per hour and works 30 hours per week, 35 weeks per year.
9. Leroy Johnson earns $1950 per month. Oprah Johnson makes $11.80 per hour, 30 hours per week, 48 weeks per year.
10. The Buscemis earn $42,800 per year. Should they buy a house selling for $126,600?
11. Vanessa averages $825 per week in wages and tips. She works 50 weeks per year. Should she buy a townhouse for $85,000?
12. Pat Merriman is paid $1245 biweekly. Should she buy a condominium selling for $102,000?

Talk Math

13. Other than price, discuss some other factors to consider when renting an apartment or buying a house.
14. A condominium costs $135,000. What is the minimum gross yearly earnings for a family which is considering buying it?

Mixed Skills Review

1. $1\frac{1}{8} \times \frac{20}{21}$	2. 50,736 + 29,468	3. 6.8 − 6.213
4. 1.93 + 6.274	5. 632 × 36	6. 669.44 ÷ 52.3
7. $12 - 4\frac{4}{5}$	8. 18,111 − 7387	9. $4\frac{2}{3} + 3\frac{3}{4}$
10. 431,616 ÷ 512	11. 12.18 × 4.5	12. $2\frac{2}{7} \div 8$

15-2 Renting

Objective
Find housing rental costs.

When renting an apartment it is important to consider extra costs such as utilities, deposits, and advanced rental payments.

Ravenview Apts
Jacuzzi — Air-Conditioned
Pool — Fireplace
1 BR — $425, plus utilities
2 BR — $500, plus utilities
Open daily, 9–5

Understand the Situation

- What are some extra features at Ravenview Apartments?
- What is the cost of a 2-bedroom apartment?
- What are some extra costs to be considered when renting an apartment?

Example

1. Utilities at Ravenview cost $55 per month. Find the monthly cost for a 1-bedroom apartment.

Rent + Utilities = Monthly Cost $425 + $55 = $480

Try This

Find the monthly cost.

a. Versailles: 2 bedroom, utilities average $65 per month

b. The Oaks: 1 bedroom, utilities average about $58 per month

Versailles
1 BR – $325
2 BR – $480
Fireplace — Carpeted
Patio or Balcony

The Oaks
1 Bedroom — $460
2 baths
Close to shops
Air-Conditioned

Your landlord may ask for certain deposits or fees. Some fees and deposits are refundable. Others are not. Thus, it is important to know what is in your rental agreement.

Example

2. Find the move-in cost.

First month's rent	$320	Electricity	$65
Security deposit	$320	Pet deposit	$75
Cleaning fee	$45	Total Move-in Cost	$825

Try This

c. Find the move-in cost.

First month's rent	$515	Application fee	$25
Last month's rent	$515	Electric bill	$75
Cleaning fee	$50		

Exercises

Ruxton Towers
1 Bedroom
Furnished $510
Unfurnished $375
2 Bedrooms
Furnished $625
Unfurnished $530

Riverview Arms
1 Bdr $410
2 Bdr $500
Add $100 for Furnished
Fireplace, Balcony
Excellent View

Notchcliff Acres
1 Bdr $310
2 Bdr $450
3 Bdr $600
2 Bathrooms
Close to shops

Windwood Way
1 Bdr $290
2 Bdr $365
Add $75 for Furnished
Large Lawn, Patio, Clubhouse, Pool

Find the monthly costs.

1. Ruxton Towers: 2 bdr, furn, util average \$70/mo.
2. Riverview Arms: 1 bdr, unfurn, util average \$65/mo.
3. Notchcliff Acres: 3 bdr, util included in rent.
4. Windwood Way: 2 bdr, furn, util average \$55/mo.

Find the total move-in cost.

5. Ruxton Towers—1 bdr, unfurn: first month's rent, cleaning dep of \$30, application fee of \$25, and security dep of \$200.
6. Notchcliff Acres—2 bdr: first month's rent, last month's rent, security dep of \$300, key dep of \$18, and cleaning dep of \$45.
7. Windwood Way—1 bdr, furn: first and last months' rent, security dep of \$300, application fee of \$50, cleaning dep of \$50, and key dep of \$28.
8. Riverview Arms—2 bdr, unfurn: first month's rent, application fee of \$45, security dep of \$350, cleaning dep of \$125.

Problem Solving

9. The Hillcrest Apartments rent units for \$365 per month, \$415 per month, \$500 per month, and \$625 per month. Find the mean and median cost of a unit at Hillcrest.
10. What percent of the move-in cost is the security deposit of \$300 for the apartment described in Exercise 7?

Mixed Skills Review

1. $7614 - 858$
2. $23{,}657 + 74{,}308$
3. $2\frac{3}{5} + 7\frac{1}{2} + \frac{3}{10}$
4. 6.053×87
5. $22{,}330 \div 154$
6. $\$50.06 - \$27.54$
7. $1.9264 \div 8.6$
8. $6\frac{2}{9} \div 2\frac{2}{3}$
9. 6201×44
10. $\$6.39 + \$7.58 + \$4.95$
11. $5\frac{1}{2} - 2\frac{4}{7}$
12. $6 \times \frac{7}{18}$

15-3 Buying a House

Objective
Find down payment and buyer qualification.

Most people need a **mortgage loan** to buy a house. The amount of any **down payment** depends on the cost of the house and the terms of the loan.

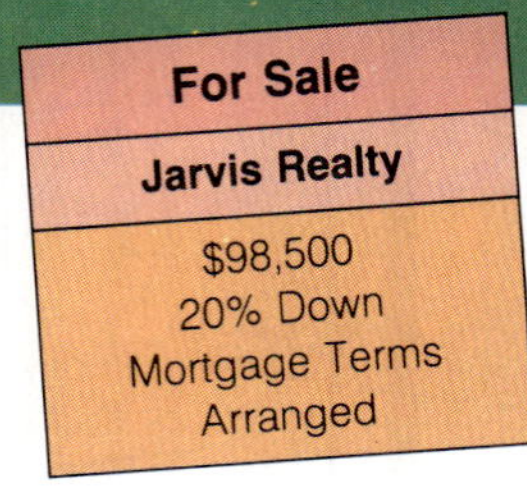

The Hivelys want to buy the house advertised on the For Sale sign.

Examples

1. Find the down payment that the Hivelys must pay.

$0.20 \times \$98{,}500 = \$19{,}700$ The Hivelys need to put $19,700 down.

2. Sarah Michael plans to buy a house costing $85,500. The bank wants a down payment of 3% of the first $25,000, 5% of the next $45,000, and 100% of the remaining cost. Find Sarah's down payment.

$0.03 \times \$25{,}000 = \$750$ down on the first $25,000
$0.05 \times \$45{,}000 = \$2250$ down on the next $45,000

$\$25{,}000 + \$45{,}000 = \$70{,}000$ $\$85{,}500 - \$70{,}000 = \$15{,}500$ remaining cost

$\$750 + \$2250 + \$15{,}500 = \$18{,}500$ Sarah must put down $18,500.

Try This

Find the down payment.

a. A house costs $63,800 with an 18% down payment.

b. A house costs $90,000 with 3% down on the first $25,000, 5% on the balance to $80,000, and 100% of the remainder.

One fourth of the difference between monthly income and terminating debts (debts that end) should exceed the monthly payment to qualify.

Example

3. Mrs. Stermer has a monthly income of $1380. Terminating debts total $210. Can she qualify for a monthly loan payment of $456?

$\$1380 - \$210 = \$1170$ $\$1170 \div 4 = \$292.50$ She does not qualify.

Try This

Does the buyer qualify for the loan? All figures are monthly amounts.

	Loan Payment	Gross Income	Terminating Debts
c.	$342	$1260	$145
d.	$422	$1850	$125

Exercises

Find the down payment.

	Purchase Price	Rate of Down Payment	Down Payment
1.	$76,800	18%	
2.	$85,700	20%	
3.	$63,900	25%	
4.	$102,000	22%	
5.	$105,000	3% on the first $10,000 plus 5% of the next $80,000 plus 100% of the balance	
6.	$126,700	3% of the first $15,000 plus 4% of the next $90,000 plus 100% of the balance	

Does the buyer qualify for the loan? All figures are monthly amounts.

	Loan Payment	Gross Income	Terminating Debts
7.	$320	$1875	$210
8.	$176	$1129	$198
9.	$325	$1400	$229
10.	$524	$2187	$356
11.	$895	$3900	$316
12.	$612	$1990	none
13.	$445	$1800	none
14.	$229	$1212	$213

15. Find the mean and median of the housing purchase prices in Exercises 1–6.

Group Activity

16. Contact at least five banks, savings and loan companies, or credit unions. Find the typical down payment that they require on a mortgage loan. Find the mean and median down payment. Draw a bar graph that compares the required down payments.

Mixed Skills Review

1. $5141.82 \div 142$ **2.** $5 \times 6 \times 18$ **3.** $12\frac{1}{4} + 7\frac{5}{8}$

4. $8\frac{1}{3} - 6\frac{1}{2}$ **5.** $3815 - 2867$ **6.** $7\frac{7}{8} \div 2\frac{1}{4}$

7. $3.2 \times 0.3 \times 2.4$ **8.** $3.06 + 2.28 + 6.34$ **9.** $1\frac{1}{4} \times \frac{3}{5} \times \frac{8}{9}$

10. $46 + 57 + 29 + 83$ **11.** $65,780 \div 253$ **12.** $0.08 - 0.0622$

15-4 Financing a House Purchase

Objective
Find mortgage payments.

If you buy a house you will probably need a mortgage loan. A loan table such as the one shown to the right is used to determine the monthly payment. Part of each monthly payment is for **principal.** The principal amount reduces the loan. The rest of the monthly payment is for interest.

Monthly Payment Per $1000			
Interest Rate	**20-Year Term**	**25-Year Term**	**30-Year Term**
7.5%	$8.06	$7.39	$6.99
8.0	8.36	7.72	7.34
8.5	8.68	8.05	7.69
9.0	9.00	8.39	8.05
9.5	9.32	8.74	8.40
10.0	9.65	9.08	8.78
10.5	9.98	9.44	9.15
11.0	10.32	9.80	9.52
11.5	10.66	10.16	9.90
12.0	11.01	10.53	10.29
12.5	11.36	10.90	10.67
13.0	11.72	11.28	11.06
13.5	12.07	11.66	11.45
14.0	12.44	12.04	11.85
14.5	12.80	12.42	12.25

Understand the Situation

- At 8.5% interest what is the amount per $1000 on a 30-year loan?
- At 10.5% interest what is the amount per $1000 on a 20-year loan?
- At 9.5% interest what is the amount per $1000 on a 25-year loan?

Examples

1. Find the monthly payment on a loan of $56,500 at 9.5% for 25 years.

Find the amount per $1000 in the table. $8.74
Write $56,500 as 56.5 thousand dollars.

Multiply to find the monthly payment.
$8.74 × 56.5 = $493.81 *Estimate: 9 × 60 = 540*

The monthly payment is $493.81.

Many monthly payments also include property taxes and insurance.

2. Find the total monthly payment.

Principal and interest	$493.81	*Estimate: 500 + 100 + 100 = 700*
Taxes	$113.80	
Insurance	$98.00	
Total monthly payment	$705.61	

Try This

Find the total monthly payment.

a. $35,700 mortgage loan at 10.5% for 20 years

b. $92,400 mortgage loan at 11% for 30 years

c. Principal and interest $678.40, taxes $132, and insurance $112

d. Principal and interest $567.60, taxes $143, and insurance $102

Exercises

Find the monthly payment. Use the table on page 348.

1. $67,000 mortgage loan at 8.5% for 20 years
2. $89,500 mortgage loan at 10.5% for 25 years
3. $54,700 mortgage loan at 11.5% for 25 years
4. $98,000 mortgage loan at 9.5% for 30 years
5. $48,000 mortgage loan at 12% for 20 years
6. $77,000 mortgage loan at 14.5% for 25 years

Find the total monthly payment.

7. Principal and interest $568.70, taxes $114.98, and insurance $89
8. Principal and interest $715.80, taxes $203.78, and insurance $139.90
9. Principal and interest $623.90, taxes $117.89, and insurance $98.76
10. Principal and interest $495.60, taxes $97.73, and insurance $89.56

A typical payment table is shown on page 463 in the Data Bank. It shows how much of each payment goes to interest and how much to principal. The table shows that most of the early payments go toward interest. As time passes the amount paid toward principal gradually increases.

11. Find the amount to interest and the amount to principal on Payment 4.
12. Find the amount to interest and the amount to principal on Payment 358.
13. How much of the loan is reduced by the first three payments?
14. How much of the loan is reduced by the last three payments?
15. By how much does the principal amount change for Payments 2 and 3?
16. By how much does the interest amount change for Payments 2 and 3?
17. Estimate the percent of Payment 1 that goes toward interest.
18. Estimate the percent of Payment 356 that goes toward interest.

Group Activity

19. A mortgage loan of $65,000 is taken at the rate of 13% for 25 years. Find the total amount of interest paid on this loan.

Mixed Skills Review

1. 60,324 − 15,935	2. $3\frac{1}{8} - \frac{1}{2}$	3. 14,625 + 26,256
4. 554 × 217	5. 4.25 + 12 + 0.6	6. 16.32 × 18.4
7. 7.32 − 6.444	8. $4\frac{4}{5} \div 18$	9. 282.5 ÷ 7.3
10. $5\frac{1}{3} \times 4\frac{1}{3}$	11. $\frac{3}{8} + 1\frac{1}{2} + 2\frac{1}{4}$	12. 5941 ÷ 65

15-5 Closing Costs

Objective
Find total housing costs.

The Santana family is buying a house. To close the deal they must pay **closing costs,** which can include various fees and taxes.

The total cost of the Santanas' $72,778 home is the sum of the down payment, the closing costs, and the mortgage payments.

Loan #43–598	$64,800
Deed recording fee	$45
Escrow fee	$520
Lawyer's fee	$275
Prepaid insurance	$375
Credit report fee	$55
Prepaid taxes	$1100
Loan fee (1% of loan)	
Total Closing Costs	

Examples

1. Find the total closing costs for the Santana's mortgage loan #43-598.

Find the loan fee. $0.01 \times \$64\,800 = \$648$

Add the costs to find the total.
$\$45 + \$520 + \$275 + \$375 + \$55 + \$1100 + \$648 = \$3018$

The total closing costs are $3018.

2. The Santanas' house purchase required a 10% down payment. They paid the down payment and closing costs in cash. Loan #43-598 is for the balance of the purchase price and will run 25 years at 11.5%.

Find the down payment, cash paid, and monthly payment.
$64,800 is 90% of the purchase price. $\$64{,}800 \div 0.9 = \$72{,}000$
The difference is the down payment. $\$72{,}000 - \$64{,}800 = \$7{,}200$

down payment	+	loan amount	=	purchase price
10%	+	90%	=	100%
$7,200	+	$64,800	=	$72,000

down payment	+	closing costs	=	total cash
$7,200	+	$3,018	=	$10,218

From the payment table on page 348, the monthly payment per $1000 at 11.5% for 25 years is $10.16. There are 64.8 thousands in a $64,800 loan.

$64.8 \times \$10.16 = \$658.37$ per month

Try This

a. Find the total closing costs on a $75,000 loan for an $88,235 home.

Loan fee	2% of the loan	Prepaid insurance	$540	Prepaid taxes	$1200
Deed recording fee	$65	Lawyer's fee	$375	Credit report fee	$85

Escrow fee $430 (half paid by the seller)

b. Use the loan in **a.** with a down payment of 15%. Find the down payment, cash paid, and monthly payment. The loan will run for 30 years at 10.5%.

Exercises

Loan Option **A:** Loan fee—1.5% of loan, interest rate—11%, length of loan—20 years
Loan Option **B:** Loan fee—1% of loan, interest rate—11%, length of loan—30 years

	Deed Recording Fee	Escrow Fee	Prepaid Insurance	Lawyer's Fee	Prepaid Taxes	Credit Report Fee
A	\$55	\$500	\$400	\$325	\$1,000	\$65
B	\$55	\$400*	\$500	\$325	\$1,000	\$70

*half escrow fee paid by seller

1. The Weirs are buying the house across the street from the Santanas. They plan on taking out a \$90,000 mortgage loan. They have found two different loan options. Find the closing costs for Loan Option A.
2. Find the closing costs for Loan Option B. Which loan has lower closing costs?
3. The Weirs are going to pay the required 12% down payment and the closing costs in cash. The loan will be for the balance of the purchase price. Find the purchase price, cash amount paid, and the monthly mortgage payment for Loan Option A.

4. Find the cash amount paid and the monthly mortgage payment for Loan Option B.
5. The Weirs must decide which loan option to accept. Based on the formula closing costs + total interest paid (over life of loan), which loan option would you suggest they accept?
6. Use Loan Option A with a 10% down payment and the purchase price found in Exercise 3. Find the cash amount paid.

Mixed Skills Review

1. $104{,}000 - 36{,}162$
2. $42 + 18 + 763$
3. $2\frac{1}{2} \times 2\frac{1}{2}$
4. $5\frac{3}{5} \div \frac{12}{25}$
5. $3\frac{1}{4} - 2\frac{9}{16}$
6. $7276 \div 428$
7. $18.4 - 11.77$
8. $552.6 \div 9.21$
9. $12\frac{7}{8} + 6\frac{5}{6}$
10. 136×73
11. 32×8.536
12. $8.09 + 5.06 + 9.3$

15-6 Utility Costs

Objective
Calculate utility costs.

Electrical energy is measured by the kilowatt hour (kWh). One kWh is 1000 watts of power used for one hour. For instance, 20 fifty-watt light bulbs burning for 1 hour use 1 kWh.

Electric meters measure the amount of electricity used. If the pointer is between two numbers we read the lower number. However, if the pointer is between 0 and 9, we read 9.

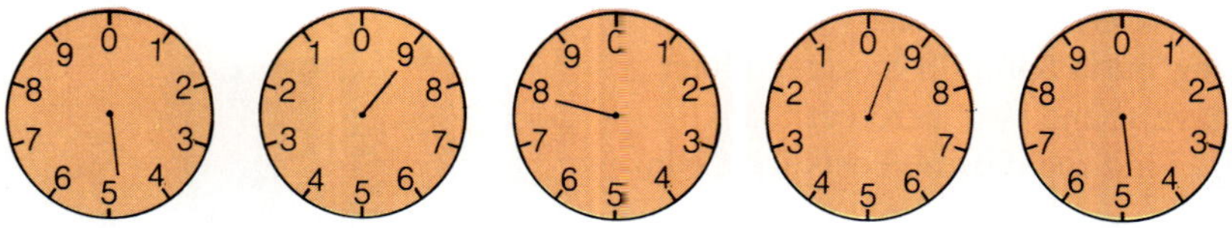

Dial 1 Dial 2 Dial 3 Dial 4 Dial 5

Understand the Situation

- Which dial would be read 8?
- Would any dial be read 9?
- What is the reading of the dial on the right?

Example

1. Read the meter above from left to right. The meter reads 48,794.

Try This

a. Read this meter.

The cost of electricity can be found by multiplying the number of kWh used by the cost per kWh. The cost varies in different geographical regions.

Example

2. Find the cost at the rate of $0.062 per kWh.
present meter reading: 67,562 previous meter reading: 65,988

Find the number of kWh used.
$67{,}562 - 65{,}988 = 1574$

Find the cost. Round to the nearest cent.
$\$0.062 \times 1574 = \$97.588$ $97.59

Try This

b. Find the cost at the rate of $0.071 per kWh.
present meter reading: 78,741 previous meter reading: 78,674

Exercises

Read each meter.

1. 

2. 

Find the cost.

3. Present meter reading: 65,984
   Previous meter reading: 62,802
   Rate: $0.0685 per kWh

4. Present meter reading: 78,061
   Previous meter reading: 76,995
   Rate: $0.072 per kWh

5. Present meter reading: 55,489
   Previous meter reading: 54,906
   Rate: $0.065 per kWh

6. Present meter reading: 49,887
   Previous meter reading: 47,997
   Rate: $0.068 per kWh

Find the estimated yearly cost.

7. A toaster uses about 4.2 kWh per year. The rate is $0.0701 per kWh.
8. A water heater uses about 4421 kWh per year at $0.0659 per kWh.
9. An electric blanket uses about 154 kWh per year at $0.071 per kWh.

The unit for water is 1000 gallons. Find the cost.

10. 78 units of water at $0.69 per unit
11. 128 units of water at $0.56 per unit

The unit for natural gas is usually 100 cubic feet. Find the cost.

12. 312 units at $0.48 per unit
13. 210 units at $0.62 per unit

Mixed Skills Review

1. $912 + 836 + 1474$
2. $158{,}148 \div 382$
3. $6.1 + 3.6 + 12.5 + 7.8$
4. $82.13 - 54.361$
5. 256×318
6. What percent is 16 of 40?
7. $2\frac{1}{5} \div 4\frac{2}{5}$
8. $17{,}374 - 6565$
9. $50 \times \frac{7}{90}$
10. 18% of 171
11. $\frac{3}{5} + 2\frac{1}{2} + 1\frac{3}{4}$
12. $3\frac{3}{8} - 1\frac{7}{24}$

15-7 Homeowners and Renters Insurance

Objective *Find homeowner and renter insurance costs.*

Homeowners and renters buy insurance to cover possible losses to personal property. Homeowners also need to insure the house itself. The **premium** is the amount of money paid for the insurance. It is based on either current value or replacement cost.

Homeowners Insurance

Replacement Cost	Typical Annual Premium
$50,000	$143
60,000	153
70,000	161
80,000	176
90,000	194
100,000	213
110,000	239
120,000	264
130,000	284

Example

1. Find the monthly premium to the nearest cent for an $80,000 house.

Annual premium: $176 Monthly premium: $\$176 \div 12 = \$14.67$

Try This

a. Find the semiannual premium for a $120,000 house.

Renters may insure personal property against certain losses. The amount of coverage is based on the replacement cost, useful life, and age of the item.

Example

2. Find the amount covered by personal property insurance.

Item Lost	Replacement Cost	Useful Years	Age (yr)	Deductible	Amount Covered
Sofa	$835	15	3	$100	$568.04

Find the cost per useful year. $\$835 \div 15 = \$55.67$ per useful year

Find the current value. $15 - 3 = 12$ useful years remaining

$\$55.67 \times 12 = \$668.04$ current value

Subtract the deductible. $\$668.04 - \$100 = \$568.04$

The amount covered by insurance is $568.04.

Try This

Find the amount covered by personal property insurance.

	Item Lost	Replacement Cost	Useful Years	Age (yr)	Deductible	Amount Covered
b.	Computer	$3120	5	1	$200	
c.	Desk	$750	20	5	$50	

Exercises

Use the table on page 354. Find the indicated insurance premium.

1. Replacement cost $90,000; monthly
2. Replacement cost $60,000; quarterly
3. Replacement cost $50,000; monthly
4. Replacement cost $70,000; semiannual
5. Replacement cost $80,000; quarterly
6. Replacement cost $110,000; semiannual
7. Jessie Cho owns a house valued at $102,000 and insured for 85% of its value. Her insurance company allows 55% coverage on personal property. What is the amount of coverage for personal property?
8. The Traceys' $125,000 home is insured for 90% of its value. Their insurance company allows them a 12% coverage for a garage and storage shed. What is the amount of coverage for the garage and shed?
9. Mel Friedman owns a home valued at $145,000. It is insured for 80% of its value. The insurance company allows Mel 20% coverage to pay expenses while living away from home during needed repairs. What is the amount allowed for these loss-of-use expenses?
10. Lou DiPietro's home is insured for $95,000. The insurance company allows Lou a 45% coverage for liability and medical expenses in case someone is injured on Lou's property. What is the amount allowed?
11. Find the amount covered by personal property insurance.

Item Lost	Replacement Cost	Useful Years	Age (yr)	Deductible	Amount Covered
TV	$940	10	4	$200	
Chair	$690	15	10	$100	
Clock	$210	10	8	$50	
Desk	$825	20	12	$200	
Computer	$2995	3	2	$100	
VCR	$595	5	1	$50	

Mixed Skills Review

1. $8.006 - 2.384$
2. $6\frac{2}{5} \div \frac{2}{5}$
3. $4892 + 3562 + 4368$
4. $6\frac{1}{4} + 2\frac{7}{8}$
5. $81,630 - 12,647$
6. 47% of what number is 282?
7. $\frac{1}{2}$ of $2\frac{2}{3}$
8. $242,892 \div 702$
9. 883×353
10. $7\frac{1}{2} - 2\frac{7}{12}$
11. 6.03×434.2
12. $\$1625.58 + \$18 + \$964.93$

15-8 Energy Conservation

Objective
Find savings from energy conservation measures.

Heating, cooling, and hot water are the greatest utility expenses for most houses. There are many ways to conserve energy. These range from something as simple as setting the thermostat lower in winter and higher in summer to installing solar heating.

Examples

1. The Wineholts' bill for hot water averages \$57 per month. They install an insulating blanket on their hot water heater as shown. About how long will it take for the savings to pay for the cost of the improvement?

Find the monthly saving. $0.10 \times \$57 = \$5.70$

Divide to find how long it will take for the savings to pay for the cost of the improvement.
$\$35 \div \$5.70 = 6.14$ or about 7 months

After about 7 months the savings will have paid for the improvement.

2. Three college students share an apartment. Their monthly expenses for heating and cooling average \$67. By setting the thermostat to 68 in the winter and 78 in the summer, they estimate that they can save 25%. What is their yearly saving?

Find the yearly cost. $12 \times \$67 = \$804$
Find the yearly saving. $0.25 \times \$804 = \$201$

Their yearly saving is about \$201.

Try This

a. The average monthly heating and cooling bill for Cherie and Bill Zachmeier is \$98. By changing filters on their furnace once a month, they estimate that they can save 8%. The filters will cost \$12 per year. Find their yearly saving.

b. Al and Ilene Ramirez have hot water expenses of about \$72 per month. By installing solar panels for producing hot water, they estimate that they can save 75% of the expenses. The cost of the solar panels is \$3600. About how long will it take for the savings to pay for the cost of the improvement?

Exercises

Energy Saving Options

Option	Saving	Cost
A. Close draperies on sunny side in summer and open in winter.	11%	None
B. Clean refrigerator coils twice a year.	7%	None
C. Lower hot water heater setting from 165° to 120°.	12%	None
D. Take showers instead of baths.	33%	None
E. Operate dishwasher only when full.	5%	None
F. Set thermostat 5° higher in summer and 5° lower in winter.	19%	None
G. Fix leaking hot water faucet (do-it-yourself).	20%	$2.50
H. Install wood burning stove.	80%	Stove $1000 Wood $200/yr

Find the yearly savings.

1. Gary and Diane Masimore's average monthly heating and cooling cost is $94. They choose Option A.
2. Ron and Betsy Levine's hot water bill averages $43 per month. They use Option D.
3. The Weimers' average monthly heating bill is $85. They choose Option F.
4. Freda Jasper's average monthly cost for running her refrigerator is $9.40. She tries Option B.
5. Neil and Lee's hot water bill averages $63 per month. They decide to use Option E.
6. The Costains pay $31 extra during June, July, August, and September for air conditioning. They choose Option F.
7. Art Micozzi's hot water bill averages $61 per month. He tries Option C.
8. Hwan's hot water bill has been averaging $37 per month. He discovers that he has a leak in the hot water faucet and uses Option G.
9. The Goldblooms' average heating bill is $73 per month. They decide to try Option H. How long will it take for the savings to pay for the improvement?

Mixed Skills Review

1. 275×59
2. $1\frac{5}{9} \div \frac{2}{3}$
3. $2\frac{1}{5} + 7\frac{2}{3}$
4. $9126 + 7812 + 6810$
5. $3\frac{1}{10} - 2\frac{1}{2}$
6. $16.562 \div 0.338$
7. $97{,}524 \div 86$
8. $27\frac{1}{2}\%$ of $134
9. $0.29 + $0.49 + $0.53 + $0.73
10. $1\frac{3}{5} \times 1\frac{1}{2} \times 1\frac{2}{3}$
11. $9600 - 2792$
12. $6.01 − $3.72

15-9 Enrichment: Pascal's Triangle

Objective
Discover patterns of Pascal's Triangle.

The triangle bearing Blaise Pascal's name was known before his time. Pascal made some unique discoveries relating the arithmetical triangle to probabilities.

first row	1
second row	1 1
third row	1 2 1
fourth row	1 3 3 1
fifth row	1 4 6 4 1

Follow the pattern to continue the sixth row.

The first number is 1. The next number is the sum of the two numbers above it, $1 + 4 = 5$. You may have noted how the second number in each row is in the pattern 1, 2, 3, 4. The third number in row six is the sum of the two numbers above it, $4 + 6 = 10$. Continue the pattern, ending with 1. Note that the numbers in the rows read the same forward and backward, as in palindrome numbers (p. 214). This is the sixth row.

1 5 10 10 5 1

Exercises

Follow the pattern to continue.

1. seventh row
2. eighth row
3. ninth row
4. tenth row
5. eleventh row
6. twelfth row

Give the sum of the numbers in each row.

7. second row
8. third row
9. fourth row
10. fifth row
11. sixth row
12. seventh row
13. eighth row
14. ninth row
15. tenth row
16. eleventh row
17. twelfth row
18. thirteenth row

Use Pascal's Triangle to complete the following math sentences.

19. 2^2
20. 2^3
21. 2^4
22. 2^5
23. 2^6
24. 2^7
25. 2^8
26. 2^9
27. 2^{10}
28. 2^{11}
29. 2^{12}
30. 2^{13}

Give the next three numbers in each pattern. Can you find these patterns in Pascal's Triangle?

31. 1, 3, 6, 10, ___, ___, ___
32. 1, 4, 10, ___, ___, ___
33. 1, 5, 15, ___, ___, ___
34. 36, 28, 21, ___, ___, ___

15-10 Computer: Housing Costs

Objective
Use the computer to find housing costs.

The Johnsons are buying a house for $96,000. The bank requires that they put 20% down. The mortgage rate is 9.5% for a 30-year loan. Annual expenses include $408 for insurance and $1374 for property taxes. The following program will calculate:

- down payment
- loan amount
- monthly payment
- amount repaid
- cost per month

```
10  REM Program for calculating monthly mortgage payments
20  CLS:INPUT "What is the price of the house";P
30  INPUT "What percent down payment is required";D
40  INPUT "How many years to repay the loan";Y
50  INPUT "What is the payment per $1000";T
60  INPUT "What is the annual amount of property tax";X
70  INPUT "What is the annual insurance premium";I
80  L = P * (1 - .01 * D): M = INT(T * .1 * L + .5)/100: A = M * Y * 12
90  PRINT "Selling"; TAB(12) "Down"; TAB(24) "Loan"; TAB(36)
    "Monthly"; TAB(48) "Amount"; TAB(60) "Monthly"
100 PRINT "Price"; TAB(12) "Payment"; TAB(24) "Amount"; TAB(36)
    "Payment"; TAB(48) "Repaid"; TAB(60) "Cost"
110 PRINT
    "$"P;TAB(12)"$"P-L;TAB(24)"$"L;TAB(36)"$"M;TAB(48) "$" A + P - L;
    TAB(60) "$" M + INT((X+I)/12 * 100 + .5)/100
120 INPUT "Do you want to do another problem";A$
130 IF A$ = "Y" OR A$ = "y" THEN 20 ELSE END
```

The following output is for the example above.

Selling Price	Down Payment	Loan Amount	Monthly Payment	Amount Repaid	Monthly Cost
$96,000	$19,200	$76,800	$645.12	$251,443.20	$793.62

Exercises

Use the program for the following house sales.

	Selling Price	Down Payment	Rate of Mortgage	Number of Years	Annual Insurance	Annual Taxes
1.	$87,600	20%	12.5%	25	$386	$1158
2.	$129,400	15%	9.0%	30	$562	$1634
3.	$102,000	25%	13.5%	20	$453	$1395

CHAPTER 15 REVIEW

Choose the one that does not belong in each group.

1. **A.** mortgage payment
   B. property taxes
   C. homeowners insurance
   D. principal

2. **A.** escrow fee
   B. deed recording fee
   C. credit report fee
   D. replacement cost

3. **A.** rate
   B. gallon
   C. kWh
   D. cubic foot

4. **A.** personal property
   B. premium
   C. deductible
   D. loan fee

5. **A.** down payment
   B. security deposit
   C. cleaning fee
   D. pet deposit

6. **A.** insulation
   B. woodburning stove
   C. furnace filter
   D. prepaid insurance

7. Arrange these meter readings in order from least to greatest.

Match the item in column A with the correct amount in column B.

Column A	Column B
8. an 18% down payment on a house costing $76,800	**A.** $527.28
9. the monthly payment on a mortgage loan of $65,500 at 8.5% for 25 years	**B.** $14,654.00
	C. $13,824.00
10. closing costs on an $85,000 loan: 2% loan fee; prepaid insurance, $580 prepaid taxes, $1300; deed recording fee, $80	**D.** $534.90
	E. $3660.00

CHAPTER 15 TEST

How much should be budgeted for monthly housing costs?

1. Harvey Hattfield earns $1800 per month. His wife, Karen, earns $7.15 per hour and works 25 hours per week.

Estimate the maximum amount to pay for a house.

2. Michael Blevins makes $2700 per month. His wife, Elsie, makes an annual salary of $22,300.

Find the monthly costs.

3. 3 bdr, unfurn $750, util average $73/mo

Find the total move-in costs for Waterview Apartments.

4. $650 per month, first and last month plus cleaning fee of $50, application fee of $20, key deposit of $15

Find the down payment.

5. purchase price $112,000, rate of down payment 25%

6. purchase price $112,000, 3% on the first $25,000, 5% on the next $75,000, 100% of the rest

Does the buyer qualify for the loan? All figures are monthly amounts.

7. loan payment $420, gross income $1950, terminating debts $275

Find the monthly payment. Use the table on p. 348.

8. $62,500 mortgage loan at 12.5% for 30 years

Find the total monthly payment.

9. principal and interest $637.50, taxes $123.53, and insurance $92

Find the cost.

10. electricity rate $0.0724 per kwh
present meter reading 34,038, previous meter reading 32,603

11. 372 gallons of heating oil at $0.61 per gallon

Use the table on p. 354. Find the indicated insurance premium.

12. replacement cost $90,000, quarterly

Find the yearly saving. Use the Energy Saving Option table on p. 357.

13. The Drake family has an average monthly water bill of $19. They choose Option D.

CHAPTER 15 CUMULATIVE REVIEW

1. Give the percent. Use a fraction.
18% of 65
   A. 12
   B. $11\frac{7}{10}$
   C. $11\frac{1}{2}$
   D. $10\frac{7}{10}$

2. Find the simple interest.
$58,750 loan, 30 years, 10.7%
Use $i = prt$.
   A. $22,068.75
   B. $21,868.75
   C. $219,686.50
   D. $220,687.50

3. Find the monthly costs at Jefferson Arms.
3 bdrm. unfurn $375/mo, util. average $35/mo
   A. $415
   B. $400
   C. $410
   D. $310

4. Find the increase in annual liability premium.
100/300/50 coverage $246, driver rating changed from 1.8 to 2.7
   A. $664.20
   B. $221.40
   C. $442.80
   D. $220.40

5. Alex had 156 baseball cards. He sold $\frac{1}{4}$ of them for 75¢ each. How much did he make from the sale?
   A. $29.25
   B. $28.25
   C. $30.75
   D. $29.75

6. Find the down payment on this house.
purchase price $68,750, rate of down payment 22%
   A. $15,025
   B. $14,025
   C. $13,914
   D. $15,125

7. Find the amount of tax withheld.
waitress, single, 2 exemptions, 45 hours per week, $6.50 per hour
   A. $30
   B. $34
   C. $41
   D. $29

Single Persons—Weekly Payroll Period

And the wages are–		And the number of withholding allowances claimed is–			
		0	1	2	3
At least	But less than	The amount of income tax to be withheld shall be–			
110	115	14	8	3	0
115	120	14	9	4	0
120	125	15	10	4	0
125	130	16	10	5	1
130	135	17	11	6	1
270	280	38	33	27	22
280	290	40	34	29	23
290	300	41	36	30	25
300	310	43	37	32	26
310	320	44	39	33	28

8. Find the total monthly payment.
principal and interest $1005.47, taxes $166.50, and insurance $76.80
   A. $1237.77
   B. $1248.77
   C. $1148.77
   D. $1227.77

9. Find the pulse rate. Use mental math.
15 seconds, 18 beats
A. 72
B. 54
C. 62
D. 108

10. Find the total earnings for an insulation contractor.
40 hours, $5.40 per hour, 25% commission on labor
A. $216
B. $270
C. $54
D. $269

11. Divide.
$8\frac{1}{4} \div 2\frac{3}{4}$
A. 3
B. 22.7
C. $\frac{1}{3}$
D. $3\frac{1}{4}$

12. In the United States in 1970 there were 2.7 million farms with an average of 374 acres per farm. What was the total acreage farmed?
A. 1009.8 million acres
B. 138.5 million acres
C. 10,098 million acres
D. 100.98 million acres

13. Budget one week's income for housing. Terry earns $12.20 per hour, 40 hours per week. How much should he budget for housing?
A. $378
B. $460
C. $488
D. $498

14. Find the electricity cost.
present reading 169,341, previous reading 168,590, rate $0.068 per kwh
A. $57.87
B. $51.07
C. $50.68
D. $50.07

15. Ms. Sanchez says the odds in favor of getting a B on the test are 3 to 1. What is the probability Joe will get a B?
A. $\frac{1}{3}$
B. $\frac{1}{4}$
C. $\frac{3}{1}$
D. $\frac{3}{4}$

16. Find the closing costs on this mortgage loan.
$54,000 loan, loan fee 2%, deed recording fee $45, escrow fee $237, credit report $75
A. $357
B. $1337
C. $1437
D. $1427

17. Find the amount of insurance coverage for personal property.
a $75,600 house insured for 85% of its value, personal property insured for 60%
A. $64,260
B. $38,556
C. $45,360
D. $38,555

18. Find the median daily mileage.
miles driven per day 450, 375, 382, 425
A. 382
B. 375
C. 403.5
D. 408

19. Tomás rode his scooter 193 miles in 3 days. The first day he rode $\frac{3}{4}$ as far as the second day. The second day he rode 66 miles. How far did he ride the third day?
A. 143.5 mi
B. 49.5 mi
C. 115.5 mi
D. 77.5 mi

20. Find the yearly saving. A monthly hot water bill averaged $57. A shower adaptor reduced this by 33%.
A. $684
B. $458.28
C. $225.72
D. $214.62

Chapter 16 Overview

Key Ideas

- estimating measurements
- measuring for accuracy
- converting units of length
- estimating materials for home improvements
- finding costs of home improvements
- using cost factors
- comparing do-it-yourself costs with contract and retail costs

Key Terms

- inch
- gallon
- square yard
- installation charge
- rise
- foot
- quart
- square foot
- labor costs
- hip roof
- pitch
- yard
- square
- cost of materials
- span

Key Skills

Find the answer.

1. 173×36
2. $248.5 \div 3.5$
3. $\frac{1}{2} \times 8 \times 12$
4. $8\frac{1}{7} \div 3$
5. 20% of 386
6. $\$1432.02 \div 29$
7. $16 \times 5\frac{3}{4}$
8. $414 \div 36$
9. 3×138.47
10. $\$84.37 - \$46.53$
11. $430 \div 16$
12. 43% of 1920
13. $\$438.16 + \$86.57 + \$36.82$
14. $\$12.93 + \$18.51 + \$27.08 + \$3.54$

Solve.

15. Find the sale price of a garden house, regularly $10.99, now 15% off.
16. Find the selling price of a swimming pool with an invoice price of $2618 and a 22% markup.
17. Find the cost of $26\frac{1}{2}$ yards of fencing at $5 per yard.
18. Find the per gallon cost of 7 gallons of latex paint for $61.18.
19. Find the annual expense of a gardening service at $8.75 per week.
20. Find the cost of carpet cleaning at $11.95 per room for 7 rooms.

HOME IMPROVEMENTS

16

16-1 Consumer Measurements

Objective
Measure line segments.
Change units of length.

When repairing, maintaining, or decorating your home or apartment, it is often necessary to take measurements. A tape or ruler marked in inches can be used to measure wood, carpet, fabric, wallpaper, etc. The edge of this picture frame can be measured to the nearest $\frac{1}{8}$ in.

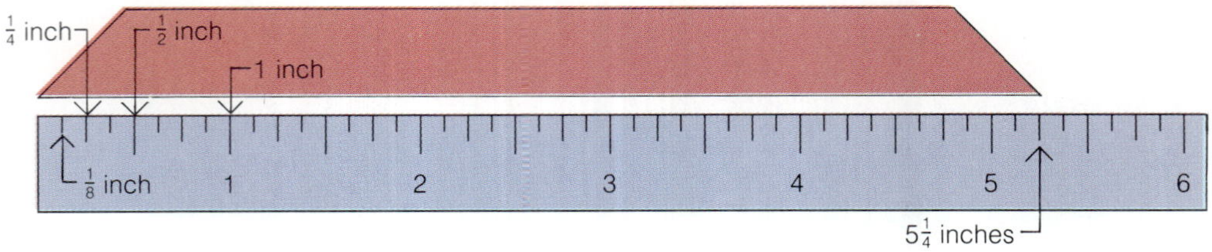

The end of the piece of frame is closer to $5\frac{1}{4}$ in. than to $5\frac{3}{8}$ in. The length of the piece is $5\frac{1}{4}$ in. to the nearest $\frac{1}{8}$ in.

Example Measure the edge of the line segment to the nearest $\frac{1}{8}$ in.

1. ______________________________

The measure is just beyond $4\frac{7}{8}$ in. To the nearest $\frac{1}{8}$ in. the measure is $4\frac{7}{8}$ in.

Try This Measure each segment to the nearest $\frac{1}{8}$ in.

a. ______________________ **b.** ______________

c. ______________________________

The inch, foot, and yard are common units of length used around the house.

Multiply if changing from a larger to smaller unit. Divide if changing from a smaller to larger unit.

Relations Among Units

1 foot (ft) = 12 inches (in.)
1 yard (yd) = 3 feet (ft)
1 (yd) = 36 inches (in.)

Examples

2. Change 11 feet to inches. $11 \times 12 = 132$ 11 ft = 132 in.

3. Change 19 feet to yards. $19 \div 3 = 6\frac{1}{3}$ 19 ft = $6\frac{1}{3}$ yd, or 6 yd 1 ft

Try This Change these units of length.

d. 13 feet to inches **e.** 23 feet to yards **f.** 8 yards to inches

Exercises

Change these units of length.

1. 9 feet to inches **2.** 26 feet to yards **3.** 14 inches to feet

4. 18 yards to feet **5.** 226 inches to feet **6.** 17 feet to inches

7. 14 feet to yards **8.** 68 feet to yards **9.** 39 inches to feet

10. 17 yards to feet **11.** 37 inches to feet **12.** 314 inches to yards

Use your inch ruler. Draw segments with the following lengths.

13. 9 in. **14.** $1\frac{1}{2}$ in. **15.** $4\frac{1}{4}$ in. **16.** $7\frac{3}{4}$ in.

Estimation

Sketch line segments whose lengths are as follows. Use a ruler to see how close your sketches are.

17. 2 in. **18.** 5 in. **19.** $8\frac{1}{2}$ in. **20.** 4 in.

Number Sense

Arrange these lengths in order from longest to shortest.

21. 23 in. 1 yd 14 ft 6 yd $\frac{1}{2}$ in. 14 yd

22. 3 ft 3 yd $\frac{1}{4}$ in. 130 in. 1200 in. 0.35 in.

23. 27 ft 14 in. 4 yd 2500 in. 6 ft 0.5 yd

Work with a Partner

Estimate the following. Compare your estimates. Then measure to see whose estimate was closer.

24. your height in inches

25. the length of this book to the nearest $\frac{1}{8}$ in.

26. the width of the classroom door in inches

27. the length of a dollar bill to the nearest $\frac{1}{4}$ in.

28. the diameter of a nickel to the nearest $\frac{1}{16}$ in.

Mixed Skills Review

1. 24,080 − 16,357 **2.** $2\frac{1}{6} + 5\frac{1}{4} + 2\frac{2}{3}$ **3.** 12.034 + 6.292 + 0.58

4. $2\frac{1}{5} \times 3\frac{1}{3}$ **5.** 63,728 + 41,590 **6.** 60,900 ÷ 420

7. 8.6 − 8.138 **8.** 21.6 × 3.18 **9.** $4\frac{2}{3} - 2\frac{8}{9}$

10. 413.58 ÷ 22.6 **11.** $12 \div 5\frac{1}{3}$ **12.** 846 × 59

16-2 Painting

Objective
Find perimeter and area.

Debbie Levine plans to repaint her family room. The room is presently painted dark blue. Debbie plans to use Spanish white to cover the dark blue. Debbie uses a table like the one at the right to help her decide how much paint to buy.

Color of Paint	Coverage per Gallon (ft^2)	Coverage per Quart (ft^2)
Same color or dark over light	400	100
Light over dark (needs 2 coats)	200	50
Deduct 15 ft^2 from the total area for each door or window.		

Understand the Situation

- In which row of the table should Debbie look for help?
- How much should Debbie deduct for each door or window?

Examples

1. How many gallons of paint are needed to cover 1500 ft^2 with two coats of paint?

Since two coats are needed, the coverage per gallon is 200 ft^2.
Divide to find the number of gallons needed. $1500 \div 200 = 7.5$

The number of gallons needed is 7.5, or 7 gallons and 2 quarts.

2. Debbie's family room is $12' \times 14'$ and has an $8'$ ceiling. There are three windows and two doors. How much paint should Debbie buy?

Draw a picture of the room.

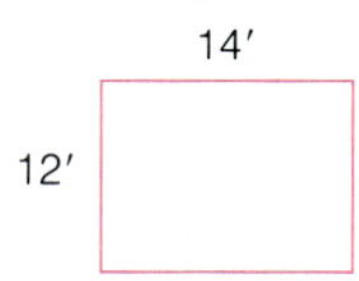

Find the area of the ceiling.
$12' \times 14' = 168$ ft^2

Find the perimeter.
$12 + 14 + 12 + 14 = 52$ ft

Multiply to find the area of the walls.
$52 \times 8 = 416$ ft^2

Find the total area.
$416 + 168 = 584$

Find the area of the windows and doors.
2 doors, 3 windows $5 \times 15 = 75$ ft^2

Find the area to be painted.
$584 - 75 = 509$ ft^2

Find the amount of paint needed.
$509 \div 200 = 2.545$ gal

It will take 2 gallons plus 3 quarts to cover the walls and ceilings.

Try This

a. How many gallons of paint will be needed to cover 1800 ft^2 with one coat?

b. Estimate the amount of paint needed to cover an $18'$ by $20'$ living room. The ceiling is $8'$ high. The room has 2 doors and 4 windows. The paint is light over light.

Exercises

Use the table on p. 368.

1. How many gallons of paint are needed to cover 1600 ft^2 with one coat?
2. How many gallons of paint are needed to cover 2500 ft^2 with two coats?
3. How many gallons of paint are needed to cover 10,000 ft^2 with one coat?
4. How many gallons of paint are needed to cover 4550 ft^2 with two coats?

Estimation About how many gallons of paint are required for each room?

	Room	Ceiling Height (ft)	Dimensions (ft)	Paint	Windows and Doors	Estimate
5.	Kitchen	8	8 × 10	Dark over light	1 door 1 window	
6.	Family room	9	16 × 18	Light over dark	2 doors 3 windows	
7.	Bedroom	8	14 × 16	New—2 coats	2 doors 2 windows	
8.	Dining room	8	12 × 14	Light over light	2 doors 3 windows	
9.	Living room	8	14 × 18	Light over dark	3 doors 4 windows	

10. When Debbie purchases the paint for her family room she finds that it costs \$15 per gallon and \$8.50 per quart. Should she buy 2 gallons and 2 quarts as the table suggests or buy 3 gallons?
11. Debbie calls two painters who say it will take about 12 hours to paint her family room. What will the labor cost be at \$9 per hour for each painter?

Group Activity

12. Find the area of the walls and ceilings in your classroom. Visit a local paint store and select a color of paint for the classroom. Find the cost of a paint and determine how much it would cost to paint the classroom. Be sure to account for windows and doors.

Mixed Skills Review

1. $5\frac{3}{8} \div \frac{9}{16}$
2. 4726×48
3. $23.2 - 16.345$
4. $64.026 + 18.598$
5. $2\frac{5}{6} \times \frac{15}{17}$
6. $3\frac{11}{15} - 1\frac{4}{5}$
7. $20{,}996 \div 58$
8. $2\frac{1}{2} + 3\frac{3}{4} + 3\frac{5}{8}$
9. 106.04×15.6
10. $3610 - 964$
11. $16.554 \div 2.67$
12. $147 + 1211 + 453$

16-3 Decorating

Objective
Find areas for wallpapering and carpeting.

Hanging wallpaper is a good way to redecorate and improve a room or a whole house or apartment. Wallpaper usually comes in 27-in.-wide rolls. This is called a **width.** The length of a single roll is 16 feet.

Finding the Number of Rolls of Wallpaper

1. Find the perimeter of the room in inches.
2. Find the number of widths needed. Divide by 27 inches.
3. Find the total length of the widths. Multiply by the height in feet.
4. Find the number of rolls. Divide by 16.
5. Subtract one roll for every two openings (doors, windows).

Example

1. Find the number of rolls of wallpaper needed for a room 18 ft wide by 21 ft long by 8 ft high. There are two doors and four windows.

Find the perimeter in inches.	$18 + 18 + 21 + 21 = 78$ ft, or 936 in.
Find the number of widths needed.	$936 \div 27 = 34.7$, or 35 widths
Find the total length of the 35 widths.	$35 \times 8 = 280$ ft
Find the number of rolls needed.	$280 \div 16 = 17.5$, or 18 rolls
Deduct 3 rolls for the openings.	$18 - 3 = 15$ rolls are needed.

Try This

Find the number of rolls of wallpaper needed for each room.

a. Room: 14 ft × 18 ft × 8 ft high, 3 openings

b. Room: 19 ft × 22 ft × 9 ft high, 5 openings

New carpet can improve the look of any house. It is sold by the square yard, so it may be necessary to change area measures to yards.

Example

2. Find the cost of installing carpet in a room 16 feet wide and 12 feet long. Carpet costs \$15.90 per yd^2, padding costs \$2.95 per yd^2, and installation costs are \$6 per yd^2.

Find the area. $16 \text{ ft} \times 12 \text{ ft} = 192 \text{ ft}^2$

Change to yd^2. $192 \text{ ft} \div 9 \text{ ft}^2/\text{yd}^2 = 21.3 \text{ yd}^2$

Carpet	Padding	Installation
$21.3 \times \$15.90 = \$338.67$	$21.3 \times \$2.95 = \$62.84$	$21.3 \times \$6 = \$127.80$

Find the total cost. $\$338.67 + \$62.84 + \$127.80 = \$529.31$

Try This

c. Find the cost of installing carpet in a room 16 feet wide and 21 feet long. Carpet costs \$17.60 per yd^2, padding costs \$3.05 per yd^2, and installation costs are \$6.50 per yd^2.

Exercises

Find the number of rolls of wallpaper needed. Use 27-inch widths.

1. Wall 9 ft × 9 ft high, no openings
2. Wall 10 ft 7 in. × 8 ft high, no openings
3. Wall 13 ft × 11 ft high, no openings
4. Wall 8 ft 7 in. × 10 ft 2 in. high, 1 opening
5. Wall 17 ft × 12 ft high, 2 openings
6. Wall 23 ft 7 in. × 8 ft high, 2 openings
7. Room 12 ft × 12 ft, 8 ft height, 2 openings
8. Room 13 ft × 16 ft, 8 ft height, 3 openings
9. Room 24 ft × 20 ft, 9 ft height, 7 openings
10. Room 14 ft × 16 ft, 8 ft height, 8 openings
11. Room 30 ft × 40 ft, 10 ft height, 10 openings

Find the cost of installing carpet in these rooms.

	Room	Carpet Cost	Padding Cost	Installation Price	Installation Cost
12.	18′ × 21′	$10.50/yd^2	$2.50/yd^2	$4.75/yd^2	
13.	10′ × 12′	$11.95/yd^2	$2.70/yd^2	$5.10/yd^2	
14.	15′ × 13′	$21.00/yd^2	$3.10/yd^2	$6.00/yd^2	
15.	16′ × 14′	$19.75/yd^2	$2.90/yd^2	$5.25/yd^2	
16.	18′ × 24′	$14.90/yd^2	$3.75/yd^2	$4.60/yd^2	

17. When decorating her living room Carolyn Fritz selected three possible carpet colors, five possible wallpaper patterns, and four matching wallpaper border patterns. From how many different combinations of carpet, wallpaper, and border can Carolyn choose?

Data Hunt

18. Find the number of rolls of wallpaper needed to paper your room. Estimate the costs of installing wallpaper purchased locally.

Mixed Skills Review

1. 233,280 ÷ 144
2. 5383 − 4444
3. $1\frac{7}{9} \div 8$
4. 634 × 508
5. 6.34 + 2.18 + 5.93
6. 20 − 0.014
7. $\frac{5}{8} + \frac{3}{4} + 4\frac{1}{4}$
8. $\frac{3}{8} \times \frac{12}{13} \times \frac{2}{3}$
9. 0.9328 ÷ 17.6
10. $1\frac{2}{3} + 4\frac{7}{12}$
11. 39 + 27 + 64 + 55
12. 8.2 × 4.4 × 3

16-4 Adding on to the House

Objective
Solve problems involving area.

Bill and Phyllis Knaub are planning to add a new family room and patio to their house. The family room is to be 16′ × 18′. The patio is to be 16′ × 8′. They learn that they can estimate the cost of the additions by using a **cost factor.** A cost factor is a cost per square foot.

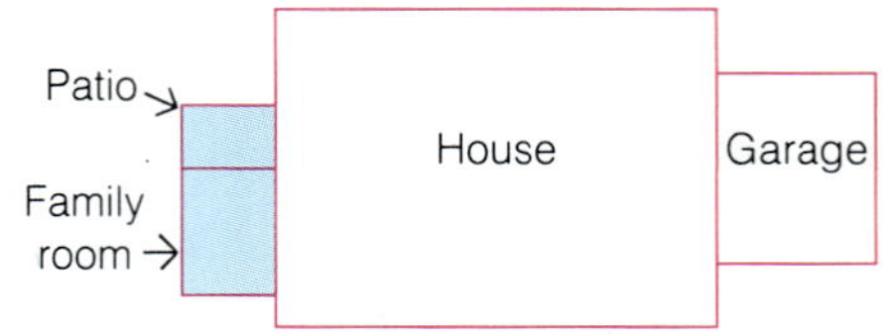

Examples

1. An average cost factor for the addition of the family room is \$48 per square foot.

Estimate the cost of the addition. $18 \text{ ft} \times 16 \text{ ft} \approx 20 \text{ ft} \times 15 \text{ ft} \approx 300 \text{ ft}^2$

The cost factor is about \$50 per ft^2 so \$50 per $\text{ft}^2 \times 300 \text{ ft}^2$ is about \$15,000 for the family room.

The new family room will have aluminum siding. A **square** of siding covers 100 square feet. Deduct 10% to allow for doors and windows.

2. These are views of the outside walls for the new family room. Find the number of squares of siding needed.

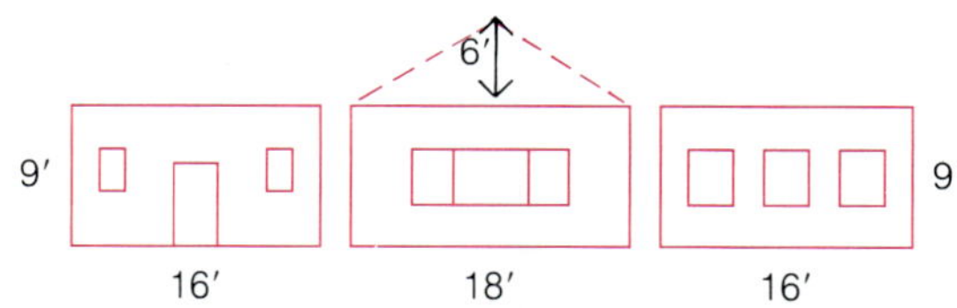

Find the area of the walls.

a. There are two 9′ × 16′ walls. $2 \times 9 \text{ ft} \times 16 \text{ ft} = 288 \text{ ft}^2$

b. The third wall is 9′ × 18′ with an 18′ × 6′ triangle.
Area of rectangle: $18 \text{ ft} \times 9 \text{ ft} = 162 \text{ ft}^2$
Area of triangle: $18 \text{ ft} \times 6 \text{ ft} \div 2 = 54 \text{ ft}^2$
The area of the third wall is $162 \text{ ft}^2 + 54 \text{ ft}^2$, or 216 ft^2.

c. The total area of the outside walls is $288 \text{ ft}^2 + 216 \text{ ft}^2 = 504 \text{ ft}^2$.
Deduct 10% for windows and doors. $0.9 \times 504 \approx 454$
Find the number of squares of siding needed. $454 \div 100 = 4.54$
About 4.54 squares of siding will be needed for the outside walls.

Try This

a. The patio cost factor is \$36 per ft^2. Estimate the cost of the patio.

b. Find the number of squares of siding needed for this wall.

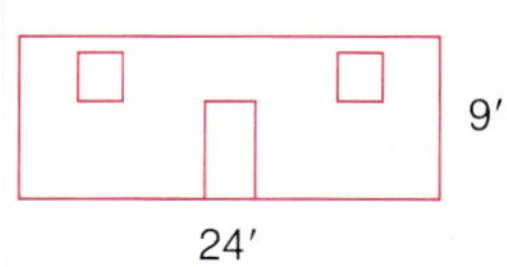

Exercises

Estimate the cost of each additional room. Use the given cost factor.

1.

2.

3.

4.

5.

6.
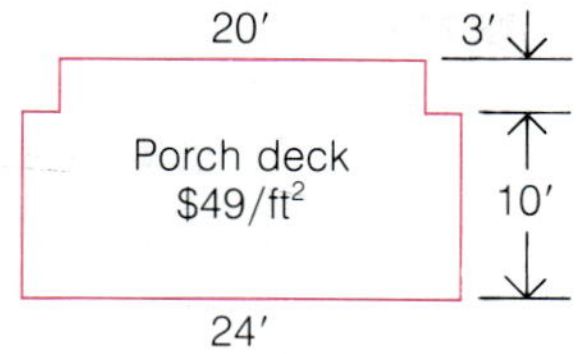

Find the number of squares of aluminum siding needed.

7.
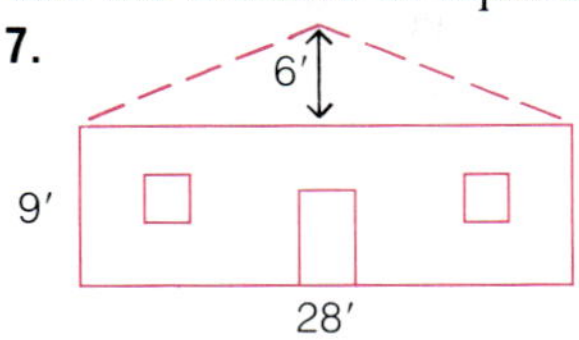

8.
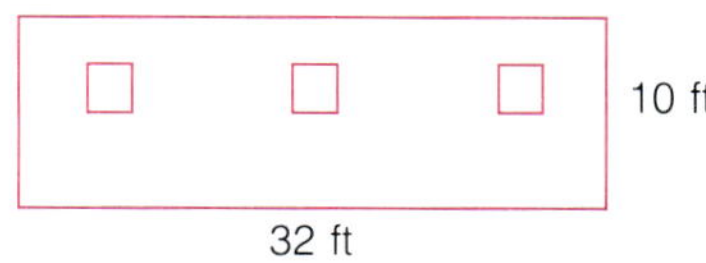

Suppose

9. Suppose that a new family room of 300 ft^2 is being planned. A contractor being considered for the job has just completed a similar room of 350 ft^2, which cost $15,750. Use a proportion to estimate the cost of the new family room.

Data Hunt

10. Contact a local contractor to find out a current cost factor that can be used to estimate the cost of adding on a room. Use this cost factor to estimate the cost of the additions described in Exercises 1–6.

Mixed Skills Review

1. $196 + 284 + 416$ **2.** $3987.2 \div 280$ **3.** $3\frac{2}{5} + \frac{7}{10} + 2\frac{1}{2}$

4. $21,870 - 6882$ **5.** $12.6 + 8 + 0.634$ **6.** 5403×19

7. 16.5×0.46 **8.** $33 \div 5\frac{1}{2}$ **9.** $2.1602 - 0.8453$

10. $74,698 \div 13$ **11.** $1\frac{19}{20} - 1\frac{4}{5}$ **12.** $18 \times \frac{3}{16}$

16-5 Landscaping

Objective *Find the costs of landscaping.*

The Taymans family has planned to have a portion of their yard landscaped. Here is a proposal given to them by Wanner's Landscape Gardening Service.

R. A. Wanner	Landscape Gardener			
Quantity	**Description**	**Size**	**Price**	**Amount**
1	Serviceberry tree	6′	\$120	
5	Bayberry	24″	\$23	
5	Yucca	4′	\$7	
7	Goldtip juniper	15″	\$26	
15	Pink astilbe	2′	\$4.60	
	Peat moss & fertilizer			\$50
	2-rail splitrail fence	44″	Labor	\$150

Understand the Situation

- How many bayberry bushes are proposed?
- What is the labor charge for installing the splitrail fence?
- What is the price for each pink astilbe?

Example

1. Find the cost of the seven goldtip junipers.

The junipers cost \$26 each.
$7 \times \$26 = \$182$ Use mental math. $(7 \times 20) + (7 \times 6)$

The cost of the seven junipers is \$182.

Try This

a. Find the cost of the pink astilbes.

b. Find the cost of the five bayberry bushes.

Example

2. Andy Wanner recommends that four spotlights be included in the project for the Taymans family. The spotlights cost \$32.85 each. Find the total cost.

$4 \times \$32.85 = \$131.40$ Estimate. $4 \times \$30 = \$120$

The spotlights will cost \$131.40.

Try This

c. A transformer costing \$69.20 and 100 feet of cable costing \$38.45 are needed to install the spotlights. What is the cost of these items?

Exercises

1. Find the cost of the five yucca plants for the Taymans' project.
2. Find the total cost of materials for the Taymans' project proposal.
3. Andy Wanner charges $975 labor for designing and planting the Taymans' project. He spent about 25 hours on the project. What is his labor charge per hour?
4. To maintain a new landscape project, Joe Taymans orders two 50-lb bags of 3-2-2 fertilizer at $6.25 per bag, one 50-lb bag of phosphate at $3.75 per bag, and two 50-lb bags of Nitro-10 costing $8.75 per bag. Find the total cost of the fertilizer products.

Find the cost per bulb.

	Description	Quantity	Cost	Cost per Bulb
5.	Fall-flowering crocus	50	$10	
6.	Autumn colchicums	5	$8	
7.	Hybrid tulips	50	$24	
8.	Sunburst narcissus	9	$8.50	
9.	Garden lilies	10	$34	
10.	Pink anemone	25	$5.50	

11. Andy Wanner installs a sod lawn covering 2400 ft^2. The sod costs $0.115 per ft^2. Find the cost for the lawn.

12. Mulch Products offers a discount of $1 per yd^3 for April delivery. Find the cost of 20 yd^3 of Nutri-mulch.

Mulch Products Co.	
Shredded hardwood	$15/cubic yard
Nutri-mulch	$16/cubic yard
Pulp/bark	$14/cubic yard
Glacto-lite	$14/cubic yard

13. Find the cost for a May delivery of 15 yd^3 of shredded hardwood. What is the average cost per yd^3 of the various mulch products?

Mixed Skills Review

1. $\frac{3}{8}$ of 28
2. $14.62 + $18.63 + $9.08
3. 627×82
4. $563 + 26 + 1285$
5. $4\frac{4}{9} \div 8\frac{1}{3}$
6. $1{,}665{,}660 \div 204$
7. $12.08 - 8.243$
8. $7010 - 4695$
9. $3\frac{4}{5} + 4\frac{3}{8}$
10. $127.63 × 24
11. $9\frac{3}{4} - 8\frac{5}{6}$
12. $68.143 \div 8.21$

16-6 Tiling

Objective
Find the cost of installing floor tile.

Lennay Watts has decided to have her game room retiled. At Dulaney Floor Coverings she learns that tile comes in cartons that cover 45 ft^2. The cost varies with the quality of the tile.

Tile Cost

Type of Tile	Cost/Carton
Standard	\$23.50
Deluxe	\$30.95
Super Deluxe	\$42.00

Example

The game room is to be tiled with the Deluxe tile. At \$30.95 for the first 10 cartons and a discounted price of \$28.95 for over 10 cartons, find the cost of the tile.

To find the area of the room, think of the room as two rectangles.

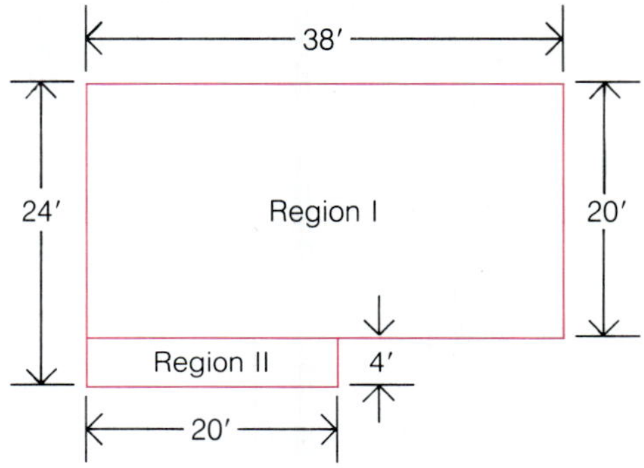

One rectangle is 20′ × 38′ and the other is 4′ × 20′.

38 × 20 = 760 ft^2 20 × 4 = 80 ft^2 760 ft^2 + 80 ft^2 = 840 ft^2

Find the number of cartons needed.   840 ÷ 45 = 18.7, or 19 cartons

Find the cost of the cartons.

for the first 10	for the remaining cartons	Find the total cost.
10 × \$30.95 = \$309.50	9 × \$28.95 = \$260.55	\$309.50 + \$260.55 = \$570.05

The tile will cost \$570.05.

Try This

Find the cost of the tile for this room. Super Deluxe tile is to be used at the regular cost on the first 10 cartons and at a discounted price of \$40 per carton on each over 10 cartons.

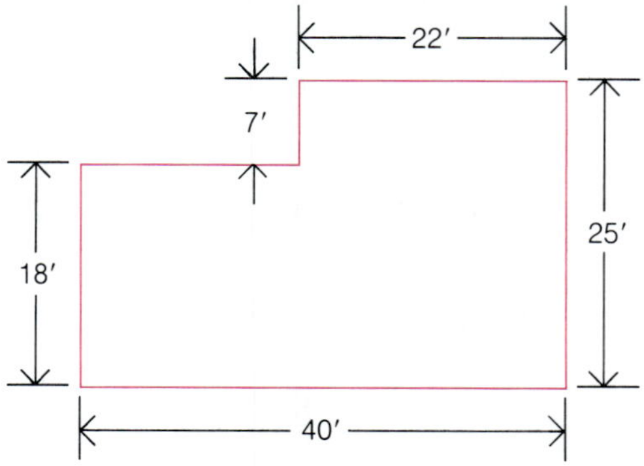

Exercises

Find the cost of tiling each room. Use the costs from the table on p. 376. Deduct \$2 per carton for each carton beyond 10.

	Room	Dimensions (in ft)	Type of Tile	Cost
1.	Kitchen	16 × 38	Standard	
2.	Laundry room	6 × 15	Standard	
3.	Bathroom	10 × 14	Deluxe	
4.	Clubroom	34 × 22	Super Deluxe	
5.	Workshop	26 × 28	Deluxe	
6.	Porch	24 × 18	Deluxe	

7. Kitchen, deluxe

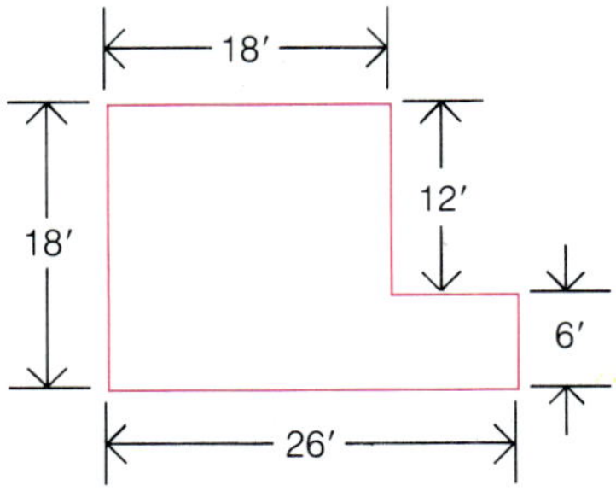

8. Laundry room, standard

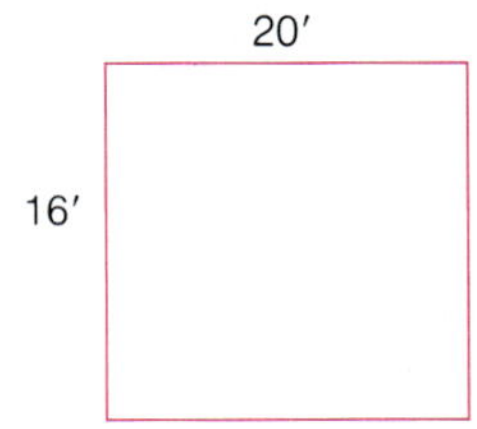

9. Bathroom, super deluxe

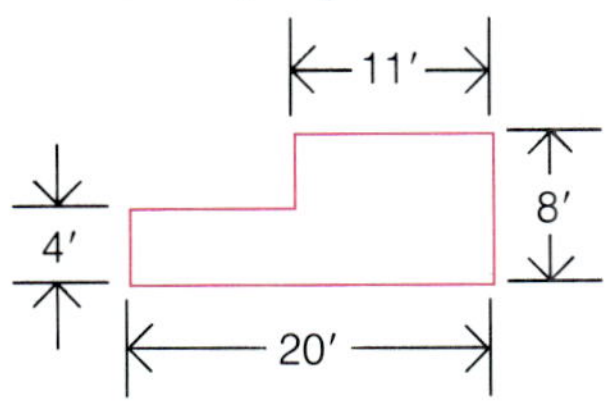

10. Clubroom, deluxe

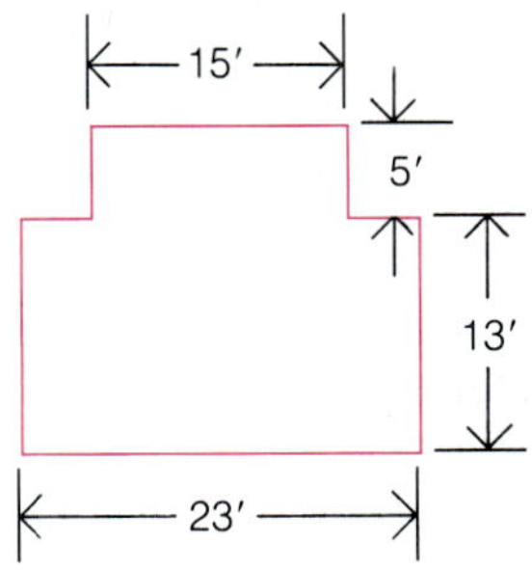

Mixed Skills Review

1. $163.139 \div 17.3$

2. 127×534

3. $260{,}034 - 26{,}165$

4. $61.03 + 8.6 + 9.242$

5. $3\frac{6}{7} + 1\frac{2}{3}$

6. $18{,}216 \div 396$

7. $4\frac{1}{2} \times 5\frac{1}{2}$

8. $72 - 15.96$

9. $3424 + 4938 + 962$

10. $\frac{5}{8} \div 40$

11. 20.62×11.8

12. $8\frac{1}{2} - 5\frac{7}{10}$

16-7 Roofing

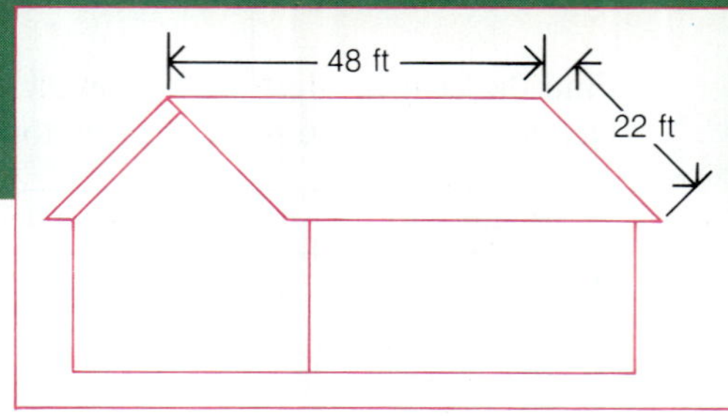

Objective
Find roofing costs.

The roof on Jim's home is old and needs to be replaced. Jim finds that roofing shingles are sold by the **square.** A square of shingles covers 100 ft^2.

Examples

1. Jim selects a style of shingle that costs $40 per square. The labor charge for the roofer that he hires is $35 per square. Find the cost of a new roof.

Find the area of the roof. Each side is a rectangle.
$2 \times 48 \text{ ft} \times 22 \text{ ft} = 2112 \text{ ft}^2$

Find the number of squares. Round up to the next square.
$2112 \div 100 = 21.12$, or 22 squares

Find the total cost per square. $\$40 + \$35 = \$75$
The total cost per square is $75.

Find the cost of the roof. $22 \times \$75 = \$1650$
The roof will cost about $1650.

2. Bill Fritz has a garage with a hip roof. He selects a shingle style costing $39 per square. The roofer charges $42 per square. Find the cost of the new roof.

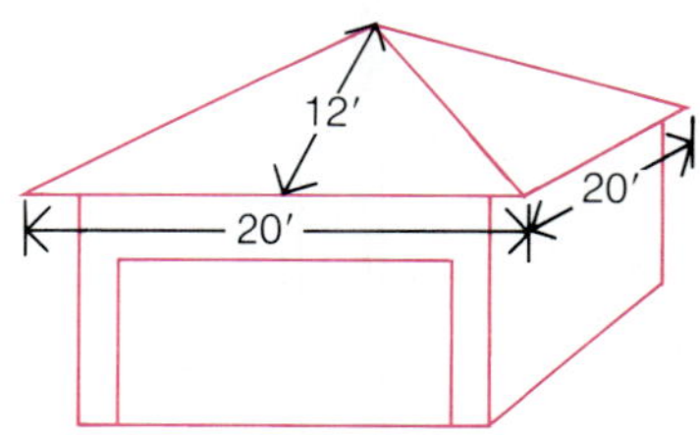

Find the area of the roof. Each section of roof is shaped like a triangle. There are 4 equal sections.

$4 \times \frac{1}{2} \times 20 \text{ ft} \times 12 \text{ ft} = 2 \times 240 = 480 \text{ ft}^2$

The area of the roof is 480 ft^2.

Find the number of squares needed.
$480 \div 100 = 4.8$, or 5 squares

Find the total cost. The cost per square is $39 + $42, or $81.
$5 \times \$81 = \$405$. The roof will cost about $405.

Try This Find the approximate cost of each new roof.

a.

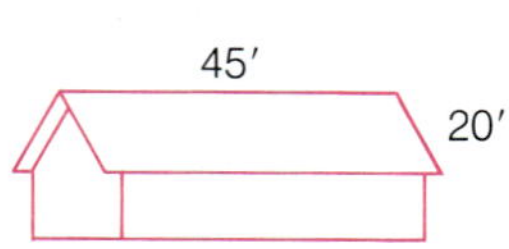

Shingle cost is $34 per square.
Labor cost is $30 per square.

b.

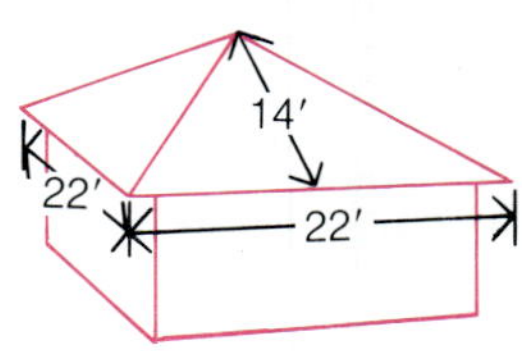

Shingle cost is $38 per square.
Labor cost is $37 per square.

Exercises

Find the approximate cost of each new roof.

1.

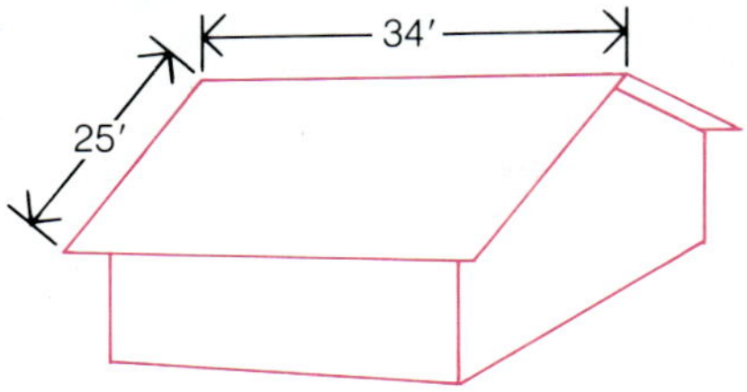

Shingle cost is \$36 per square.
Labor cost is \$31 per square.

2.

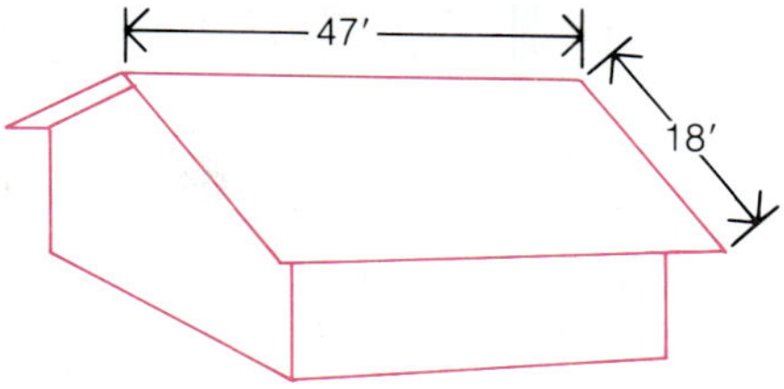

Single cost is \$37 per square.
Labor cost is \$34 per square.

3.

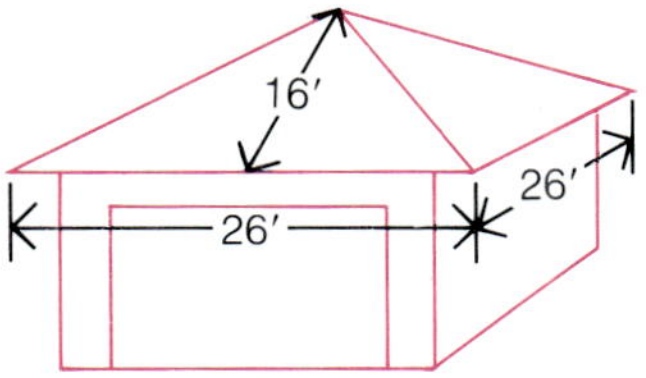

Shingle cost is \$39 per square.
Labor cost is \$38 per square.

4.

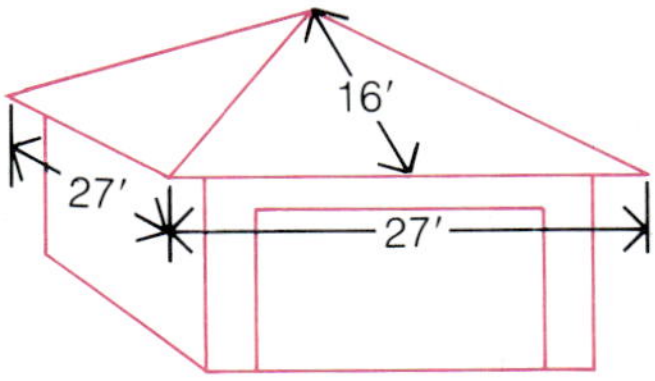

Shingle cost is \$33 per square.
Labor cost is \$31 per square.

Group Activity

5. Contact a local building supplier and find the cost of five different styles of roofing shingles. Make a bar graph to display the comparative costs.

The width of a building is called the **span.** The distance from the top of the ceiling to the peak of the roof is called the **rise.** The ratio of the rise to the span is called the **pitch** of the roof. Find the pitch of these roofs.

6. Rise: 8 ft
Span: 16 ft

7. Rise: 12 ft
Span: 20 ft

8. Rise: 10 ft
Span: 18 ft

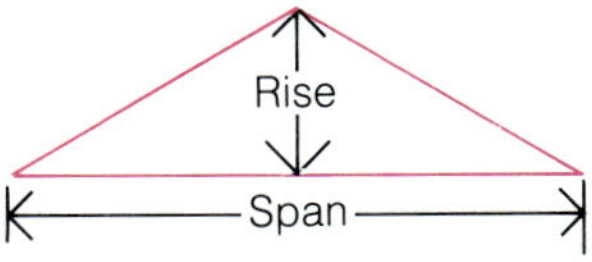

Mixed Skills Review

1. $8\frac{1}{3} - 6\frac{1}{2}$

2. 2.79×79.2

3. \$504 + \$1620 + \$884 + \$1027

4. $1.414 + 2.653 + 0.59$

5. $18{,}075 - 16{,}368$

6. $\frac{15}{16} \div 2\frac{7}{10}$

7. $386{,}568 \div 413$

8. $2\frac{3}{4} + 8\frac{3}{16}$

9. $4 \times 6 \times 8 \times 5$

10. $\frac{9}{14} \times \frac{7}{24}$

11. $0.026 - 0.0094$

12. $553.656 \div 2.006$

16-8 Do It Yourself

Objective
Find the savings in do-it-yourself projects.

Sometimes money can be saved by "doing it yourself." If you have the time and interest you can often save money by making your own clothes, finishing your own furniture, or growing and canning your own vegetables and fruits.

Canning Supplies	
Pint jars	\$4.95/dozen
Vinegar	\$2.00/gallon
Salt	\$0.34/pound
Sugar	\$0.35/pound
Peaches	\$14/bushel
Tomatoes	\$12/bushel
Cucumbers	\$9/bushel

Understand the Situation

- What does "do it yourself" mean?
- What is the cost of a dozen canning jars?
- How much will you pay for a bushel of peaches?

Examples

1. The food market sells pint jars of peaches for \$1.15 each. How much would you save by canning a bushel of peaches that yields about 48 jars? You need 3 pounds of sugar to can the peaches.

Find the cost of canning the peaches.

Cost of peaches	\$14.00	
Cost of jars	\$19.80	(4 × \$4.95)
Cost of sugar	\$ 1.05	(3 × \$0.35)
Total cost	\$34.85	

Find the cost of buying 48 jars of peaches. 48 × \$1.15 = \$55.20

Find the difference in costs. \$55.20 − \$34.85 = \$20.35

The amount saved by canning is \$20.35.

2. Fabric, patterns, and other sewing supplies to make a skirt cost \$14.95. At a clothing store a similar skirt costs \$27. How many skirts would have to be made to pay for a sewing machine costing \$346?

Find the savings per skirt.
\$27 − \$14.95 = \$12.05, or about \$12 per skirt

Divide to find the number of skirts needed to pay for the sewing machine.
\$346 ÷ \$12 ≈ 28.8, or 29 skirts to pay for the sewing machine.

Try This

a. A bushel of tomatoes will yield about 36 home-canned pint jars of tomatoes. About \$0.10 worth of salt is needed. A food market charges \$0.98 per pint can. How much can be saved by canning?

b. The materials for knitting a sweater cost \$12.80. A similar sweater in a store costs \$55.95. How much can be saved by knitting the sweater?

Exercises

1. A bushel of cucumbers will yield about 18 home-canned pint jars of dill pickles. Pickling spices cost \$1.00 and 2 quarts of vinegar are needed. A grocery store charges \$1.10 per pint jar. How much can be saved by canning instead of buying?
2. A dress shirt costs \$22.50 at a store. The costs of making a similar shirt are \$4.24 per yd for 2 yards of material, \$3.25 for a pattern, 8 buttons at \$0.30 each, and thread costing \$0.90. How much less would be spent by making the shirt?
3. An unfinished cabinet costs \$97.75. The costs to build a similar cabinet include \$30 for wood, \$8.30 for nails and glue, and \$9.45 for hinges. What is the difference in costs?
4. A pattern for making a jacket costs \$3.10. The fabric costs \$14.75 with buttons, thread, etc. costing \$7.15. A similar jacket at a clothing store costs \$48.95. What is the difference in costs?
5. A finished bookcase at a furniture store costs \$145.95. An unfinished bookcase costs \$83.50. A can of walnut stain costs \$5.10 and a can of varnish costs \$7.95. Sandpaper costs \$1.10 and brushes cost \$6.50. How much can be saved by finishing the bookcase instead of buying it already finished?
6. You buy an old coffee table at an auction for \$120. To refinish the coffee table you spend \$6.25 on paint remover, \$2.75 for sandpaper, \$3.95 for stain, \$7.95 for varnish, and \$4.50 on brushes. A new coffee table of the same style costs \$325. How much less would you spend by buying the old table and refinishing it?

Suppose

7. Suppose that you spend about \$150 less on summer clothes by making them yourself. If the time that it took you to make the clothes was 20 hours you can think of the savings as money that you earned by doing it yourself. How much did you earn per hour by making the summer clothes?

Mixed Skills Review

1. $2.8 - 1.8042$
2. $7\frac{1}{3} \div \frac{2}{15}$
3. 19% of what number is 47.5?
4. $104{,}040 \div 85$
5. $\$12.13 + \$20.48 + \$28.63$
6. $5\frac{13}{15} - 5\frac{4}{5}$
7. 52% of \$65
8. $4\frac{5}{6} + 7\frac{5}{18}$
9. 3218×47
10. $9 + 3 + 18 + 7$
11. $\frac{7}{12}$ of 21
12. $1100 - 951$

16-9 Enrichment: Painting a Room

Objective
Calculate area problems.

Before painting a room, you must estimate the amount of paint needed to complete the job.

How much paint do you need to paint this bedroom?

Find the area of the walls by multiplying the distance around the room by the height of the rooms. You may wish to use a calculator.

Color of Paint	Coverage/ Gallon	Coverage/ Quart
Same color or dark over light	400 square feet	100 square feet

Bedroom

length (l) = 14 ft
width (w) = 12 ft
height (h) = 8 ft

length of the 4 walls × height = (14 + 12 + 14 + 12) × 8
= 52 × 8 = 416 sq ft

One gallon of paint covers 400 square feet. You will need one quart of paint to cover the remaining 16 square feet of wall area.

Exercises

Find the area and the amount of same-color paint needed.

1. l = 13 ft
w = 13 ft
h = 9 ft
Area = □ sq ft
paint needed = □ gal □ qt

2. l = 19 ft
w = 15 ft
h = 8 ft
Area = □ sq ft
paint needed = □ gal □ qt

3. l = 11 ft
w = 14 ft
h = 8 ft
Area = □ sq ft
paint needed = □ gal □ qt

4. l = 14 ft
w = 17 ft
h = 8 ft
Area = □ sq ft
paint needed = □ gal □ qt

5. l = 15 ft
w = 17 ft
h = 8 ft
Area = □ sq ft
paint needed = □ gal □ qt

6. l = 18 ft
w = 24 ft
h = 9 ft
Area = □ sq ft
paint needed = □ gal □ qt

7. l = 22 ft
w = 27 ft
h = 7 ft
Area = □ sq ft
paint needed = □ gal □ qt

8. l = 16 ft
w = 15 ft
h = 8 ft
Area = □ sq ft
paint needed = □ gal □ qt

16-10 Computer: Carpeting Cost

Objective
Use the computer to find the cost of carpeting.

Many retail stores have computers that help salespeople find the cost of some products. The program below finds the cost of installing carpet and padding.

```
10  REM This program finds the area of a room to be carpeted,
20  REM the cost of installation, padding, carpet, and total cost
30  INPUT "HOW MANY FEET LONG IS THE ROOM" ;L
40  INPUT "HOW MANY FEET WIDE IS THE ROOM" ;W
50  LET A = INT(L*W/9*10+.5)/10
60  INPUT "HOW MUCH DOES 1 SQUARE YARD OF CARPET COST" ;C
70  INPUT "HOW MUCH DOES 1 SQUARE YARD OF PADDING COST" ;P
80  INPUT "HOW MUCH DOES INSTALLATION COST PER SQ YD" ;I
90  CR=INT(C*A*100+.5)/100:PR=INT(P*A*100+.5)/100:IR=INT(I*A*100+.5)/100
100 T = CR+PR+IR
110 PRINT "AREA";
120 PRINT TAB(15) "---------------------COST---------------------"
130 PRINT "SQ YD", "CARPET", "PADDING", "INSTALL.", "TOTAL"
140 PRINT A, "$"CR, "$"PR, "$"IR, "$"T
150 END
```

The program was run for the following case.

room, 15 ft × 11 ft	area sq yd, 18.3
cost carpet, $12.99 per sq yd	cost carpet, $237.72
cost padding, $3.05 per sq yd	cost padding, $55.81
cost installation, $5.50 per sq yd	cost installation, $100.65
	total, $394.18

Exercises

1. How many square yards of carpet are needed for each room?

   A. 9 ft × 12 ft   B. 15 ft × 18 ft

2. Use the program to find the cost of installing the following:

Room Size	Cost Carpet (per sq yd)	Cost Padding (per sq yd)	Cost Installation (per sq yd)	Total
8 ft × 14 ft	$ 9.49	$2.89	$5.49	
12 ft × 17 ft	$12.39	$2.79	$6.25	
15 ft × 29 ft	$11.49	$3.25	$6.50	
11 ft × 13 ft	$13.29	$3.05	$5.89	

3. The area of the room is calculated in line 50 of the program. Explain why L*W is divided by 9 (/9).

CHAPTER 16 REVIEW

Choose the correct answer from A–F.

1. Find the total cost of a sprinkler system with 6 sprinkler heads at \$2.49 each, 8 pipes at \$3.16 each, and a timer for \$46.38.
2. How much do you save? Movie tickets and popcorn for 2 cost \$5.50 per ticket and \$1.75 per bucket of popcorn. Compare the cost of renting 2 movies at \$2.25 per movie.
3. Find the total cost to cover a 9′ × 15′ floor with Deluxe tile. Use the chart on p. 376.
4. Estimate the cost of adding a 22′ × 15′ room to a home. The cost factor is \$42 per square foot.
5. Find the total cost of a new roof if its 2 sections are 37′ × 28′. Materials cost \$42 per square. The roofer charges \$35 per square.
6. Carpet costs \$23.60 per square yard. Padding costs \$3.45 per square yard. Installation costs \$6 per square yard. What is the total cost for installing carpet and padding in a 24′ × 21′ room?

A. \$10
B. \$1617
C. \$92.85
D. \$86.60
E. \$1850.80
F. \$13,860

Change these units of measure.

7. 16 ft = □ in.
8. 520 in. = □ yd
9. 34 ft = □ yd
10. 774 ft^2 = □ yd^2
11. 345 ft^2 = □ yd^2
12. 562 ft^2 = □ yd^2

Solve.

13. Find the amount of same-color paint needed to cover a 16′ × 12′ room with a 10′ ceiling and 2 windows and 2 doors. Use the table on p. 368.
14. Find the number of rolls of wallpaper needed for a 15′ × 15′ room with 1 window, 1 door, and an 8′ ceiling. Use the chart on p. 370.

Complete each sentence using each of these words once:

area   perimeter   square   square feet   yards

15. To find the amount of carpet needed, you find the _______ of the floor.
16. To tile a laundry room floor, find the number of _______.
17. The length of an office building is measured in _______.
18. The distance around a garden is its _______.
19. Roofing shingles are sold by the _______.

CHAPTER 16 TEST

Make these changes.

1. 4 ft = _______ in.	**2.** 18 in. = _______ ft
3. 38 ft = _______ yd	**4.** 13 yd = _______ ft
5. 5 yd = _______ in.	**6.** 279 in. = _______ yd

Use the table on p. 368.

7. How many gallons of paint are needed to cover 7550 sq ft with two coats of paint?

8. Estimate the amount of paint needed to cover a 16 ft × 21 ft room with 8.5 ft ceilings. The room has 1 door and 5 windows. Only one coat of paint is needed.

Find the number of rolls of wallpaper needed. Use 27-in. widths.

9. wall 14 ft × 8 ft, 2 openings **10.** room 23′ × 15′, 9′ height, 5 openings

11. Find the cost of installing carpet in a room 18 ft × 14 ft. Carpet costs $13.95 per sq yd, padding costs $2.79 per sq yd, and installation costs $6.50 per sq yd.

12. Estimate the cost of a 12 ft × 15 ft kitchen, cost factor $45/sq ft.

13. Two of the outside walls of a new family room are 10′ × 18′. The third wall includes a 10′ × 22′ rectangle and a triangle with a base of 22′ and a height of 8′. Find the number of squares of siding needed.

14. Use the proposal given by Wanner's Landscape Gardening Service on p. 374. Find the total cost of 4 goldtip junipers, 18 pink astibles, 4 bayberrys, 6 spotlights at $28.49 each, 8 pole lights at $52.98 each, a transformer costing $75.50, and 140 feet of cable at $0.32 per foot.

15. A game room having dimensions 22′ × 36′ is to be tiled. Otto's Flooring sells cartons of tile covering 60 sq ft at $38.50 for the first six cartons and $34.50 for those over six. How many cartons are needed? How much will the tile cost?

16. Mamie Lawson decides to replace the roof of her house. Both sections of the roof are rectangles with dimensions 21 ft × 68 ft. The shingles cost $36 per square. The roofer charges $42 per square. How much will the new roof cost?

17. Joe's garage charges $18.50 to change the oil in your car. You can buy 4 quarts of oil at $2.49 per quart. How much can you save by changing the oil yourself?

CHAPTER 16 CUMULATIVE REVIEW

1. Multiply.
$\frac{1}{3}$ of \$57.81
- **A.** \$19.27
- **B.** \$38.54
- **C.** \$173.43
- **D.** \$18.90

2. Change 25 feet to yards.
- **A.** $\frac{2}{3}$ yd.
- **B.** $2\frac{1}{2}$ yd
- **C.** $8\frac{1}{3}$ yd
- **D.** $8\frac{2}{3}$ yd

3. Translate to a number sentence.
325 is what percent of 750?
- **A.** $n\% = 325 \times 750$
- **B.** $750 = n\% \times 325$
- **C.** $325 = n\% \times 750$
- **D.** $n = 750$

4. Taxi fare is \$1.35 for the first 0.35 mile and \$0.55 for each additional mile. What is the fare for a 4.25-mile trip?
- **A.** \$3.50
- **B.** \$3.69
- **C.** \$5.65
- **D.** \$5.75

5. Find the annual salary at the end of the given period.
recreation leader \$3800,
3.5% increase every year,
2 years
- **A.** \$4070.66
- **B.** \$3933
- **C.** \$4213.13
- **D.** \$3969.66

6. Find the amount deducted from a weekly paycheck.
annual life insurance premium \$380
- **A.** \$31.67
- **B.** \$7.31
- **C.** \$6.21
- **D.** \$7.03

7. Find the cost for a new sod lawn covering 3100 sq ft. Sod costs \$0.155 per sq ft.
- **A.** \$480.50
- **B.** \$379.50
- **C.** \$370.50
- **D.** \$485

8. Find the sale price of a microwave oven.
list price \$245,
$\frac{1}{5}$ off
- **A.** \$186
- **B.** \$49
- **C.** \$190
- **D.** \$196

9. Find the mean.
prices of shoes:
\$31, \$29, \$32, \$31, \$56, \$72, \$40
- **A.** \$31
- **B.** \$30.50
- **C.** \$41.57
- **D.** \$72

10. Estimate the cost of an additional room. Use \$50 per sq ft cost factor.
new room 10 ft × 12 ft
- **A.** \$2000
- **B.** \$600
- **C.** \$1200
- **D.** \$6000

Use.
First 22 pounds = 1000 cal
Next 22 pounds = 500 cal
Each additional 22 pounds = 200 cal
Active athlete = 500 cal
Lose/gain weight ±500 cal

11. Find the number of calories to gain weight. Sam Chu plays baseball at 165 pounds. Use the chart to the left.
- **A.** 3500 cal
- **B.** 2600 cal
- **C.** 3100 cal
- **D.** 3600 cal

12. Find the number of rolls of wallpaper needed. Use 27-in. widths, 16 ft length. wall 14 ft 6 in. × 8 ft, no openings
Use $P = 2l + 2w$.
A. 10 rolls
B. 13 rolls
C. 11 rolls
D. 15 rolls

13. Find the amount in the account after 2 years. $1800 principal, 10.7% interest, compounded semiannually
Use $i = prt$.
A. $2318.23
B. $2216.23
C. $2007.23
D. $2217.23

14. Estimate the maximum amount to pay for a house. Use 2.5 times annual gross pay.
monthly salary of $2010
A. about $24,000
B. about $60,000
C. about $120,000
D. about $80,000

15. Find the cost of tiling the room. A carton that covers 45 sq ft costs $23.50.
family room 13 ft × 16 ft
A. $108.10
B. $108.57
C. $117.50
D. $141

16. FICA deductions are 7.15% of the first $42,000 of gross income. Find the monthly deduction on an annual salary of $31,200.
A. $185.90
B. $184.16
C. $2230.80
D. $194.16

17. Find the approximate cost of a new roof.
roof 1836 sq ft, shingle $37 per square, labor $34 per square
A square covers 100 sq ft.
A. $1278
B. $1249
C. $1378
D. $1349

18. Find the car rental cost.
rental per day $45.30, number of days 8, cost per mile $0.35, number of miles 245, discount 50 free miles
A. $448.15
B. $430.65
C. $398.15
D. $429.55

19. A new upholstered chair costs $178. An old chair is sold for $50. Materials to repair it cost $35 for fabric, $15 for tacks and tools, and $20 for stuffing. How much can be saved by reupholstering the old chair?
A. $58
B. $68
C. $108
D. $48

20. How many gallons of paint are needed to cover 2150 sq ft? One gallon covers 400 sq ft.
A. $5\frac{1}{2}$ gal
B. $4\frac{3}{4}$ gal
C. $5\frac{3}{8}$ gal
D. 53 gal

Chapter 17 Overview

Key Ideas

- finding routes and distances
- renting and operating a recreational vehicle
- using a travel guide
- using currency exchange rates
- finding distances on water
- using formulas and tables
- using golf scoring terms
- making scuba calculations
- finding camera settings

Key Terms

- recreational vehicle
- travel guide
- facilities
- lodging
- currency
- exchange rate
- yen
- peso
- pound sterling
- guilder
- franc
- mark
- nautical mile
- knot
- par
- bogie
- birdie
- pressure
- dive time
- atmosphere
- aperture
- f-stop
- shutter

Key Skills

Find the answer.

1. 92 + 114 + 216 + 84 + 110
2. 85 + 199 + 213 + 308 + 74
3. \$1264 + \$218 + \$1915
4. \$412 + \$605.19 + \$206.19
5. $512 \div 45$
6. $610 \div 55$
7. $494 \div 50$
8. $612 \div 28$
9. $495 \div 31$
10. $586 \div 29$
11. 60×139
12. 49×236
13. 60×157
14. 0.11×1396
15. 0.12×1375
16. 0.14×1421
17. 0.5947×1400
18. 0.4857×1600
19. 0.6637×1500
20. $926 \div 60$
21. $860 \div 60$
22. $1050 \div 60$
23. $\frac{1}{2} \times \frac{1}{3}$
24. $\frac{1}{2} \times \frac{1}{2}$
25. $\frac{1}{4} \times \frac{1}{2}$

Solve.

26. What is 8% of \$92?
27. What is 9% of \$65?
28. n = 8% of \$162
29. 16% of \$126 = n

TRAVEL AND LEISURE

17

17-1 Planning a Trip

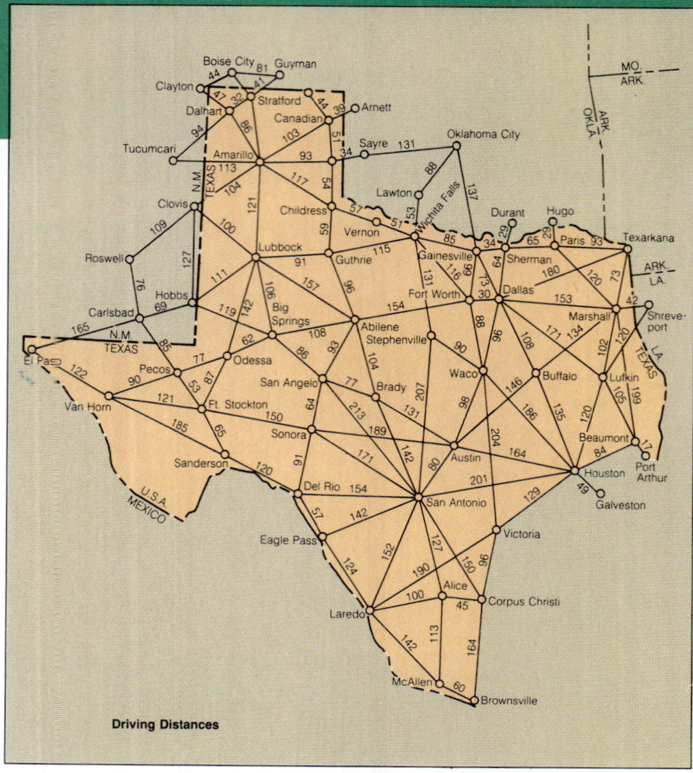

Objective
Determine routes and distances on a map.

The Tyburski family lives in Waco, Texas. They use this map to help plan a vacation trip to Clovis, New Mexico. The map shows the distances between cities.

Understand the Situation

- What is the distance between Waco and Fort Worth?
- What is the distance between Fort Worth and Wichita Falls?
- What is the shortest distance between Wichita Falls and Clovis?

Example

Mr. Tyburski plans to average about 45 miles per hour.
Allowing one hour for lunch, about how long will the trip take?
Mr. Tyburski expects to average 29 miles per gallon.
About how many gallons of gasoline will he use?

Add to find the distance to Clovis.
$88 + 154 + 157 + 100 = 499$ miles

Divide to find the driving time. $499 \div 45 \approx 11.1$ hours

Add 1 hour for lunch. $11.1 + 1 = 12.1$ hours

Divide to find the number of gallons of gasoline.
499 miles $\div$ 29 mpg $\approx$ 17.2 gallons

The trip will take about 12.1 hours and about 17.2 gallons of gasoline.

Try This

The Packer family plans to make a trip from San Angelo to San Antonio in one day via Brady and Austin. Mr. Packer estimates that they will average 45 miles per hour. They also plan a 1.5-hour rest stop including lunch. About how long will the trip take? Mr. Packer expects to get 25 miles per gallon. About how many gallons of gasoline will they use?

Exercises

Find the shortest distance. Use the map on p. 390.

1. Lubbock, TX, to Del Rio, TX

2. Waco, TX, to Shreveport, LA

3. Roswell, NM, to Wichita Falls, TX

4. Odessa, TX, to Laredo, TX

5. Port Arthur, TX, to Durant, OK

6. Sonora, TX, to Sayre, OK

7. Lufkin, TX, to Fort Stockton, TX

8. Abilene, TX, to Carlsbad, NM

9. El Paso, TX, to Vernon, TX

10. Tucumcari, NM, to Marshall, TX

11. Andy Dotterweich plans to average 45 miles per hour on a trip from Pecos, TX, to Buffalo, TX, via Big Springs and Brady. Allowing two hours for lunch and rest, about how long will the trip take? Andy knows that his car averages 27 miles per gallon. About how many gallons of gasoline will he use on the trip?

12. Laura Talbott and her sister are planning a trip from Corpus Christi, TX, to Abilene via San Antonio, Dallas, and Fort Worth. Laura plans to average 45 miles per hour. Allowing three hours for lunch, rest, and sightseeing, about how long will the trip take? Laura's car averages 31 miles per gallon. About how many gallons of gasoline will she need?

Estimate the total cost for each of these trips.

13. Preston Fischer's 7-day trip to the Gettysburg battlefield

Motel	\$420
Food	\$250
Gasoline	\$120
Miscellaneous	\$150

14. The Rowe family's 10-day trip to Marco Island, Florida

Campgrounds	\$150
Food	\$300
Gasoline	\$160
Entertainment	\$100
Miscellaneous	\$150

Mental Math

15. Estimate the cost per day of the trip to Marco Island described in Exercise 14.

Group Activity

16. Obtain a map of your state and the surrounding area. Plan several trips. Estimate the distances involved, gasoline needs, and expenses.

Mixed Skills Review

1. $4\frac{2}{5} \div 2$

2. $1\frac{1}{3} \times 1\frac{1}{9} \times 1\frac{4}{5}$

3. $5.096 + 12.17 + 21.84$

4. $59{,}630 \div 335$

5. $12\frac{3}{8} - 7\frac{3}{4}$

6. $12.13 - 5.65$

7. $2043 - 1638$

8. $1056.64 \div 83.2$

9. 184×416

10. $1\frac{3}{8} + 6\frac{1}{2} + 5\frac{1}{4}$

11. $\$8.53 \times 207$

12. $166 + 552 + 19 + 487$

17-2 Recreational Vehicles

Objective
Find the costs of renting and operating a recreational vehicle.

Michelle and Keith and their children enjoy traveling and camping. This year they rented a recreational vehicle (RV) for their camping trips.

Understand the Situation

- What is the weekly rental fee?
- What is the daily rental fee?
- What is the mileage charge?

Examples

1. The Joneses choose the weekly rental for a three-week, 1400-mile trip. Find the cost of the RV for the trip.

Multiply to find the RV cost.
$\$395 \times 3 = \$1185$ *Estimate: $3 \times \$400 = \$1200$*

Multiply to find the mileage cost.
$\$0.11 \times 1400 = \$154$ *Estimate: $\$0.10 \times 1400 = \$140$*

Find the total cost.
$\$1185 + \$154 = \$1339$ *Estimate: $\$1200 + \$140 = \$1340$*

The cost of the RV will be $1339.

2. Al and Martha Tartal own a pop-up camping trailer. Because of the trailer's extra weight, they estimate that it takes an extra gallon of gasoline every 100 miles to tow it. About how many extra gallons of gasoline can they expect to use on a 1416-mile trip from Big Springs, Texas, to the Grand Canyon?

Use mental math to divide 1416 by 100. Move the decimal point two places to the left.
$1416 \div 100 = 14.16$

They can expect to use about 14 extra gallons of gasoline.

Try This

a. Vaughn and Eloise Weimer rent an RV at the daily rate (see the table above). Find the cost of the RV for a four-day round trip of about 470 miles from Cheyenne, Wyoming, to the Black Hills of South Dakota.

b. The Jordan family owns a small travel trailer. They estimate that to tow the trailer 100 miles they use about two extra gallons of gasoline. About how many extra gallons of gasoline can they expect to use on a 750-mile round trip from Rochester, Minnesota, to Independence, Missouri?

Exercises

1. Steve and Jestina Held rent an RV for $345 per week plus $0.13 per mile. Find the cost of the RV for a two-week, 1125-mile round trip from White Hall, Maryland, to Bar Harbor, Maine.
2. The Besche family rents an RV for $396 per week plus $0.11 per mile. Find the cost of the RV for a three-week, 2350-mile round trip from Grand Island, Nebraska, to Yosemite National Park.
3. The Yinglings visit Woodland Way Campground. The cost is $13 per night. They plan to stay seven days and nights. They estimate that food will cost them $14 per meal, two meals per day. Not including gasoline, how much will the trip cost?
4. Paul and Linda Leimbach and their children camp at Oceanview Campgrounds. The cost is $15 per day. They plan to stay 11 days and nights. Food costs are estimated at $30 per day. Not including gasoline, how much will the trip cost?

Data Hunt

5. Contact a local agency which rents RVs. Find out the cost of renting various types of RVs. Plan a weekend trip using an RV to some nearby site and find the cost for the trip.

Use the vacation graph to answer these questions about when U.S. residents take vacations.

6. In which month are most vacations taken? Estimate the number of vacations taken in this month.
7. In which month are the least vacations taken? Estimate the number of vacations taken in this month.
8. Estimate the number of vacations taken in the three summer months of June, July, and August. Find the average number of vacations taken during the summer months.

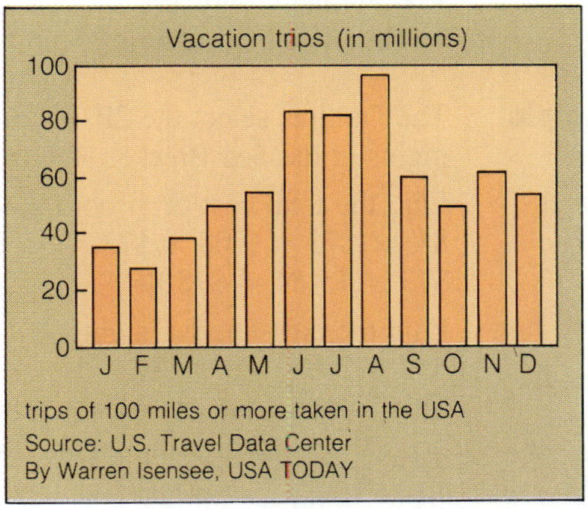

Mixed Skills Review

1. $4.59 \div 2.55$
2. $7\frac{2}{9} + 8\frac{1}{3}$
3. 12.04×16.8
4. $4\frac{1}{2} - 1\frac{2}{3}$
5. $294{,}722 \div 86$
6. $\$32.08 + \$163.49 + \$88.52$
7. $5\frac{1}{3} \times 1\frac{1}{9}$
8. $46{,}101 - 9352$
9. $4\frac{1}{2} \div 5\frac{2}{5}$
10. 2956×28
11. $2043 - 164.8$
12. $194.7 + 8 + 15.06 + 10.5$

17-3 Hotels and Motels

Objective
Find cost of lodging using a travel guide.

Gloria and Phil Donlin are planning a vacation at the beach. They use a travel guide to select a motel. The travel guide contains information about costs, room sizes, and other motel or hotel facilities.

Sea Breeze	Rate Guaranteed	Motel	♦♦♦

All year 1P 61.00–65.00 2P/1B 73.00–77.00 2P/2B 69.00–73.00 XP 5 F
Senior discount. 289 units. Tax 8%. On SR 4 at Newtown Rd exit: 1 mi e of jct I-6. A/C; C/CATV; movies; radios; phones. Suites. Htd indoor pool, sauna; whirlpool; health spa. AE, DI, MC, VI. ● Dining rm & coffeeshop: 6:30 am-10 pm; $7-$17 Ⓢ Ⓓ ♿ (804) 555-4400

Understand the Situation

- What are the rates for two persons with one bed (2P/1B)?
- What are the rates for two persons with two beds (2P/2B)?
- What are the rates for one person (1P)?
- What is the room tax rate?
- How many units are available?
- What is the cost for an extra person (XP)?

Example

The Donlins select the 2P/2B room at the $73 rate. They plan to spend five nights at the Sea Breeze. How much will they spend on lodging?

Find the total cost of a room including tax.
$8\% \times \$73 = 0.08 \times \$73 = \$5.84$
$\$73 + \$5.84 = \$78.84$

Find the cost for five nights.
$5 \times \$78.84 = \$394.20$
This can be done using mental math.
To multiply by 5 using mental math, multiply by 10 and then take $\frac{1}{2}$.

$10 \times 78.84 = 788.40, \frac{1}{2} \times 788.40 = 394.20$

The cost of lodging will be $394.20.

Try This

a. Dan and Jane Sides spend six nights at the Sea Breeze. They choose the 2P/1B room at the $77 rate. What is the total cost of lodging?

b. Jerry and Marian Davis plan to spend four nights at the Sea Breeze. They choose the 2P/2B room at a rate of $73. They also pay the extra-person rate for their daughter. What is the total cost of lodging?

Exercises

Use the travel guide table for Exercises 1–9.

The Breakers					Rates Subject to Change				Motel		♦♦♦
7/20-8/3	1P	....	51.00	2P/1B	.... 51.00	2P/2B	....	51.00	XP	7	F
6/3-7/19 & 8/4-9/2	1P	40.00-	46.00	2P/1B	.... 40.00	2P/2B	43.00-	46.00	XP	6	F
9/3-11/30 & 4/1-6/2	1P	31.00-	38.00	2P/1B	.... 31.00	2P/2B	36.00-	38.00	XP	5	F
12/1-3/31	1P	29.00-	35.00	2P/1B	.... 29.00	2P/2B	33.00-	35.00	XP	5	F

Senior discount. 51 units. Tax 5%. A/C; C/CATV; phones. Pool, wading pool. Pets. AE, CB, DI, MC, VI.

1. What is the room rate range for 2P/1B if the room is rented during July?
2. What is the room rate range for 2P/2B if the room is rented on May 26?
3. What is the room rate range for 1P during December?
4. What is the sales tax rate?
5. What is the extra-person rate on July 4?
6. How many units does this motel have?
7. Jack and Darla McArthur rent a 2P/2B room for three nights from July 12–14. The rate is $45. What is the total cost of lodging?
8. Esteban Rodriguez attends a convention at the beach. He stays at the Breakers for four nights during March. His room rate is $35 per night. What is the total cost of lodging?
9. Jim Garretson and his wife Nancy rent a 2P/1B room during mid-August. Their room rate is $40 per night and they stay seven nights. What is the total cost of lodging?

Group Activity

10. Obtain some travel guides for a vacation site. Select a motel or hotel and determine the lodging cost for a six-night stay.

The Data Bank on p. 463 contains a copy of a hotel bill for a five-night stay in Anaheim, CA.

11. What is the rate per night for the room?
12. Find the tax rate.
13. What other charges appear on the bill?
14. What is the total hotel bill?

Mixed Skills Review

1. $8\frac{2}{5} - 6\frac{2}{3}$
2. $6358 + 7194 + 1956$
3. $36.108 - 35.326$
4. $86{,}052 - 79{,}459$
5. $1\frac{5}{7} \div 1\frac{1}{7}$
6. $650{,}592 \div 216$
7. $4\frac{2}{7} \times 1\frac{1}{15}$
8. 0.193×461
9. $4\frac{5}{6} + 7\frac{1}{3}$
10. $86.268 \div 316$
11. $\$7.53 + \$8.29 + \$4.74$
12. 1096×57

17-4 Exchanging Currency

Objective
Use currency exchange rates.

Bob and Carolyn Cox are planning a trip to London. For their trip they will need to convert some of their cash to British pounds. This table shows exchange rates for various currencies. These rates change daily. In the table the British pound is 0.5947. This means that each U.S. dollar will be exchanged for £0.5947.

New York Rates	Foreign Currency per Dollar	New York Rates	Foreign Currency per Dollar
Australian dollar	1.3820	Israeli shekel	1.5732
Belgian franc	36.92	Italian lira	1291.00
Brazilian cruzado	30.99	Japanese yen	140.20
British pound	0.5947	Mexican peso	1231.00
Canadian dollar	1.3395	Norwegian krone	6.6075
Dutch guilder	2.0075	Spanish peseta	124.40
French franc	5.9545	Soviet ruble	0.2200
Greek drachma	132.40	Swedish krona	6.2265
Hong Kong dollar	7.7980	Swiss franc	1.4610
Irish punt	0.6637	West German mark	1.7808

Examples

1. Bob and Carolyn exchange \$1200 for British pounds. How many pounds will they receive?

Since each dollar will be exchanged for 0.5947 pounds, multiply 1200 by 0.5947.

$0.5947 \times 1200 = 713.64$ *Estimate:* $0.6 \times 1200 = 720$

Bob and Carolyn will receive 713.64 pounds.

2. After returning from their trip Bob and Carolyn have 126 pounds to exchange for dollars. How many dollars will they receive?

Since every dollar is worth 0.5947 pounds, divide 126 by 0.5947.

$126 \div 0.5947 \approx 211.87$ *Estimate:* $120 \div 0.6 = 200$

Bob and Carolyn will receive \$211.87.

Try This

a. For a vacation trip to Canada, Gary Bauer exchanged \$700 for Canadian dollars. How many Canadian dollars will he receive?

b. After his Canadian trip Gary had 238 Canadian dollars left. He exchanged them for U.S. dollars. How many U.S. dollars did he receive?

Exercises

Use the exchange rate table on p. 396 for Exercises 1–6.

1. Jane and Dan Sides are planning a trip to Spain. They exchange $1800 for Spanish pesetas. How many pesetas will they receive?
2. After their Spanish vacation Jane and Dan have 2356 pesetas left. How many U.S. dollars will they receive when the pesetas are exchanged?
3. Joan Guitterrez must make a business trip to Japan. She exchanges $750 for Japanese yen. How many Japanese yen will she receive?
4. After returning from her business trip Joan has 8645 yen left. How many U.S. dollars will she receive when the yen are exchanged?

Mental Math

5. Mike Rader and his family plan to vacation in Mexico. They exchange $1000 for Mexican pesos. Use mental math to find how many pesos they will receive.
6. Josh ordered a book from Australia. The price of the book was 18 Australian dollars. How many U.S. dollars must Josh spend to send a money order for 18 Australian dollars?

Estimation

7. the number of French francs received for $1200
8. the number of Dutch guilders received for $2500

Some people invest in foreign currencies. The table shows U.S. dollar exchange rates for four currencies a year ago and for the same currencies now. If $10,000 had been invested in each currency last year, what would the investment be worth at today's rate?

	Currency	Last Year's Rate	Last Year's Investment	Current Rate	Current Value
9.	Guilder (Netherlands)	2.2845	$10,000	2.0075	
10.	Franc (France)	7.0190	$10,000	5.9545	
11.	Franc (Switzerland)	1.8425	$10,000	1.4610	
12.	D-Mark (Germany)	2.2045	$10,000	1.7808	

Mixed Skills Review

1. $13.02 - 7.451$
2. $66{,}138 \div 73$
3. $\frac{1}{3} + 6\frac{1}{2} + 2\frac{1}{3}$
4. 263×258
5. $8 + 3 + 17 + 9 + 3$
6. $\frac{24}{35} \times \frac{7}{8}$
7. $\frac{16}{21} \div 4\frac{2}{7}$
8. $12 - 2\frac{3}{11}$
9. $132.104 \div 19.6$
10. $2900 - 1848$
11. $5.8 \times 2.7 \times 0.3$
12. $0.6 + 0.84 + 0.07 + 11.35$

17-5 Boating

Objective
Find distances on water using rates and tables.

Al Naeny enjoys boating as a leisure activity. Two terms that are frequently used in boating are the **nautical mile** and the **knot.**

Various forms of the distance formula $d = rt$ can be used when working with nautical miles and knots.

---

1 nautical mile = 6076 ft or 1858 meters
1 knot is a speed of one nautical mile per hour.

$d = \frac{rt}{60}$ $\quad r = \frac{60d}{t}$ $\quad t = \frac{60d}{r}$

d = distance in nautical miles
r = rate or speed in knots
t = time in minutes

Examples

1. Al takes 1 hour and 42 minutes to sail a distance of 14.8 nautical miles. To the nearest tenth of a knot, what was his speed?

Convert the time to minutes. $\quad t = 1\text{ h }42\text{ min} = 102\text{ min}$
Use the formula to find the speed.

$$r = \frac{60d}{t} = \frac{60 \times 14.8}{102} = \frac{888}{102} = 8.705 \text{ or about } 8.7 \text{ knots}$$

$$\frac{60 \times 15.0}{100} = \frac{900}{100} = \text{about 9 knots}$$ *Estimate: Use compatible numbers.*

Al's average speed was 8.7 knots.

2. Jean Kuhlman owns a small power boat. She cruises at an average speed of 9.8 knots for 1 hour and 25 minutes. To the nearest tenth of a nautical mile, how far has Jean cruised?

Use the formula $d = \frac{rt}{60}$. The time is 85 minutes.

$$d = \frac{rt}{60} = \frac{9.8 \times 85}{60} = \frac{833}{60} = 13.883 \text{ or about } 13.9$$

$$\frac{840}{60} = \text{about 14 nautical miles}$$ *Estimate: Use compatible numbers.*

Jean has cruised about 13.9 nautical miles.

Try This

a. Wayne took 47 minutes to go 5.1 nautical miles in a sailboat. To the nearest tenth of a knot, what was his speed?

b. Stacey drove her motor boat at a speed of 11.6 knots for 1 h 12 minutes. To the nearest tenth of a nautical mile, how far has Stacey traveled?

Exercises

Solve.

1. Jan took 2.7 hours to cruise 19.8 nautical miles in her small power boat. To the nearest tenth of a knot, what was her speed?
2. In his sailboat George Gabriel took about 4.7 hours to sail 23.9 nautical miles. To the nearest tenth of a knot, what was his speed?
3. In a large chartered motorboat it took 1.9 hours to cruise 18.3 nautical miles. What was the speed to the nearest tenth of a knot?
4. Nancy Sershan sailed for 56 minutes at a speed of 3.2 knots. To the nearest tenth of a nautical mile, how far did she sail?

The horizon is the apparent boundary between the earth and sky. At sea, objects beyond the horizon are hidden from view. Taller objects can be seen from greater distances. A table such as the one to the right can be used to find distances for objects at different heights.

Visual Distances at Sea

Height in Feet	Distance to the Horizon in Nautical Miles
1	1.1
2	1.7
3	2.0
4	2.3
5	2.5
10	3.6
15	4.4
20	5.1
25	5.7
30	6.3

5. Lee is sitting in a seat on a chartered fishing boat. Her line of sight is about 10 feet above the water. About how far is it to a point on the horizon?
6. You are sitting in a rowboat and your eyes are 3 feet above the water. You see the mast of a familiar boat. You know that the mast is about 15 feet high. About how far away is the other boat?
7. From the flying bridge of a large power boat you can see the top of a lighthouse just above the horizon. Your eyes are about 15 feet above the water. The height of the lighthouse is 30 feet. How far away is the lighthouse?

Mixed Skills Review

1. $9.116 + 2.481 + 5.493$
2. $3006 - 789$
3. $13{,}257.56 \div 86.2$
4. 2161.7×0.28
5. $\frac{6}{63} \div \frac{8}{9}$
6. $\$386 + \$851 + \$418$
7. $2\frac{1}{4} + 4\frac{2}{3}$
8. $3\frac{3}{5} \times \frac{1}{2} \times 1\frac{1}{9}$
9. $61.09 - 58.782$
10. $42{,}412 \div 922$
11. $51 \times 12 \times 7$
12. $9\frac{1}{8} - 3\frac{1}{2}$

17-6 Golf

Objective
Find golf scores and par ratings.

Golf Terms
Eagle — 2 strokes less than par on a hole
Birdie — 1 stroke less than par on a hole
Bogie — 1 stroke more than par on a hole
Double bogie — 2 strokes more than par on a hole
Triple bogie — 3 strokes more than par on a hole

Jean Sullivan plays golf for exercise and recreation. Jean explains that a round of golf consists of 18 holes. The object is to complete the 18 holes with the fewest possible number of strokes of the club at the ball. Total scores are compared to **par,** which is the number of strokes normally expected of an expert player on the course. The term par is also used to tell the number of strokes expected on a particular hole.

Understand the Situation

- Jean shot a bogie on a par 5 hole. What was her number of strokes?
- Jean shot a birdie on a par 4 hole. What was her number of strokes?
- If Jean shot par on a par 5 hole and a par 3 hole, how many strokes did she take?

Examples

1. On a par 70 course Jean shot 3 bogies, 2 double bogies, 4 triple bogies, and 1 birdie. Find her total number of strokes.

Add the strokes in bogies to the par 70.
$(3 \times 1) + (2 \times 2) + (4 \times 3) = 3 + 4 + 12 = 19$
$70 + 19 = 89$

Subtract the strokes for birdies. $89 - 1 = 88$

Jean had a total of 88 strokes.

Positive and negative numbers are used to show a golfer's score compared to par. The score -3 means 3 strokes under par. The score $+7$ means a score 7 strokes over par.

2. Jim Kleman finishes at $+8$ on a par 71 course. Find the total number of strokes.

$71 + 8 = 79$

Jim Kleman had 79 total strokes.

Try This

a. Neal shoots 3 birdies, 4 bogies, and 1 triple bogie on a par 72 course. Find his total score.

b. Carly finishes at -2 on a par 73 course. Find her total score.

Exercises

Find the number of strokes.

1. Terry shoots a double bogie on a par 5 hole.
2. Kenra shoots an eagle on a par 5 hole.
3. Louisa shoots a birdie on a par 3 hole.
4. Tyrone shoots a triple bogie on a par 4 hole.
5. Sally shoots 2 birdies, 5 bogies, 1 eagle, and 2 double bogies on a par 73 course.
6. Larry shoots 4 birdies, 2 bogies, and 1 triple bogie on a par 74 course.
7. Janine shoots 7 bogies, 1 birdie, 4 double bogies, and 3 triple bogies on a par 70 course.
8. Marietta shoots 3 birdies, 4 bogies, and 3 double bogies on a par 72 course.
9. Jake Charles shoots +7 on a par 69 course.
10. Orlando Hitzig shoots −3 on a par 72 course.
11. Denise Dunlap shoots +18 on a par 73 course.
12. Gary Saperstein shoots −4 on a par 73 course.
13. Jean Sullivan recorded these scores for the month of June.
    86, 89, 90, 78, 91, 85, 82, 79, 80, 89, 91, 83, 86, 84
    Find the mean and median of her scores.

The formula $P = \frac{L}{200} + 38.25$ is used to assign a par rating to golf courses. P stands for the par rating and L stands for the length of the course in yards. Find the par rating of these golf courses.

14. Mt. Pleasant 7160 yards
15. Longview 6978 yards
16. Ocean Dunes 7208 yards
17. Bon Air 7153 yards

Data Hunt

18. Check with some local golf courses. For each course find the length in yards and the par rating. Find the par rating using the formula above and compare it with the given par rating.

Mixed Skills Review

1. $1500 \times \$0.36$
2. $2\frac{4}{5} - 1\frac{3}{4}$
3. $272.232 \div 0.38$
4. $3\frac{9}{10} + 5\frac{4}{5}$
5. $8 - 0.116$
6. 306×841
7. $775{,}866 \div 637$
8. $5\frac{4}{9} \times \frac{3}{14}$
9. $5706 + 1496 + 7358$
10. $112{,}823 - 105{,}786$
11. $\frac{11}{20} \div 8\frac{1}{4}$
12. $\$5318.62 + \$94.26 + \$347.58$

17-7 Scuba Diving

Objective *Calculate diving time and air consumption during scuba diving.*

Kim DeSoto is taking a course to become certified as a scuba diver. Kim learns that certain formulas can be used to find underwater pressure and maximum dive time.

Scuba Formulas

Underwater Pressure

$P = 1 + \frac{d}{34}$ fresh water

$P = 1 + \frac{d}{33}$ salt water

P stands for pressure measured in atmospheres (1 atmosphere = 14.7 lb per sq in.)
d = depth of the water in feet

Maximum Dive Time

$T = \frac{v}{kP}$

T = maximum time in minutes
v = volume of tank in cu ft
k = air-use rate in cu ft/min
P = pressure in atmospheres

Understand the Situation

- What do P and d stand for in the underwater pressure formulas?
- How is pressure measured?
- What do v and k stand for in the maximum dive time formula?
- How is T measured?

Example Kim's tank contains 35 cubic feet of air and he uses the air at a rate of 0.9 cubic feet per minute. What is the maximum dive time on a salt-water dive to a depth of 60 feet?

Find P, the underwater pressure. Use the salt-water formula.

$$P = 1 + \frac{d}{33} = 1 + \frac{60}{33} = 1 + 1.818 = 2.818 \text{ atmospheres}$$

Find the maximum dive time.

$$T = \frac{v}{kP} = \frac{35}{0.9 \times 2.818} = \frac{35}{2.536} = 13.8 \text{ minutes}$$

Round down as a safety factor. The dive should last no more than 13 minutes.

Try This Find the underwater pressure.

a. 40 feet in salt water **b.** 60 feet in fresh water

Find the maximum dive time.

c. You dive to a depth of 65 feet in fresh water. The air tank holds 35 cubic feet and the air-use rate is 0.7 cubic feet per minute.

Exercises

Find the underwater pressure.

1. 40 feet in fresh water **2.** 65 feet in fresh water

3. 90 feet in salt water **4.** 75 feet in salt water

5. 75 feet in fresh water **6.** 100 feet in salt water

7. 35 feet in fresh water **8.** 78 feet in fresh water

Find the maximum dive time for each of these dives.

	v, Air in Tank (cu ft)	k, Air-Use Rate (cu ft/min)	d, Depth of Dive (feet)	Type of Water	Dive Time
9.	60	0.85	60	salt	
10.	35	1.4	35	fresh	
11.	42	0.95	50	salt	
12.	55	1.00	95	fresh	
13.	80	0.90	25	salt	

Number Sense

14. Look at the underwater pressure formulas. What effect does changing the denominator of 34 in the fresh-water formula to 33 for the salt-water formula have on the pressure?

Talk Math

15. Why is it safer for a diver to round the dive time down rather than apply the usual rules for rounding?

Mixed Skills Review

1. $11\frac{1}{4} \times \frac{20}{27}$ **2.** $0.083 + 0.025 + 0.204$ **3.** 7162×36

4. $7\frac{3}{4} - 3\frac{5}{6}$ **5.** 311.6×4.5 **6.** $2\frac{8}{9} \div 4\frac{1}{3}$

7. $155{,}782 \div 194$ **8.** $74{,}059 + 6973 + 248$ **9.** $2201.331 - 1844.648$

10. $20{,}104 - 6375$ **11.** $2452.28 \div 40.4$ **12.** $4\frac{3}{5} + 7\frac{1}{6}$

17-8 Photography

Objective *Find aperture and shutter speed settings.*

Mark Goldstein's hobby is photography. Mark can control the amount of light entering his camera by setting the **aperture.**

The aperture settings are called **f-stops.** The smallest aperture is f 32. Increasing the aperture by one stop to the left doubles the light admitted. Decreasing the aperture by one stop to the right halves the light admitted.

Example

1. Mark changes the aperture opening from f 8 to f 3.8. What happens to the amount of light admitted?

Changing from f 8 to f 3.8 opens the aperture two stops. Each opening doubles the amount of light admitted. Since $2 \times 2 = 4$, there is 4 times the amount of light admitted.

Try This

What happens to the amount of light admitted when Mark changes the aperture opening?

a. from f 11 to f 8 **b.** from f 16 to f 32

The **shutter speed** controls the length of time that light is admitted. Shutter speeds are measured in fractions of one second. For instance, a setting of 30 means $\frac{1}{30}$ of a second. The next setting, 60, means $\frac{1}{60}$ of a second. Each setting approximately halves the length of time of the setting to the left. Likewise, each setting approximately doubles the setting to the right.

B 1 2 4 8 15 30 60 125 250 500 1000

Example

2. What happens if the shutter speed is changed from 30 to 250?

The number of settings from 30 to 250 is 3.
Each setting decreases the time by $\frac{1}{2}$. $\frac{1}{2} \times \frac{1}{2} \times \frac{1}{2} = \frac{1}{8}$
Changing speed from 30 to 250 means that the shutter is open $\frac{1}{8}$ as long.

Try This

c. What happens if the shutter speed is changed from 60 to 250?

d. What happens if the shutter speed is changed from 500 to 250?

Exercises

For each aperture opening change tell what happens to the amount of light admitted.

1. from f 16 to f 11
2. from f 11 to f 3.8
3. from f 22 to f 5.6
4. from f 16 to f 3.8
5. from f 3.8 to f 5.6
6. from f 8 to f 16
7. from f 11 to f 22
8. from f 3.8 to f 16

Tell what happens when the following shutter speed changes are made.

9. from 15 to 30
10. from 60 to 500
11. from 2 to 15
12. from 125 to 500
13. from 500 to 125
14. from 1000 to 250
15. from 60 to 8
16. from 125 to 30

The aperture and shutter speed can be adjusted together. If you increase the aperture by one f-stop and decrease the shutter speed by one setting, the exposure is the same.

Suppose

17. The aperture opening is changed from f 5.6 to f 8 while the shutter speed is 60. What new shutter speed setting should be selected to keep the exposure the same?
18. The aperture opening is changed from f 8 to f 3.8 while the shutter speed is 125. What new shutter speed setting should be selected to keep the exposure the same?

Number Sense

19. Most cameras with fixed shutter speeds have a speed of $\frac{1}{75}$ second. Copy the shutter speed scale and place a 75 where this speed would be located.
20. If a camera had another shutter speed faster than 1000, what would it be?
21. How long does the aperture stay open if the shutter speed is set at 1?
22. What shutter speed would be between B and 1?

Mixed Skills Review

1. 9440×53
2. $21.2 - 14.6433$
3. $2\frac{4}{5} \div 3\frac{3}{20}$
4. $224{,}455 \div 583$
5. $1\frac{17}{25} \times 5\frac{5}{6}$
6. $592.4 + 1.673 + 52.761$
7. $4 + 6508 + 363 + 27$
8. $4\frac{3}{10} - 2\frac{1}{4}$
9. $8006 - 488$
10. $747.86 \div 0.613$
11. 5.294×65
12. $2\frac{3}{16} + 1\frac{5}{8} + 3\frac{1}{2}$

17-9 Enrichment: Travel Itinerary

Objective
Use an itinerary to find travel time.

The Ellison family kept a record of their summer trip through the eastern part of the United States and Canada.

Departed			Arrived		
City	**Time**	**Date**	**City**	**Time**	**Date**
Williamsburg	8:30 A.M.	6/21	Washington, D.C.	1:00 P.M.	6/21
Washington, D.C.	2:00 P.M.	6/25	Gettysburg	5:20 P.M.	6/25
Gettysburg	7:30 A.M.	6/27	New London	4:00 P.M.	6/27
New London	11:15 A.M.	6/28	Provincetown	3:00 P.M.	6/28
Provincetown	9:00 A.M.	6/30	Boston	12:15 P.M.	6/30
Boston	8:30 A.M.	7/3	Bangor	4:30 P.M.	7/3
Bangor	9:00 A.M.	7/5	Quebec City	4:30 P.M.	7/5
Quebec City	10:00 A.M.	7/8	Montreal	2:00 P.M.	7/8

How long did they take to drive from Williamsburg to Washington, D.C.?

Use mental math. Add 12:00 hours to afternoon times.
1:00 p.m. + 12:00 a.m. = 13:00

Subtract the departure time from the arrival time.
13:00 p.m. − 8:30 a.m. = 12:60 − 8:30 = 4:30

It took 4 hours 30 minutes to drive from Williamsburg to Washington, D.C.

Exercises

Solve.

1. How long did the Ellisons take to drive from Gettysburg to New London?
2. Where did the Ellisons spend about 4 days?
3. How many hours did the Ellisons spend in Bangor?
4. In which city did the Ellisons spend the least amount of time?
5. How many hours long was the drive from Provincetown to Boston?
6. How many hours did the Ellisons spend in New London?
7. About how many days did the Ellisons spend in Boston?
8. How much longer did the trip from Boston to Bangor take than the trip from Provincetown to Boston?
9. How many days did the Ellisons take to get to Montreal from Williamsburg?
10. How many hours did the Ellisons take to get from Williamsburg to Montreal?

17-10 Calculator: Population and Area

Objective
Compare areas. Find and compare population densities.

The Five Largest States			The Five Smallest States		
State	Area (km^2)	Population	State	Area (km^2)	Population
Alaska	1,518,717	400,481	New Jersey	20,294	7,364,158
Texas	692,364	14,228,383	Hawaii	16,705	965,000
California	410,991	23,668,562	Connecticut	12,973	3,107,576
Montana	381,065	786,690	Delaware	5,327	595,225
New Mexico	315,096	1,299,968	Rhode Island	3,144	947,154

This table lists states with the largest and smallest areas and also gives their populations from a recent census. You can use a calculator to compare their areas.

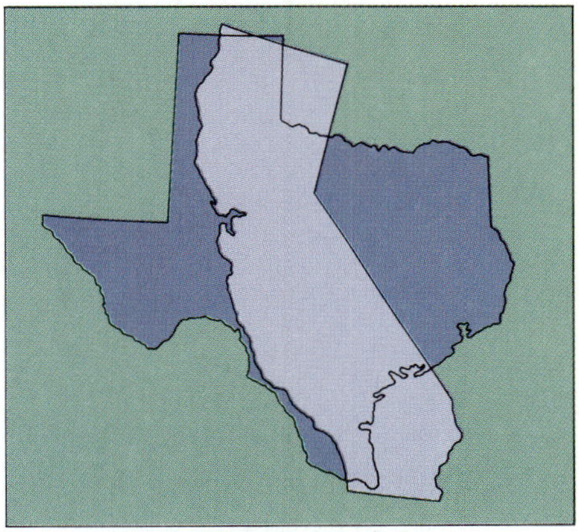

How many times larger in area is Texas than California?

Divide the larger area by the smaller area. Round the answer to hundredths.
692364 [÷] 410991 $\approx$ 1.70

Texas is more than $1\frac{1}{2}$ times larger than California.

What is the population density of Texas?

Population density is the number of people living per square kilometer of area. Divide the population by the area.
14228383 [÷] 692364 $\approx$ 20.55 people per km^2

The population density of Texas is about 21 people per square kilometer.

Exercises

Solve. Use your calculator. Round to the nearest hundredth whenever necessary.

1. Is the population of Texas more or less than the combined populations of New Jersey and Connecticut? How much more or less?
2. Is Alaska large enough to contain the combined areas of all the states listed except Montana?
3. What is the population density of California?
4. Which state listed has the smallest population density?
5. Which state listed has the greatest population density?
6. What are some reasons why a smaller state like Rhode Island has a greater population density than a larger state like Texas?

CHAPTER 17 REVIEW

Use the data provided in Lessons 17-1 to 17-8.

1. Find the number of gallons of gasoline needed for a 615-mile trip in a car which averages 28 mpg.
2. Find the distance from Dallas to Odessa via Ft. Worth, Abilene, and Big Springs.
3. Find the cost of an RV at the daily rate for a 775-mile, 6-day trip from St. Louis, Missouri, to Wichita, Kansas.
4. Find the cost of a 4-night stay at the Ebb Tide in a 2P/2B room at the $69 rate.
5. Find the number of British pounds received in exchange for $900.
6. Find the number of Spanish pesetas received in exchange for $1200.
7. Find the speed, to the nearest tenth of a knot, of a boat that takes 1 hour 52 minutes to sail 15.2 nautical miles.
8. Find the distance a boat can cruise in 2 hours 10 minutes at an average speed of 10.2 knots.
9. Find the distance to a point on the horizon from a person whose line of sight is 5 feet above the water.
10. Find the score of a golfer on a par 72 course who shoots 3 birdies, 4 bogies, and 1 triple bogie.
11. Find the maximum dive time in 40 ft of fresh water of a diver using a tank holding 35 cu ft of air and using air at a rate of 0.9 cu ft per minute.
12. Find the number of times longer that the shutter remains open when the shutter speed is changed from 500 to 60.

Choose the one that does not belong. Explain why.

13. aperture
    dive time
    shutter speed
    f-stop
14. exchange rate
    knot
    nautical mile
    horizon
15. bogie
    par
    RV
    eagle
16. atmosphere
    dive time
    underwater pressure
    stroke
17. frame
    peseta
    currency
    ruble
18. 2P/1B
    XP
    V
    2P/2B

CHAPTER 17 TEST

1. Peter Bassford is planning to drive from Corpus Christi, TX, to Arnett, OK, via San Antonio, Brady, Abilene, Guthrie, Childress, Shamrock, and Canadian. Peter will average 45 miles an hour. He plans to spend 2 hours for lunch and rest stops. How long will the trip take? Peter's car averages 33 miles per gallon. About how many gallons of gasoline will he use?
2. Tom and Barbara Shanks rent an RV for $355 per week plus $0.15 per mile. Find the cost of the RV for a 2-week, 1425-mile camping trip.
3. Jack and Maggie Sipple stay at the Edgecliff Lodge for 5 nights. The rate is $47. What is the total cost of the lodging?

Use the exchange rate table on p. 396 for Exercises 4 and 5.

4. Joe and Marilyn Dolan are going to Greece on their honeymoon. They exchange $450 for Greek drachmas. How many drachmas do they receive?
5. Upon returning from a trip to Japan, Harold Smith exchanges 3500 Japanese yen for U.S. dollars. How many U.S. dollars does he receive?
6. John took 1 hour 20 minutes to cruise to an island that is 6.3 nautical miles from shore. What was his speed to the nearest tenth of a knot?
7. Jack and Debbie Williams cruised in their power boat at 5.8 knots for 45 minutes. How far did they travel?

Find the number of strokes.

8. Nadine shoots a double bogie on a par 4 hole.
9. Alfred shoots 1 eagle, 2 birdies, 3 bogies, 2 double bogies, and 1 triple bogie on a par 71 course.

Find the underwater pressure.

10. 80 feet in salt water
11. 95 feet in fresh water

Find the maximum dive time for the dives.

	v, Air in Tank (cubic feet)	k, Air-Use Rate (cu ft/min)	d, Depth of Dive (feet)	Type of Water
12.	75	1.2	85	fresh
13.	68	0.90	90	salt

For each aperture change tell what happens to the amount of light admitted.

14. from f 1.8 to f 5.6
15. from f 16 to f 8

Tell what happens when the following shutter speed changes are made.

16. from 4 to 30
17. from 500 to 30

CHAPTER 17 CUMULATIVE REVIEW

1. Estimate the percent. Use compatible numbers.
   32 is what percent of 478?
   A. 100%
   B. 106%
   C. 60%
   D. 6.7%

2. Robbie is driving 267 miles to Clear Lake. He can average 50 miles per hour. He will take 1 hour for lunch. About how long will the trip take?
   A. about 6 h
   B. about 5 h
   C. about 10 h
   D. about 4 h

3. Find the selling price of the car.
   luxury $11,270, air-conditioning $800, power door locks $225, tinted glass $135, pin striping $60
   A. $11,390
   B. $12,490
   C. $12,380
   D. $11,380

4. Compare the pay for a weekly period.
   Job A: $17,592 per year
   Job B: $9.60 per hour for 40-hour week
   A. Job A, $48.69 more
   B. Job B, $48.69 less
   C. Job A, $38.59 less
   D. Job B, $48.69 more

5. Find the cost of the camping trip.
   campground $12 per night, food $25 per day, for 8 days and nights
   A. $296
   B. $308
   C. $496
   D. $286

6. Compare the weekly transportation costs.
   14 miles one way
   Driving: $0.22/mi,
   Bus: $1.50 one way
   A. Bus saves $4.66.
   B. Bus saves $15.80.
   C. Car saves $2.20.
   D. Car saves $0.22.

7. Find the income tax.
   adjusted income $13,766, tax 4.96%
   A. $736.80
   B. $655.80
   C. $682.80
   D. $671.70

8. Find the underwater pressure. Use $P = 1 + \frac{d}{33}$.
   67 feet in salt water
   A. 0.03
   B. 1.03
   C. 2.03
   D. 3.03

9. Camera shutter speeds—
   B 1 2 4 8 15 30
   What happens when the speed is changed from 30 to 4?
   A. 8 times as long
   B. $\frac{1}{8}$ as long
   C. 6 times as long
   D. $\frac{1}{4}$ as long

10. Find the annual cost for the insured. Biweekly family coverage deduction per pay is $31.73.
    A. $1689.96
    B. $380.76
    C. $824.98
    D. $190.38

11. Carl exchanges $150 for German marks at a rate of 1.78 marks per dollar. How many marks will he receive?
    A. 270
    B. 83.14
    C. 84.24
    D. 267

12. Find the catalog cost.
    1 dress $18.75, 2 blouses $9.50 each, 3 scarves $4.15 each, weight 7 lb 6 oz, shipping cost 5–8 lb $4.15
    A. $36.55
    B. $54.35
    C. $50.20
    D. $62.10

13. The RDA of protein is 55.6 grams. One serving of cereal provides 5% RDA of protein. How many grams of protein are in a serving?

A. 2.78 g
B. 11.12 g
C. 0.278 g
D. 1.11 g

14. Find the total move-in cost.
monthly rent $275,
cleaning deposit $30,
application fee $15,
security deposit $150

A. $755
B. $350
C. $320
D. $470

15. Find the speed of a power boat going 25.5 nautical miles in 2.5 hours.

A. 6.4 knots
B. 10.2 knots
C. 5 knots
D. 12.8 knots

16. Baseball Scores

	Won	Lost
Sequoias	67	70
Alamos	71	65

How many games back are the Sequoias? Use $G = \frac{dW + dL}{2}$.

A. 2
B. $4\frac{1}{2}$
C. 6
D. $5\frac{1}{2}$

17. Find the par rating. Use $P = \frac{L}{200} + 38.25$.
Aberdeen Hills
7178 yards

A. 74.14
B. 35.89
C. 75.14
D. 34.89

18. In Heidi's coin collection there are 6 German coins, 4 Swiss coins, and 8 Canadian coins. She takes out one coin. What is the probability it is Swiss?

A. $\frac{2}{3}$
B. $\frac{1}{2}$
C. $\frac{2}{9}$
D. $\frac{3}{18}$

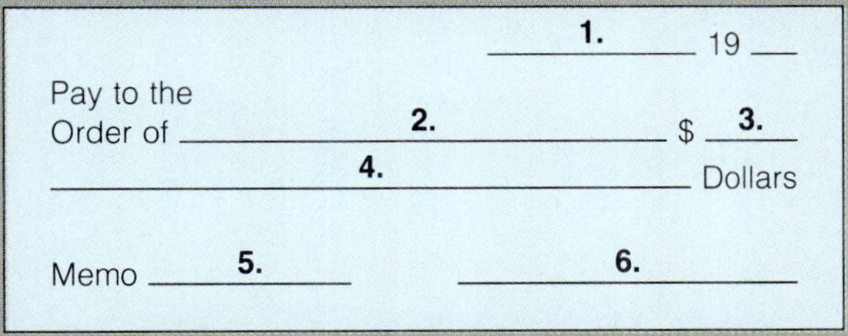

19. On which line do you write the purpose of the check?

A. 5
B. 3
C. 6
D. 2

20. Find the room rate for a family of four sharing a room.
1P 38-40 2P/2B 43-48
XP 6

A. $43–$48
B. $55–$60
C. $52
D. $152

21. Compare the weekly transportation costs.
52 mi one way
Driving: $0.18 per mile
Bus: $4.50 one way

A. Bus saves $98.60.
B. Bus saves $45.
C. Bus saves $53.60.
D. Bus saves $48.60.

22. Find the cost of tile for a 13 in. × 17 in. storeroom.
Tile: $34.80 per carton, covers 45 ft^2

A. $386.70
B. $221
C. $174
D. $139.20

Chapter 18 Overview

Key Ideas

- finding stock prices
- finding bond values
- finding mutual fund values
- finding value of collectibles
- comparing compound interest
- finding social security benefits
- finding contribution to individual retirement plans
- comparing group retirement benefits

Key Terms

- stock
- dividend
- share
- bond
- pass value
- maturity date
- market value
- mutual fund
- net asset value
- loading charge
- no-load fund
- collectible
- compound interest
- social security
- retirement plans
- Keogh plan

Key Skills

Find the answer.

1. $160 × 39.5
2. $8.47 + $11.55 + $6.94
3. $6 ÷ 24
4. $0.68 + $0.36 + $0.88
5. $34.99 × 0.045
6. $42 − $28.36
7. $3016.88 ÷ 4.3
8. $3111.84 − $2692.57
9. $157.42 + $687.93
10. $1063.50 − $612.98
11. $181.50 ÷ 0.75
12. $12.54 × 40

Find the percent of increase or decrease.

13. $192 to $124.80
14. $67.50 to $78.30
15. $125 to $287.50
16. $106.30 to $31.89
17. $0.64 to $1.60
18. $52.17 to $34.78

Find the mean.

19. $36.53, $42.08, $41.36, $39.52, $40.11
20. $827, $846, $850, $853, $864, $874, $882, $893, $901, $911
21. $16.32, $9.38, $11.85, $30.71, $19.93, $25.08, $21.84, $18.12
22. $106.43, $124.84, $245.78, $131.52, $168.83, $153.46, $200.17

Solve.

23. 11% of $76
24. 6% of $32.50
25. 18% of $439.18
26. 3.5% of $1036
27. 108% of $14
28. $5\frac{1}{2}$% of $175

INVESTMENTS AND RETIREMENT

18

18-1 Buying Stocks

Objective
Find stock prices.

When you buy **stock**, you buy part of a company. The part that you buy is called a **share.** Profits to stockholders are **dividends.** The financial page of the newspaper gives information about many stocks.

Sales 100s is the number of hundreds of shares sold that day.

Dividend on AMI is $0.72 per share.

Abbreviation for the stock's name

High-Low is highest price and lowest price paid this year per share in dollars.

Close is the final price of the day in dollars.

Net chg. is the difference between yesterday's close and today's close.

	Div.	Sales 100s	Year High	Year Low	Close	Net Chg.
Amrtch	6.60	1751	102¾	99⅞	102¾	+2
AlnGrp	.44	1018	108⅜	106¾	107¼	− ¾
AMI	.72	5349	20⅜	20¼	20½	− ⅛
AmMat		1399	3	2¾	3	+ ⅛
APresd	s.50	226	19	18¾	18⅞	.....
ASLFla		570	8⅞	8⅛	8½	− ⅛
ASLFl	pf2.19	122	17⅜	17⅛	17⅛	− ⅛
AShip	.80	60	11⅝	11½	11½	− ⅛
AmStd	1.60	1521	37½	36⅞	37½	+ ¾
AmStor	.84	268	64	63½	64	+ ¼
AStr	pfA4.38	27	75¼	74½	74½	.....
AStr	pfB6.80	13	u59¾	59	59½	+ ½
AT&T	1.20	21106	23⅝	22¾	23⅛	− ½
AT&T	pf3.64	93	41⅞	41⅝	41⅝	.....
AT&T	pf3.74	1224	42¾	42⅝	42¾	.....

Examples

1. What dividend is paid on AShip?
Find AShip in the stock column. In the *Div.* column opposite AShip the dividend is .80. AShip pays a dividend of $0.80 per share.

2. How many hundreds of shares of AMI were sold?
Find AMI in the stock column. In the *Sales 100s* column, the number of hundreds of shares sold is 5349. There were 534,900 shares sold.

Try This

a. What dividend was paid on AmStor?

b. How many hundreds of shares of ASLFla were sold?

Prices in the *Close* column are given in dollars. The $18\frac{7}{8}$ beside APresd means that a share of APresd stock sold for 18 and $\frac{7}{8}$ of a dollar at the end of the day. Since $\frac{7}{8} = 0.875$, APresd sold for $18.875 per share.

Example

3. What is the cost of 300 shares of AT&T at $41\frac{5}{8}$ per share?
$41\frac{5}{8}$ means $41.625 per share. Multiply to find the total cost.
Multiply the cost per share by the number of shares.
$\$41.625 \times 300 = \$12,487.50$

The cost of the 300 shares of AT&T is $12,487.50.

Try This

c. What is the cost of 100 shares of IBM at $156\frac{1}{2}$?

d. What was the closing cost of AlnGrp?

Exercises

Use this chart for Exercises 1–6.

	Div.	Sales 100s	High	Low	Close	Net Chg.
FinCoa		4448	10¼	9⅞	10¼	+ ¼
FinCo	pf6.08e	20	37⅜	37	37⅛	− ¼
FnsBar		71	8¾	8⅛	8⅝	+ ⅛
Firefd	n.30	2042	33¾	33	33¼	+ ¼
Firestn	.80	1128	22⅞	22	22⅞	+ ½
FtBkSv	1.60	811	u43⅞	41¼	43⅜	+2⅛
FBost $	1	1376	43	42	42¾	− ½
FstChic	1.32	379	28⅞	28⅜	28½	+ ½
FChi	pf4.92e	29849	½	49⅛	49⅛	− ¼
FtBTex	.60	679	13½	12⅞	13⅛	− ½
FtBTx	pf5.32e	12	38¼	37	37⅞	+ ¼
FtBTx	pf5.07e	9	35	35	35	+ ⅜
FtCity		158	7⅜	7¼	7⅜	− ⅛
FFB	3.12	235	59⅞	58¾	59	+ ½

1. What was the lowest cost per share for Firefd?
2. What dividend is paid by Firestn?
3. How many hundred shares of FBost were sold?
4. What was the high cost per share for FFB?
5. What was the closing cost of FChi?
6. What dividend is paid by FinCoa?
7. What is the cost of 400 shares of Gearht at $8\frac{1}{8}$?
8. What is the cost of 250 shares of LAC at $28\frac{1}{4}$?
9. What is the cost of 180 shares of Huffy at $11\frac{1}{8}$?
10. What is the cost of 360 shares of MONY at $8\frac{7}{8}$?

Group Activity

11. Consult the financial section of your newspaper. Choose a stock to follow for a period of two weeks. Make a line graph that shows the closing price of the stock for each of the ten days.

Data Hunt

12. Consult a local stockbrokerage firm. For the stock chosen in Exercise 11 find the commission that would be charged on 100 shares. Then find the total cost, including commission, for the 100 shares.

Use this stock table for Exercises 13 and 14.

13. What dividend will be paid on 100 shares of Carlisle?
14. What dividend will be paid on 270 shares of CascNG?

Stock	Div.	Yld.	PE	Sales 100s	High	Low	Last	Chg.
CdnPac	.48	...	20	3091	18¼	17⅞	18⅛	− ⅛
CanonG		...	8	1441	4¾	4½	4¾	+ ⅛
CapCits	.20	.1	30	326	u380¼	375½	278	+1½
CapHld	.88	2.8	7	840	31⅝	30⅞	31	...
Carlisle	1.10	3.0	17	74	37¼	37	37	...
CaroFl	.50	1.5	13	172	33½	32¾	33½	+ ⅜
CarPw	2.76	7.5	9	1661	36⅞	36⅝	36⅝	− ¼
CarPir	.70	1.4	23	891	49½	49¼	49¼	− ½
CartHw	1.22	1.9	47	596	64⅜	63½	64⅛	+ ½
CartWl	.80	.8	24	258	103¼	101¼	101½	−2¾
CartSv n	.07r	.6	5	121	12¼	11¾	12⅛	+ ¼
CascNG	1.28	8.3	82	29	15⅝	15⅜	15½	− ⅛
CastlCk		...	37	274	25⅜	25⅛	25⅜	...
Catlyst		...	30	281	23¼	22⅝	22⅝	...
Caterp	.50	.9	...	2309	54⅜	53½	54¼	+ ⅜
Cedr F n		...	...	368	9⅞	9¾	9⅞	+ ⅛
Cengy		...	28	116	9¼	9⅛	9¼	− ⅛
Centel	2.50	3.9	17	399	65	63⅞	64	− ¾
CentEn	2.56	15.6	5	1432	16⅝	16¼	16⅜	− ¼

Mixed Skills Review

1. $2\frac{7}{10} - 1\frac{1}{2}$
2. $47{,}270 \div 326$
3. $\$50 - \$12.67$
4. $10.6 + 3.18 + 7.26$
5. $17{,}206 + 36{,}895$
6. $2\frac{1}{4} \div 1\frac{1}{14}$
7. 2434×18
8. $3\frac{2}{3} \times \frac{6}{7}$
9. $5\frac{3}{8} + 2\frac{3}{4}$
10. 3.6×9.02
11. $8102 - 1576$
12. $52.728 \div 15.6$

18-2 Buying Bonds

Objective
Find bond values.

Governments and corporations can raise money by selling **bonds.** The **par value** of a bond is the amount that will be paid to the holder on the **maturity date,** which may be 20 years in the future.

Information about bonds is given in the financial pages of the newspaper.

Bonds	Cur Yld	Vol	High	Low	Close	Net Chg.
Savin 11 3/8 98	3.	23	50 3/8	47 1/8	50	.....
Savin 14s00	8.	105	51	48 3/4	49 1/4	+ 1/2
Seafst 9 1/4 01	1.	15	83 1/4	83 1/4	83 1/4	.....
Seagrm zr06	..	511	24	23 1/2	23 3/4	− 1/4
Sears 6 3/8 93	.6	11	85 1/2	83 5/8	83 5/8	−1 7/8
Sears 8 5/8 95	.0	29	96	96	96	+ 7/8
Sears 8s06	.8	10	82	82	82	.....
Sears 10 1/4 88	.9	10	103 3/8	103 3/8	103 3/4	+ 3/8
Sears 11 7/8 14	1.	25	109	109	109	−1 1/2
SearA 8 3/8 86	.4	59	29	29	29	.....
SLR 15 1/4 90	4.	20	108	108	108	+1 7/8
ShellO 8 1/2 00	.7	3	87 1/4	87 1/4	87 1/4	− 1/4
Signi 8.85s94	.5	11	93 1/4	93	93	.....
Signi 8s09	cv	62	88 1/2	88	88	.....
Singer 8s99	.9	30	81	81	81	+ 7/8
Singer 9s06	cv	36	120	118	118	+ 1/2
SoestPS 11 7/8 98	6.	7	73	72	72	− 3/4
SoCG 17 3/8 01	5.	5	117	117	117	−1
SNET 8 1/2 08	.9	5	81 3/4	81 3/4	81 3/4	+ 3/4

Current yield

Year of maturity

Rate of interest

Abbreviation for company's name

High-Low is the highest price and lowest price shown as a percent of the par value.

Close is the final price of the day as a percent of the par value.

Net Chg. is the difference between yesterday's close and today's close.

The **market value** of a bond is the current selling price.

Example

1. Suppose that the par value of Seafst is \$1000. What is the market value?

The closing value is $83\frac{1}{4}\%$ of the par value. So, the market value is $0.8325 \times \$1000$ or \$832.50.

Try This

a. What is the year of maturity for ShellO?

b. Suppose the par value of SNET is \$1000. What is the market value?

Interest is paid on the par value of each bond.

Example

2. The Hagerstown City Park bonds pay 5.25% interest. What is the annual interest paid on a bond of \$1000 par value?

5.25% of $\$1000 = 0.0525 \times \$1000 = \$52.50$

The annual interest paid is \$52.50.

Try This

c. Cecil County is selling bonds that pay 6.75%. What is the annual interest paid on a bond of \$1000 par value?

Exercises

For Exercises 1–18 use the table to the right. Suppose par value is $1000.

Bonds		Cur Yld	Vol	High	Low	Close	Net Chg.
NortS	6s98	9.4	12	63 5/8	63 5/8	63 5/8	+ 1/4
Oakin	10 1/2 02	cv	14	68	67	67	+ 3/4
OcciP	16s89	15.	25	109 1/2	108 1/4	108 1/4	− 1/4
OccP	9.65s94	11.	100	90 3/8	90 1/4	90 1/4	−1 1/4
OhBIT	7 1/2 11	10.	4	75	75	75	.....
OhBIT	12 5/8 20	12.	8	109 1/8	109 1/8	109 1/8	−1 1/4
OhEd	9 1/2 06	11.	6	88	88	88	.....
OhEd	8 1/2 06	11.	5	79	79	79	− 1/2
OhEd	11 7/8 10	12.	27	101 7/8	101 3/4	101 7/8	− 1/8
OklGE	4 1/2 87	4.7	9	96 3/4	96 3/4	96 3/4	− 1/4
Orion	11s98	14.	20	80	79 7/8	80	.....
OwCor	12s10	11.	10	104 1/2	104 1/2	104 1/2	+ 1/2
Ownill	7 5/8 01	9.6	10	79 3/8	79 3/8	79 3/8	− 7/8
PSA	11 1/8 04	cv	10	110	110	110	.....

1. What is the maturity date of NortS?
2. Which bond has a maturity date of 2001?
3. What was the low cost of OwCor?
4. What was the high cost of Orion?
5. What was the closing cost of PSA?
6. What was the closing cost of Oakin?
7. What rate of interest does OwCor pay?
8. What rate of interest does NortS pay?
9. Which bond in the table pays the highest rate of interest?
10. Which bond in the table pays the lowest rate of interest?
11. What is the market value of PSA?
12. What is the market value of OwCor?
13. How much interest does Oakin pay per year?
14. How much interest does OccP pay per year?
15. What was the closing cost of OklGE yesterday?
16. What was the closing cost of Orion yesterday?
17. What was the market value of Ownill yesterday?
18. What was the market value of OcciP yesterday?

Data Hunt

19. One column in the bond table of the financial page shown above is headed *Cur Yld*. Find out what **current yield** is. Determine the current yield of a bond that pays 6.8% interest on a par value of $1000 and has a market value of $850.

Group Activity

20. Consult the financial page of your local newspaper. Select a bond and record the closing cost of the bond for 10 days. Make a line graph that shows the changes in the closing cost over the 10-day period.

Mixed Skills Review

1. 637×446
2. $0.061 + 0.473 + 2.08$
3. $\frac{2}{5} \div \frac{3}{25}$
4. $30{,}451 - 6714$
5. $212.338 \div 0.523$
6. $3\frac{2}{3} + 4\frac{1}{2}$
7. $132 + 1638 + 845 + 74$
8. $8.2 - 6.54$
9. 16.4×16.23
10. $\frac{7}{9} \times 1\frac{13}{14}$
11. $8\frac{4}{9} - 6\frac{2}{3}$
12. $52.728 \div 15.6$

18-3 Investing in Mutual Funds

Objective
Find mutual fund values.

Many people invest in mutual funds. A **mutual fund** pools the money of many investors to buy stocks, bonds, or other investments. The cost of a share of a mutual fund is called the **net asset value (NAV).** Some mutual funds charge commissions and fees called a **loading charge.** Funds that do not have a loading charge are called **no-load** funds.

Mutuals

Mutual Funds Closing Quotations

	Sell	Buy	Chg
AARP Invst:			
CapGr	26.30	NL	+.07
GiniM	15.52	NL	+.01
GenBd	15.32	NL	+.01
Gthinc	24.48	NL	+.14
TxFBd	15.82	NL	+.02
TxFSh	15.43	NL	...
ABT Midwest:			
Emry	10.76	11.30	−.02
Fl Gth	16.71	17.54	+.06
Gthinc	14.12	14.82	+.03
IntGv	10.36	10.57	+.01
Secinc	12.25	12.86	−.03
TFLtd	10.26	10.47	+.01
Utillnc	14.82	15.56	...
AdsnCa	16.25	16.75	+.06
ADTEK	13.87	NL	+.04
Social	26.60	27.85	+.16
TxF Lt	10.58	NL	...
TxF Lg	15.33	NL	+.03
US Gov	14.58	15.26	...
WshA r	21.10	NL	−.17
Carnegie Funds:			
Govt	9.78	10.24	...
CapGr	15.72	16.46	+.09
CapTR	11.31	11.84	+.01
Cardnl	17.12	18.71	+.09
CrdnlGv	9.09	9.54	−.01
Cnt Shs	20.03	NL	+.16
Chp Dir	12.55	NL	+.06
Chestnut	84.75	NL	+.56
CIGNA Funds:			
Agrsv	14.15	14.89	+.03
Grwth	16.12	16.97	+.07
HiYld	10.45	11.00	+.02
Incom	7.83	8.24	+.01
MuniB	7.41	7.80	+.01
Value	14.40	15.16	+.03

Understand the Situation

- What is the closing cost of AARP TxFBd?
- Find a no-load (NL) fund in the list.
- What was the change (Chg) in AdsnCa?

Example

1. June Danaher invested \$2000 in a mutual fund with a 6.8% loading charge. How much of her money was actually invested?

After the loading charge is deducted, the difference is invested.
If 6.8% is deducted for the loading charge, then 93.2% was invested.
93.2% = 0.932   0.932 × \$2000 = \$1864

There is \$1864 left to be invested.

Try This

a. Frank Kupres invested \$4500 in a mutual fund with a loading charge of 8.1%. How much money was actually invested?

The number of shares of a mutual fund that can be purchased with a given amount of money depends on the net asset value and the loading charge.

Example

2. Chip Herbold invested \$5000 in 20th Century Select, a no-load fund with a NAV of \$37.89. How many shares did he buy?

Since the fund is no-load, the entire \$5000 will go to the investment.

Divide \$5000 by the NAV.   \$5000 ÷ \$37.89 = 131.96

There were 131.96 shares bought.

Try This

b. Donna Mandel invested \$7000 in the Price New Era no-load fund with a NAV of \$21.39. How many shares did she buy?

Exercises

1. Mandy Gill invested \$2500 in a mutual fund with a 4.9% loading charge. How much of her money was actually invested?
2. Ted Zelinski invested \$10,000 in a mutual fund with a 7.8% loading charge. How much of his money was actually invested?
3. Sally Neutzel invested \$4500 in a mutual fund with a 6.2% loading charge. How much of her money was actually invested?
4. Gary Levine invested \$6000 in a mutual fund with a NAV of \$16.26 and a loading charge of 7.8%. How many shares did he buy?
5. Jessie Fisher invested \$2800 in a no-load mutual fund with a NAV of \$4.98. How many shares did she buy?
6. Fred Driver invested \$7500 in a mutual fund with a NAV of \$22.78 and an 8.9% loading charge. How many shares did he buy?
7. Shannon Pew bought 40 shares of a mutual fund with an NAV of \$12.87. Two years later he sold his shares and made a profit of \$250. What was the net asset value when he sold his shares?
8. A mutual fund has a total market value of \$29,760,800. The fund also has 18,000,000 shares. What is the NAV?
9. A mutual fund has a total market value of \$32,986,560. The fund also has 14,500,000 shares. What is the NAV?
10. A mutual fund has a total market value of 3.2 million dollars. It also has 2.2 million shares. What is the NAV?

Group Activity

11. Consult the financial section of your newspaper. Choose a mutual fund to follow for a period of two weeks. Make a line graph that shows changes in the net asset value of the fund for each of the ten days.

Consider the 20th Century funds shown in the table.

12. Which of the no-load funds has the greatest NAV?
13. Which of the funds had the greatest change?
14. What was the previous NAV for 20th Century Select?
15. Which 20th Century fund had a loss?

20th Century:			
Gift r	9.01	9.05	+.15
Grwth	19.40	NL	+.08
LTBnd	94.50	NL	+.02
Select	39.88	NL	+.18
TxEln	97.46	NL	...
TxELT	94.94	NL	−.02
Ultra r	12.47	12.53	+.09
USGv	96.68	NL	...
Vista r	8.09	8.13	+.23

Mixed Skills Review

1. $3.07 - 1.6$
2. $130{,}642 \div 83$
3. $4.217 + 2 + 8.79$
4. $6\frac{1}{3} + 1\frac{2}{5}$
5. $10{,}135 - 8746$
6. $3\frac{15}{16} \div 1\frac{1}{8}$
7. 1.42×73
8. $4\frac{4}{5} \times \frac{15}{16}$
9. $162{,}635 + 352{,}747$
10. $3386.6 \div 236$
11. 846×359
12. $26 - 17\frac{3}{7}$

18-4 Investing in Collectibles

Objective
Find values of collectible items.

Many people collect valuable items such as coins, stamps, antiques, dolls, and baseball cards. Later they may sell them for a profit. Rarity, age, quality, or popularity may increase the value of a collectible.

Examples

1. Chris Bess bought a 1921 Peace dollar in fine condition for \$48. After 10 years, a coin journal listed the value at \$56. What was the increase in value, to the nearest percent?

Find the difference in value. $\$56 - \$48 = \$8$

Divide the difference by the original cost to find the percent of increase.
$\$8 \div \$48 \approx 0.167 \approx 17\%$

2. Gerry Bloomberg bought 10 Eddie Murray rookie baseball cards in 1979 for \$2.50 each. In 1986 he sold them all at an auction for \$170. The commission for the auctioneer was 15%. What percent of profit did Gerry make?

Find the commission. 15% of $\$170 = 0.15 \times \$170 = \$25.50$

Find the amount of money received for the cards.
$\$170 - \$25.50 = \$144.50$

The cost of the 10 cards was \$25. Subtract to find the amount of profit.
$\$144.50 - \$25 = \$119.50$

Divide to find the percent of profit.
$\$119.50 \div \$25 = 4.78 \qquad 4.78 = 478\%$

Gerry's percent of profit was 478%.

Try This

a. Jay Weimer bought an 1889 CC Morgan silver dollar at an auction for \$370. Several years later he sold the coin to a collector for \$410. What was the percent of increase?

b. Tess Foster sold 5 George Brett baseball cards. She had paid \$1.90 for each card. The 5 cards sold for a total of \$96 at a baseball card shop. The shop charged 12% commission. Find the percent of profit.

Exercises

Find each answer to the nearest percent.

1. Fred Weiss bought a 1901 Pan-American 10¢ stamp from a collector for \$175. Later he sold the stamp at an auction for \$235. He paid a commission of 15% to the auctioneer. What was his percent of profit?
2. A 1930 Graf Zeppelin \$2.60 airmail stamp was valued at \$1800. What was the percent of increase in the value?
3. Chris Wild bought a 1909S VDB wheat penny for \$290. He later sold the coin for \$386. What was his percent of profit?
4. A 1916 D Mercury dime in mint condition was valued at \$3000. What was the percent of increase in the value?
5. Seth Isaacs bought a set of Franklin half dollars for \$986. He later sold the set at auction for \$895. The auctioneer charged a commission of 15%. What was the percent of loss?
6. Marion Reesey bought a 1924 St. Gaudens double eagle \$20 gold coin for \$280. When the price of gold rose, Marion sold the coin for \$1020. What was the percent of profit?

7. Pat Kilduff found a Dwight Gooden card in a 25¢ pack of bubble gum. She sold the card to a collector for \$5.00. What was her percent of profit?
8. Ted Hertz owns Ted's Nostalgia, a sports memorabilia shop. Ted says that a Hank Aaron card purchased in 1954 for \$0.01 is now worth \$125. What is the percent increase in value of this card?
9. A 1952 Mickey Mantle card was once valued as high as \$2500. One of these cards was sold at a baseball card show for \$1800. What was the percent of decrease in value?

Data Hunt

10. Check with a baseball card shop or baseball card magazine. Find the current value of a Pete Incaviglia rookie card and the original cost of this card. Find the percent of increase in value.

Mixed Skills Review

1. $2\frac{2}{5} \div 5\frac{1}{3}$
2. $97 + 64 + 82 + 76 + 39$
3. 1037×608
4. $5\frac{1}{8} - 2\frac{3}{4}$
5. 214.8×0.34
6. $3\frac{5}{6} + 4\frac{1}{2}$
7. $29.736 \div 47.2$
8. $1432.51 - 650.826$
9. $15,000 - 3263$
10. $2\frac{5}{8} \times \frac{4}{9}$
11. $333,792 \div 608$
12. \$18.03 + \$6.53 + \$8.78

18-5 Using Compound Interest

Objective
Find compound interest.

By using the ''magic'' of compound interest, small but regular savings can become large sums of money. This table shows the amount you could earn by investing only 50¢ per day for five days per week.

Saving 50¢ Per Day—5 Days Per Week (Compounded Monthly)

	5%	10%
10 years	$1559	$2065
20 years	$4127	$7656
30 years	$8357	$22,793
40 years	$15,323	$63,767
50 years	$26,297	$174,687

Understand the Situation

- How much is saved per day?
- About how much is saved per month?
- After 40 years invested at 5%, how much would this be worth?

Example

1. Estimate the amount earned at 10% after 15 years.
15 is halfway between 10 and 20.

Estimate the amount halfway between $2065 and $7656.
$\$2065 \approx \$2000 \qquad \$7656 \approx \$8000$
The number 5000 is halfway between $2000 and $8000.

About $5000 would be earned after 15 years.

Try This

a. Estimate the amount earned at 5% after 25 years.
b. Estimate the amount earned at 10% after 35 years.

This graph shows how $100 invested at the beginning of each month will grow at various rates compounded monthly for 30 years.

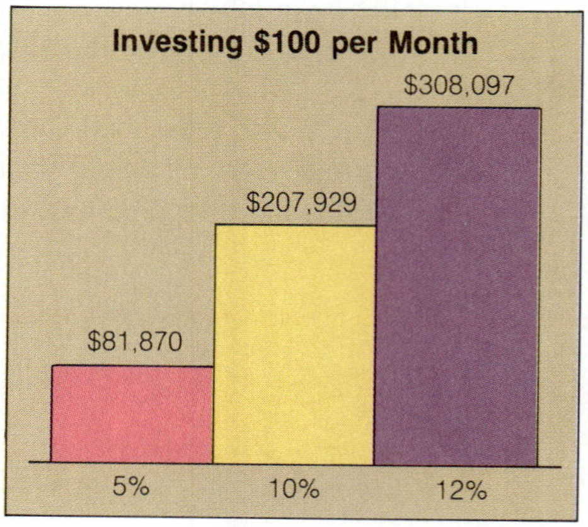

Example

2. Estimate how many times more money is earned at 12% than at 5%.

We can use compatible numbers.
$80,000 for $81,870
$320,000 for $308,097
$\$320,000 = 4 \times \$80,000$

So, about 4 times as much money is earned at 12% than at 5%.

Try This

c. Estimate how many times more money is earned at 10% than at 5%.

Exercises

Use this table for Exercises 1–6.

Saving $1 per Day—5 Days per Week (Compounded Monthly)		
	5%	10%
10 years	$3118	$4131
20 years	$8254	$15,313
30 years	$16,714	$45,586
40 years	$30,647	$127,535
50 years	$53,595	$349,375

1. How much more money is earned at 10% than at 5% after 20 years?
2. How much more money is earned at 10% than at 5% after 30 years?
3. How much more money is earned at 5% after 30 years than after 20 years?
4. Estimate the amount earned at 10% after 15 years.
5. Estimate the amount earned at 5% after 25 years.
6. Estimate the amount earned at 10% after 35 years.

Use this graph for Exercises 7–10.

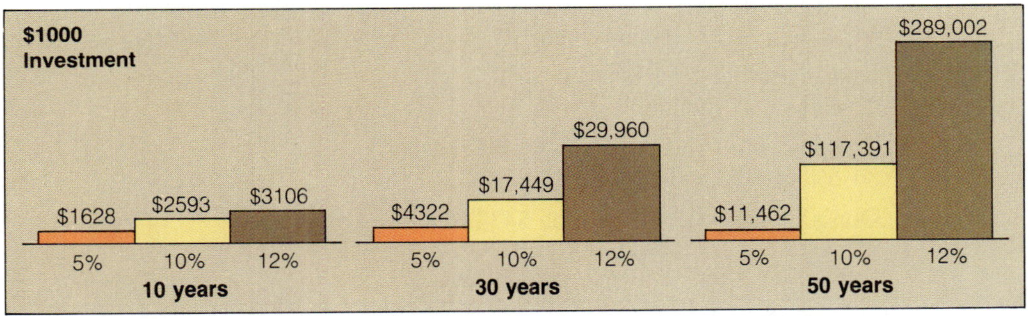

7. How much more will $1000 earn in 30 years at 12% than at 10%?
8. How much more will $1000 earn at 10% for 50 years than at 12% for 30 years?
9. Estimate how many times more money the $1000 will earn at 12% for 30 years than at 5% for 30 years.
10. Estimate how many times more money the $1000 will earn at 12% for 50 years than at 12% for 30 years.

Mixed Skills Review

1. 0.17×364.3
2. $3\frac{5}{9} \div 1\frac{1}{3}$
3. $28{,}842 + 25{,}586$
4. $0.2 - 0.128$
5. $18{,}291 \div 67$
6. $\$64 + \$248 + \$316 + \$27$
7. $4\frac{4}{9} \times 3\frac{3}{5}$
8. $7.9156 \div 3.08$
9. $0.7 + 0.26 + 0.8 + 1.63$
10. $4354 - 2638$
11. 78×458
12. $6\frac{5}{12} - 2\frac{5}{8}$

18-6 Social Security

Objective
Find the amount of Social Security benefits.

Social Security is a federal insurance program that provides retirement benefits. The monthly retirement benefit depends on average annual income and age at retirement. The present retirement age is 65. Early retirement can begin at 62 with a penalty. The following chart shows how early retirement affects benefits. It is based on **reduction months.**

How Reduction Months Affect Benefits

Reduction Months	Percent Payable	Reduction Months	Percent Payable
1	99.4	19	89.4
2	98.8	20	88.8
3	98.3	21	88.3
4	97.7	22	87.7
5	97.2	23	87.2
6	96.6	24	86.6
7	96.1	25	86.1
8	95.5	26	85.5
9	95.0	27	85.0
10	94.4	28	84.4
11	93.8	29	83.8
12	93.3	30	83.3
13	92.7	31	82.7
14	92.2	32	82.2
15	91.6	33	81.6
16	91.1	34	81.1
17	90.5	35	80.5
18	90.0	36	80.0

Example

1. Don Webb plans to retire at age 63. His basic monthly benefit is at \$500. What will be his monthly benefit?

Don is retiring 2 years early.
2 years = 24 months

The percent of full benefit is 86.6%.
86.6% of 500 = $0.866 \times 500 = 433$

Don's monthly benefit will be \$433.

Try This

a. Maynard Keadle retires at age 62. His basic monthly benefit is calculated at \$725. What will be his monthly benefit?

To be eligible for Social Security you need 40 work credits. A work credit is earned for each 3 months of work.

Example

2. Jason Jarre has worked for 5 years 2 months. Use mental math to find how many more work credits he needs.

Express work in months. 5 years × 12 months per year = 60 months

Divide by 3 months per credit to find how many credits he has earned.
60 months ÷ 3 months per credit = 20

Subtract the work credits from the minimum of 40. $40 - 20 = 20$

Jason needs 20 more work credits.

Try This

Use mental math to find how many more work credits are needed.

b. Stuart Monk has worked for 6 years 9 months.

Exercises

Use the table on p. 424.

1. Chuck Raynor plans to retire 14 months before his 65th birthday. His basic monthly benefit is \$589. What will be his monthly benefit?
2. Myrl Dunty plans to retire 21 months before her 65th birthday. Her basic monthly benefit is \$538. What will be Myrl's monthly benefit?
3. Nick Souris retires at age 64. His basic monthly benefit is calculated at \$715. What will be Nick's monthly benefit?
4. Debra Felak retires at age 62. Her basic monthly benefit is \$695. What will be Debra's monthly benefit?
5. Freda Jasper plans to retire 9 months before her 65th birthday. Her basic monthly benefit is \$672. What will be Freda's monthly benefit?
6. Frank Diehl retires at age 67. Social Security pays a bonus of 0.25% for each month that someone works between the ages of 65 and 70. Frank's basic monthly benefit is \$725. What will be his monthly benefit?
7. Lisa Godfrey plans to work until age 70. Her basic monthly benefit is \$765. What will be her monthly benefit? Use data from Exercise 6.

Use mental math to find how many more work credits are needed.

8. Esteban Puentes has worked 4 years 3 months.
9. Anna Krebs has worked 9 years 2 months.

Mixed Skills Review

1. 146,672 ÷ 712
2. $8\frac{3}{8} + 3\frac{5}{16}$
3. 87,842 + 25,586
4. 9.006 × 38
5. $9\frac{1}{2} - \frac{7}{8}$
6. 0.228 ÷ 38
7. 60,016 − 42,563
8. $8\frac{1}{3} \div 3\frac{8}{9}$
9. 694 × 317
10. \$19 − \$12.23
11. $2\frac{2}{5} \times 1\frac{5}{8}$
12. 4.834 + 9.6 + 13.027

18-7 Individual Retirement Plans

Objective
Find contributions to individual retirement plans.

There are several types of retirement plans available to individuals. Each plan may be subject to certain restrictions.

Retirement Plans

Type	Contribution
IRA (Individual Retirement Account)	Up to $2000 per year
Simplified Employee Pension—IRA (SEP-IRA)	Up to 15% or $30,000, whichever is less
KEOGH	Up to 25% or $30,000, whichever is less
403b TSA	Up to 20% (only available for employees of nonprofit organizations)

Examples

1. Paul Upton has a SEP-IRA plan. His income in 1989 was $64,320. He contributes the maximum allowed to his plan. What was his contribution?

Paul can contribute 15%.
15% of $64,320 = 0.15 × $64,320 = $9648

Paul contributes $9648 to his SEP-IRA plan.

2. Jan Elliot works at the museum and has a 403b TSA plan. Her weekly pay is $438. She contributes 16% to the 403b TSA. How much is deducted weekly from her paycheck?

16% of $438 = 0.16 × $438 = $70.08

Jan has $70.08 deducted from her weekly pay.

3. Two years ago, Jesse Greenriver started an IRA and contributed the maximum. Last year he made $18,000 and contributed 10%. This year he made $31,000. What is the most he could have in his plan all together?

first year $2000
second year 10% of $18,000 = 0.10 × $18,000 = $1800
third year $2000
$2000 + $1800 + $2000 = $5800 Jesse can have saved $5800.

Try This

a. Betsy Ness is self-employed. She has a Keogh plan. In a year in which her income was $72,584 she contributed 18%. How much did Betsy contribute?

b. Randy Vogel earns $32,860 per year and has no other retirement plan. He contributes the maximum to an IRA. How much income is left?

Exercises

1. Denise Graham has an annual income of \$47,880. She contributes the maximum to her SEP-IRA plan. What is her contribution?
2. Sam Calleri plays the violin with the symphony. He is eligible for a 403b TSA plan. Sam contributes 12% of his \$39,800 annual salary. How much does he contribute?

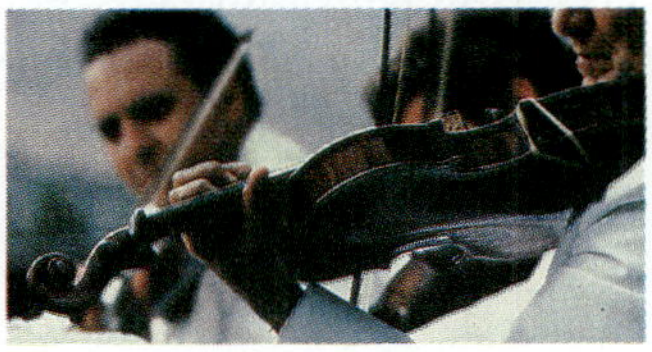

3. Mitsu Sasajima, an author, has a Keogh plan. In a year when his royalties were \$123,862 he contributed the maximum allowed. How much did he contribute?
4. Dale Kief is self-employed as a seamstress. One year she earned \$26,750. She contributed 12% to her Keogh and also made the maximum allowed contribution to her IRA. How much did she contribute to her retirement plans?
5. Harper Martin is an officer of a wildlife preservation organization. Last year he made \$78,596 and contributed 11% to his 403b TSA plan. How much was deducted from each of Harper's biweekly paychecks?
6. Margo Montez has a SEP-IRA. In a year when she earned \$44,625 she contributed \$5801.25 to her plan. Was her contribution within the allowed percentage?

Mental Math Use mental math to find the contributions.

	Income	Retirement Plan	% Contribution	Amount
7.	\$80,596	SEP-IRA	10%	
8.	\$56,000	Keogh	20%	
9.	\$90,000	403b TSA	15%	
10.	\$36,000	IRA	5%	
11.	\$85,000	Keogh	10%	

Data Hunt 12. What are profit-sharing and money-purchase Keogh plans?

Mixed Skills Review

1. $1\frac{9}{16} \times 1\frac{1}{5}$
2. \$392 + \$126 + \$706
3. $8 \div 2\frac{2}{7}$
4. $4\frac{7}{15} - 3\frac{2}{3}$
5. 6.06×539
6. $8411 - 839$
7. 1538×67
8. $20.003 - 6.86$
9. $21{,}528 \div 552$
10. $6\frac{4}{5} + 8\frac{1}{2}$
11. $146.63 \div 0.215$
12. $163.6 + 2.051 + 18.367$

18-8 Group Retirement Plans

Objective
Find and compare group retirement benefits.

Some people receive retirement benefits from company pension plans. Pension benefits are usually based on years of service and some type of average salary.

Examples

1. Jake Liman worked 32 years for ACME Manufacturing. His pension plan is based on the average of his 3 highest annual salaries: $38,500, $40,100, and $40,800. His annual pension is 1.5% of the product of his salary average and years of service. Find his annual pension.

Find the average of the 3 highest annual salaries. $\frac{\$38{,}500 + \$40{,}100 + \$40{,}800}{3} = \$39{,}800$

Multiply the average by the total years of service. $\$39{,}800 \times 32 = \$1{,}273{,}600$

Multiply the result by the company's pension rate. $0.015 \times 1{,}273{,}000 = \$19{,}104$

Jake's annual pension is $19,104.

2. ACME also has a pension option that uses this formula.

$$\frac{\text{Years of Service}}{65} \times \text{Average of 3 Highest Annual Salaries} = \text{Annual Pension}$$

Find Jake's annual pension under this plan.
How does it compare to the benefits in Example 1?

Divide years of service by 65 and round to the nearest hundredth.
$32 \div 65 = 0.49$
Multiply by the salary average. $0.49 \times \$39{,}800 = \$19{,}502$

Jake's annual pension under this plan is $19,502.
Compare to $19,104. $\$19{,}502 - \$19{,}104 = \$398$

This plan pays $398 more per year.

Try This

a. Erika Copenhauser worked 36 years for the Binde Corporation. Her pension plan is based on the average of her 3 highest annual salaries, which were $29,500, $32,600, and $35,800. The company's pension rate is 1.8%. Find Erika's annual pension.

b. Binde Corporation has another pension plan that uses this formula

$$\frac{\text{Years of Service}}{55} \times \text{Average of 3 Highest Annual Salaries} = \text{Annual Pension}$$

Find Erika's annual pension under this plan. Compare it to **a.**

Exercises

Use this formula for Exercises 1, 3, 4, 6, and 7.
Annual pension = Pension Rate × High Salary Average × Years of Service

1. James Whitmore worked 29 years for R & D Sales. His pension rate is 1.6%. His 3 highest annual salaries were \$26,800, \$30,500, and \$32,600. Find his annual pension.
2. Find James' annual pension using the formula:

   $$\frac{\text{Years of Service}}{55} \times \text{High Salary Average} = \text{Annual Pension}$$

   How does it compare to the benefits in Exercise 1?
3. Theresa Garvey worked 31 years for the Tucker Manufacturing Company. Her 4 highest annual salaries were \$36,550, \$37,000, \$38,600, and \$40,000. The company's pension rate is 1.5%. Find her annual pension.
4. Bill Kitchens worked 35 years as a machinist for Glen Arm Tool Company. His 4 highest annual salaries were \$36,800, \$37,000, \$38,750, and \$40,000. The company's pension rate is 1.4%. Find Bill's annual pension.
5. Find Bill's annual pension using the formula:

   $$\frac{\text{Years of Service}}{65} \times \text{High Salary Average} = \text{Annual Pension}$$

   How does it compare to the benefits in Exercise 4?
6. Don Martinez plans to retire with 30 years of service. His retirement benefits are based on his three highest annual salaries, which were \$42,600, \$45,000, and \$47,700. His company's pension rate is 1.6%. Don elects to take only 80% of his pension so that his family will have survivor benefits. Find Don's annual pension.
7. Donna Vitale worked 20 years for S & L Transit Corporation. Her three highest annual salaries were \$28,500, \$29,000, and \$29,800. The company's pension rate is 1.55%. Find her annual pension.

Write Your Own Problem

8. Contact a company or industry in your community. Ask about their pension plans. Write several problems based on these plans.

Mixed Skills Review

1. $9348 + 3056 + 479$
2. $3\frac{9}{11} \times \frac{3}{14}$
3. $4\frac{3}{16} - 1\frac{5}{8}$
4. $573.93 \div 91.1$
5. $\$4.36 + \$10.08 + \$0.93$
6. $9\frac{3}{5} \div 4$
7. $1 - 0.351$
8. $279{,}027 \div 903$
9. $11{,}264 - 6827$
10. $\$12.46 \times 85$
11. $7\frac{2}{3} + 2\frac{1}{4}$
12. 527×83

18-9 Enrichment: Stock Quotations

Objective
Find the difference between stock quotes. Write stock quotes as dollars and cents.

Stock quotations listed in the newspaper appear as mixed numbers. A quotation of $13\frac{5}{8}$ means that one share of stock sells for $13\frac{5}{8}$ dollars.

Write a quotation of $21\frac{5}{8}$ as dollars and cents.

Write the fraction as a decimal. $\frac{5}{8} = 0.625$
Add the whole number. $21 + 0.625 = 21.625 \approx \$21.63$

A stock quotation of $21\frac{5}{8}$ is about \$21.63.

Stockbrokers use the highest and lowest price per share in the past year and the current price per share to evaluate stocks.

Find the difference between the 52-week high and low in dollars and cents for AmTT stock.

Write $22\frac{3}{4}$ and $16\frac{5}{8}$ as decimals. $\$22\frac{3}{4} = \$22.75 \quad 16\frac{5}{8} = \$16.625$
Subtract the low from the high. $\$22.75 - \$16.625 = 6.125 \approx \$6.13$

Exercises

Find the difference between the 52-week high and low. Express the current price in dollars and cents. You may wish to use a calculator.

	52 Week High	Low	Difference	Stock	Current Price	Current Price in Dollars and Cents
1.	$15\frac{1}{2}$	$12\frac{3}{8}$		BaxTr	$13\frac{1}{4}$	
2.	$3\frac{5}{8}$	$1\frac{7}{16}$		ESSp	$2\frac{3}{16}$	
3.	$59\frac{1}{8}$	$33\frac{1}{2}$		ExO	$48\frac{1}{2}$	
4.	40	$18\frac{7}{8}$		FMC	$27\frac{3}{8}$	
5.	$158\frac{1}{4}$	$136\frac{3}{4}$		HnvBk	$119\frac{7}{8}$	
6.	$8\frac{1}{16}$	$5\frac{7}{8}$		NovaPh	$3\frac{9}{16}$	
7.	$100\frac{1}{2}$	$95\frac{5}{8}$		WeMD	$95\frac{7}{16}$	
8.	$63\frac{3}{8}$	$29\frac{1}{4}$		ReAir	$54\frac{1}{8}$	
9.	15	$11\frac{3}{8}$		CrFr	$11\frac{7}{8}$	
10.	$26\frac{1}{4}$	$22\frac{1}{2}$		FnInc	$24\frac{5}{8}$	
11.	$113\frac{1}{8}$	$99\frac{3}{4}$		GLC	$108\frac{3}{8}$	
12.	$70\frac{7}{8}$	$64\frac{1}{2}$		ZebK	$68\frac{1}{4}$	
13.	$136\frac{3}{16}$	$108\frac{3}{4}$		CrnN	$125\frac{5}{16}$	
14.	$89\frac{1}{2}$	$72\frac{1}{2}$		RMCi	$86\frac{1}{8}$	
15.	$33\frac{3}{8}$	$17\frac{1}{4}$		KoLN	$24\frac{5}{8}$	
16.	$56\frac{7}{8}$	$51\frac{1}{2}$		TnMn	$55\frac{1}{2}$	
17.	$100\frac{7}{16}$	$82\frac{5}{8}$		FePR	$97\frac{1}{8}$	

18-10 Calculator: Stock Prices

Objective
Find stock prices.

Because stock prices are usually listed as whole numbers and fractions, you will want to write the fraction part as a decimal.

Use mental math or your calculator to write the following fractions as decimals.

$\frac{1}{8}$ $\frac{1}{4}$ $\frac{3}{8}$ $\frac{1}{2}$ $\frac{5}{8}$ $\frac{3}{4}$ $\frac{7}{8}$

Find the price of 50 shares of Cine Tech stock at the closing price. Use your calculator.

Multiply the closing price by the number of shares.

\$81.375 ⊠ 50 ⊟

The price of 50 shares of Cine Tech stock is \$4068.75.

Stock	Year High	Year Low	Close	Net Change
Cine Tech	$88\frac{1}{2}$	$37\frac{1}{2}$	$81\frac{3}{8}$	$-1\frac{1}{2}$
Acevedo	$25\frac{3}{8}$	$9\frac{7}{8}$	$24\frac{7}{8}$	$-\frac{1}{4}$
Tog Corp	16	$10\frac{1}{4}$	$15\frac{1}{8}$	$+\frac{1}{4}$
Yaz Kal	$27\frac{1}{2}$	$11\frac{3}{8}$	$23\frac{3}{4}$	$+\frac{1}{4}$
Rep NY	$50\frac{7}{8}$	$22\frac{3}{4}$	$39\frac{3}{4}$	$-\frac{1}{2}$

Find the difference between the high and low prices for the year of 50 shares of Tog Corp. Then multiply that figure by the number of shares.

16 ⊟ 10.25 ⊠ 50 ⊟

The difference between the high and low prices is \$287.50.

Exercises

Write the closing prices for these stocks as dollars and cents.

1. Cine Tech **2.** Acevedo **3.** Tog Corp
4. Yaz Kal **5.** Rep NY

Find the price of the following stocks at the closing price. Use your calculator.

6. Yaz Kal, 100 shares **7.** Rep NY, 27 shares
8. Acevedo, 250 shares **9.** Tog Corp, 200 shares
10. Cine Tech, 42 shares **11.** Rep NY, 75 shares

Find the difference in price between the high and low prices for the following stocks.

12. Acevedo, 300 shares **13.** Yaz Kal, 160 shares
14. Rep NY, 420 shares **15.** Cine Tech, 280 shares

CHAPTER 18 REVIEW

Choose the one that does not belong. Explain why.

1. A. no load
   B. net asset value
   C. reduction month
   D. load charge
2. A. dividend
   B. closing cost
   C. share
   D. SEP-IRA
3. A. share
   B. par value
   C. maturity date
   D. market value
4. A. rarity
   B. dividend
   C. age
   D. popularity
5. A. net asset value
   B. work credit
   C. FICA
   D. reduction months
6. A. 4036 TSA
   B. SEP-IRA
   C. no load
   D. Keogh

Match each item in column A with the correct item from column B.

	Column A	Column B
7.	The maximum contribution to a SEP-IRA plan on an income of $58,920.	A. $67.50
8.	The number of Social Security work credits still needed by Sandy who has worked 6 years 3 months.	B. 32%
9.	The market value of a $1000 bond with a closing value of $86\frac{3}{4}$% of the par value.	C. $4705
10.	The amount actually available for investing when $5000 is invested in a mutual fund with a 5.9% loading charge.	D. 92.59
11.	The cost of 200 shares of IBM at $157\frac{3}{4}$.	E. $867.50
12.	The interest paid at 6.75% on a bond with par value of $1000.	F. $31,550
13.	The number of shares of a no-load fund with NAU of $32.40 bought with an investment of $3000.	G. $8838
14.	The percent of profit on a cow bought for $52 and later sold for $65.	H. 25%
		I. 15
		J. 34%
		K. $5803
		L. 84.76

Complete each sentence using each of these words once.

market value
bonds
dividends
no load

15. Profits to stockholders are called _______.
16. Mutual funds which do not charge commissions or fees are called _______ funds.
17. Governments and corporations sell _______ in order to raise money.
18. The current selling price of a bond is its _______.

CHAPTER 18 TEST

1. What is the cost of 550 shares of Bugle Stock at $12\frac{3}{4}$?
2. AutoClub paid a dividend of $1.12 per share. Anna May owns 224 shares of AutoClub stock. How much did she receive in dividends?
3. What is the annual interest paid on a bond with par value of $1000, market value of $950, and an 11.125% interest rate?
4. Jane Okuma invested $6500 in a mutual fund with an 8.5% loading charge. How much of her money was actually invested?
5. Carl Wright invested $4300 in a no-load mutual fund with a NAV of $35.86. How many shares did he buy?
6. Al Smith bought a model train for $150 in 1968. In 1981, he sold it at an auction for $395. He paid a 12% commission to the auctioneer. What was his profit?
7. Sharon had her grandmother's gold ring appraised. Its current value is $2900. Her grandfather paid $500 for the ring. What is the percent of increase in the value?

Refer to the Compounded Monthly table on p. 422 for questions 9 and 10.

8. Estimate how much more is earned at 10% for 25 years than at 5%?
9. Estimate the amount earned at 10% after 35 years.

Refer to the Reduction Months table on p. 424 for questions 11 and 12.

10. Della Gmeindel retires at age 62. Her basic monthly benefit is calculated at $623. What will Della's monthly benefit be?
11. George Smeller retires at age 68. Social security pays a bonus of 0.25% for each month that you work between 65 and 70. George's basic monthly benefit is calculated at $715. What will his monthly benefit be?

Refer to the Retirement Plans table on p. 426 for questions 12 and 13.

12. Michael Chan is a freelance writer. One year he made $48,000. He contributed 12% to his Keogh and made the maximum contribution allowed to his IRA. How much did he contribute to his retirement plans?
13. Pat Ireton worked 33 years as a stockbroker. His pension is based on a rate of 1.8% and the average of his highest 3 years. His highest 3 years' salaries were $38,900, $41,588, and $45,734. Find his annual pension.

CHAPTER 18 CUMULATIVE REVIEW

1. What is the cost of 150 shares of IBM at $160\frac{1}{4}$?
   A. $24,027.50
   B. $23,027.50
   C. $24,037.50
   D. $23,037.50

2. Add.

   $$\begin{array}{r} 87.024 \\ 43.228 \\ +21.59 \\ \hline \end{array}$$

   A. 142.832
   B. 151.842
   C. 152.842
   D. 141.832

3. Write 0.7512 as a percent.
   A. 75.12%
   B. 7.512%
   C. 0.07512%
   D. 0.7512%

4. The probability that all 12 people invited will come to a party is about $\frac{3}{4}$. How many people can be expected?
   A. about 6
   B. about 8
   C. about 9
   D. about 10

5. Subtract.

   $8\frac{1}{5} - 3\frac{5}{6}$

   A. $5\frac{2}{3}$
   B. $4\frac{11}{30}$
   C. $4\frac{2}{15}$
   D. $4\frac{13}{15}$

6. Find the total contribution to retirement plans.
   total earnings $38,600, 11% to Keogh, $2000 to IRA
   A. $4246
   B. $4646
   C. $6446
   D. $6246

7. Find the property tax on $89,720 at $5 per $100 of the assessed value.
   A. $4,486
   B. $1,794,40
   C. $4,055
   D. $1,684.40

8. $1500 was invested in a mutual fund with a 2% loading charge. How much is the actual balance?
   A. $1,300
   B. $1,370
   C. $1,200
   D. $1,470

9. Compare the health plan benefits. Plan A pays 80% of doctor's visit. Plan B pays up to $60 for a doctor's visit. The doctor charged $75.
   A. Plan A, $20 more
   B. Plan A, $20 less
   C. They are equal.
   D. Plan B, $15 less

10. Find the gross pay.
    regular hours 37, regular rate $10.20, overtime hours 3, overtime rate $11.50
    A. $408
    B. $411.90
    C. $460
    D. $410.80

11. Find the electricity cost.
    present meter reading 12227, previous meter reading 11979, rate $0.0675 per kWh
    A. $15.73
    B. $14.64
    C. $16.74
    D. $16.64

12. Cheryl has a monthly income of $1150. She budgeted $97.75 for personal items. What percent of her budget is this?
    A. 8.5%
    B. 8.0%
    C. 7.5%
    D. 7.0%

13. Mattie bought an antique vase for $125. She later sold it for $119. What was the percent of decrease in value?
A. 1.5%
B. 2.1%
C. 5.0%
D. 4.8%

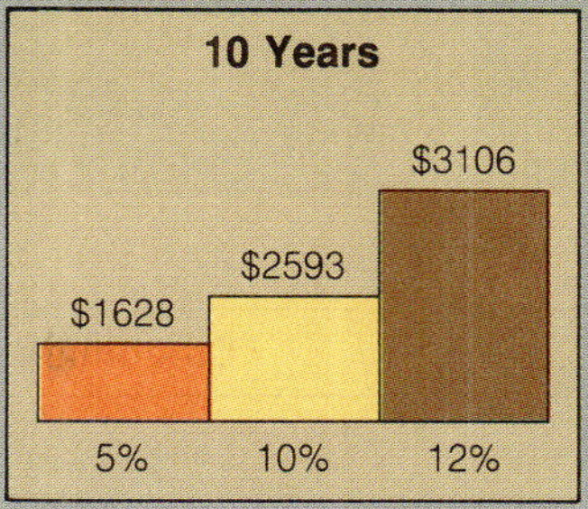

14. How much more will $1000 earn at 12% than at 10%?
A. $513
B. $1478
C. $965
D. $493

15. It is 404 miles from Cairo to Freeport. If Sue's car aveıages 21 mpg how much gas will be used on the trip?
A. 18.2 gal
B. 20 gal
C. 19.2 gal
D. 19.3 gal

16. Name the food group for the following foods: butter, salad oil, margarine.
A. dairy
B. meat
C. vegetable
D. fats

17. There are about 78,600 registered voters. A poll of 50 voters showed 28 voters preferred Candidate B. Predict the number of votes for Candidate B.
A. 35,108
B. 44,016
C. 89,604
D. 105,324

18. What rate of interest does OhEd pay?
A. 8.8%
B. $9\frac{1}{2}\%$
C. 11%
D. 6%

Bonds	Cur Yld	Vol	High	Low	Close	Net Chg.
OhEd	$9\frac{1}{2}06$	11.	88	88	88	...

19. Julius called Red Bluff for a total of 20 minutes. The rate for the call is $0.70/$0.80. What was the cost of the call?
A. $5.60
B. $6.02
C. $14
D. $5.32

20. Find the annual pension. Annual Pension = Pension Rate × High Salary Average × Years of Service
25 years, high salary average = $35,000, pension rate 1.4%
A. $10,750
B. $10,850
C. $11,970
D. $12,250

21. Compare the cost of 15 gallons of unleaded gas: $1.05 per gallon self-service or $1.57 per gallon full-service.
A. save $11.78
B. save $15.75
C. save $7.80
D. save $23.55

DATA BANK

Measurement Tables

Metric Measures

Length

1 kilometer (km)	=	1000 meters (m)
1 meter (m)	=	100 centimeters (cm)
1 meter	=	10 decimeters (dm)
1 meter	=	1000 millimeters (mm)
1 decimeter	=	10 centimeters (cm)
1 centimeter	=	10 millimeters (mm)

Area

100 square millimeters (mm^2)	=	1 square centimeter (cm^2)
10,000 square centimeters	=	1 square meter (m^2)
100 square meters	=	1 acre (a)
10,000 square meters	=	1 hectare (ha)

Volume

1000 cubic millimeters (mm^3)	=	1 cubic centimeter (cm^3)
1000 cubic centimeters	=	1 cubic decimeter (dm^3)
1,000,000 cubic centimeters	=	1 cubic meter (m^3)

Capacity

1 liter (L)	=	1000 milliliters (mL)

Mass

1 metric ton (t)	=	1000 kilograms (kg)
1 kilogram (kg)	=	1000 grams (g)
1 gram (g)	=	1000 milligrams (mg)

Temperature

0°C	=	freezing point of water
37°C	=	normal body temperature
100°C	=	boiling point of water

Customary Measures

Length

12 inches (in.)	=	1 foot (ft)
3 feet	=	1 yard (yd)
36 inches	=	1 yard
5280 feet	=	1 mile
1760 yards	=	1 mile
6076 feet	=	1 nautical mile

Area

144 square inches (sq in.)	=	1 square foot (sq ft)
9 square feet	=	1 square yard (sq yd)
4840 square	=	1 acre (A)

Volume

1728 cubic inches (cu in.)	=	1 cubic foot (cu ft)
27 cubic feet	=	1 cubic yard (cu yd)

Capacity

8 fluid ounces (fl oz)	=	1 cup (c)
2 cups	=	1 pint (pt)
2 pints	=	1 quart (qt)
4 quarts	=	1 gallon (gal)

Weight

16 ounces (oz)	=	1 pound (lb)
2000 pounds	=	1 ton (T)

Temperature

32°F	=	freezing point of water
98.6°F	=	normal body temperature
212°F	=	boiling point of water

Geometric Formulas

Perimeter

Rectangle	$P = 2l + 2w$, or $2(l + w)$

Area

Rectangle	$A = lw$
Square	$A = s^2$
Parallelogram	$A = bh$
Triangle	$A = \frac{1}{2}bh$
Trapezoid	$A = \frac{1}{2}h(a + b)$
Circle	$A = \pi r^2$

Circumference

Circle	$C = \pi d$ or $C = 2\pi r$

Volume

Rectangular prism	$V = lwh$
Cube	$V = s^3$
Cylinder	$V = \pi r^2 h$
Pyramid	$V = \frac{1}{3}lwh$
Cone	$V = \frac{1}{3}\pi r^2 h$
Sphere	$V = \frac{4}{3}\pi r^3$

Common Fractions and Their Decimal and Percent Equivalents

Fraction	$\frac{1}{1000}$	$\frac{1}{100}$	$\frac{1}{50}$	$\frac{1}{25}$	$\frac{1}{20}$	$\frac{1}{10}$	$\frac{1}{8}$	$\frac{1}{6}$	$\frac{1}{5}$	$\frac{1}{4}$	$\frac{1}{3}$
Decimal	0.001	0.01	0.02	0.04	0.05	0.1	0.125	0.1666	0.2	0.25	0.333
Percent	0.1%	1%	2%	4%	5%	10%	12.5%	16.6%	20%	25%	33.3%

Fraction	$\frac{3}{8}$	$\frac{2}{5}$	$\frac{1}{2}$	$\frac{3}{5}$	$\frac{5}{8}$	$\frac{2}{3}$	$\frac{3}{4}$	$\frac{4}{5}$	$\frac{5}{6}$	$\frac{7}{8}$	1
Decimal	0.375	0.4	0.5	0.6	0.625	0.666	0.75	0.8	0.833	0.875	1
Percent	37.5%	40%	50%	60%	62.5%	66.6%	75%	80%	83.3%	87.5%	100%

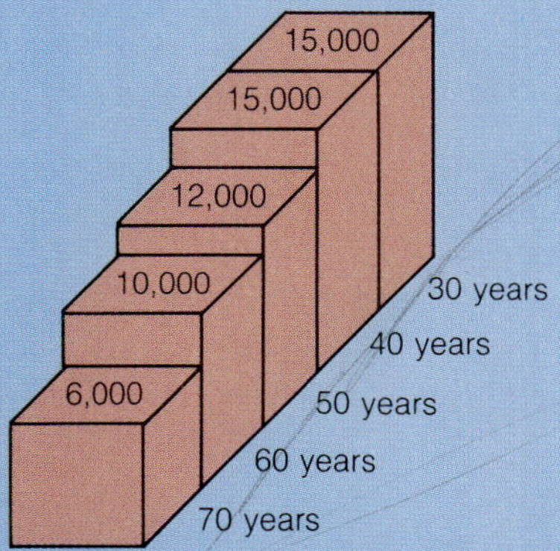

Hearing
(highest frequency in cycles per second)

Recommended Daily Dietary Allowances

	Age (yr)	Weight (kg)	Weight (lb)	Height (cm)	Height (in.)	Energy (cal)	Protein (g)
Infants	0.0–0.5	6	13	60	24	kg × 115	kg × 2.2
	0.5–1.0	9	20	71	28	kg × 105	kg × 2.0
Children	1–3	13	29	90	35	1300	23
	4–6	20	44	112	44	1700	30
	7–10	28	62	132	52	2400	34
Boys	11–14	45	99	157	62	2700	45
	15–18	66	145	176	69	2800	56
Men	19–22	70	154	177	70	2900	56
	23–50	70	154	178	70	2700	56
	51+	70	154	178	70	2400	56
Girls	11–14	46	101	157	62	2200	46
	15–18	55	120	163	64	2100	46
Women	19–22	55	120	163	64	2100	44
	23–50	55	120	163	64	2000	44
	51+	55	120	163	64	1800	44

Calorie Count Table

Kinds of Food	Amount	Food Energy (Calories)
Beverages, Milk, Fats		
Butter or margarine ($\frac{1}{4}$ lb)	$\frac{1}{2}$ cup	815
Butter or margarine	1 pat	25
Buttermilk	1 cup	100
Club soda (unsweetened)	12 oz	0
Cream, half-and-half	1 tbsp	20
Cream, sour	1 tbsp	25
Fats, vegetable	1 tbsp	110
Ginger ale	12 oz	115
Lard	1 tbsp	115
Malted milk	1 cup	235
Milk, skim	1 cup	85
Milk, whole	1 cup	150
Oils, salad or cooking	1 tbsp	120
Soda, fruit-flavored	12 oz	170
Breads, Cereals, Desserts, Pastas, Snacks		
Bagel (egg)	1	165
Biscuit	1	105
Bread:		
Raisin	1 slice	65
Rye	1 slice	60
White	1 slice	70
Whole-wheat	1 slice	60
Breakfast cereals:		
Bran flakes	1 cup	105
Corn flakes	1 cup	95
Oatmeal	1 cup	130
Puffed rice	1 cup	60
Puffed wheat	1 cup	55
Shredded wheat	1 piece	90
Corn muffin	1	125
Crackers, graham	4	110
Crackers, saltine	4	50
Fig bar	1	50
Gelatin dessert	1 cup	140
Honey	1 tbsp	65
Jams and preserves	1 tbsp	55
Macaroni, plain	1 cup	190
Macaroni w/cheese	1 cup	430
Muffin	1	120
Noodles	1 cup	200
Pancake (4″ diam.)	1	50
Popsicle	1	70
Pretzel, thin twisted	1	25
Roll, frankfurter	1	120
Sherbet	1 cup	260
Spaghetti, plain	1 cup	155
Spaghetti, tomato & cheese	1 cup	260
Sugar, white granular	1 tbsp	45
Syrup	1 tbsp	60
Waffle (7″ diam.)	1	205
Whipped cream	1 tbsp	10
Fruits and Fruit Juices		
Apple, raw ($2\frac{3}{4}$″ diam.)	1	80
Apple juice	1 cup	120
Applesauce, sweetened	1 cup	230
Apricots, raw	3	55
Apricots, canned in syrup	1 cup	220
Apricots, dried, uncooked	1 cup	340
Avocado, raw	1	370
Banana, raw	1	100
Blackberries, raw	1 cup	85
Blueberries, raw	1 cup	90
Cantaloupe, raw	$\frac{1}{2}$	80
Cherries, canned	1 cup	105
Cranberry juice, canned	1 cup	165
Cranberry sauce, sweet	1 cup	405
Dates, pitted	1 cup	490
Fruit cocktail, canned	1 cup	195
Grapefruit, raw	$\frac{1}{2}$	50
Grapefruit, canned	1 cup	180
Grapefruit juice	1 cup	95
Grape juice, bottled	1 cup	165
Grapes, raw	1 cup	70
Lemon, raw	1	20
Lemonade	1 cup	105
Limeade	1 cup	100
Orange, raw	1	65
Orange juice	1 cup	110
Peach, raw	1	40
Peaches, canned in syrup	1 cup	200
Pear, raw	1	100
Pears, canned in syrup	1 cup	195
Pineapple, raw, diced	1 cup	80
Pineapple, canned in syrup	1 cup	190
Pineapple juice	1 cup	140
Plum, raw	1	30
Prunes, cooked	1 cup	255
Prune juice	1 cup	195
Raisins ($\frac{1}{2}$-oz package)	1 pkg	40
Raspberries, red, raw	1 cup	70
Rhubarb, cooked with sugar	1 cup	380
Strawberries, raw	1 cup	55
Watermelon, wedge 4″ × 8″	1	110

Calorie Count Table

Kinds of Food	Amount	Food Energy (Calories)
Meat, Cheese, Eggs, Poultry, Fish		
Bacon (slices)	2	85
Beef, dried or chipped	$2\frac{1}{2}$ oz	145
Beef, hamburger	3 oz	235
Beef, roast (lean only)	3 oz	210
Beef, steak, broiled	3 oz	220
Beef potpie $\left(4\frac{1}{4}''\text{ diam.}\right)$	1	515
Bologna (slices)	2	170
Cheese:		
American	1 oz	105
American spread	1 oz	82
Blue or Roquefort	1 oz	105
Cheddar	1 oz	115
Cottage	1 cup	235
Cream	1 oz	100
Parmesan	1 oz	130
Yogurt	1 cup	140
Chicken, broiled	3 oz	115
Chicken, fried	3 oz	215
Chicken potpie (9″ diam.)	$\frac{1}{3}$	545
Chili con carne w/beans	1 cup	340
Corned beef (canned)	3 oz	185
Egg, scrambled in fat	1	95
Egg, whole (boiled)	1	80
Fish and seafood:		
Clams, raw	3 oz	65
Fish sticks	3 oz	150
Haddock, fried	3 oz	140
Lobster, canned	3 oz	80
Oysters, raw	1 cup	160
Salmon, canned	3 oz	120
Sardines, canned	3 oz	175
Tuna, canned	3 oz	170
Ham, baked	3 oz	185
Hot dog (frankfurter)	2 oz	170
Lamb chop (broiled)	3.1 oz	360
Lamb, leg (roasted)	3 oz	235
Liver, beef (fried)	3 oz	195
Luncheon ham (canned)	slice	175
Pork, chop	2.7 oz	305
Pork, roast	3 oz	310
Pork, sausage links	2	120
Salami	3 oz	450
Veal, cutlet	3 oz	185
Veal, roast	3 oz	230

Kinds of Food	Amount	Food Energy (Calories)
Vegetables and Nuts		
Almonds, shelled, chopped	1 cup	775
Asparagus	1 cup	30
Beans, green	1 cup	30
Beans, lima	1 cup	260
Beans, navy	1 cup	210
Beans and pork (canned)	1 cup	310
Beets	1 cup	55
Broccoli	1 cup	40
Brussels sprouts	1 cup	55
Cabbage, cooked	1 cup	30
Cabbage, raw	1 cup	15
Carrot, raw (7″ long)	1	30
Carrots, cooked	1 cup	50
Cashew nuts	1 cup	785
Cauliflower, cooked	1 cup	31
Celery, diced, raw	1 cup	20
Coconut, grated	1 cup	275
Collards	1 cup	65
Corn, canned (whole kernel)	1 cup	175
Corn, sweet	1 ear	70
Cucumber, peeled	1	20
Lettuce, iceberg (head)	1	70
Mushrooms, raw	1 cup	20
Onion, cooked	1 cup	60
Onion, raw $\left(2\frac{1}{2}''\text{ diam.}\right)$	1	40
Parsnips	1 cup	100
Peanut butter	1 tbsp	95
Peanuts, roasted	1 cup	840
Peas, green, canned	1 cup	150
Pecan halves	1 cup	810
Pepper, green, raw	1	15
Pickle, sweet	1	20
Potato, baked	1 large	145
Potato, boiled	1 med.	105
Potato, french-fry pieces	10	135
Potato, mashed	1 cup	135
Radishes, raw	10	10
Rice, cooked	1 cup	225
Sauerkraut, canned	1 cup	40
Spinach	1 cup	40
Sweet potato, baked	1	160
Tomato, raw (3″ diam.)	1	40
Tomato catsup	1 tbsp	15
Tomato juice	1 cup	45
Tomatoes, canned	1 cup	50
Turnip greens	1 cup	30
Turnips, diced	1 cup	35
Walnuts, black	1 cup	785

Calorie Count Table

Kinds of Food	Amount	Food Energy (Calories)	Kinds of Food	Amount	Food Energy (Calories)
Salad Dressings, Sauces, and Soups			Salad dressings:		
Barbecue sauce	1 cup	230	Blue cheese	1 tbsp	75
Bean with pork soup	1 cup	170	French	1 tbsp	65
Beef broth	1 cup	30	Mayonnaise	1 tbsp	100
Bouillon cube	1	5	Thousand Island	1 tbsp	80
Cream of chicken soup	1 cup	180	Tartar sauce	1 tbsp	75
Cream of mushroom soup	1 cup	215	Tomato soup	1 cup	90
Cream of tomato soup	1 cup	175	Vegetarian soup	1 cup	80
			White sauce	1 cup	405

Sewing Pattern

Girls' Pull-on Pants or Shorts and Shirt Loose-fitting top-stitched pants V.1 and shorts V.2 have triple elastic waistline casing, and side seam pockets. Top-stitched buttoned front shirt has back inverted pleat, notched collar, buttoned patch pockets, and short sleeves with notch detail.

Fabrics Cotton types: broadcloth, chambray, chintz, oxford cloth, batik. Pants and shorts also in lightweight denim, poplin. Extra fabric needed to match plaids and stripes. One-way design fabrics: extra fabric may be needed. Not suitable for obvious diagonal fabrics.

Body Measurements							
Chest		26	27	$28\frac{1}{2}$	30	32	Inches
Waist		23	$23\frac{1}{2}$	$24\frac{1}{2}$	$25\frac{1}{2}$	$26\frac{1}{2}$	Inches
Hip		27	28	30	32	34	Inches
Back-neck to waist		$11\frac{1}{2}$	12	$12\frac{3}{4}$	$13\frac{1}{2}$	$14\frac{1}{2}$	Inches
Sizes		7	8	10	12	14	
View 1 Pants	44"/45"***	$1\frac{3}{4}$	$1\frac{3}{4}$	2	$2\frac{1}{8}$	$2\frac{1}{4}$	Yards
	58"/60"*	$1\frac{1}{8}$	$1\frac{1}{4}$	$1\frac{1}{4}$	$1\frac{1}{2}$	$1\frac{3}{4}$	Yards
View 2 Shorts	44"/45"*	1	1	$1\frac{1}{8}$	$1\frac{1}{4}$	$1\frac{3}{8}$	Yards
	58"/60"*	$\frac{3}{8}$	$\frac{7}{8}$	$\frac{7}{8}$	$\frac{7}{8}$	$\frac{7}{8}$	Yard
View 2 Shirt	Even lengthwise striped or plain fabric						
	35"/36"*	$1\frac{1}{2}$	$1\frac{1}{2}$	$1\frac{5}{8}$	$1\frac{5}{8}$	$1\frac{3}{4}$	Yards
	44"/45"*	$1\frac{1}{4}$	$1\frac{1}{4}$	$1\frac{1}{4}$	$1\frac{3}{8}$	$1\frac{3}{8}$	Yards
	Interfacing: $\frac{1}{2}$ yd of 2" to 36" woven or non-woven						
Pants side length V.1		$26\frac{1}{4}$	$27\frac{3}{4}$	31	$34\frac{1}{4}$	$35\frac{1}{2}$	Inches
Pants leg width V.1		$17\frac{1}{4}$	$17\frac{1}{2}$	18	$18\frac{1}{2}$	19	Inches
Shorts side length V.2		$12\frac{3}{4}$	$13\frac{1}{4}$	14	$14\frac{3}{4}$	$15\frac{1}{2}$	Inches

Notions: Thread. Pants or shorts: Two pkg. $\frac{3}{8}$" wide elastic. Shirt: $\frac{3}{8}$" buttons.

*without nap **with nap ***with or without nap

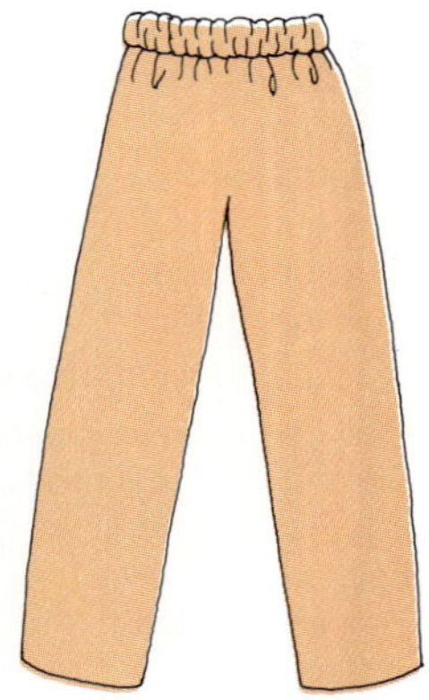

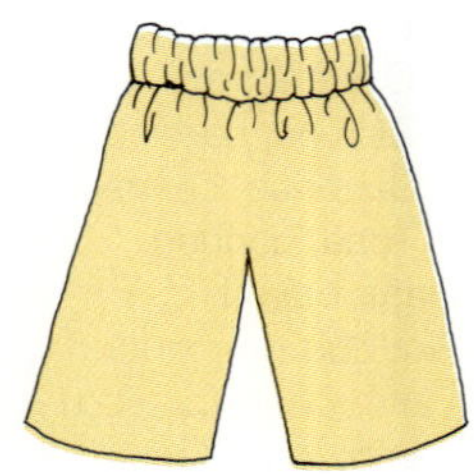

Formulas

1. $P = IE$ (Watts of power = Amps × Volts)
2. Sound travels at the rate of 1460 meters per second.
   Distance = 1460 meters/second × time (in seconds)
3. Vital capacity (women)
   $V = 0.041h - 0.018a - 2.69$
   V = vital capacity in liters
   h = height in cm
   a = age in years
4. Maximum pulse rate
   $P = 0.8 \times (220 - a)$
   P = maximum pulse rate
   a = age in years
5. Number of liquid gallons per cu in. $= \frac{\pi r^2 h}{231} = \frac{\text{volume in cu in.}}{231}$

Resumé

You submit your resumé to prospective employers or include it with your job application. It lists your education and work experience. A good resumé is concise, correct, and neatly typed.

Resume of Susan Juarez
1834 W. Angeles
Dallas, Texas 75216
(214) 555-6543

Objective — To get a job as a park recreation leader

Education
1984–1988 — Kennedy High School

Experience
1986–present — YWCA Water Safety Instructor
5700 Moraga Blvd.
Dallas, Texas 75243

1984–1987 — Math tutor, Wilson Junior High School
8484 N. Hunter St.
Dallas, Texas 75201

Activities
1987–1988 — Captain, pitcher for varsity softball team
1985–1988 — Spanish Club
Honor Society

References

Mrs. Alison Roberti
Principal, Kennedy High School
3860 Rossmond St.
Dallas, Texas 75201

Ms.Valerie Grant
Director, YMCA
5700 Moraga Blvd.
Dallas, Texas 75243

Mr. Peter Richmond
Vice Principal, Wilson Junior High School
8484 N. Hunter St.
Dallas, Texas 75201

Social Security Number Application

HOW TO APPLY FOR AN ORIGINAL OR REPLACEMENT SOCIAL SECURITY NUMBER CARD

WHERE TO APPLY Mail or take your evidence documents and application to the nearest Social Security office. However, you must apply in person at a Social Security office if:

1. You are age 18 or older and have never had a Social Security number card before, or
2. You are an alien whose immigration documents should not be mailed.

EVIDENCE YOU WILL NEED Social Security law requires that you furnish evidence of your age, identity, and U.S. citizenship or lawful alien status. Find one of the four categories below which applies to you and read the instructions carefully.

1. U.S. Citizen — Born in the U.S. — Applying for original Social Security number

Please submit one document from list A. **and** at least one from list B.

List A. Evidence of Age and Citizenship

One of the following records established before your 5th birthday:

- Public birth certificate (**This is the preferred document.** You should submit it if at all possible.)
- Religious record showing age or date of birth (To establish citizenship, it must have been recorded within 3 months of birth.)
- Hospital record of birth

If your birth was never recorded or the public record no longer exists, submit documents from list B for evidence of age and citizenship. At least one document must show your name, age or date of birth, and your place of birth. It should be at least one year old.

List B. Evidence of Identity

A public or hospital birth record is not considered evidence of identity. An identity document should contain enough information to identify you. Some items we look for on the document are name, age, address, signature, physical description and picture. Examples are:

- Identity card
- Driver's license
- Adoption record
- Insurance policy
- Youth organization record
- School record
- Medical record
- Work badge or building pass
- Military record
- Court order for name change
- Marriage or Divorce record
- U.S. Passport
- U.S. Citizen ID card
- Religious record (If not submitted as evidence of age)
- Any other document providing identifying data sufficient to establish proper identity.

2. U.S. Citizen — Born outside the U.S. — Applying for Original Social Security Number

If you were born outside the U.S. but are a U.S. citizen you should submit either:

a. U.S. consular report of birth (FS-240 or FS-545) **and** evidence of your identity (see category 1, list B); OR
b. Your foreign birth certificate (if not readily available, another document showing age or date of birth) **and** one of the following: • U.S. citizen ID card • Citizenship certificate • U.S. passport • Certificate of Naturalization

3. Alien — Not a U.S. Citizen Applying for Original Social Security Number

If you are an alien living in or visiting the U.S., you should bring in your foreign birth certificate (if available) or passport **and** one or more of the following:

- Alien Registration Receipt Card I-151 or I-551.
- U.S. Immigration Form I-94 (and I-201D, if applicable), I-181a, I-185, I-186, I-210, or I-586 (and I-444, if applicable)

4. Any Applicant Requesting a Replacement Card (Duplicate or Correction)

You must present at least evidence of your identity such as one or more of the documents in categories 1 (List B), 2 or 3. In addition, if you are foreign-born (or U.S.-born but no longer a U.S. citizen), you must submit evidence of your current U.S. citizenship or lawful alien status. After we examine your documents, you may be asked for additional evidence of your age, or U.S. citizenship or lawful alien status.

If you are changing your name, you must provide evidence identifying you under both your **old** and **new** names. We will accept one document if it shows both the old and new names or multiple documents if each shows only one name. If the name change is due to marriage or divorce, your marriage certificate or divorce decree is usually sufficient if it shows both your previous and new names.

DEPARTMENT OF HEALTH AND HUMAN SERVICES
SOCIAL SECURITY ADMINISTRATION

Form Approved
OMB No. 0960-006

FORM SS-5 — APPLICATION FOR A SOCIAL SECURITY NUMBER CARD (Original, Replacement or Correction)

Unless the requested information is provided, we may not be able to issue a Social Security Number (20 CFR 422-103(b))

INSTRUCTIONS TO APPLICANT ▶ Before completing this form, please read the instructions on the opposite page. Type or print, using pen with dark blue or black ink. Do not use pencil. SEE PAGE 1 FOR REQUIRED EVIDENCE.

NAA	NAME TO BE SHOWN ON CARD	First	Middle	Last
NAB 1	FULL NAME AT BIRTH (IF OTHER THAN ABOVE)	First	Middle	Last
ONA	OTHER NAME(S) USED			
STT 2	MAILING ADDRESS	(Street/Apt. No., P.O. Box, Rural Route No.)		
CTY	CITY (Do not abbreviate)	STE STATE	ZIP ZIP CODE	

CSP 3 CITIZENSHIP (Check one only)
- ☐ a. U.S. citizen
- ☐ b. Legal alien allowed to work
- ☐ c. Legal alien not allowed to work
- ☐ d. Other (See instructions on Page 2)

SEX 4 SEX
- ☐ MALE
- ☐ FEMALE

ETB 5 RACE/ETHNIC DESCRIPTION (Check one only) (Voluntary)
- ☐ a. Asian, Asian-American or Pacific Islander (Includes persons of Chinese, Filipino, Japanese, Korean, Samoan, etc., ancestry or descent)
- ☐ b. Hispanic (Includes persons of Chicano, Cuban, Mexican or Mexican-American, Puerto Rican, South or Central American, or other Spanish ancestry or descent)
- ☐ c. Negro or Black (not Hispanic)
- ☐ d. Northern American Indian or Alaskan Native
- ☐ e. White (not Hispanic)

DOB 6	DATE OF BIRTH ▶	MONTH	DAY	YEAR
AGE 7	PRESENT AGE			
PLB 8	PLACE OF BIRTH ▶	CITY (Do not abbreviate)	STATE OR FOREIGN COUNTRY (Do not abbreviate)	FCI ☐
MNA 9	MOTHER'S NAME AT HER BIRTH	First	Middle	Last (Her maiden name)
	FATHER'S NAME	First	Middle	Last

PNO 10 a. Has a Social Security number card ever been requested for the person listed in item 1? ☐ YES(2) ☐ NO(1) ☐ Don't know(1)

b. Was a card received for the person listed in item 1? ☐ YES(3) ☐ NO(1) ☐ Don't know(1)

▶ IF YOU CHECKED YES TO A OR B, COMPLETE ITEMS C THROUGH E; OTHERWISE GO TO ITEM 11.

SSN c. Enter the Social Security number assigned to the person listed in item 1. ☐☐☐–☐☐–☐☐☐☐

NLC d. Enter the name shown on the most recent Social Security card issued for the person listed in item 1.

PDB e. Date of birth correction (See instruction 10 on page 2) ▶ MONTH | DAY | YEAR

DON 11 TODAY'S DATE ▶ MONTH | DAY | YEAR

12 Telephone number where we can reach you during the day. Please include the area code. ▶ HOME | OTHER

ASD **WARNING: Deliberately furnishing (or causing to be furnished) false information on this application is a crime punishable by fine or imprisonment, or both.**

IMPORTANT REMINDER: WE CANNOT PROCESS THIS APPLICATION WITHOUT THE REQUIRED EVIDENCE.

13 YOUR SIGNATURE

14 YOUR RELATIONSHIP TO PERSON IN ITEM 1 ☐ Self ☐ Other (Specify) ________

WITNESS (Needed only if signed by mark "X")

WITNESS (Needed only if signed by mark "X")

DO NOT WRITE BELOW THIS LINE (FOR SSA USE ONLY)

DTC (SSA RECEIPT DATE)	NPN		DOC
NTC	CAN	BIC	IDN
ITV	☐ MANDATORY IN PERSON INTERVIEW CONDUCTED		
TYPE(S) OF EVIDENCE SUBMITTED	SIGNATURE AND TITLE OF EMPLOYEE(S) REVIEWING EVIDENCE AND/OR CONDUCTING INTERVIEW		
	DATE		
	DCL	DATE	

Form SS-5 (11-86) 3
5/84, 1/85 and 8/85 editions may be used until supply is exhausted.

Employment Application Form

Name ______________________________ Social Sec. No. ______________
LAST FIRST MIDDLE

Address ______________________________ Telephone No. ______________
NUMBER AND STREET CITY STATE ZIP

Education

High School ______________ Location ______________

Year Graduated ________ Course of Study ______________

College 1. ______________ Location ______________

No. of Years _____ Year Graduated _____ Degree _____ Major _____ Minor _____ G.P.A. _____

2. ______________ Location ______________

No. of Years _____ Year Graduated _____ Degree _____ Major _____ Minor _____ G.P.A. _____

Other Degrees:
College ______________ Degree ________ Area of Specialization ______________

Employment Record

Name of Employer ______________ Kind of Business ______________

Position ________ Employment Dates: From _____ To _____ Salary at Termination ________

Description of Work Performed ______________

Reason for Leaving ______________

Name of Employer ______________ Kind of Business ______________

Position ________ Employment Dates: From _____ To _____ Salary at Termination ________

Description of Work Performed ______________

Reason for Leaving ______________

I certify that the above information is correct to the best of my knowledge and belief, and I am aware that proof of falsification of this application on my part may be cause for dismissal.

Signed ______________

Aptitude Test

A. Verbal

Find the word with the same meaning.

1. Boring:
- **a.** devoted
- **b.** dull
- **c.** current
- **d.** interesting

2. Blissful:
- **a.** dreaded
- **b.** regular
- **c.** ecstatic
- **d.** poor

Find the word with the opposite meaning.

3. Curt:
- **a.** gesture
- **b.** cruel
- **c.** happy
- **d.** friendly

4. Spry:
- **a.** disabled
- **b.** promotion
- **c.** commit
- **d.** athletic

Find the misspelled word.

5.
- **a.** jovial
- **b.** unabel
- **c.** occurrence
- **d.** computer

6.
- **a.** consaquently
- **b.** collective
- **c.** equation
- **d.** language

B. Numerical

1. Multiply 62 × 15.
- **a.** 77
- **b.** 930
- **c.** 1020
- **d.** None of the above

2. Add 76 + 27.
- **a.** 115
- **b.** 93
- **c.** 103
- **d.** None of the above

3. Subtract 186 − 32.
- **a.** 53
- **b.** 218
- **c.** 51
- **d.** None of the above

4. A customer buys a pair of shoelaces for $1.32 and a bottle of shampoo for $3.89. How much change should be given from a $20 bill?
- **a.** $14.89
- **b.** $14.79
- **c.** $13.67
- **d.** None of the above

5. If Anna and Michael spend $15.79 on lunch, find the 15% tip that they should leave for the waitress.
- **a.** $2.34
- **b.** $3.24
- **c.** $1.87
- **d.** None of the above

C. Clerical Perception

1. Which names are the same?
- **a.** Boothe Inc. – Boothe Co.
- **b.** Jennifer Strup – Jennifer Strap
- **c.** Adams, R. A. – Abrams, R. A.
- **d.** Farra & Co. – Farra & Co.

2. Which numbers are the same?
- **a.** 0.7682 – 0.7683
- **b.** 68.32765 – 68.32765
- **c.** 3.76112 – 3.67112
- **d.** 117.87913 – 117.87813

3. Which set is incorrectly alphabetized?
- **a.** slither, roll, ventricle
- **b.** mandatory, optical, produce
- **c.** banana, crisp, effort
- **d.** fright, gorilla, helicopter

D. Spatial

What can be made from the object on the left?

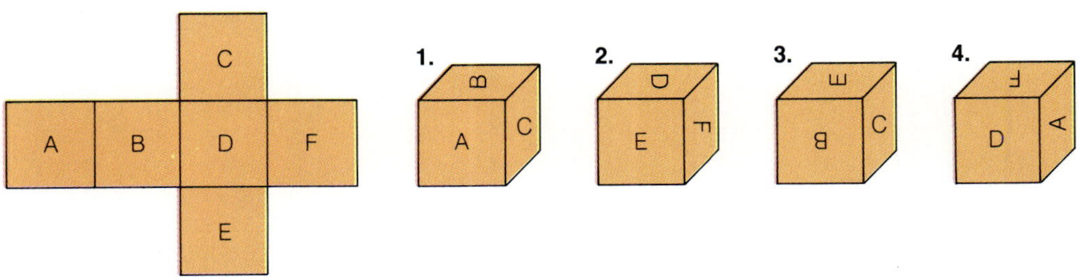

E. Form Perception

Which of the figures on the right is exactly the same as the figure on the left?

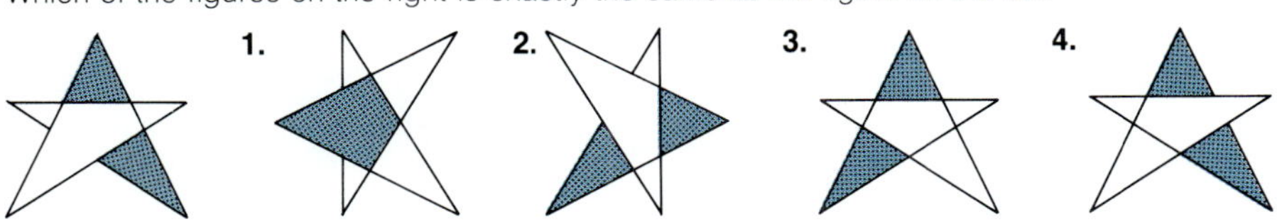

Aptitude Test

F. Motor Coordination

As quickly as possible, create the same pattern in the blank triangles on the right.

G. Finger Dexterity

Braid and knot the pieces of rope as quickly as possible.

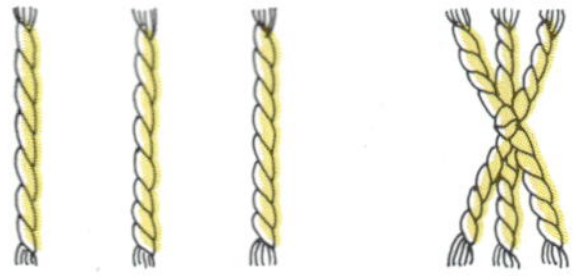

H. Manual Dexterity

Move the pegs to the lower part of this board as fast as you can.

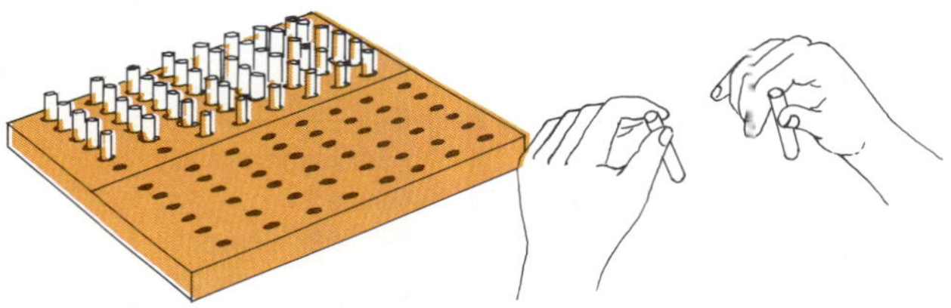

Do parts A–E. Think about parts F–H.
Which parts do you think you would do the best?
What kinds of jobs does this imply.

Compound Interest Table

Number of Periods	2%	2.5%	3%	3.5%	4%	4.5%	5%	6%	7%	8%	9%	10%
1	1.0200	1.0250	1.0300	1.0350	1.0400	1.0450	1.0500	1.0600	1.0700	1.0800	1.0900	1.1000
*												
*												
6	1.1262	1.1597	1.1941	1.2293	1.2653	1.3023	1.3401	1.4186	1.5007	1.5869	1.6771	1.7716
7	1.1487	1.1887	1.2299	1.2723	1.3159	1.3609	1.4071	1.5036	1.6058	1.7138	1.8280	1.9487
8	1.1717	1.2184	1.2668	1.3168	1.3686	1.4221	1.4775	1.5938	1.7182	1.8059	1.9926	2.1436
9	1.1951	1.2489	1.3048	1.3629	1.4233	1.4861	1.5513	1.6895	1.8385	1.9990	2.1719	2.3579
10	1.2190	1.2801	1.3439	1.4106	1.4802	1.5530	1.6289	1.7908	1.9672	2.1589	2.3674	2.5937
11	1.2434	1.3121	1.3842	1.4600	1.5395	1.6229	1.7103	1.8983	2.1049	2.3316	2.5804	2.8531
12	1.2682	1.3449	1.4258	1.5111	1.6010	1.6959	1.7959	2.0122	2.2522	2.5182	2.8127	3.1384
13	1.2936	1.3785	1.4685	1.5640	1.6651	1.7722	1.8856	2.1329	2.4098	2.7196	3.0658	3.4523
14	1.3195	1.4130	1.5126	1.6187	1.7317	1.8519	1.9799	2.2609	2.5785	2.9372	3.3417	3.7975
15	1.3459	1.4483	1.5580	1.6753	1.8009	1.9353	2.0789	2.3966	2.7590	3.1722	3.6425	4.1772
16	1.3728	1.4845	1.6047	1.7340	1.8730	2.0224	2.1829	2.5404	2.9522	3.4259	3.9703	4.5950
17	1.4002	1.5216	1.6528	1.7947	1.9479	2.1134	2.2920	2.6928	3.1588	3.700	4.3276	5.0545
18	1.4282	1.5597	1.7024	1.8575	2.0258	2.2085	2.4066	2.8543	3.3799	3.9960	4.7171	5.5599
19	1.4568	1.5987	1.7535	1.9225	2.1068	2.3079	2.5270	3.0256	3.6165	4.3157	5.1417	6.1159
20	1.4895	1.6386	1.8061	1.9898	2.1911	2.4117	2.6533	3.2071	3.8697	4.6610	5.6044	6.7275

Shipping and Handling Charges

Zone	1	2	3	4	5	6	7
Destination	DC DE MD VA	NJ NY NC PA WV	CT, GA IL, IN KY, ME MA, MI NH, OH RI, SC TN, VT	AL FL, AR IA, LA MN, MS MO, WI	KS NB ND OK SD TX	CO NM UT WY	AK HI, MT AZ, CA ID, NV OR, WA VI, PR
Weight							
1–5 lb	2.75	2.90	3.00	3.25	3.50	3.75	4.25
6–10 lb	3.75	3.90	4.00	4.25	5.00	5.35	6.70
11–15 lb	4.40	4.60	4.80	5.00	6.25	7.00	8.40
16–20 lb	4.75	5.50	5.60	6.40	7.80	9.00	10.60
21–25 lb	5.25	6.20	6.40	7.40	9.40	10.90	12.90
26–30 lb	5.75	6.70	7.40	8.50	10.90	12.75	15.00
31–35 lb	6.25	7.50	8.60	9.90	12.50	14.50	17.30
36–40 lb	6.75	8.15	9.60	11.40	13.90	16.40	19.40
41–45 lb	7.25	8.80	10.60	12.50	15.20	18.20	21.50
46–50 lb	7.50	9.40	11.40	13.50	16.70	19.90	23.60
51–55 lb	7.75	9.80	11.80	13.75	17.40	20.90	24.70
56–60 lb	8.00	10.00	12.05	14.00	17.70	21.10	25.30
61–65 lb	8.25	10.20	12.25	14.70	18.10	21.50	25.80
66–70 lb	8.40	10.30	12.45	15.00	18.40	22.20	26.25

If any items in your order weigh over 70 lb, or have a combined height and girth measurement of 108 inches, these items must be shipped by truck. We will notify you in advance and will not ship without your permission.

All Truck Shipments	30.00	35.00	45.00	50.00	55.00	60.00	65.00

Credit Card Statement

New Balance	Minimum Payment Due	Closing Date	Payment Due Date	Credit Limit	Credit Available
400.64	12.00	06/01/89	06/25/89	1000	599.36

Date	Reference Number	Transaction	Amount
05/10	061399461	Open Parachutes, Inc.	189.06
05/13	062043845	Sky Dives	68.00
05/20	506432988	Payment	232.91 –
05/23	075409113	Army-Navy Surplus	35.49
05/25	063850912	Anthony Pharmacy	6.13

Previous Balance	Payments & Credits	Purchases & Charges	Cash Advances	Finance Charge	New Balance
332.91	232.91	298.68	.00	1.96	400.64

Some of the World's Longest Suspension Bridges

Bridge	Location	Year Completed	Span (feet)
Brooklyn	East River, New York City	1883	1595
Miampimi	Mexico	1900	1030
Bear Mountain	Hudson River, New York	1924	1632
Benjamin Franklin	Philadelphia	1926	1750
Ambassador	Detroit-Canada	1929	1850
Mid-Hudson	Poughkeepsie, New York	1930	1500
George Washington	New York City	1931	3500
San Francisco-Oakland Bay	San Francisco	1936	2310[1]
Golden Gate	San Francisco	1937	4200
Bronx-Whitestone	New York City	1939	2300
Lion's Gate	Burrard Inlet, Brit. Columbia	1939	1550
Tacoma Narrows II	Puget Sound, Washington	1950	2800
Wm. Preston Lane Memorial	Sandy Point, Maryland	1952	1600[2]
Mackinac	Mackinac Straits, Michigan	1957	3800
Walt Whitman	Philadelphia	1957	2000
Throgs Neck	Long Island Sound	1961	1800
Verrazano Narrows	New York City	1964	4260
Forth Road	Queensferry, Scotland	1964	3300
Vincent Thomas	Los Angeles Harbor	1964	1500
Amu-Darja	U.S.S.R.	1964	1278
Salazar	Lisbon, Portugal	1966	3323
Severn	Beachley, England	1966	3240
Angostura	Ciudad Bolivar, Venezuela	1967	2336
Second Delaware Memorial	Wilmington, Delaware	1968	2150
Pierre Laporte	Quebec City	1970	2190
Little Belt	Denmark	1970	1968
Bosphorus	Istanbul, Turkey	1973	3520
Kanmon Straits	Shimonosaki, Japan	1973	2350
Humber Estuary	nr. Hull, England	1980	4626
Akashi-Kaikyo	Honshu to Shikoku, Japan	1988[3]	5840

[1]Two spans, each 2310 feet

[2]A second parallel bridge was completed in 1973.

[3]Expected completion

Number of Kennedy Half Dollars Minted

Year	Half Dollars Minted
1967	295,046,978
1968	246,951,930
1969	129,881,880
1970	2,150,000

Food Nutrient Chart

Food	Measure	Calories	Carbohydrates (g)	Protein (g)	Fats (g)
Apple, raw	1 medium	70	18	tr	tr
Bacon, fried	2 slices	100	1	5	8
Beans, green, canned	1 c	45	10	2	tr
Beef, roast	3 oz	245	0	23	16
Bran, raisin	$\frac{2}{3}$ c	99	22	2	tr
Bread					
white	1 slice	60	12	2	1
whole wheat	1 slice	55	11	2	1
Brussels sprouts	1 c	45	8	5	1
Carrots, raw	1 c	45	11	1	tr
Cheese, processed	1 oz	105	1	7	9
Chicken, fried	$\frac{1}{2}$ breast	155	tr	25	5
Coffee, unsweetened	1 c	0	0	0	0
Corn flakes	1 oz	110	24	2	tr
Eggs, scrambled	1 medium	110	1	7	8
Jams and jellies	1 tbsp	55	14	tr	tr
Lamb chop	4.8 oz	400	0	25	33
Lettuce leaves	2 large	10	2	1	tr
Macaroni and cheese	1 c	470	44	18	24
Margarine	1 pat	50	tr	tr	6
Milk					
whole	1 c	160	12	9	9
nonfat, skim	1 c	90	12	9	tr
Orange					
navel	1 medium	60	16	2	tr
juice	1 c	110	27	2	tr
Pancakes	1, 4″ diameter	60	9	2	2
Peanut butter	1 tbsp	95	3	4	8
Pork					
chop	3.5 oz	260	0	16	21
boiled ham	2 oz	135	0	11	10
Potatoes					
baked	1 medium	90	21	3	tr
mashed	1 c	185	24	4	8
Raisins	1 c	460	124	4	tr
Rice, white	1 c	185	26	2	tr
Spaghetti	1 c	260	37	9	9
Sugar, granulated	1 tbsp	45	12	0	0
Syrup, table blends	1 tbsp	60	15	0	0
Tea, unsweetened	1 c	0	0	0	0
Watermelon	1, 4″ × 8″ wedge	115	27	2	1

tr = trace

U.S. Recommended Daily Dietary Allowances for Teenagers (15–18)

			Vitamins					
Protein	Calcium	Iron	A	B1	B2	Niacin	C	D
mg	mg	mg	IU	mg	mg	mg	mg	IU
85	1400	15	5000	1.4	2	22	80	400

IU = International Units

Yield Rates of Various Produce

Raw Produce	Measure	Weight	Amount Needed for One Quart
Apples	1 bushel	48 lb	$2\frac{1}{2}$ to 3 lb
Apricots	1 box	22 lb	2 to $2\frac{1}{2}$ lb
Beans, green	1 bushel	30 lb	$1\frac{1}{2}$ to $2\frac{1}{2}$ lb
Beets	1 bushel	52 lb	2 to $3\frac{1}{2}$ lb
Carrots	1 bushel	50 lb	2 to 3 lb
Cherries	1 lug	22 lb	2 to $2\frac{1}{2}$ lb
Corn	1 bushel	35 lb	3 to 6 lb
Peaches	1 lug	22 lb	2 to 3 lb
Pears	1 box	35 lb	2 to 3 lb
Peas	1 bushel	30 lb	3 to 6 lb
Plums	1 lug	24 lb	$1\frac{1}{2}$ to $2\frac{1}{2}$ lb
Tomatoes	1 lug	30 lb	$2\frac{1}{2}$ to $3\frac{1}{2}$ lb

World Production of Food Crops

Food Crop	Metric Tons (billions)
Cereal grains	1.70
Roots and tubers	0.56
Sugar cane and sugar beets	1.15
Legumes and oil seeds	0.11
Vegetables, fruits, berries, and nuts	0.68
Total	4.20

Agriculture: World Wheat, Rice, and Corn Production

Country	Wheat	Rice	Corn
World, total	**521,682**	**469,959**	**449,255**
Afghanistan	2850[3]	466[3]	800[3]
Argentina	13,000	476	9500
Australia	18,580	635	210
Austria	1501	NA	1542
Bangladesh	1200	21,500[3]	1[3]
Belgium[2,4]	1330	NA	39[3]
Brazil	1830	9023	21,174
Bulgaria	3600[3]	74[3]	3000[1]
Burma	191	14,500[3]	360
Canada	21,199	NA	7024
Chile	988	165	721
China	87,682[3]	181,028[3]	72,690[3]
Colombia	76	1696	874[1]
Cuba	NA	555	97[3]
Czechoslavakia	6170	NA	940
Denmark[4]	2446	NA	NA
Ecuador	24[1]	470[1]	300[1]
Egypt	1815	2230[1]	3600
Ethiopia	675[3]	NA	1275[3]
Finland	478	NA	NA
France[4]	32,884	42	10,321
German Dem. Rep.	4110[3]	NA	20[1]
Germany, Fed. Rep.[4]	10,223	NA	1026
Greece	2646	91	1992
Hungary	7300[1]	30[1]	6700[1]
India	45,148	91,000[3]	7750[3]
Indonesia	NA	NA	4000[1]
Iran	5500[3]	1230[3]	50[3]
Iraq	300[3]	95[3]	25[3]
Ireland[4]	660	NA	NA
Israel	130	NA	26
Italy[4]	10,005	1027	6781
Japan	741	14,848	3[1]
Kampuchea	NA	1300[3]	75
Korea, D. Peo. Rep.[1]	600[3]	5400[3]	2580[3]
Korea, Republic of	17	7970	133
Laos	NA	1322	40[3]
Madagascar	NA	1755	141
Malaysia	NA	2132	22
Mexico	4262	1755	14,050
Nepal	634	635	751
Netherlands[4]	1133	NA	2[3]
New Zealand	294	NA	172
Pakistan	11,053	5009	1100[3]
Panama	NA	175[1]	80[3]
Peru	100[1]	1134	576
Philippines	NA	8280[1]	3400[3]
Poland	6010	NA	57
Portugal	475	134	530[1]
Romania	7900[1]	75[1]	13,000[1]
South Africa	2150	3[3]	4400
Soviet Union	76,000[3]	2500[3]	13,000[3]
Spain	6044	437	2505
Sri Lanka	NA	2270[1]	40[3]
Sweden	1776	NA	NA
Switzerland	577	NA	143
Syria	1051	NA	40[3]
Thailand	NA	19,200[3]	4150
Turkey	17,235	280	1500
United Kingdom[4]	14,960	NA	1[3]
United States	70,638	6216	194,475
Uruguay	450[3]	340	120
Venezuela	NA	408	547
Vietnam	NA	15,416	475[3]
Yugoslavia	5596	36	11,265[1]

(1) FAO estimate.
(2) Includes Luxembourg.
(3) Unofficial figure.
(4) European Economic Community.
(NA) Not available.

Source: UN Food and Agriculture Organization (thousands of metric tons)

HMO Plans

World Health Plan

Annual deductible, $50/person
Maximum benefit, $300,000/person
Service available from physicians and hospitals worldwide.

Professional Services

Surgery	Full coverage
Doctor's visits	
Hospital	Full coverage
Office	80%
Home	80%
Hospital Services	
Room and board	Full coverage
Emergency room	100% of first $300, then 80%

Other Services

X-ray, lab tests	100% of first $300, then 80%
Prescriptions	80%
Routine physical	Not covered
Eye examination	Not covered

Capital Care Medical Plan

No deductibles
No maximums
Services available from local physicians and hospitals.

Professional Services

Surgery	Full coverage
Doctor's visits	
Hospital	Full coverage
Office	$4/visit
Home	$10/visit
Hospital Services	
Room and board	Full coverage
Emergency room	$10/visit

Other Services

X-ray, lab tests	Full coverage
Prescriptions	Full coverage after $2 copayment
Routine physical	Full coverage after $4 copayment
Eye examination	Full coverage after $4 copayment

Some Human Physiological Data

Measure	Women	Men
Weight of brain		
small	37.04 oz	38.8 oz
average	44.98 oz	49.38 oz
large	54.68 oz	60 oz
Weight of heart	8 oz	10 oz
Quantity of blood	0.875 gal	1.5 gal
Surface area of skin	1.93 sq yd	2.21 sq yd
Water: % of body weight	54%	60%
Muscle: % of body weight	36%	42%
Fat: % of body weight	28%	18%
Bone: % of body weight	18%	18%
Average length of spine	24 in.	28 in.
Total lung capacity at age 25		
small	3.3 qt	4.5 qt
average	4.4 qt	6.8 qt
large	5.7 qt	9.5 qt
Number of breaths per minute (at rest)	20–22	14–18
Intake of air per breath		
resting	0.36 qt	0.79 qt
light work	0.91 qt	1.77 qt
heavy work	0.93 qt	2.15 qt
deepest possible intake *(vital capacity)* at age 25	3.17 qt	5.18 qt
Number of red blood cells in millions per cubic millimeter *(blood count)* – U.S. average	4.2–5.4	4.6–6.2

SOURCE: Alan E. Nourse, *The Body*. New York: Time-Life Books, 1968.

Insurance Rates per $1000

Age Purchased		Term		Straight	Limited Pay		Endowment		Universal
M	F	5 yr	10 yr		20 yr	30 yr	20 yr	30 yr	
15	17			$7.60	$9.50	$7.40	$11.80	$7.50	$26.50
20	22	$2.70	$2.90	9.40	11.60	9.10	14.30	9.20	28.00
25	27	2.80	3.00	11.40	13.90	10.80	17.20	11.00	31.00
30	32	2.90	3.30	14.00	17.10	13.30	21.30	13.60	36.00
35	37	3.50	4.50	17.60	21.70	16.90	27.10	17.30	42.00

6% Sales Tax Schedule

Transaction	Tax	Transaction	Tax	Transaction	Tax	Transaction	Tax
.01– .10	.00						
.11– .22	.01	5.09– 5.24	.31	10.09–10.24	.61	15.09–15.24	.91
.23– .39	.02	5.25– 5.41	.32	10.25–10.41	.62	15.25–15.41	.92
.40– .56	.03	5.42– 5.58	.33	10.42–10.58	.63	15.42–15.58	.93
.57– .73	.04	5.59– 5.74	.34	10.59–10.74	.64	15.59–15.74	.94
.74– .90	.05	5.75– 5.91	.35	10.75–10.91	.65	15.75–15.91	.95
.91–1.08	.06	5.92– 6.08	.36	10.92–11.08	.66	15.92–16.08	.96
1.09–1.24	.07	6.09– 6.24	.37	11.09–11.24	.67	16.09–16.24	.97
1.25–1.41	.08	6.25– 6.41	.38	11.25–11.41	.68	16.25–16.41	.98
1.42–1.58	.09	6.42– 6.58	.39	11.42–11.58	.69	16.42–16.58	.99
1.59–1.74	.10	6.59– 6.75	.40	11.59–11.74	.70	16.59–16.74	1.00
1.75–1.91	.11	6.75– 6.91	.41	11.75–11.91	.71	16.75–16.91	1.01
1.92–2.08	.12	6.92– 7.08	.42	11.92–12.08	.72	16.92–17.08	1.02
2.09–2.24	.13	7.09– 7.24	.43	12.09–12.24	.73	17.09–17.24	1.03
2.25–2.41	.14	7.25– 7.41	.44	12.25–12.41	.74	17.25–17.41	1.04
2.42–2.58	.15	7.42– 7.58	.45	12.42–12.58	.75	17.42–17.58	1.05
2.59–2.74	.16	7.59– 7.74	.46	12.59–12.74	.76	17.59–17.74	1.06
2.75–2.91	.17	7.75– 7.91	.47	12.75–12.91	.77	17.75–17.91	1.07
2.92–3.08	.18	7.92– 8.08	.48	12.92–13.08	.78	17.92–18.08	1.08
3.09–3.24	.19	8.09– 8.24	.49	13.09–13.24	.79	18.09–18.24	1.09
3.25–3.41	.20	8.25– 8.41	.50	13.25–13.41	.80	18.25–18.41	1.10
3.42–3.58	.21	8.42– 8.58	.51	13.42–13.58	.81	18.42–18.58	1.11
3.59–3.74	.22	8.59– 8.74	.52	13.59–13.74	.82	18.59–18.74	1.12
3.75–3.91	.23	8.75– 8.91	.53	13.75–13.91	.83	18.75–18.91	1.13
3.92–4.08	.24	8.92– 9.08	.54	13.92–14.08	.84	18.92–19.08	1.14
4.09–4.24	.25	9.09– 9.24	.55	14.09–14.24	.85	19.09–19.24	1.15
4.25–4.41	.26	9.25– 9.41	.56	14.25–14.41	.86	19.25–19.41	1.16
4.42–4.58	.27	9.42– 9.58	.57	14.42–14.58	.87	19.42–19.58	1.17
4.59–4.74	.28	9.59– 9.74	.58	14.59–14.74	.88	19.59–19.74	1.18
4.75–4.91	.29	9.75– 9.91	.59	14.75–14.91	.89	19.75–19.91	1.19
4.92–5.08	.30	9.92–10.08	.60	14.92–15.08	.90	19.92–20.08	1.20

Withholding Tax Table

Single Persons—Weekly Payroll Period
(For Wages Paid After December 1986)

And the wages are–		And the number of withholding allowances claimed is–										
At least	But less than	0	1	2	3	4	5	6	7	8	9	10
		The amount of income tax to be withheld shall be–										
$0	$16	$0	$0	$0	$0	$0	$0	$0	$0	$0	$0	$0
16	18	1	0	0	0	0	0	0	0	0	0	0
18	20	1	0	0	0	0	0	0	0	0	0	0
20	22	1	0	0	0	0	0	0	0	0	0	0
22	24	1	0	0	0	0	0	0	0	0	0	0
24	26	1	0	0	0	0	0	0	0	0	0	0
26	28	2	0	0	0	0	0	0	0	0	0	0
28	30	2	0	0	0	0	0	0	0	0	0	0
30	32	2	0	0	0	0	0	0	0	0	0	0
32	34	2	0	0	0	0	0	0	0	0	0	0
34	36	2	0	0	0	0	0	0	0	0	0	0
36	38	3	0	0	0	0	0	0	0	0	0	0
38	40	3	0	0	0	0	0	0	0	0	0	0
40	42	3	0	0	0	0	0	0	0	0	0	0
42	44	3	0	0	0	0	0	0	0	0	0	0
44	46	4	0	0	0	0	0	0	0	0	0	0
46	48	4	0	0	0	0	0	0	0	0	0	0
48	50	4	0	0	0	0	0	0	0	0	0	0
50	52	4	0	0	0	0	0	0	0	0	0	0
52	54	5	0	0	0	0	0	0	0	0	0	0
54	56	5	1	0	0	0	0	0	0	0	0	0
56	58	5	1	0	0	0	0	0	0	0	0	0
58	60	6	1	0	0	0	0	0	0	0	0	0
60	62	6	1	0	0	0	0	0	0	0	0	0
62	64	6	2	0	0	0	0	0	0	0	0	0
64	66	7	2	0	0	0	0	0	0	0	0	0
66	68	7	2	0	0	0	0	0	0	0	0	0
68	70	7	2	0	0	0	0	0	0	0	0	0
70	72	7	2	0	0	0	0	0	0	0	0	0
72	74	8	3	0	0	0	0	0	0	0	0	0
74	76	8	3	0	0	0	0	0	0	0	0	0
76	78	8	3	0	0	0	0	0	0	0	0	0
78	80	9	3	0	0	0	0	0	0	0	0	0
80	82	9	4	0	0	0	0	0	0	0	0	0
82	84	9	4	0	0	0	0	0	0	0	0	0
84	86	10	4	0	0	0	0	0	0	0	0	0
86	88	10	4	0	0	0	0	0	0	0	0	0
88	90	10	5	0	0	0	0	0	0	0	0	0
90	92	10	5	1	0	0	0	0	0	0	0	0
92	94	11	5	1	0	0	0	0	0	0	0	0
94	96	11	6	1	0	0	0	0	0	0	0	0
96	98	11	6	1	0	0	0	0	0	0	0	0
98	100	12	6	1	0	0	0	0	0	0	0	0
100	105	12	7	2	0	0	0	0	0	0	0	0
105	110	13	7	2	0	0	0	0	0	0	0	0
110	115	14	8	3	0	0	0	0	0	0	0	0
115	120	14	9	4	0	0	0	0	0	0	0	0
120	125	15	10	4	0	0	0	0	0	0	0	0
125	130	16	10	5	1	0	0	0	0	0	0	0
130	135	17	11	6	1	0	0	0	0	0	0	0

Withholding Tax Table

Single Persons—Weekly Payroll Period
(For Wages Paid After December 1986)

And the wages are–		And the number of withholding allowances claimed is–										
At least	But less than	0	1	2	3	4	5	6	7	8	9	10
		The amount of income tax to be withheld shall be–										
$135	$140	$17	$12	$6	$2	$0	$0	$0	$0	$0	$0	$0
140	145	18	13	7	2	0	0	0	0	0	0	0
145	150	19	13	8	3	0	0	0	0	0	0	0
150	160	20	15	9	4	0	0	0	0	0	0	0
160	170	22	16	11	5	1	0	0	0	0	0	0
170	180	23	18	12	7	2	0	0	0	0	0	0
180	190	25	19	14	8	3	0	0	0	0	0	0
190	200	26	21	15	10	4	0	0	0	0	0	0
200	210	28	22	17	11	6	1	0	0	0	0	0
210	220	29	24	18	13	7	2	0	0	0	0	0
220	230	31	25	20	14	9	3	0	0	0	0	0
230	240	32	27	21	16	10	5	0	0	0	0	0
240	250	34	28	23	17	12	6	1	0	0	0	0
250	260	35	30	24	19	13	8	3	0	0	0	0
260	270	37	31	26	20	15	9	4	0	0	0	0
270	280	38	33	27	22	16	11	5	1	0	0	0
280	290	40	34	29	23	18	12	7	2	0	0	0
290	300	41	36	30	25	19	14	8	3	0	0	0
300	310	43	37	32	26	21	15	10	4	0	0	0
310	320	44	39	33	28	22	17	11	6	1	0	0
320	330	46	40	35	29	24	18	13	7	2	0	0
330	340	47	42	36	31	25	20	14	9	3	0	0
340	350	50	43	38	32	27	21	16	10	5	0	0
350	360	53	45	39	34	28	23	17	12	6	2	0
360	370	55	46	41	35	30	24	19	13	8	3	0
370	380	58	48	42	37	31	26	20	15	9	4	0
380	390	61	51	44	38	33	27	22	16	11	5	1
390	400	64	54	45	40	34	29	23	18	12	7	2
400	410	67	56	47	41	36	30	25	19	14	8	3
410	420	69	59	49	43	37	32	26	21	15	10	4
420	430	72	62	52	44	39	33	28	22	17	11	6
430	440	75	65	55	46	40	35	29	24	18	13	7
440	450	78	68	57	47	42	36	31	25	20	14	9
450	460	81	70	60	50	43	38	32	27	21	16	10
460	470	83	73	63	53	45	39	34	28	23	17	12
470	480	86	76	66	55	46	41	35	30	24	19	13
480	490	89	79	69	58	48	42	37	31	26	20	15
490	500	92	82	71	61	51	44	38	33	27	22	16
500	510	95	84	74	64	54	45	40	34	29	23	18
510	520	97	87	77	67	56	47	41	36	30	25	19
520	530	100	90	80	69	59	49	43	37	32	26	21
530	540	103	93	83	72	62	52	44	39	33	28	22
540	550	107	96	85	75	65	55	46	40	35	29	24
550	560	110	98	88	78	68	57	47	42	36	31	25
560	570	114	101	91	81	70	60	50	43	38	32	27
570	580	117	104	94	83	73	63	53	45	39	34	28
580	590	121	108	97	86	76	66	56	46	41	35	30
590	600	124	111	99	89	79	69	58	48	42	37	31
600	610	128	115	102	92	82	71	61	51	44	38	33
610	620	131	118	106	95	84	74	64	54	45	40	34

Withholding Tax Table

Single Persons—Weekly Payroll Period
(For Wages Paid After December 1986)

And the wages are–		And the number of withholding allowances claimed is–										
At least	But less than	0	1	2	3	4	5	6	7	8	9	10
		The amount of income tax to be withheld shall be–										
$620	$630	$135	$122	$109	$97	$87	$77	$67	$57	$47	$41	$36
630	640	138	125	113	100	90	80	70	59	49	43	37
640	650	142	129	116	103	93	83	72	62	52	44	39
650	660	145	132	120	107	96	85	75	65	55	46	40
660	670	149	136	123	110	98	88	78	68	58	47	42
670	680	152	139	127	114	101	91	81	71	60	50	43
680	690	156	143	130	117	105	94	84	73	63	53	45
690	700	159	146	134	121	108	97	86	76	66	56	46
700	710	163	150	137	124	112	99	89	79	69	58	48
710	720	166	153	141	128	115	102	92	82	72	61	51
720	730	170	157	144	131	119	106	95	85	74	64	54
730	740	173	160	148	135	122	109	98	87	77	67	57
740	750	177	164	151	138	126	113	100	90	80	70	59
750	760	180	167	155	142	129	116	103	93	83	72	62
760	770	184	171	158	145	133	120	107	96	86	75	65
770	780	187	174	162	149	136	123	110	99	88	78	68
780	790	191	178	165	152	140	127	114	101	91	81	71
790	800	194	181	169	156	143	130	117	105	94	84	73
800	810	198	185	172	159	147	134	121	108	97	86	76
810	820	201	188	176	163	150	137	124	112	100	89	79
820	830	205	192	179	166	154	141	128	115	102	92	82
830	840	208	195	183	170	157	144	131	119	106	95	85
840	850	212	199	186	173	161	148	135	122	109	98	87
850	860	215	202	190	177	164	151	138	126	113	100	90
860	870	219	206	193	180	168	155	142	129	116	104	93
870	880	222	209	197	184	171	158	145	133	120	107	96
880	890	226	213	200	187	175	162	149	136	123	111	99
890	900	229	216	204	191	178	165	152	140	127	114	101
900	910	233	220	207	194	182	169	156	143	130	118	105
910	920	236	223	211	198	185	172	159	147	134	121	108
920	930	240	227	214	201	189	176	163	150	137	125	112
930	940	243	230	218	205	192	179	166	154	141	128	115
940	950	247	234	221	208	196	183	170	157	144	132	119
950	960	250	237	225	212	199	186	173	161	148	135	122
960	970	254	241	228	215	203	190	177	164	151	139	126

Withholding Tax Table

Married Persons—Weekly Payroll Period
(For Wages Paid After December 1986)

And the wages are–		And the number of withholding allowances claimed is–										
At least	But less than	0	1	2	3	4	5	6	7	8	9	10
		The amount of income tax to be withheld shall be–										
$0	$40	$0	$0	$0	$0	$0	$0	$0	$0	$0	$0	$0
40	42	1	0	0	0	0	0	0	0	0	0	0
42	44	1	0	0	0	0	0	0	0	0	0	0
44	46	1	0	0	0	0	0	0	0	0	0	0
46	48	1	0	0	0	0	0	0	0	0	0	0
48	50	1	0	0	0	0	0	0	0	0	0	0
50	52	2	0	0	0	0	0	0	0	0	0	0
52	54	2	0	0	0	0	0	0	0	0	0	0
54	46	2	0	0	0	0	0	0	0	0	0	0
56	58	2	0	0	0	0	0	0	0	0	0	0
58	60	3	0	0	0	0	0	0	0	0	0	0
60	62	3	0	0	0	0	0	0	0	0	0	0
62	64	3	0	0	0	0	0	0	0	0	0	0
64	66	3	0	0	0	0	0	0	0	0	0	0
66	68	3	0	0	0	0	0	0	0	0	0	0
68	70	4	0	0	0	0	0	0	0	0	0	0
70	72	4	0	0	0	0	0	0	0	0	0	0
72	74	4	0	0	0	0	0	0	0	0	0	0
74	76	4	0	0	0	0	0	0	0	0	0	0
76	78	5	1	0	0	0	0	0	0	0	0	0
78	80	5	1	0	0	0	0	0	0	0	0	0
80	82	5	1	0	0	0	0	0	0	0	0	0
82	84	5	1	0	0	0	0	0	0	0	0	0
84	86	5	1	0	0	0	0	0	0	0	0	0
86	88	6	2	0	0	0	0	0	0	0	0	0
88	90	6	2	0	0	0	0	0	0	0	0	0
90	92	6	2	0	0	0	0	0	0	0	0	0
92	94	6	2	0	0	0	0	0	0	0	0	0
94	96	7	2	0	0	0	0	0	0	0	0	0
96	98	7	3	0	0	0	0	0	0	0	0	0
98	100	7	3	0	0	0	0	0	0	0	0	0
100	105	8	3	0	0	0	0	0	0	0	0	0
105	110	8	4	0	0	0	0	0	0	0	0	0
110	115	9	4	0	0	0	0	0	0	0	0	0
115	120	10	5	1	0	0	0	0	0	0	0	0
120	125	11	6	2	0	0	0	0	0	0	0	0
125	130	11	6	2	0	0	0	0	0	0	0	0
130	135	12	7	3	0	0	0	0	0	0	0	0
135	140	13	7	3	0	0	0	0	0	0	0	0
140	145	14	8	4	0	0	0	0	0	0	0	0
145	150	14	9	4	0	0	0	0	0	0	0	0
150	160	16	10	5	1	0	0	0	0	0	0	0
160	170	17	12	6	2	0	0	0	0	0	0	0
170	180	19	13	8	3	0	0	0	0	0	0	0
180	190	20	15	9	4	0	0	0	0	0	0	0
190	200	22	16	11	5	1	0	0	0	0	0	0
200	210	23	18	12	7	3	0	0	0	0	0	0
210	220	25	19	14	8	4	0	0	0	0	0	0
220	230	26	21	15	10	5	1	0	0	0	0	0
230	240	28	22	17	11	6	2	0	0	0	0	0

Withholding Tax Table

Married Persons—Weekly Payroll Period
(For Wages Paid After December 1986)

And the wages are–		And the number of withholding allowances claimed is–										
At least	But less than	0	1	2	3	4	5	6	7	8	9	10
		The amount of income tax to be withheld shall be–										
$240	$250	$29	$24	$18	$13	$7	$3	$0	$0	$0	$0	$0
250	260	31	25	20	14	9	4	0	0	0	0	0
260	270	32	27	21	16	10	5	1	0	0	0	0
270	280	34	28	23	17	12	6	2	0	0	0	0
280	290	35	30	24	19	13	8	3	0	0	0	0
290	300	37	31	26	20	15	9	4	0	0	0	0
300	310	38	33	27	22	16	11	6	1	0	0	0
310	320	40	34	29	23	18	12	7	3	0	0	0
320	330	41	36	30	25	19	14	8	4	0	0	0
330	340	43	37	32	26	21	15	10	5	1	0	0
340	350	44	39	33	28	22	17	11	6	2	0	0
350	360	46	40	35	29	24	18	13	7	3	0	0
360	370	47	42	36	31	25	20	14	9	4	0	0
370	380	49	43	38	32	27	21	16	10	5	1	0
380	390	50	45	39	34	28	23	17	12	6	2	0
390	400	52	46	41	35	30	24	19	13	8	3	0
400	410	53	48	42	37	31	26	20	15	9	4	0
410	420	55	49	44	38	33	27	22	16	11	6	2
420	430	56	51	45	40	34	29	23	18	12	7	3
430	440	58	52	47	41	36	30	25	19	14	8	4
440	450	59	54	48	43	37	32	26	21	15	10	5
450	460	61	55	50	44	39	33	28	22	17	11	6
460	470	62	57	51	46	40	35	29	24	18	13	7
470	480	64	58	53	47	42	36	31	25	20	14	9
480	490	65	60	54	49	43	38	32	27	21	16	10
490	500	67	61	56	50	45	39	34	28	23	17	12
500	510	68	63	57	52	46	41	35	30	24	19	13
510	520	70	64	59	53	48	42	37	31	26	20	15
520	530	71	66	60	55	49	44	38	33	27	22	16
530	540	73	67	62	56	51	45	40	34	29	23	18
540	550	74	69	63	58	52	47	41	36	30	25	19
550	560	76	70	65	59	54	48	43	37	32	26	21
560	570	77	72	66	61	55	50	44	39	33	28	22
570	580	79	73	68	62	57	51	46	40	35	29	24
580	590	81	75	69	64	58	53	47	42	36	31	25
590	600	84	76	71	65	60	54	49	43	38	32	27
600	610	87	78	72	67	61	56	50	45	39	34	28
610	620	90	80	74	68	63	57	52	46	41	35	30
620	630	93	82	75	70	64	59	53	48	42	37	31
630	640	95	85	77	71	66	60	55	49	44	38	33
640	650	98	88	78	73	67	62	56	51	45	40	34
650	660	101	91	81	74	69	63	58	52	47	41	36
660	670	104	94	83	76	70	65	59	54	48	43	37
670	680	107	96	86	77	72	66	61	55	50	44	39
680	690	109	99	89	79	73	68	62	57	51	46	40
690	700	112	102	92	82	75	69	64	58	53	47	42
700	710	115	105	95	84	76	71	65	60	54	49	43
710	720	118	108	97	87	78	72	67	61	56	50	45
720	730	121	110	100	90	80	74	68	63	57	52	46
730	740	123	113	103	93	83	75	70	64	59	53	48

Withholding Tax Table

Married Persons—Weekly Payroll Period
(For Wages Paid After December 1986)

And the wages are–		And the number of withholding allowances claimed is–										
At least	But less than	0	1	2	3	4	5	6	7	8	9	10
		The amount of income tax to be withheld shall be–										
$740	$750	$126	$116	$106	$96	$85	$77	$71	$66	$60	$55	$49
750	760	129	119	109	98	88	78	73	67	62	56	51
760	770	132	122	111	101	91	81	74	69	63	58	52
770	780	135	124	114	104	94	84	76	70	65	59	54
780	790	137	127	117	107	97	86	77	72	66	61	55
790	800	140	130	120	110	99	89	79	73	68	62	57
800	810	143	133	123	112	102	92	82	75	69	64	58
810	820	146	136	125	115	105	95	84	76	71	65	60
820	830	149	138	128	118	108	98	87	78	72	67	61
830	840	151	141	131	121	111	100	90	80	74	68	63
840	850	154	144	134	124	113	103	93	83	75	70	64
850	860	157	147	137	126	116	106	96	85	77	71	66
860	870	160	150	139	129	119	109	98	88	78	73	67
870	880	163	152	142	132	122	112	101	91	81	74	69
880	890	165	155	145	135	125	114	104	94	84	76	70
890	900	168	158	148	138	127	117	107	97	86	77	72
900	910	171	161	151	140	130	120	110	99	89	79	73
910	920	175	164	153	143	133	123	112	102	92	82	75
920	930	178	166	156	146	136	126	115	105	95	85	76
930	940	182	169	159	149	139	128	118	108	98	87	78
940	950	185	173	162	152	141	131	121	111	100	90	80
950	960	189	176	165	154	144	134	124	113	103	93	83
960	970	192	180	167	157	147	137	126	116	106	96	86
970	980	196	183	170	160	150	140	129	119	109	99	88
980	990	199	187	174	163	153	142	132	122	112	101	91
990	1,000	203	190	177	166	155	145	135	125	114	104	94
1,000	1,010	206	194	181	168	158	148	138	127	117	107	97
1,010	1,020	210	197	184	171	161	151	140	130	120	110	100
1,020	1,030	213	201	188	175	164	154	143	133	123	113	102
1,030	1,040	217	204	191	178	167	156	146	136	126	115	105
1,040	1,050	220	208	195	182	169	159	149	139	128	118	108
1,050	1,060	224	211	198	185	173	162	152	141	131	121	111
1,060	1,070	227	215	202	189	176	165	154	144	134	124	114
1,070	1,080	231	218	205	192	180	168	157	147	137	127	116
1,080	1,090	234	222	209	196	183	170	160	150	140	129	119

Annual Fuel Costs Chart

For 1986 Model Car/Based on 15,000 Miles per Year

	Dollars Per Gallon					
Est. MPG	$1.55	$1.45	$1.35	$1.25	$1.15	$1.05
50	465	435	405	375	345	315
49	474	444	413	382	352	321
48	484	452	421	390	359	328
47	495	463	431	399	367	335
46	505	472	439	407	375	342
45	516	483	450	416	383	350
44	528	494	460	426	392	358
43	542	507	472	437	401	366
42	553	518	482	446	411	375
41	567	531	494	458	421	384
40	581	544	506	469	431	394
39	595	557	518	480	442	404
38	611	572	533	493	454	414
37	628	587	547	506	466	426
36	646	605	563	521	479	438
35	665	622	579	536	493	450
34	684	639	595	551	507	463
33	704	659	614	568	523	477
32	725	679	632	585	539	492
31	751	703	654	606	556	508
30	774	724	674	624	575	525
29	802	750	699	647	595	543
28	830	776	723	669	616	563
27	860	805	749	694	639	583
26	895	837	780	722	663	606
25	930	870	810	750	690	630
24	970	907	844	782	719	656
23	1011	946	881	816	750	685
22	1058	990	921	853	784	716
21	1107	1035	964	892	821	750
20	1162	1088	1012	938	862	788
19	1223	1144	1065	986	908	829
18	1293	1209	1126	1042	958	875
17	1367	1279	1191	1102	1015	926
16	1453	1359	1266	1172	1078	984
15	1551	1451	1351	1251	1150	1050
14	1660	1553	1446	1339	1232	1125
13	1788	1673	1557	1442	1327	1212
12	1937	1812	1687	1562	1438	1313
11	2113	1977	1841	1704	1568	1432
10	2325	2175	2025	1875	1725	1575
9	2583	2416	2250	2083	1917	1750

For those who purchase gasoline by the liter, 3.785 liters equals one gallon.

Mileage Chart

	1 5000	5001 10000	10001 15000	15001 20000	20001 25000	25001 30000	30001 35000	35001 40000	40001 45000	45001 50000	50001 55000	55001 60000	60001 65000	65001 70000	70001 75000	75001 80000	80001 85000	85001 90000	90001 UP
1987	120	80	70	140	210	280	350	420	490	560	630	630	630	630	630	630	630	630	630
1986	240	180	120	60	70	140	210	280	350	420	490	560	630	630	630	630	630	630	630
1985	360	300	240	180	120	60	70	140	210	280	350	420	490	560	630	630	630	630	630
1984	480	ADD		240	180	120	60	0	70	140	210	280	350	DEDUCT			630	630	630
1983	540			360	300	240	180	120	60	70	140	210	280				560	630	630
1982	540	540	480	420	360	300	240	180	120	60	0	70	140	210	280	350	420	490	560
1981	540	540	540	480	420	360	300	240	180	120	0	0	70	140	210	280	350	420	490
1980	540	540	540	540	480	420	360	300	240	180	0	0	0	70	140	210	280	350	420

Equal Monthly Payment Factors—Without Insurance

	Terms in Months									
APR	12	18	24	30	36	42	48	60	72	84
10.00	.087915	.060057	.046144	.037811	.032267	.028316	.025362	.021247	.018525	.016601
10.25	.088032	.060172	.046260	.037927	.032384	.028435	.025482	.021370	.018652	.016730
10.50	.088148	.060287	.046376	.038044	.032502	.028554	.025603	.021493	.018778	.016860
10.75	.088265	.060402	.046491	.038161	.032620	.028674	.025724	.021617	.018906	.016991
11.00	.088381	.060518	.046607	.038278	.032620	.028793	.025845	.021742	.019034	.017122
11.25	.088498	.060634	.046723	.038395	.032857	.028913	.025967	.021867	.019167	.017254
11.50	.088615	.060750	.046840	.038512	.032976	.029034	.026089	.021992	.019291	.017386
11.75	.088731	.060865	.046956	.038630	.033095	.029154	.026211	.022118	.019420	.017519
12.00	.088848	.060982	.047073	.038748	.033214	.029275	.026333	.022244	.019550	.017652
12.25	.088965	.061098	.047190	.038866	.033333	.029396	.026456	.022370	.019680	.017786
12.50	.089082	.061214	.047307	.038984	.033453	.029518	.026579	.022497	.019811	.017921
12.75	.089200	.061331	.047424	.039102	.033573	.029640	.026703	.022625	.019942	.018056
13.00	.089317	.061447	.047541	.039221	.033693	.029762	.026827	.022753	.020074	.018191
13.25	.089434	.061564	.047659	.039340	.033814	.029884	.026951	.022881	.020206	.018328
13.50	.089552	.061681	.047777	.039459	.033935	.030007	.027076	.023009	.020338	.018464
13.75	.089669	.061798	.047894	.039578	.034056	.030130	.027201	.023138	.020472	.018602
14.00	.089787	.061915	.048012	.039698	.034177	.030253	.027326	.023268	.020605	.018740
14.25	.089904	.062032	.048130	.039818	.034299	.030376	.027452	.023398	.020739	.018878
14.50	.090022	.062149	.048249	.039938	.034420	.030500	.027577	.023528	.020874	.019017
14.75	.090140	.062267	.048367	.040058	.034543	.030624	.027704	.023658	.021009	.019156
15.00	.090258	.062384	.048486	.040178	.034665	.030749	.027830	.023789	.021145	.019296
15.25	.090376	.062502	.048605	.040299	.034787	.030873	.027957	.023921	.021281	.019437
15.50	.090494	.062620	.048724	.040419	.034910	.030998	.028084	.024053	.021417	.019578
15.75	.001612	.062738	.048843	.040540	.035033	.031123	.028212	.024185	.021554	.019719
16.00	.090730	.062856	.048963	.040662	.035157	.031249	.028340	.024318	.021691	.019862
16.25	.090849	.062974	.049082	.040783	.035280	.031375	.028468	.024451	.021829	.020004
16.50	.090967	.063093	.049202	.040905	.035404	.031501	.028597	.024584	.021968	.020147
16.75	.091086	.063211	.049322	.041025	.035528	.031627	.028725	.024718	.022106	.020291
17.00	.091204	.063330	.049442	.041143	.035652	.031754	.028855	.024852	.022246	.020435
17.25	.091323	.063448	.049562	.041271	.035777	.031881	.028984	.024987	.022385	.020580
17.50	.091442	.063567	.049682	.041393	.035902	.032008	.029114	.025122	.022526	.020725
17.75	.091561	.063686	.049803	.041515	.036027	.032136	.029244	.025257	.022666	.020871

Equal Monthly Payment Factors—Without Insurance

	Terms in Months									
APR	12	18	24	30	36	42	48	60	72	84
18.00	.091679	.063805	.049924	.041639	.036152	.032264	.029374	.025393	.022807	.021017
18.25	.091799	.063925	.050044	.041762	.036277	.032392	.029505	.025529	.022949	.021164
18.50	.091918	.064044	.050166	.041885	.036403	.032520	.029636	.025666	.023091	.021311
18.75	.092037	.064163	.050287	.042009	.036529	.032649	.029768	.025803	.023233	.021459
19.00	.092156	.064283	.050408	.042132	.036656	.032778	.029900	.025940	.023376	.021608
19.25	.092275	.064403	.050530	.042256	.036782	.032907	.030032	.026078	.023520	.021756
19.50	.092395	.064523	.050651	.042380	.036909	.033037	.030164	.026216	.023663	.021906
19.75	.092514	.064643	.050773	.042505	.037036	.033167	.030297	.026354	.023808	.027055
20.00	.092604	.064763	.050895	.042629	.037163	.033297	.030430	.026493	.023952	.022206
20.25	.092754	.064883	.051018	.042879	.037291	.033427	.030563	.026633	.024097	.022356
20.50	.092873	.065003	.051140	.042879	.037418	.033558	.030697	.026772	.024243	.022508
20.75	.092993	.065174	.051262	.043004	.037546	.033689	.030831	.026912	.024389	.022659

Time Zones

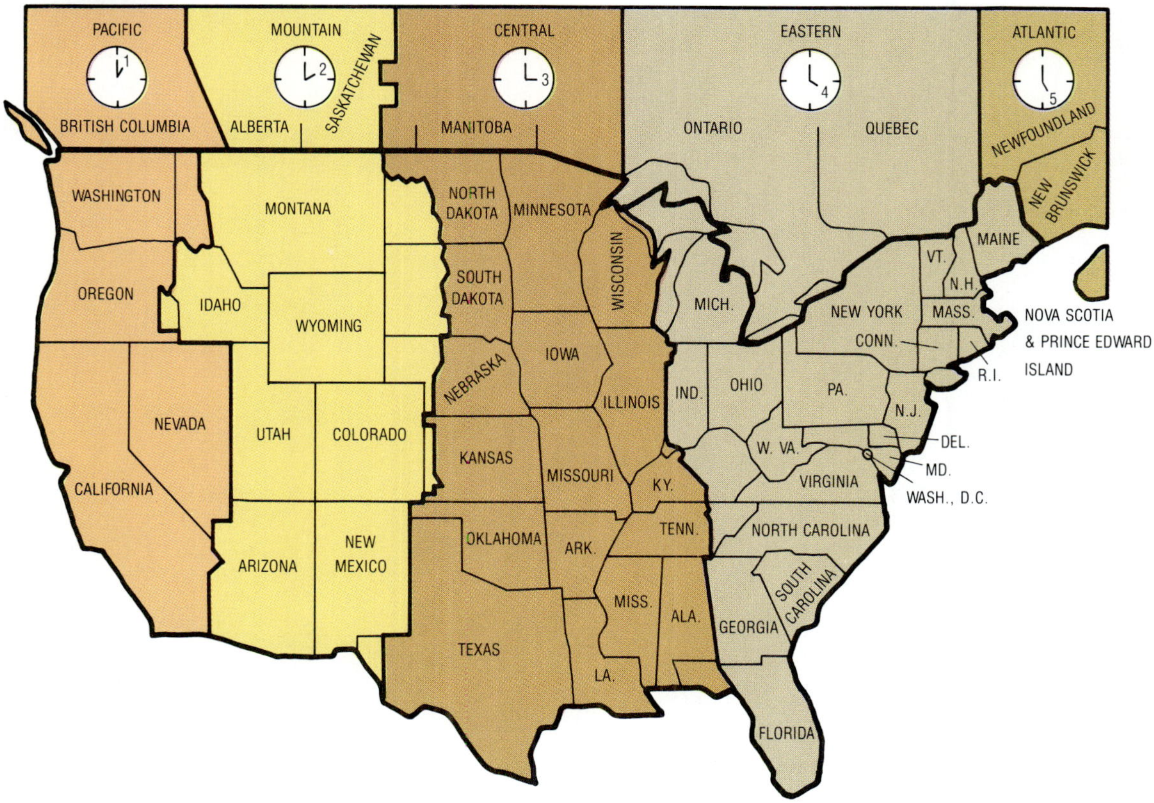

Airfare Chart

				Percent of Base Fare and 8% Tax				
Total	8% Tax	Base Fare	50%	8% Tax	$66\frac{2}{3}$%	8% Tax	75%	8% Tax
371.00	27.48	343.52	171.76	13.74	229.02	18.32	257.64	20.61
372.00	27.56	344.44	172.22	13.78	229.64	18.37	258.33	20.67
373.00	27.63	345.37	172.69	13.81	230.26	18.42	259.03	20.72
374.00	27.70	346.30	173.15	13.85	230.88	18.47	259.72	20.78
375.00	27.78	347.22	173.61	13.89	231.49	18.52	260.42	20.83
376.00	27.85	348.15	174.07	13.93	232.11	18.57	261.11	20.89
377.00	27.93	349.07	174.54	13.96	232.73	18.62	261.81	20.94
378.00	28.00	350.00	175.00	14.00	233.34	18.67	262.50	21.00
379.00	28.07	350.93	175.46	14.04	233.96	18.72	263.19	21.06
380.00	28.15	351.85	175.93	14.07	234.58	18.77	263.89	21.11
381.00	28.22	352.78	176.39	14.11	235.20	18.82	264.58	21.17
382.00	28.30	353.70	176.85	14.15	235.81	18.87	265.28	21.22
383.00	28.37	354.63	177.31	14.19	236.43	18.91	265.97	21.28
384.00	28.44	355.56	177.78	14.22	237.05	18.96	266.67	21.33
385.00	28.52	356.48	178.24	14.26	237.67	19.01	267.36	21.39
386.00	28.59	357.41	178.70	14.30	238.28	19.06	268.06	21.44
387.00	28.67	358.33	179.17	14.33	238.90	19.11	268.75	21.50
388.00	28.74	359.26	179.63	14.37	239.52	19.16	269.44	21.56
389.00	28.81	360.19	180.09	14.41	240.14	19.21	270.14	21.61
390.00	28.89	361.11	180.56	14.44	240.75	19.26	270.83	21.67
391.00	28.96	362.04	181.02	14.48	241.37	19.31	271.53	21.72
392.00	29.04	362.96	181.48	14.52	241.99	19.36	272.22	21.78
393.00	29.11	363.89	181.94	14.56	242.60	19.41	272.92	21.83
394.00	29.19	364.81	182.41	14.59	243.22	19.46	273.61	21.89
395.00	29.26	365.74	182.87	14.63	243.84	19.51	274.31	21.94
396.00	29.33	366.67	183.33	14.67	244.46	19.56	275.00	22.00
397.00	29.41	367.59	183.80	14.70	245.07	19.61	275.69	22.06
398.00	29.48	368.52	184.26	14.74	245.69	19.66	276.39	22.11
399.00	29.56	369.44	184.72	14.78	246.31	19.70	277.08	22.17
400.00	29.63	370.37	185.19	14.81	246.93	19.75	277.78	22.22
401.00	29.70	371.30	185.65	14.85	247.54	19.80	278.47	22.28
402.00	29.78	372.22	186.11	14.89	248.16	19.85	279.17	22.33
403.00	29.85	373.15	186.57	14.93	248.78	19.90	279.86	22.39
404.00	29.93	374.07	187.04	14.96	249.40	19.95	280.56	22.44
405.00	30.00	375.00	187.50	15.00	250.01	20.00	281.25	22.50
406.00	30.07	375.93	187.96	15.04	250.63	20.05	281.94	22.56
407.00	30.15	376.85	188.43	15.07	251.25	20.10	282.64	22.61
408.00	30.22	377.78	188.89	15.11	251.86	20.15	283.33	22.67
409.00	30.30	378.70	189.35	15.15	252.48	20.20	284.03	22.72
410.00	30.37	379.63	189.81	15.19	253.10	20.25	284.72	22.78
411.00	30.44	380.56	190.28	15.22	253.72	20.30	285.42	22.83
412.00	30.52	381.48	190.74	15.26	254.33	20.35	286.11	22.89
413.00	30.59	382.41	191.20	15.30	254.95	20.40	286.81	22.94
414.00	30.67	383.33	191.67	15.33	255.57	20.45	287.50	23.00
415.00	30.74	384.26	192.13	15.37	256.19	20.49	288.19	23.06
416.00	30.81	385.19	192.59	15.41	256.80	20.54	288.89	23.11
417.00	30.89	386.11	193.06	15.44	257.42	20.59	289.58	23.17
418.00	30.96	387.04	193.52	15.48	258.04	20.64	290.28	23.22
419.00	31.04	387.96	193.98	15.52	258.65	20.69	290.97	23.28
420.00	31.11	388.89	194.44	15.56	259.27	20.74	291.67	23.33

Airfare Chart

				Percent of Base Fare and 8% Tax				
Total	8% Tax	Base Fare	50%	8% Tax	$66\frac{2}{3}$%	8% Tax	75%	8% Tax
421.00	31.19	389.91	194.91	15.59	259.89	20.79	292.36	23.39
422.00	31.26	390.74	195.37	15.63	260.51	20.84	293.06	23.44
423.00	31.33	391.67	195.83	15.67	261.12	20.89	293.75	23.50
424.00	31.41	392.59	196.30	15.70	261.74	20.94	294.44	23.56
425.00	31.48	393.52	196.76	15.74	262.36	20.99	295.14	23.61
426.00	31.56	394.44	197.22	15.78	262.98	21.04	295.83	23.67
427.00	31.63	395.37	197.69	15.81	263.59	21.09	296.53	23.72
428.00	31.70	396.30	198.15	15.85	264.21	21.14	297.22	23.78
429.00	31.78	397.22	198.61	15.89	264.83	21.19	297.92	23.83
430.00	31.85	398.15	199.07	15.93	265.45	21.24	298.61	23.89

Interest/Principal Payment Table

Loan Table					
Smith Loan—$75,000 at 11.75% for 30 years—Monthly Payment $757.06					
Payment Number	Loan Balance	Payment Amount	Interest Amount	Principal Amount	New Loan Balance
1	$75,000.00	$757.06	$734.38	$22.68	$74,977.32
2	74,977.32	757.06	734.15	22.91	74,954.41
3	74,954.41	757.06	733.93	23.13	74,931.28
4	74,931.28	757.06	733.70	23.36	74,907.92
356	3676.56	757.06	36.00	721.06	2955.50
357	2955.50	757.06	28.94	728.12	2227.38
358	2227.38	757.06	21.81	735.25	1492.13
359	1492.13	757.06	14.61	742.45	749.68
360	749.68	757.06	7.38	749.68	0.00

Hotel Bill

Date	Reference	Charges	Credits	Balance Due
04/05	Restaurant 5150	21.97		
04/05	Local Call 0678-Dir	.75		
04/05	Selfpark	3.00		
04/05	Coffee Shop 7705	8.50		
04/05	Movies	5.95		
04/05	Room 725, 1	87.00		
04/05	Rm. Tx 725, 1	6.96		
04/06	Restaurant 5268	14.93		
04/06	Coffee Shop 2924	1.00		
04/06	Room 725, 1	87.00		
04/06	Rm. Tx 725, 1	6.96		
04/07	Room 725, 1	87.00		
04/07	Rm. Tx 725, 1	6.96		
04/08	Room 725, 1	87.00		
04/08	Rm. Tx 725, 1	6.96		
04/09	Room 725, 1	87.00		
04/09	Rm. Tx 725, 1	6.96		

EXTRA PRACTICE

Chapter 1

Add.

1. $\begin{array}{r}165\\82\\464\\+\ 73\\\hline\end{array}$ **2.** $\begin{array}{r}143.8\\61.34\\+\ 94.6\\\hline\end{array}$ **3.** $\begin{array}{r}43,285\\16,176\\824\\+\quad 352\\\hline\end{array}$ **4.** $\begin{array}{r}\$130.68\\19.24\\36.80\\+507.29\\\hline\end{array}$ **5.** $\begin{array}{r}8565\\4716\\9302\\+2588\\\hline\end{array}$

6. $\begin{array}{r}0.38\\0.846\\0.76\\+0.429\\\hline\end{array}$ **7.** $\begin{array}{r}1274\\168.32\\54.74\\+2605.8\\\hline\end{array}$ **8.** $\begin{array}{r}\$\ 3.81\\0.16\\24.95\\+23\\\hline\end{array}$ **9.** $\begin{array}{r}93.08\\136\\48.726\\+\quad 7.658\\\hline\end{array}$ **10.** $\begin{array}{r}\$467.39\\528.54\\291.03\\+746.68\\\hline\end{array}$

Subtract.

11. $\begin{array}{r}628\\-452\\\hline\end{array}$ **12.** $\begin{array}{r}19.24\\-12.88\\\hline\end{array}$ **13.** $\begin{array}{r}428\\-144.62\\\hline\end{array}$ **14.** $\begin{array}{r}\$1503\\-\ 846.54\\\hline\end{array}$ **15.** $\begin{array}{r}2952\\-1968\\\hline\end{array}$

16. $\begin{array}{r}0.307\\-0.289\\\hline\end{array}$ **17.** $\begin{array}{r}2400\\-1603\\\hline\end{array}$ **18.** $\begin{array}{r}\$982\\-295.73\\\hline\end{array}$ **19.** $\begin{array}{r}2070\\-\ 473.84\\\hline\end{array}$ **20.** $\begin{array}{r}\$6256.18\\-2448.35\\\hline\end{array}$

Multiply.

21. $\begin{array}{r}56\\\times\ 4\\\hline\end{array}$ **22.** $\begin{array}{r}0.37\\\times\ \ 58\\\hline\end{array}$ **23.** $\begin{array}{r}83\\\times 64\\\hline\end{array}$ **24.** $\begin{array}{r}\$127.45\\\times\quad 92\\\hline\end{array}$ **25.** $\begin{array}{r}863\\\times 479\\\hline\end{array}$

26. $\begin{array}{r}3.8\\\times 0.8\\\hline\end{array}$ **27.** $\begin{array}{r}\$15.08\\\times\ \ 2.4\\\hline\end{array}$ **28.** $\begin{array}{r}639\\\times\ 25\\\hline\end{array}$ **29.** $\begin{array}{r}2.84\\\times\ 5.5\\\hline\end{array}$ **30.** $\begin{array}{r}\$394.25\\\times\ \ 56.4\\\hline\end{array}$

Divide.

31. 448 ÷ 8 **32.** 42.9 ÷ 0.3 **33.** 460.8 ÷ 36 **34.** $287 ÷ 3.5 **35.** 1363 ÷ 47

36. 27,328 ÷ 64 **37.** 3759.28 ÷ 28 **38.** $1734 ÷ 0.85 **39.** 163.029 ÷ 9.3 **40.** $19.72 ÷ 58

Estimate, then find the exact answer.

41. 47 × 63 **42.** $973 06 − $428.83 **43.** 3.58 + 0.7 + 4.43

44. $3.64 + $47 + $62.94 **45.** 271.92 ÷ 10.3 **46.** 74 + 18 + 96 + 53

47. 34.76 − 8.93 **48.** 72.6 × 37.3 **49.** 1620 ÷ 36

50. $790.56 ÷ 18 **51.** 2064 − 653 **52.** $58.16 × 37

53. 134.82 ÷ 96.3 **54.** $1183 × 0.5 **55.** 3624 − 2756

Chapter 2

Add.

1. 78,263 + 56,385 + 92,012 + 3,657

2. $1,028,563 + 463,926 + 782,354 + 3,608,164

3. 2.0638 + 0.8452 + 2.934 + 11.66

4. 372.838 + 1684 + 201.08 + 53.719

5. 978 + 73,126 + 8,632 + 52,054

6. $ 267.82 + 540.94 + 316.55 + 1791.86

7. 0.3658 + 0.927 + 0.0654 + 0.5218

8. $1412.65 + 0.89 + 362.48 + 53.74

9. 4609 + 273.62 + 18 + 9385.094

10. $ 904.16 + 5632.82 + 50.58 + 743.19

Subtract.

11. $37,625 − 28,756

12. 80,036 − 59,245

13. 63.34 − 60.7432

14. 0.83924 − 0.82969

15. $1006.43 − 468.35

16. 1,607,806 − 463,837

17. $27,806 − 19,388.47

18. 17 − 12.70614

19. 9206.3 − 8756.32

20. 62,125 − 50,684

Multiply.

21. 483 × 79

22. $568.57 × 84

23. 3026 × 15.7

24. 7254 × 630

25. $4278 × 3.75

26. 198 × 9.4

27. $2280.63 × 75

28. 564 × 378

29. 291.9 × 0.53

30. 369 × 248

Divide.

31. 67,536 ÷ 42
32. 1128.864 ÷ 4.8
33. 35.424 ÷ 0.82
34. 245,088 ÷ 37

35. 997.36 ÷ 28
36. 51.584 ÷ 16
37. 2653.92 ÷ 0.97
38. 8154 ÷ 135

39. 11,550.6 ÷ 621
40. $34.04 ÷ 74

Estimate, then find the exact answer.

41. 5236 − 4710
42. 280.8 ÷ 23.4
43. $493.27 × 8.6

44. 2.03 + 2.71 + 1.84
45. 555 × 628
46. 158.76 ÷ 12.6

47. 36.3 × 485
48. $6483 − $3028
49. $0.82 + $0.38 + $2.24

50. $73.91 ÷ 3.89
51. 8 + 4 + 7 + 3
52. 18,034 − 4692

53. 89.63 + 26.08
54. 1495 ÷ 65
55. 23.24 − 16.5

56. $0.43 × 123
57. $17.68 + $24.27
58. 48.3 × 0.15

Chapter 3

Add.

1. $\frac{5}{8} + \frac{1}{4}$
2. $\frac{3}{5} + \frac{2}{3}$
3. $1\frac{3}{10} + \frac{2}{5}$
4. $\frac{5}{8} + \frac{3}{4} + \frac{5}{16}$
5. $2\frac{4}{9} + 4\frac{11}{18}$
6. $6\frac{5}{8} + 7\frac{1}{2}$
7. $3\frac{7}{12} + 5\frac{1}{4}$
8. $7\frac{2}{7} + 4\frac{2}{3}$
9. $3\frac{3}{4} + 2\frac{5}{6}$
10. $6\frac{3}{8} + 4\frac{7}{12}$
11. $9\frac{1}{6} + 5\frac{4}{9}$
12. $8\frac{3}{8} + 7\frac{2}{3}$

Subtract.

13. $\frac{5}{9} - \frac{1}{3}$
14. $\frac{7}{8} - \frac{1}{2}$
15. $2\frac{3}{4} - \frac{5}{16}$
16. $9\frac{3}{10} - 9\frac{1}{5}$
17. $4 - 3\frac{6}{11}$
18. $6\frac{2}{3} - 4\frac{5}{8}$
19. $8\frac{1}{2} - 6\frac{2}{3}$
20. $7\frac{4}{5} - 3\frac{19}{20}$
21. $6\frac{1}{4} - 2\frac{1}{3}$
22. $6 - 1\frac{3}{5}$
23. $7\frac{5}{6} - 3\frac{8}{9}$
24. $4\frac{3}{14} - 2\frac{13}{21}$

Multiply.

25. $\frac{3}{7} \times \frac{5}{6}$
26. $\frac{2}{9} \times \frac{3}{4}$
27. $\frac{4}{5} \times \frac{2}{3}$
28. $\frac{7}{15} \times \frac{5}{14}$
29. $1\frac{1}{15} \times \frac{5}{12}$
30. $2\frac{2}{3} \times \frac{7}{8}$
31. $1\frac{7}{9} \times 6\frac{2}{3}$
32. $2\frac{10}{11} \times 4\frac{1}{8}$
33. $\frac{5}{6} \times \frac{9}{10} \times \frac{4}{5}$
34. $3\frac{2}{3} \times \frac{2}{3}$
35. $3\frac{3}{7} \times 1\frac{5}{16}$
36. $10\frac{4}{5} \times 1\frac{1}{12}$

Divide.

37. $\frac{3}{8} \div \frac{3}{4}$
38. $\frac{7}{12} \div \frac{1}{2}$
39. $18 \div \frac{6}{11}$
40. $2\frac{1}{4} \div 6$
41. $2\frac{2}{5} \div 1\frac{1}{2}$
42. $3\frac{1}{9} \div 4\frac{2}{3}$
43. $2\frac{4}{21} \div 1\frac{9}{14}$
44. $7\frac{1}{8} \div 9\frac{1}{2}$
45. $2\frac{5}{8} \div 7$
46. $26 \div \frac{13}{18}$
47. $4\frac{4}{9} \div 4\frac{2}{7}$
48. $1\frac{3}{7} \div 4\frac{2}{3}$

Write in lowest terms.

49. $\frac{17}{51}$
50. $\frac{24}{60}$

Write the missing numeral.

51. $\frac{7}{8} = \frac{\square}{24}$
52. $\frac{8}{13} = \frac{32}{\square}$

Write as a fraction.

53. $7\frac{3}{4}$
54. $5\frac{5}{8}$

Write as a mixed number or a whole number.

55. $\frac{15}{8}$
56. $\frac{24}{8}$

Estimate, then find the exact answer.

57. $\frac{24}{35} \times 6$
58. $\frac{19}{40} \times 52$
59. $\frac{1}{12} \times \$71$
60. $\frac{5}{21} \times \$12$
61. $\frac{4}{13} \times 26.8$
62. $\frac{20}{27} \times \$72$
63. $\$2 \div \frac{8}{79}$
64. $124 \div \frac{16}{65}$
65. $13 \div \frac{15}{26}$
66. $32 \div \frac{3}{10}$
67. $7 \div \frac{3}{5}$
68. $2 \div \frac{17}{73}$

Chapter 4

Solve.

1. $\$346.59 + \$885.43$
2. $1.6 + 2.4 + 8.3$
3. $46 + 97 + 38$
4. $\frac{13}{20} + \frac{3}{4}$
5. $3\frac{2}{3} + 6\frac{3}{8}$
6. $5784 + 4808 + 13$
7. $2.08 + 2.4 + 8.57$
8. $\$1706.18 + \$426.54$
9. $1975 + 727 + 65$
10. $\$6.82 + \$11.95$
11. $4\frac{7}{12} + 4\frac{5}{8}$
12. $18.47 + 13.542 + 6.7$
13. $2.053 + 8.3794$
14. $\frac{3}{4} + \frac{2}{3} + \frac{1}{6}$
15. $\$0.89 + \$0.36 + \$7$
16. $8646 + 7295 + 18$
17. $1\frac{1}{2} + \frac{2}{5} + \frac{5}{8}$
18. $0.365 + 0.828$
19. $1754 + 4267$
20. $\$8579 + \$6895$
21. $\frac{7}{9} - \frac{3}{5}$
22. $7326 - 5464$
23. $\$19.35 - \$6.79$
24. $163.082 - 5.78$
25. $\$12 - \$10.38$
26. $6\frac{3}{16} - 4\frac{7}{16}$
27. $22.5 - 16.48$
28. $10{,}637 - 8829$
29. $1443 - 856$
30. $10.64 - 7.0836$
31. $2\frac{1}{2} - \frac{5}{7}$
32. $\$904.18 - \$726.09$
33. $6 - 3.452$
34. $\$3.01 - \$1.52$
35. $2922 - 1367$
36. $6010 - 4234$
37. 74×56
38. $\$26.19 \times 82$
39. $\frac{5}{9} \times 12$
40. 56.3×9
41. $2\frac{3}{4} \times \frac{2}{9}$
42. 259×18
43. 7.26×8.4
44. $\$0.43 \times 236$
45. 103.7×0.65
46. $6\frac{1}{8} \times 2\frac{2}{3}$
47. $\$163.04 \times 40.5$
48. $1\frac{11}{16} \times 1\frac{20}{21}$
49. $\$4696 \times 5.06$
50. 4384×168
51. $5\frac{1}{10} \times \frac{7}{12}$
52. 12.007×826
53. $16{,}848 \div 18$
54. $19{,}584 \div 96$
55. $20.91 \div 0.34$
56. $\$9116.64 \div 52$
57. $36 \div 0.0072$
58. $36{,}960 \div 55$
59. $\$3009 \div 1.5$
60. $2\frac{2}{3} \div \frac{1}{2}$
61. $144 \div 2\frac{2}{5}$
62. $\$254.88 \div \$0.27$
63. $0.3108 \div 3.7$
64. $5\frac{2}{3} \div 6$
65. $6\frac{6}{7} \div 2\frac{2}{9}$
66. $23.7 \div 75$
67. $3114 \div 18$

Find the missing percent.

68. $36\frac{1}{2}\% + 18\frac{3}{4}\% + \square = 100\%$

Write as a decimal.

69. 6%
70. 0.8%
71. 143%

Write as a fraction or mixed number.

72. 70%
73. $62\frac{1}{2}\%$

Write as a percent.

74. $\frac{7}{5}$
75. 0.0034

Compare. Use <, >, or =.

76. $\frac{1}{4} \square 27\%$
77. $0.6\% \square 0.006$
78. $\frac{5}{3} \square 1.33$
79. $47\% \square \frac{7}{5}$

Find each percent.

80. What percent of 80 is 16?

Find the percent of increase or decrease.

81. from 16 to 60

Chapter 5

Solve.

1. $37 + 13$
2. $89 + 3$
3. $34 + 333$
4. $12 + 8$
5. $17 + 457$
6. $0.098 + 7.002$
7. $1398 + 3456$
8. $0.976 + 2.87$
9. $2.07 + 0.13$
10. $\frac{2}{7} + \frac{13}{3}$
11. $4\frac{5}{9} + 6\frac{5}{18}$
12. $234 + 432$
13. $27 + 3$
14. $1902 + 379$
15. $3.4 + 7.6$
16. $1.82 + 0.08$
17. $2.76 + 1.75$
18. $\frac{1}{6} + 3\frac{1}{3}$
19. $8971 + 333$
20. $7\frac{4}{5} + 3\frac{3}{10}$
21. $57 - 24$
22. $2789 - 944$
23. $29 - 13$
24. $2034 - 696$
25. $3\frac{12}{13} - 2\frac{7}{13}$
26. $0.373 - 0.037$
27. $12\frac{7}{32} - 2\frac{1}{8}$
28. $7\frac{1}{3} - 1\frac{7}{8}$
29. $9.98 - 2.57$
30. $26 - 12$
31. $807 - 17$
32. $32 - 8$
33. $4\frac{8}{7} - 2\frac{7}{8}$
34. $432 - 398$
35. $\frac{13}{27} - \frac{4}{9}$
36. $10.63 - 2.44$
37. $91.01 - 30.09$
38. $89 - 3$
39. $207 - 178$
40. $29047 - 3222$
41. 13×3
42. 5×13
43. 34×7
44. 8.8×2
45. $\frac{1}{2} \times \frac{1}{12}$
46. 0.0202×6
47. $4\frac{3}{4} \times \frac{1}{7}$
48. 27×37
49. 67×8
50. 9128×3
51. 56×9
52. 2.7×9.06
53. 13×4
54. 8.6×5.6
55. 76×56
56. 9×9
57. $12\frac{1}{3} \times 6\frac{1}{4}$
58. $\frac{111}{24} \times \frac{3}{5}$
59. 6×9
60. $\frac{1}{32} \times 3\frac{1}{2}$
61. $108 \div 9$
62. $7.2 \div 12$
63. $135 \div 45$
64. $54 \div 12$
65. $36.09 \div 3$
66. $128 \div 4$
67. $51 \div 17$
68. $1656 \div 12$
69. $656 \div 164$
70. $\frac{5}{6} \div \frac{1}{8}$
71. $\frac{3}{5} \div \frac{2}{5}$
72. $\frac{72}{32} \div \frac{3}{4}$
73. $\frac{1}{7} \div \frac{2}{7}$
74. $8\frac{1}{4} \div 33$
75. $56 \div 6$
76. $7\frac{3}{4} \div 4$
77. $2.25 \div 3$
78. $4758 \div 6$
79. $7 \div 8$
80. $20097 \div 87$
81. \$29,600 − \$27,853
82. \$1370 − \$1146
83. \$5230 − \$4853.18
84. \$6461.08 − \$5385.70
85. $\frac{1}{2}$ of \$17.48
86. $\frac{2}{3}$ of \$525.18
87. $\frac{1}{4}$ of \$94.36
88. $\frac{3}{8}$ of \$1152.64
89. 6% of \$9.38
90. 15% of \$34.82
91. 25% of \$753.08
92. $33\frac{1}{3}$% of \$62 16
93. \$930 + 11%
94. \$1807 + 4%
95. \$2762.48 + 3%
96. \$24,025 + $5\frac{1}{2}$%
97. \$0.035 × 460
98. \$0.065 × 1250
99. \$200.60 ÷ \$0.85
100. \$7.05 ÷ 94

Chapter 6

Solve.

1. $2 + 39$
2. $89 + 45$
3. $16 + 13$
4. $18 + 24$
5. $41 + 52 + 55$
6. $68 + 14 + 4$
7. $132 + 3 + 86$
8. $12 + 8 + 777$
9. $3\frac{1}{4} + 5\frac{3}{8}$
10. $1\frac{2}{17} + 6\frac{1}{2}$
11. $4\frac{7}{16} + 8\frac{3}{4}$
12. $\frac{1}{3} + 5\frac{1}{6}$
13. $4\frac{3}{4} + 8\frac{7}{16}$
14. $2\frac{1}{3} + 3\frac{1}{6}$
15. $9\frac{2}{3} + 1\frac{1}{3}$
16. $8\frac{3}{5} + 3\frac{1}{3}$
17. $43.2 + 16.1$
18. $31.7 + 22.1$
19. $4.99 + 33$
20. $12.22 + 8.01$
21. $22 - 9$
22. $29 - 14$
23. $960 - 13$
24. $542 - 189$
25. $61 - 52 - 5$
26. $98 - 14 - 4$
27. $32 - 23 - 9$
28. $123 - 8 - 77$
29. $5\frac{2}{5} - 5\frac{1}{5}$
30. $1\frac{8}{9} - \frac{2}{3}$
31. $4\frac{3}{4} - 2\frac{1}{16}$
32. $\frac{1}{12} - \frac{1}{32}$
33. $3.2 - 1.1$
34. $1.7 - 0.9$
35. $3.99 - 0.87$
36. $12.20 - 12.01$
37. $4.1 - 2.9 - 1$
38. $18.9 - 0.7 - 17$
39. $4.32 - 3.99 - 0.33$
40. $2.22 - 1.9 - 0.3$
41. 2×43
42. 69×15
43. 999×3
44. 8×24
45. $1 \times 55 \times 55$
46. $78 \times 4 \times 4$
47. $132 \times 3 \times 86$
48. $2 \times 8 \times 707$
49. $9\frac{1}{4} \times 5\frac{1}{2}$
50. $1\frac{2}{7} \times 6\frac{1}{14}$
51. $2\frac{1}{4} \times 2\frac{3}{5}$
52. $\frac{1}{9} \times \frac{1}{32}$
53. 1.2×8.9
54. 0.9×0.14
55. 9.68×1.3
56. 5.42×1.11
57. $7.7 \times 0.2 \times 3$
58. $7.8 \times 0.14 \times 2$
59. $32.11 \times 2.3 \times 0$
60. $23.9 \times 0.5 \times 2$
61. $32 \div 2$
62. $29 \div 9$
63. $960 \div 12$
64. $99 \div 33$
65. $1 \div 52 \div 5$
66. $78 \div 4 \div 2$
67. $102 \div 6 \div 3$
68. $23 \div 8 \div 5$
69. $3\frac{2}{7} \div 3\frac{1}{7}$
70. $\frac{4}{9} \div \frac{2}{3}$
71. $10\frac{1}{4} \div 2\frac{1}{6}$
72. $\frac{1}{2} \div 2$
73. $1 \div \frac{7}{22} \div 5$
74. $7\frac{8}{9} \div \frac{1}{4} \div \frac{2}{3}$
75. $33\frac{1}{2} \div 6 \div 3$
76. $23 \div \frac{1}{8} \div 3\frac{6}{7}$
77. $3.2 \div 2$
78. $0.09 \div 1$
79. $87.2 \div 1.2$
80. $1.6 \div 0.6$
81. \$37.16 + \$82.86
82. \$1126.38 − \$257.63
83. \$85.24 + \$71.98
84. \$324.08 − \$105.68
85. 8% of \$1920
86. 35% of \$2172
87. 16.2% of \$1500
88. 12.5% of \$3800
89. \$2500 × 8% × 3
90. $\$1600 \times 6.5\% \times 1\frac{1}{2}$
91. $\$900 \times 12\% \times \frac{3}{4}$
92. \$23,500 × 11% × 5
93. $72 \div 5$
94. $72 \div 10$
95. $72 \div 18$
96. $72 \div 6.5$
97. $\frac{\$175}{\$2000} = n\%$
98. $\frac{\$860}{\$2250} = n\%$
99. $\frac{\$235}{\$1840} = n\%$
100. $\frac{\$48.50}{\$2150} = n\%$

Chapter 7

Solve.

1. $406 + 188$
2. $389 + 44$
3. $12 + 14 + 34$
4. $9022 + 679$
5. $24 + 457 + 29$
6. $0.14 + 0.702$
7. $9.74 + 3.56$
8. $2.075 + 2.87$
9. $335 + 76$
10. $\frac{18}{3} + \frac{13}{3}$
11. $15\frac{6}{5} + 6\frac{9}{5}$
12. $1121 + 27$
13. $4.75 + 3.875$
14. $4.3 + 22.87$
15. $3040 + 2020$
16. $11.03 + 0.14$
17. $96 + 22$
18. $4\frac{3}{4} + 3\frac{7}{8}$
19. $43 + 76 + 29$
20. $6\frac{1}{7} + 33\frac{5}{14}$
21. $0.11 - 0.011$
22. $2200 - 1600$
23. $20212 - 421$
24. $3.75 - 1.125$
25. $11\frac{2}{3} - 2\frac{3}{8}$
26. $323 - 241$
27. $7\frac{12}{5} - 6\frac{5}{12}$
28. $7\frac{14}{3} - \frac{11}{9}$
29. $103 - 37$
30. $112 - 54$
31. $2819 - 1009$
32. $46 - 38$
33. $\frac{36}{7} - 4$
34. $5932 - 4781$
35. $\frac{7}{9} - \frac{2}{5} - \frac{1}{8}$
36. $4098 - 3287$
37. $0.07 - 0.065$
38. $23 - 17$
39. $257 - 43$
40. $57.5 - 3.012$
41. 2044×3
42. 4×75
43. 87×7
44. 4.5×2
45. $12 \times \frac{13}{4}$
46. 3.045×2
47. $6\frac{1}{9} \times 4\frac{1}{2}$
48. $65 \times .0203$
49. 0.4×8.07
50. 9128×3
51. 56×9
52. 2.7×9.06
53. $222 \times 7 \times 6$
54. 12.7×6.25
55. 184×0.002
56. 12×7
57. $\frac{15}{32} \times \frac{6}{7}$
58. $\frac{1}{3} \times \frac{1}{2}$
59. 45×0.05
60. $\frac{1}{20} \times 3\frac{1}{2}$
61. $1024 \div 32$
62. $9.9 \div 9.9$
63. $221 \div 13$
64. $1449 \div 63$
65. $27.75 \div 3$
66. $855 \div 19$
67. $4815 \div 12$
68. $10701 \div 87$
69. $100.94 \div 980$
70. $\frac{12}{45} \div \frac{7}{9}$
71. $7.6 \div \frac{2}{5}$
72. $3\frac{8}{3} \div 9\frac{6}{5}$
73. $8\frac{1}{2} \div 6\frac{1}{3}$
74. $12\frac{1}{8} \div \frac{1}{2}$
75. $116 \div 29$
76. $3\frac{4}{15} \div \frac{7}{2}$
77. $0.085 \div 17$
78. $91143 \div 369$
79. $3 \div 8$
80. $22 \div 7$
81. 9 lb 12 oz = *n* oz
82. 21 lb 3 oz = *n* oz
83. 3 lb 7 oz + 6 lb 9 oz
84. 11 lb 13 oz + 9 lb 13 oz
85. $\frac{2}{3}$ of \$460.71
86. $\frac{3}{4}$ of \$516.76
87. $12\frac{1}{2}$% of \$37.84
88. 40% of \$179.26
89. \$52.67 + \$174.58
90. \$3380.73 − \$1651.86
91. \$29.99 × 46
92. \$215.18 ÷ \$0.53
93. \$5.44 ÷ 34 oz
94. \$0.20 ÷ 8 oz
95. \$1.10 ÷ 1 lb 6 oz
96. \$81.18 ÷ 4 lb 2 oz
97. \$63.50 − 20%
98. \$240 − 3.5%
99. \$7024 − 8%
100. \$3875 − 16.5%

Chapter 8

Solve.

1. $156 + 99$	**2.** $12 + 41$	**3.** $320 + 13$	**4.** $88 + 214$
5. $90 + 2 + 65$	**6.** $43 + 224 + 21$	**7.** $47 + 7 + 76$	**8.** $2 + 85 + 100$
9. $9\frac{5}{6} + 8\frac{5}{6}$	**10.** $1\frac{2}{13} + \frac{1}{2}$	**11.** $5\frac{3}{4} + 81\frac{7}{16}$	**12.** $\frac{1}{3} + 6\frac{5}{6}$
13. $4\frac{1}{4} + \frac{3}{16}$	**14.** $12 + 3\frac{1}{23}$	**15.** $3\frac{2}{9} + 3\frac{1}{3}$	**16.** $\frac{3}{5} + 90\frac{1}{3}$
17. $89.9 + 16.1$	**18.** $1.7 + 2.1$	**19.** $0.99 + 84$	**20.** $65.04 + 13.01$
21. $876 - 9$	**22.** $39 - 12$	**23.** $60 - 33$	**24.** $52 - 9$
25. $661 - 92 - 5$	**26.** $8 - 4 - 4$	**27.** $132 - 23 - 9$	**28.** $23 - 8 - 7$
29. $7\frac{2}{5} - 3\frac{1}{5}$	**30.** $\frac{8}{9} - \frac{2}{3}$	**31.** $4\frac{1}{4} - 2\frac{1}{16}$	**32.** $2\frac{1}{64} - 1\frac{1}{32}$
33. $2.2 - 0.1$	**34.** $87.7 - 86.9$	**35.** $3.76 - 0.07$	**36.** $2.99 - 1.01$
37. $23.1 - 2.6 - 20$	**38.** $8.9 - 0.7 - 7$	**39.** $6.32 - 1.32 - 3$	**40.** $12.22 - 1.1 - 0.9$
41. 4×13	**42.** 9×55	**43.** 999×8	**44.** 81×24
45. $12 \times 5 \times 55$	**46.** $18 \times 14 \times 4$	**47.** $322 \times 3 \times 2$	**48.** $98 \times 18 \times 77$
49. $\frac{1}{4} \times 7\frac{1}{2}$	**50.** $11\frac{1}{7} \times 6\frac{3}{14}$	**51.** $2\frac{1}{4} \times \frac{3}{5}$	**52.** $\frac{1}{56} \times \frac{1}{2}$
53. 0.2×9.9	**54.** 8.9×0.44	**55.** 6.68×0.3	**56.** 88.42×1.71
57. $8.1 \times 2 \times 3$	**58.** $71.8 \times 0.1 \times 1$	**59.** $2.11 \times 12.3 \times 0$	**60.** $3.9 \times 55.5 \times 2$
61. $12 \div 2$	**62.** $19 \div 8$	**63.** $160 \div 12$	**64.** $69 \div 13$
65. $1 \div 5 \div 15$	**66.** $128 \div 2 \div 7$	**67.** $12 \div 16 \div 3$	**68.** $123 \div 8 \div 25$
69. $1\frac{4}{7} \div \frac{1}{7}$	**70.** $\frac{4}{15} \div \frac{2}{3}$	**71.** $9\frac{3}{4} \div 2\frac{1}{6}$	**72.** $\frac{1}{22} \div 4$
73. $2 \div \frac{7}{52} \div 5$	**74.** $\frac{8}{9} \div \frac{1}{5} \div \frac{2}{5}$	**75.** $3\frac{1}{9} \div 6 \div 3$	**76.** $21 \div \frac{5}{8} \div 1\frac{6}{7}$
77. $2.2 \div 7.8$	**78.** $66.7 \div 88$	**79.** $88.9 \div 5$	**80.** $771.9 \div 33.3$
81. $\frac{138 + 146 + 142}{3}$	**82.** $\frac{15 + 13 + 10 + 8}{4}$	**83.** $\frac{\$1.62 + \$14 + \$7}{3}$	**84.** $\frac{9 + 7 + 7 + 8 + 10}{5}$
85. $\frac{2.3 + 4.4 + 3.7}{4}$	**86.** $\frac{\$1.48 + \$3 + \$2.57}{3}$	**87.** $\frac{0.7 + 4.6 + 2.2}{3}$	**88.** $\frac{3 + 9 + 5 + 12}{4}$
89. $\frac{9 \times 78}{248}$	**90.** $\frac{9 \times 107}{236}$	**91.** $\frac{9 \times 124}{253}$	**92.** $\frac{9 \times 206}{276}$
93. $\frac{96}{200} = \frac{n}{15{,}000}$	**94.** $\frac{185}{250} = \frac{n}{10{,}000}$	**95.** $\frac{17}{50} = \frac{n}{600}$	**96.** $\frac{63}{150} = \frac{n}{1000}$
97. $\frac{72}{100} = \frac{n}{800}$	**98.** $\frac{51}{80} = \frac{n}{4000}$	**99.** $\frac{292}{500} = \frac{n}{11{,}000}$	**100.** $\frac{184}{400} = \frac{n}{20{,}000}$

Chapter 9

Solve.

1. 109 + 12

2. 18 + 18

3. 4024 + 1113

4. 75 + 25

5. $4\frac{4}{5} + \frac{2}{18}$

6. 0.105 + 0.9

7. 1200 + 850

8. 1053 + 387

9. 2.125 + 3.875

10. 19.3 + 7.45

11. 338 + 3040

12. 23 + 56

13. $\frac{14}{3} + \frac{17}{9}$

14. 4842 + 349

15. 63 + 76

16. 3.28 + 94.02

17. 290 + 175

18. $3\frac{1}{9} + 3\frac{1}{3}$

19. $\frac{15}{17} + \frac{6}{51}$

20. $\frac{114}{5} + \frac{1}{5}$

21. 2.16 − 0.93

22. 42 − 17

23. 109 − 38

24. 150 − 105

25. 45 − 15

26. 1956 − 1944

27. 3026 − 1599

28. $3\frac{9}{8} - 2\frac{3}{5}$

29. 4.36 − 1.23

30. 46 − 26

31. 903 − 471

32. 42 − 8

33. $12\frac{4}{3} - 2\frac{13}{9}$

34. 14 − 3

35. $\frac{116}{29} - \frac{7}{8}$

36. 12.1 − 3.4

37. 81 − 33

38. 11.025 − 3.75

39. 3089 − 48

40. 1276 − 392

41. 24 × 7

42. 34 × 92

43. 12 × 12 × 12

44. 4.5 × 124

45. $\frac{6}{2} \times 14\frac{1}{3}$

46. 2.028 × 0.75

47. $6\frac{1}{5} \times 8\frac{1}{3}$

48. 22 × 107

49. 2.25 × 7.8

50. 5300 × 0.3

51. 27 × 9

52. 6.4 × 2

53. $5\frac{3}{5} \times 6\frac{8}{7}$

54. $\frac{14}{33} \times \frac{12}{7}$

55. 7 × 47

56. $\frac{2}{3} \times 7542$

57. 19 × 6

58. 0.175 × 8

59. 46 × 7

60. 1.4 × 7.8

61. 189 ÷ 3

62. 12 ÷ 4

63. 140 ÷ 4 ÷ 7

64. 54 ÷ 9

65. 1988 ÷ 28

66. $\frac{7}{12} \div \frac{1}{9}$

67. $\frac{19}{32} \div \frac{1}{2}$

68. $\frac{81}{51} \div \frac{108}{12}$

69. 441 ÷ 21

70. 62.92 ÷ 7.15

71. 168 ÷ 3

72. 2184 ÷ 91

73. $\frac{4}{3} \div 6$

74. $6\frac{1}{4} \div 30$

75. 134 ÷ 3

76. $9\frac{1}{3} \div 3\frac{4}{3}$

77. 9.375 ÷ 3

78. 9300 ÷ 30

79. 12 ÷ 8

80. 1945 ÷ 5

81. $\frac{1}{6} \times \frac{1}{2} \times \frac{3}{5}$

82. $\frac{3}{8} \times \frac{6}{10} \times \frac{1}{2}$

83. 0.72 × 0.72

84. 0.48 × 0.65

85. 16% of 180

86. 55% of 2000

87. 63% of 30

88. 80% of 365

89. $\frac{18}{40} = n\%$

90. $\frac{6}{240} = n\%$

91. $\frac{31}{65} = n\%$

92. $\frac{74}{300} = n\%$

93. 7 × 8 × 5

94. 26 × 18 × 24

95. 6 × 12 × 8 × 4

96. 10 × 9 × 8 × 7 × 6

97. 6!

98. 4! − 2!

99. 5! + 3!

100. 4! × 3!

Chapter 10

Solve.

1. $7 + 99$	**2.** $8 + 15$	**3.** $16 + 73$	**4.** $28 + 24$
5. $1 + 992 + 55$	**6.** $58 + 4 + 3$	**7.** $731 + 13 + 86$	**8.** $82 + 22 + 707$
9. $6\frac{3}{4} + 2\frac{1}{8}$	**10.** $\frac{1}{16} + 9\frac{1}{2}$	**11.** $2\frac{3}{8} + 8\frac{1}{16}$	**12.** $\frac{2}{3} + 3\frac{1}{6}$
13. $6\frac{1}{4} + 6\frac{7}{16}$	**14.** $5\frac{9}{17} + 3\frac{1}{17}$	**15.** $\frac{2}{5} + 1\frac{1}{13}$	**16.** $4\frac{3}{5} + 3\frac{1}{3}$
17. $1.2 + 15.1$	**18.** $3.7 + 26.1$	**19.** $9.9 + 13$	**20.** $2.02 + 0.01$
21. $142 - 9$	**22.** $9 - 4$	**23.** $60 - 13$	**24.** $742 - 89$
25. $551 - 12 - 5$	**26.** $918 - 4 - 121$	**27.** $32 - 3 - 9$	**28.** $103 - 8 - 77$
29. $2\frac{1}{5} - 1\frac{1}{5}$	**30.** $2\frac{7}{9} - \frac{2}{9}$	**31.** $\frac{3}{4} - \frac{1}{16}$	**32.** $\frac{1}{2} - \frac{1}{32}$
33. $12.2 - 1.8$	**34.** $0.7 - 0.6$	**35.** $3.01 - 0.57$	**36.** $2.22 - 2.11$
37. $44.1 - 19 - 11$	**38.** $11.9 - 2.7 - 7$	**39.** $0.92 - 0.9 - 0.02$	**40.** $1.02 - 1 - 0.01$
41. 2×58	**42.** 9×45	**43.** 19×98	**44.** 8×1003
45. $68 \times 25 \times 1$	**46.** $8 \times 14 \times 44$	**47.** $12 \times 33 \times 86$	**48.** $2 \times 89 \times 70$
49. $\frac{3}{4} \times 11\frac{1}{2}$	**50.** $4\frac{2}{5} \times 1\frac{1}{15}$	**51.** $\frac{2}{3} \times 2\frac{1}{5}$	**52.** $\frac{1}{3} \times \frac{1}{32}$
53. 8.9×8.9	**54.** 5.9×0.13	**55.** 9.18×0.3	**56.** 0.42×0.11
57. $8.7 \times 0.2 \times 2$	**58.** $8.8 \times 0.4 \times 6$	**59.** $2.17 \times 2.3 \times 0$	**60.** $3.9 \times 7.4 \times 6$
61. $2 \div 2$	**62.** $16 \div 8$	**63.** $96 \div 12$	**64.** $9 \div 33$
65. $1 \div 2 \div 8$	**66.** $8 \div 1 \div 2$	**67.** $100 \div 5 \div 10$	**68.** $20 \div 8 \div 5$
69. $2\frac{1}{2} \div 3\frac{1}{7}$	**70.** $\frac{8}{9} \div \frac{1}{3}$	**71.** $11\frac{1}{6} \div 3\frac{1}{6}$	**72.** $\frac{1}{2} \div 3$
73. $4 \div \frac{1}{23} \div 5$	**74.** $\frac{8}{9} \div \frac{1}{5} \div \frac{1}{3}$	**75.** $33\frac{1}{2} \div 1 \div 3$	**76.** $3 \div \frac{1}{3} \div \frac{6}{7}$
77. $0.2 \div 0.03$	**78.** $0.89 \div 99$	**79.** $7.2 \div 9.2$	**80.** $7.66 \div 0.6$
81. 36 min = n h	**82.** 18 min = n h	**83.** 1 h 25 min = n h	**84.** 3 h 38 min = n h
85. 2% of 75 g	**86.** 10% of 102 mg	**87.** $\frac{1}{2}$% of 90 mg	**88.** 120% of 26 g
89. $\frac{5}{12} \times 660$	**90.** $\frac{3}{5} \times 540$	**91.** $1\frac{2}{3} \times 198$	**92.** $1\frac{7}{15} \times 300$
93. $\frac{20}{24} \times \$42$	**94.** $\frac{50}{24} \times \$70$	**95.** $\frac{72}{24} \times \$101$	**96.** $\frac{35}{24} \times \$58$
97. 5% of \$1.59	**98.** \$17.53 + \$18.79	**99.** $168 \div 22$	**100.** \$42.63 − \$7.94

Chapter 11

Solve.

1. $29 + 47$
2. $94 + 54$
3. $2002 + 464$
4. $128 + 74$
5. $14 + 44$
6. $1.59 + 9.05$
7. $1245 + 1235$
8. $9.6 + 7.5$
9. $29 + 6$
10. $\frac{6}{23} + 14\frac{9}{46}$
11. $6\frac{1}{8} + 6\frac{3}{4}$
12. $632 + 438$
13. $12.35 + 67.8$
14. $1309 + 427$
15. $398 + 923$
16. $459 + 948$
17. $35.7 + 13.4$
18. $0.98 + 0.72$
19. $\frac{9}{17} + \frac{12}{51}$
20. $\frac{213}{100} + \frac{45}{150}$
21. $47 - 23$
22. $2496 - 2223$
23. $49 - 14$
24. $8.6 - 2.7$
25. $19\frac{1}{4} - 12\frac{7}{8}$
26. $143 - 47$
27. $\frac{15}{13} - \frac{3}{13}$
28. $2\frac{1}{9} - 1\frac{1}{3}$
29. $5.66 - 2.87$
30. $67 - 34$
31. $1308 - 711$
32. $28 - 8$
33. $\frac{12}{5} - 2\frac{1}{9}$
34. $\frac{45}{30} - \frac{29}{100}$
35. $5\frac{1}{2} - \frac{4}{9}$
36. $8.87 - 2.5$
37. $5900 - 3009$
38. $46 - 33$
39. $147 - 88$
40. $4709 - 3800$
41. 123×6
42. 6×6
43. 24×9
44. 6.4×0.7
45. $7\frac{3}{8} \times 4\frac{5}{6}$
46. 0.67×6
47. $12\frac{1}{9} \times 1\frac{1}{3}$
48. 94×4
49. 42×9
50. 7.4×2.3
51. 19×8
52. 4023×0.006
53. 22×11
54. 5.45×12.8
55. 178×349
56. 54×98
57. $39\frac{2}{3} \times \frac{1}{3}$
58. $\frac{54}{55} \times \frac{1}{6}$
59. 7×12
60. $\frac{9}{16} \times \frac{7}{32}$
61. $0.875 \div 0.125$
62. $185 \div 5$
63. $468 \div 156$
64. $1024 \div 64$
65. $4.59 \div 9$
66. $338 \div 20$
67. $81 \div 3$
68. $50041 \div 71$
69. $968 \div 22$
70. $6\frac{2}{5} \div 8$
71. $\frac{9}{5} \div \frac{3}{14}$
72. $1\frac{1}{7} \div \frac{1}{14}$
73. $13\frac{1}{3} \div 4\frac{2}{5}$
74. $4\frac{2}{5} \div 0.11$
75. $2937 \div 881.1$
76. $\frac{14}{3} \div \frac{10}{9}$
77. $3.06 \div 0.9$
78. $1725 \div 69$
79. $23 \div 8$
80. $1400 \div 32$
81. 70% of \$1633
82. 85% of \$4781
83. 75% of \$3094
84. 90% of \$8582
85. 80% of \$15.46
86. 80% of \$78.21
87. 6% of 250
88. 32% of 400
89. 0.003×89
90. $1.36(14)(0.87)$
91. $1.36(15)(0.91)$
92. $0.80(220 - 22)$
93. $0.80 \times (220 - 55)$
94. $0.9(30) - 0.7(6) - 0.2$
95. $2.8(4) - 5(0.8) - 0.52$
96. $6.3(4) - 2.4(4) - 6.38$
97. $224(1 - 0.15)$
98. $108(1 - 0.174)$
99. $\frac{121}{1 - 0.12}$
100. $\frac{96}{1 - 0.14}$

Chapter 12

Solve.

1. $15 + 79$
2. $9 + 102$
3. $66 + 23$
4. $806 + 94$
5. $991 + 2 + 75$
6. $8 + 12 + 76$
7. $887 + 5 + 120$
8. $32 + 18 + 47$
9. $8\frac{1}{3} + 1\frac{2}{3}$
10. $1\frac{1}{7} + 14\frac{1}{2}$
11. $6\frac{3}{5} + 3\frac{7}{15}$
12. $\frac{1}{9} + \frac{1}{6}$
13. $3\frac{3}{24} + 8\frac{5}{6}$
14. $\frac{4}{7} + 3\frac{1}{3}$
15. $19\frac{2}{9} + 20\frac{1}{3}$
16. $33\frac{7}{15} + 3\frac{1}{2}$
17. $32.5 + 6.05$
18. $1.79 + 8.1$
19. $3.45 + 6$
20. $87.4 + 3.86$
21. $332 - 9$
22. $49 - 39$
23. $124 - 23$
24. $42 - 19$
25. $326 - 82 - 1$
26. $66 - 4 - 54$
27. $65 - 11 - 9$
28. $103 - 28 - 7$
29. $\frac{4}{5} - \frac{1}{5}$
30. $6\frac{2}{5} - \frac{1}{3}$
31. $14\frac{1}{4} - 3\frac{5}{16}$
32. $\frac{3}{64} - \frac{1}{32}$
33. $0.2 - 0.1$
34. $8.6 - 0.2$
35. $7.34 - 7.32$
36. $20 - 11.06$
37. $9.1 - 3.9 - 5$
38. $76.2 - 0.1 - 76$
39. $24.9 - 0.9 - 21$
40. $90 - 34.89 - 0.11$
41. 2×899
42. 9×76
43. 11×3
44. 62×26
45. $4 \times 35 \times 15$
46. $8 \times 400 \times 2$
47. $54 \times 45 \times 4$
48. $12 \times 876 \times 7$
49. $7\frac{3}{4} \times 1\frac{4}{5}$
50. $\frac{2}{5} \times \frac{1}{24}$
51. $5\frac{3}{4} \times \frac{4}{5}$
52. $\frac{7}{12} \times \frac{6}{19}$
53. 15.2×7.4
54. 0.6×0.33
55. 7.99×2.3
56. 0.42×0.6
57. $6 \times 0.7 \times 3$
58. $4.4 \times 0.6 \times 1$
59. $111 \times 1.9 \times 0$
60. $87 \times 15.5 \times 2$
61. $4 \div 2$
62. $75 \div 25$
63. $67 \div 670$
64. $122 \div 244$
65. $12 \div 3 \div 3$
66. $88 \div 2 \div 11$
67. $87 \div 75 \div 2$
68. $3 \div 8 \div 5$
69. $5\frac{1}{7} \div \frac{1}{7}$
70. $\frac{2}{3} \div \frac{2}{3}$
71. $10\frac{5}{6} \div 2\frac{1}{12}$
72. $\frac{1}{2} \div 3$
73. $7 \div \frac{7}{56} \div 4$
74. $5\frac{8}{13} \div \frac{1}{2} \div \frac{2}{3}$
75. $3\frac{1}{2} \div 5 \div 10$
76. $45 \div \frac{3}{8} \div \frac{8}{9}$
77. $0.2 \div 4$
78. $0.29 \div 0.09$
79. $1.01 \div 1.01$
80. $56 \div 0.6$
81. $4\frac{1}{2}\%$ of \$16.37
82. 6% of \$180.95
83. 5% of \$746.81
84. $6\frac{1}{2}\%$ of \$499.36
85. 7.51% of \$360.35
86. 7.51% of \$183.66
87. 7.51% of \$400.76
88. 7.51% of \$523.78
89. 18% of \$26,540
90. 5.35% of \$20,924
91. $2\frac{1}{4}\%$ of \$19,000
92. 4.96% of \$34,460
93. 1.5% of \$10,630
94. 8.34% of \$17,118
95. 6.28% of \$13,534
96. 2.84% of \$38,270
97. $\frac{\$6.85}{\$13.42} = n\%$
98. $\frac{\$127}{\$493} = n\%$
99. $\frac{\$672}{\$2320} = n\%$
100. $\frac{\$1450}{\$1850} = n\%$

Chapter 13

Solve.

1. $1144 + 667$
2. $\frac{17}{48} + 3\frac{17}{32}$
3. $4\frac{2}{3} + 8\frac{9}{10}$
4. $3.52 + 2.8$
5. $20 + 36$
6. $19.96 + 12.06$
7. $819 + 3751$
8. $16.03 + 3.7$
9. $15.68 + 4.7$
10. $57 + 93$
11. $3090 + 345$
12. $283 + 395$
13. $4.7 + 23.9$
14. $6470 + 3330$
15. $53 + 92$
16. $2601 + 5641$
17. $2.79 + 7.96$
18. $249 + 171$
19. $9\frac{1}{2} + 4\frac{1}{4}$
20. $\frac{19}{33} + \frac{37}{150}$
21. $52 - 43$
22. $5760 - 4408$
23. $36 - 1.6$
24. $3.2 - 2.7$
25. $5.05 - 2.85$
26. $654 - 332$
27. $1590 - 881$
28. $45 - 6$
29. $29\frac{1}{8} - 18\frac{1}{9}$
30. $630 - 595$
31. $4\frac{2}{5} - 3\frac{7}{10}$
32. $15\frac{3}{8} - 14\frac{7}{16}$
33. $\frac{9}{23} - \frac{15}{46}$
34. $\frac{21}{5} - 0.2$
35. $7\frac{4}{7} - \frac{19}{21}$
36. $12.5 - 6.85$
37. $548 - 379$
38. $271 - 193$
39. $78.9 - 9.07$
40. $938 - 645$
41. 47×49
42. 7×9
43. 520×3
44. 0.025×400
45. 5.44×20
46. 960×2.5
47. 173×53
48. 214×9.4
49. $3\frac{7}{15} \times 2\frac{4}{9}$
50. 9.33×5
51. $5\frac{3}{4} \times 123$
52. 8.32×4.25
53. 31×15
54. 2.25×3.4
55. 92×27
56. 11.125×8
57. $4\frac{1}{3} \times 3\frac{1}{4}$
58. $9\frac{1}{7} \times 7\frac{1}{9}$
59. 61×13
60. $\frac{101}{3} \times 2\frac{5}{11}$
61. $841 \div 29$
62. $7\frac{1}{8} \div \frac{741}{13}$
63. $\frac{12}{7} \div \frac{3}{2}$
64. $17\frac{1}{3} \div 51\frac{1}{9}$
65. $12.1 \div 22$
66. $448 \div 1120$
67. $82 \div 6$
68. $1211 \div 14$
69. $90.6 \div 151$
70. $1911 \div 13$
71. $1266 \div 211$
72. $9030 \div 300$
73. $1577 \div \frac{12}{2}$
74. $8\frac{1}{2} \div 2$
75. $1906 \div 848$
76. $96\frac{1}{3} \div 6$
77. $84.03 \div 2801$
78. $15.04 \div 32$
79. $2500 \div 31.25$
80. $1479 \div 493$
81. $741 + $83 + $263 → $\$741 + \$83 + \$263$
82. $\$87 + \$49 + \$7 + \$6$
83. $14{,}036 - 7548.6$
84. $9336.3 - 4638.5$
85. $7\frac{1}{2}\%$ of \$24,500
86. $4\frac{3}{4}\%$ of \$9062
87. 20% of \$2113
88. 70% of \$479
89. $\frac{\$7200}{\$10{,}000} = n\%$
90. $\frac{\$6264}{\$8700} = n\%$ 72
91. $\frac{\$3980}{\$6600} = n\%$
92. $\frac{\$2600}{\$5430} = n\%$
93. $3.4 \times \$207$
94. $1.9 \times \$255$
95. $3.4 \times \$81$
96. $4.15 \times \$163$
97. $\$0.44 \times 376$
98. $\$0.35 \times 129$
99. $\$0.41 \times 532$
100. $\$0.52 \times 361$

Chapter 14

Solve.

1. $3 + 19$
2. $8 + 25$
3. $86 + 93$
4. $11 + 44$
5. $28 + 2 + 95$
6. $658 + 204 + 3$
7. $32 + 10 + 76$
8. $90 + 7 + 3$
9. $8\frac{3}{4} + 1\frac{1}{4}$
10. $1\frac{2}{15} + 4\frac{1}{5}$
11. $3\frac{1}{4} + \frac{3}{8}$
12. $\frac{1}{9} + 2\frac{5}{6}$
13. $6\frac{3}{5} + 1\frac{7}{30}$
14. $2\frac{1}{6} + \frac{1}{18}$
15. $20\frac{2}{7} + 1\frac{1}{3}$
16. $\frac{1}{5} + 3\frac{1}{10}$
17. $90.8 + 6.2$
18. $11.44 + 2.99$
19. $0.87 + 12$
20. $34.8 + 2.99$
21. $32 - 2$
22. $119 - 44$
23. $1122 - 234$
24. $685 - 199$
25. $67 - 12 - 5$
26. $100 - 14 - 6$
27. $123 - 23 - 90$
28. $854 - 23 - 142$
29. $3\frac{1}{5} - 3\frac{1}{5}$
30. $3\frac{2}{3} - \frac{1}{3}$
31. $2\frac{5}{8} - 2\frac{1}{16}$
32. $\frac{1}{2} - \frac{1}{12}$
33. $0.5 - 0.4$
34. $14.7 - 11.7$
35. $6.75 - 2.65$
36. $556.6 - 333.5$
37. $10.7 - 4.22 - 2$
38. $67 - 3.4 - 10$
39. $3.32 - 2.5 - 0.82$
40. $32.6 - 0.7 - 1.3$
41. 1×2333
42. 42×5
43. 119×23
44. 18×14
45. $19 \times 5 \times 2$
46. $508 \times 4 \times 1$
47. $70 \times 2 \times 16$
48. $42 \times 23 \times 77$
49. $1\frac{3}{4} \times 3\frac{1}{2}$
50. $\frac{2}{15} \times \frac{5}{6}$
51. $12\frac{1}{2} \times 2\frac{4}{9}$
52. $\frac{2}{7} \times \frac{7}{32}$
53. 7.9×1.9
54. 4.99×0.21
55. 3.1×1.3
56. 22×9.09
57. $0.2 \times 0.1 \times 9$
58. $2.7 \times 3 \times 1.1$
59. $12.1 \times 0.3 \times 0$
60. $23.8 \times 75 \times 2$
61. $25 \div 5$
62. $2.5 \div 5$
63. $549 \div 3$
64. $9 \div 3$
65. $1 \div 2 \div 4$
66. $18 \div 4 \div 2$
67. $972 \div 6 \div 8$
68. $13 \div 1 \div 1$
69. $6\frac{1}{4} \div \frac{1}{8}$
70. $\frac{2}{9} \div \frac{2}{3}$
71. $4\frac{3}{4} \div \frac{1}{16}$
72. $\frac{1}{2} \div 4$
73. $2 \div \frac{1}{34} \div 4$
74. $\frac{1}{9} \div \frac{1}{3} \div \frac{2}{3}$
75. $\frac{1}{6} \div 6 \div 3$
76. $3 \div \frac{3}{8} \div \frac{5}{6}$
77. $0.2 \div 1$
78. $12 \div 0.61$
79. $8.2 \div 3$
80. $7 \div 0.14$
81. 18 min + 1 h 43 min
82. 36 min + 2 h 40 min
83. 90 min + 2 h 8 min
84. 6 h 29 min + 52 min
85. 8 h − 3 h 12 min
86. 1 h 43 min − 90 min
87. 2 h 4 min − 35 min
88. 3 h 18 min − 89 min
89. $\frac{2}{3} \times \$17.01$
90. $215 \times \$0.17$
91. $\frac{1}{4} \times \$37.68$
92. $129 \times \$0.23$
93. 28% × $48.50
94. 15% of $237
95. 73% of $74
96. 44% × $184
97. 8% of $363
98. 7% of $251.84
99. 6% of $158.23
100. 8% of $211.50

Chapter 15

Solve.

1. $74 + 29 + 53$
2. $46 + 81 + 98$
3. $\$136.84 + \$267.38$
4. $\$87.62 + \$420.54$
5. $15.324 + 8.077$
6. $8.9 + 16 + 13.68$
7. $1726 + 3697$
8. $\$24{,}065 + \$19{,}367$
9. $6\frac{3}{8} + 7\frac{3}{4}$
10. $\frac{8}{15} + 2\frac{2}{3}$
11. $11\frac{7}{12} + 5\frac{1}{2}$
12. $2\frac{3}{5} + 4 + 1\frac{2}{5}$
13. $0.1856 + 0.4737$
14. $3\frac{4}{7} + 8\frac{2}{3}$
15. $12\frac{3}{4} + 37.5$
16. $11 + 16\frac{5}{9} + 7\frac{1}{6}$
17. $\$6.94 + \$8.28$
18. $84 + 8 + 65 + 3$
19. $\$36.84 + \$609.55$
20. $9.6 + 17.56 + 6.35$
21. $1703 - 867$
22. $\$23 - \$15.42$
23. $518 - 339$
24. $2.4 - 1.0584$
25. $8\frac{1}{2} - 2\frac{3}{10}$
26. $63 - 18\frac{9}{11}$
27. $2006 - 948$
28. $43\frac{9}{24} - 32\frac{5}{6}$
29. $3.064 - 2.2336$
30. $\$3924 - \$1996$
31. $\$1811 - \$1256.69$
32. $96.328 - 87.086$
33. $18\frac{1}{2} - 16\frac{7}{8}$
34. $9\frac{2}{3} - 5\frac{3}{4}$
35. $\$59.32 - \$37.64$
36. $43 - 18\frac{3}{16}$
37. $\$8622.71 - \$594.83$
38. $6.3 - 0.642$
39. $1093 - 884$
40. $4364 - 3767$
41. 83×236
42. $\$54.60 \times 16.9$
43. $\$158 \times 47$
44. 16.3×94
45. $\frac{3}{8} \times \frac{2}{9}$
46. $23 \times 7\frac{1}{2}$
47. $1\frac{7}{8} \times 5\frac{1}{3}$
48. 0.047×73
49. 768×356
50. 108.4×0.64
51. $\$0.53 \times 1254$
52. 4317×604
53. $12\frac{3}{8} \times \frac{7}{22}$
54. $11\frac{2}{3} \times 3\frac{4}{7}$
55. $1\frac{4}{21} \times \frac{3}{5}$
56. $3\frac{1}{5} \times 1\frac{1}{8}$
57. $\$374.93 \times 26$
58. $\$24.56 \times 4.5$
59. 2983×86
60. 3.724×6.8
61. $931 \div 3.8$
62. $3185.04 \div 92$
63. $89.804 \div 26$
64. $6474 \div 78$
65. $10{,}200 \div 150$
66. $1\frac{5}{8} \div 7$
67. $4.984 \div 0.08$
68. $\frac{17}{20} \div 1\frac{3}{5}$
69. $0.4225 \div 6.5$
70. $\$8232 \div 56$
71. $409.6 \div 1.6$
72. $\$26.40 \div \$0.15$
73. $16\frac{2}{3} \div 2\frac{2}{9}$
74. $18 \div 4\frac{1}{2}$
75. $10\frac{1}{2} \div 8\frac{2}{5}$
76. $34{,}216 \div 94$
77. $\$519.52 \div 34$
78. $12{,}236 \div 28$
79. $\$206.28 \div \$0.36$
80. $\$962.80 \div 83$
81. $\$2166 \div 4.3$
82. $\$3248 \div 4.3$
83. $\$17.68 \times 40 \times 50$
84. $\$9.84 \times 40 \times 50$
85. $\$763 + \$280 + \$55$
86. $\$825 + \$546 + \$387$
87. $\frac{\$3164 - \$1870}{4}$
88. $\frac{\$2590 - \$1298}{4}$
89. 20% of \$126,000
90. 22% of \$163,900
91. 20% of \$94,500
92. 25% of \$108,700
93. $\$9.43 \times 93.6$
94. $\$8.92 \times 78.5$

Chapter 16

Solve.

1. $84 + 57$
2. $626 + 325 + 8$
3. $\$12.94 + \$283.83$
4. $846 + 392 + 83$
5. $\$4483 + \$1928$
6. $15.374 + 6.087$
7. $156 + 18 + 743$
8. $1907 + 17 + 364$
9. $4\frac{1}{5} + 3\frac{2}{3}$
10. $\$685.46 + \$264.75$
11. $\frac{5}{16} + \frac{3}{8}$
12. $6\frac{3}{4} + 2\frac{9}{20}$
13. $\$43 + \$84 + \$67$
14. $0.7 + 0.13 + 0.58$
15. $\$56.13 + \$73.76$
16. $164.08 + 19.93$
17. $\frac{3}{5} + \frac{2}{3} + \frac{1}{4}$
18. $6 + 3\frac{4}{9} + 2\frac{1}{3}$
19. $81.06 + 35.74$
20. $8\frac{11}{12} + 4\frac{5}{8}$
21. $\$12 - \$4.45$
22. $3\frac{1}{3} - 2\frac{1}{2}$
23. $7\frac{5}{16} - 5\frac{1}{4}$
24. $6 - 3.502$
25. $11.3 - 9.64$
26. $3536 - 2988$
27. $\$200.12 - \$83.47$
28. $1581 - 894$
29. $602.083 - 47.59$
30. $806 - 673$
31. $147.13 - 129.8$
32. $\$542.82 - \$462.87$
33. $11 - 6\frac{14}{23}$
34. $5\frac{2}{9} - 2\frac{1}{4}$
35. $1612 - 1284$
36. $13\frac{8}{11} - 7\frac{26}{33}$
37. $1675 - 1096$
38. $\$249.11 - \$229.94$
39. $\$250 - \$168.46$
40. $2.831 - 1.8256$
41. $\$13.76 \times 82$
42. $\$1734 \times 3.5$
43. 1635×76
44. 16.4×19.3
45. $\frac{3}{7} \times \frac{4}{5}$
46. 26.47×53
47. $3\frac{1}{2} \times 2\frac{1}{4}$
48. $2\frac{2}{5} \times \frac{5}{6}$
49. $13\frac{1}{8} \times 1\frac{7}{35}$
50. $\$649 \times 8.6$
51. 2.54×53.8
52. $1\frac{7}{9} \times 1\frac{5}{7}$
53. 237×426
54. 1003×94
55. 27.36×10.8
56. 0.67×0.29
57. 452×0.26
58. 18×86
59. 63×548
60. $\$432.16 \times 56$
61. $3953 \div 67$
62. $7.175 \div 0.007$
63. $16.8 \div 1.4$
64. $\$1414.50 \div 1.5$
65. $0.576 \div 0.032$
66. $8976 \div 34$
67. $3283.2 \div 9.6$
68. $\$28{,}944 \div 48$
69. $18 \div 2\frac{2}{3}$
70. $171{,}360 \div 85$
71. $\$444.88 \div 6.7$
72. $1978 \div 43$
73. $1\frac{5}{8} \div 3\frac{3}{5}$
74. $2\frac{4}{11} \div 2\frac{1}{6}$
75. $15{,}283 \div 29$
76. $\$42{,}775 \div 39$
77. $\$1.68 \div 0.16$
78. $15\frac{3}{4} \div \frac{7}{8}$
79. $1\frac{7}{8} \div 1\frac{19}{26}$
80. $0.9664 \div 16$
81. 6 ft 8 in. = n in.
82. 10 ft 10 in. = n in.
83. 13 ft = n yd
84. 25 ft = n yd
85. 4 yd = n in.
86. 2 yd 5 ft = n in.
87. 39 in. = n ft
88. 83 in. = n ft
89. $17' \times 22'$
90. $12\frac{1}{2}' \times 14'$
91. $9\frac{3}{4}' \times 10'$
92. $11\frac{1}{2}' \times 10\frac{1}{2}'$
93. $\frac{1}{2} \times 9' \times 8'$
94. $\frac{1}{2} \times 21' \times 18'$
95. $\frac{1}{2} \times 11\frac{1}{2}' \times 14'$
96. $\frac{1}{2} \times 16\frac{1}{2}' \times 9\frac{1}{2}'$
97. $\frac{52 \times 30 \times 2}{100}$
98. $\frac{43 \times 36 \times 2}{100}$
99. $\left(\frac{1}{2} \times 520\right) \div 100$
100. $\left(\frac{1}{2} \times 670\right) \div 100$

Chapter 17

Solve.

1. $16 + 38 + 53$
2. $\$47.53 + \$32.65$
3. $2\frac{4}{5} + 4\frac{3}{7}$
4. $\frac{5}{8} + \frac{7}{8}$
5. $14.3 + 8.6 + 6.7$
6. $685 + 72 + 946$
7. $\$1487.84 + \$96.09$
8. $2.34 + 6.8 + 11.04$
9. $\$7 + \$2.23 + \$578$
10. $13.3 + 9 + 6.52$
11. $\$0.74 + \$2.30$
12. $2846 + 7359$
13. $7\frac{5}{6} + 4\frac{7}{8}$
14. $\frac{1}{2} + \frac{1}{3} + \frac{3}{4}$
15. $0.462 + 20.7538$
16. $6\frac{5}{6} + 8\frac{5}{9}$
17. $8.333 + 6.787$
18. $\$89.99 + \$12.41$
19. $8361 + 4708$
20. $4 + 1388 + 525$
21. $802 - 425$
22. $63 - 17.358$
23. $\$237 - \$168$
24. $\$6.97 - \$3.89$
25. $8.3 - 0.4629$
26. $7 - 3\frac{3}{8}$
27. $360 - 184$
28. $2\frac{1}{4} - 1\frac{3}{4}$
29. $\$940.03 - \$275.38$
30. $1600 - 941$
31. $\$60 - \$48.74$
32. $36.48 - 27.626$
33. $6\frac{2}{3} - 4\frac{3}{5}$
34. $13 - 0.6073$
35. $3\frac{5}{12} - 1\frac{1}{2}$
36. $7\frac{2}{9} - 5\frac{11}{15}$
37. $20.7 - 8.86$
38. $\$51 - \$29.06$
39. $6040 - 1684$
40. $7183 - 5555$
41. $\$136.04 \times 46$
42. $\$56.39 \times 0.8$
43. $\frac{7}{38} \times \frac{19}{21}$
44. $1\frac{1}{15} \times \frac{20}{21}$
45. 0.238×54
46. 47×68
47. 29×354
48. $\$78 \times 23.06$
49. 756×320
50. $\$0.47 \times 625$
51. 163.45×42.8
52. 20.36×81.4
53. $\$8170 \times 6.4$
54. 4063×187
55. 96.2×85.7
56. 676×934
57. $1\frac{13}{15} \times 1\frac{4}{21}$
58. $\$19.54 \times 768$
59. $2\frac{2}{3} \times \frac{4}{5}$
60. $1\frac{8}{27} \times 2\frac{4}{7}$
61. $35{,}112 \div 42$
62. $\frac{4}{5} \div \frac{3}{10}$
63. $8 \div 1\frac{9}{11}$
64. $\$540 \div 3.6$
65. $53{,}575 \div 25$
66. $277.733 \div 61$
67. $\$23{,}714 \div 71$
68. $62{,}269 \div 73$
69. $0.63486 \div 0.018$
70. $\$57.80 \div \$0.85$
71. $4.416 \div 96$
72. $71{,}328 \div 96$
73. $\$6.84 \div 36$
74. $45 \div 125$
75. $352{,}176 \div 58$
76. $65.28 \div 4.8$
77. $1\frac{7}{8} \div \frac{7}{12}$
78. $3\frac{1}{2} \div 1\frac{5}{9}$
79. $\$16.80 \div \$1.20$
80. $3\frac{3}{21} \div 1\frac{11}{14}$
81. $684 \div 23$
82. $1095 \div 31$
83. $\$0.14 \times 2364$
84. $\$0.16 \times 4507$
85. $7\% \times \$78$
86. $6\frac{1}{2} \times \$211.86$
87. 1.4610×1300
88. 0.6637×2400
89. $435 \div 30.99$
90. $860 \div 0.5947$
91. $\frac{60 \times 26.3}{174}$
92. $\frac{5.4 \times 43}{60}$
93. $\frac{60 \times 18.6}{9.2}$
94. $\frac{7086}{200} + 38.25$
95. $1 + \frac{46}{30}$
96. $1 + \frac{50}{34}$
97. $\frac{35}{1.3 \times 2.267}$
98. $\frac{35}{0.8 \times 2.606}$
99. $\frac{35}{0.9 \times 2.882}$
100. $\frac{1}{2} \times \frac{1}{2} \times \frac{1}{2} \times \frac{1}{2}$

Chapter 18

Solve.

1. $826 + 876$
2. $84.6 + 19.8$
3. $\$12.09 + \$7 + \$8.38$
4. $\$19 + \$58 + \$63$
5. $\$5.42 + \$68.59$
6. $1640 + 3572$
7. $3.4 + 8.5 + 2.5$
8. $375 + 63 + 108$
9. $8\frac{7}{16} + 5\frac{3}{8}$
10. $\frac{5}{9} + \frac{2}{3}$
11. $16\frac{2}{5} + 8.4$
12. $2.68 + 5.1 + 0.34$
13. $14.251 + 3 + 18.68$
14. $12.6 + 9.37 + 2.59$
15. $\$4.16 + \$9.53$
16. $9562 + 854 + 17$
17. $26 + 32 + 59 + 84$
18. $5\frac{3}{7} + 4\frac{1}{8}$
19. $6\frac{3}{4} + 1\frac{1}{2} + 8\frac{2}{5}$
20. $\$67.24 + \$20.77$
21. $\$73 - \$28.45$
22. $5\frac{5}{12} - 3\frac{4}{9}$
23. $8.86 - 5.93$
24. $2\frac{2}{3} - 2\frac{1}{2}$
25. $63 - 28$
26. $\$9.23 - \$4.80$
27. $716 - 437$
28. $6.36 - 2.256$
29. $34 - 15.028$
30. $74.364 - 59.657$
31. $\$0.72 - \$0.56$
32. $6103 - 2944$
33. $4008 - 3237$
34. $7.1904 - 3.5435$
35. $9235 - 1689$
36. $\$144.31 - \$64.63$
37. $\$8008 - \$13.67$
38. $7\frac{2}{3} - 5\frac{5}{8}$
39. $10\frac{9}{20} - 5\frac{1}{2}$
40. $8\frac{1}{4} - 4\frac{16}{25}$
41. $\$16.54 \times 39.5$
42. 3.8×6.15
43. 782×24
44. $\$582 \times 37$
45. $\frac{4}{15} \times \frac{1}{6}$
46. $\frac{12}{13} \times 1\frac{1}{4}$
47. 12.07×37.2
48. $11\frac{1}{3} \times 2\frac{1}{4}$
49. 926×445
50. $\$7029 \times 15.6$
51. $2\frac{4}{9} \times 1\frac{5}{16}$
52. $5\frac{1}{3} \times \frac{9}{10} \times \frac{5}{12}$
53. 0.4706×265
54. 3827×16
55. $\$0.83 \times 2675$
56. 65×34
57. $\$894.37 \times 48$
58. 139.4×50.18
59. 853×607
60. 1543×6.8
61. $\frac{5}{8} \div 3\frac{1}{3}$
62. $\$16,014 \div 12$
63. $9 \div 2\frac{1}{7}$
64. $1\frac{5}{12} \div 5\frac{1}{10}$
65. $26,282 \div 34$
66. $120.4 \div 8.6$
67. $\$520.50 \div 1.5$
68. $612 \div 45$
69. $3\frac{7}{11} \div 12$
70. $6432 \div 67$
71. $2\frac{8}{9} \div 3\frac{3}{5}$
72. $\$131.41 \div 0.085$
73. $\$4.41 \div 98$
74. $1.2532 \div 2.6$
75. $11,776 \div 46$
76. $21,624 \div 53$
77. $6.6 \div 0.88$
78. $\$2591.98 \div 19$
79. $5.888 \div 92$
80. $78.455 \div 71$
81. $\$34.875 \times 400$
82. $\$11.625 \times 200$
83. $92\frac{1}{4}\% \times \$1000$
84. $78\frac{1}{2}\% \times \$1000$
85. $100\% - 7.3\%$
86. $\$3000 \div \$43.72$
87. $\$10,000 \div \$17.64$
88. $\$6000 \div \$25.08$
89. $\frac{\$7}{\$16} = n\%$
90. $\frac{\$12}{\$45} = n\%$
91. $\frac{\$64}{\$27} = n\%$
92. $\frac{\$0.40}{\$18} = n\%$
93. 91.1% of $\$592$
94. 87.2% of $\$635$
95. 15% of $\$40,839$
96. 12% of $\$36,427$

GLOSSARY

acre A measure of land equaling 43,560 sq ft, or $8\frac{1}{4}$ sq mi.

adjusted gross income The total taxable earnings and income.

all together Collectively.

amount financed The difference between the selling price and the down payment.

annual cost The amount of money paid every year.

annual percentage rate (APR) The yearly amount charged to finance a purchase.

annual premium The total cost every year for insurance.

assessed value The estimated value of anything for tax purposes.

average The sum of the values in a set divided by the number of values in that set. See *mean*.

average daily balance Average of the daily balances in a credit or bank account.

axis (plural: axes) A reference line on a graph showing the scale.

balance The amount of money in a bank account.

bank fees Charges for various bank services, for example, monthly account fee.

base price The price set by the manufacturer.

basic rate The cost based on various factors, for example, insurance.

benefit A service or monetary incentive to attract qualified employees.

blue collar job Job that uses manual skills, usually refers to industry.

bonds A type of investment in which an individual lends money to a company or a government agency that will later repay the amount of the loan with interest. See *stock*.

break apart To separate the digits of a number into place values of ones, tens, hundreds, etc., in order to do mental calculations.

budget A plan for using money.

calculator A machine that quickly computes mathematical functions.

calorie intake The amount of calories in the food you eat.

cancelled check A check that has been deposited by the payee and has been cleared through the checkwriter's bank.

carbohydrates Foods containing sugars and starches that become sugar through digestion.

carrying charge The fee added to the list price when buying on an installment plan.

check *n.* A written order to a bank to pay out money from a depositor's account.

check *v.* Reworking a problem to make sure that the answer is correct.

check notation The specific way to write amounts on checks.

check register A written record of checks to keep a running balance.

checking account A bank account against which a depositor may write checks.

circle graph A graph that uses parts of a circle to compare parts of a whole.

closing costs Costs added to the price of a house, including fees and taxes, at the completion of the sale.

clustering When given addends are close in value to the same number, using that number for each addend in order to estimate the sum.

collision insurance Pays for damage to the insured's car in case of an accident.

commission A percent of the money taken in on sales, often paid to a sales clerk or agent in addition to salary or wages.

comparisons Study of the similarities and differences of two or more things.

compatible numbers Pairs of numbers close in value to given numbers that can be used instead of the given numbers in order to make a problem easier to estimate.

compound interest Interest computed on both the principal and the unpaid interest.

compound probability The probability for more than one event.

copayment Contribution of an insurance company to the expenses of the insured.

cosigner One who agrees in writing to repay a loan if the borrower cannot.

count up/count back To find an answer by starting with the first number and adding or subtracting integer by integer until you reach the answer.

counting principle A way to find all possible outcomes for a set of related events. With m choices for one event and n choices for another event, $m \times n$ gives the number of possible outcomes.

current yield Amount of interest accumulated on an investment to date.

data Information from which conclusions can be inferred.

decreased by Minus.

deductible The amount of money that the insured pays on each claim before insurance coverage begins.

deferred payment Gradual payment for an item over time.

denominator The number of equal parts into which a whole is divided; in the fraction $\frac{a}{b}$, the denominator is b.

deposit An amount of money put into a bank account or put down as a pledge or partial payment.

difference The answer to a subtraction problem.

direct dial Making a telephone call without the help of an operator.

discount A reduction from the usual price of an item.

dividends Part of a company's profits that are paid periodically to stockholders.

down payment The initial amount paid when buying an item on an installment plan.

duplicate check An exact copy of a check used for record-keeping purposes.

earned run average (ERA) In baseball, a run charged against a pitcher, not as a result of an error.

effective annual rate Compound interest rate expressed as a simple interest rate.

energy allowance The number of calories per food group needed to maintain proper weight and good health.

equivalent fractions Fractions that name the same number, for example, $\frac{1}{2}$, $\frac{2}{4}$, and $\frac{3}{6}$.

escrow Bond or deed put into the care of a third party until certain conditions are met.

estate value For tax purposes, the net value of a person's property after his or her death.

estimation Methods of deriving an approximation of an exact answer.

exemptions Persons claimed by a taxpayer as legally dependent on the taxpayer for support.

expectation The expected outcome based upon the probability and the number of events.

factor A number that is multiplied by another number to yield a product.

factorial The product of a certain series of consecutive whole numbers that begins with a certain number and decreases to 1.

favorable outcomes The desired chances of occurrence in a probability problem.

federal excise tax Tax on certain federally regulated goods and services.

FICA Federal Insurance Contributions Act. Amount of money withheld from a person's pay for social security. See *social security*.

filing status On a tax form, the categories of single, married, or head of household.

finance charge The amount of money charged to a credit account if the balance has not been paid.

formula A mathematical rule expressed in symbols.

fraction A numeral used to name the number of objects in a set or parts in a whole; any number expressed in the form $\frac{a}{b}$ where a is the numerator, b is the denominator, and $b \neq 0$.

frequency table A table that shows how many times an event occurs.

graph A diagram that compares numerical data. Includes bar, line, circle, coordinate graphs, and pictographs.

greater than (>) Larger than.

gross income Income before deductions.

gross pay The total amount of earnings before any deductions.

gross profit Money earned before expenses are subtracted.

handling charges The cost to process an order.

horizontal scale The scale used on the horizontal (side-to-side) axis of a graph.

hourly pay Money earned every hour.

ideal body weight Individual optimum weight based on age, body structure, and lifestyle.

increased by Plus.

independent event In compound probability, an event that does not affect the outcome of other events.

installment price The sum of the down payment and the monthly payments for an item.

insurance A contract where the insurer pays the insured a certain sum for a specified loss.

interest Money paid for the use of money, usually a percent of the amount invested, loaned, or borrowed.

interval A set containing all numbers between two given numbers.

invert To reverse an object or action; when dividing fractions, to reverse the position of the numerator and the denominator. See *reciprocal*.

Keogh plan A tax-sheltered retirement plan for self-employed people.

kilowatt hour A unit of electrical energy measurement, 1000 watts of power per hour.

knot The unit of speed of one nautical mile.

labor charge The charges on a bill for physical work.

lean body weight The difference between the actual body weight and the body fat weight.

least common denominator (LCD) The smallest common multiple of the denominators of two or more fractions.

less than (<) Smaller than.

liability insurance Automobile insurance covering injury and damage caused by the insured person.

line graph A graph using line segments to show direction or change in data.

loading charge Commissions and fees charged by a stockbroker on mutual funds.

loan fee A percent of the loan charged as a fee to the lendee.

lowest terms A fraction is in *lowest terms* when the only common factor of the numerator and denominator is 1.

market value The selling price of an item on the open market.

maturity date The year when a bond is due for payment. See *bond*.

mean The average; the sum of a set of numbers divided by the number of members in the set.

median The middle number in a set of numbers given in order.

mental math Working problems without using pencil and paper or a calculator; solving a problem "in your head."

misleading Anything that might lead to an error of judgment.

mixed number A numeral that uses a whole number and a fraction to name a number, for example, $1\frac{4}{5}$. See *fraction*.

mode The value that occurs most often in a set of data.

mortgage loan Money borrowed to buy a house.

mutual fund Money from many investors brought together to buy stocks or bonds.

nautical mile A unit of distance for sea and air travel, 6076.12 feet.

net asset value The cost of a share of a mutual fund.

net deposit The amount of money deposited after receiving cash.
net income The income remaining after taxes.
net pay The amount of money earned after deductions.
net profit The amount of money earned after expenses.
number sentence An equation.
numerator The number of equal parts of a whole that are being taken; in the fraction $\frac{a}{b}$, the numerator is *a*.
nutritionist A person specializing in the study of balanced diets for better health.

odds In probability, the ratio comparing the ways an event can and cannot occur.
operator-assisted calls A telephone call made with the help of an operator.
original amount The amount before any increase or decrease.
overtime Paid work done beyond the set working hours, usually at a higher rate of pay.
oxygen pressure Factor used in determining the amount of oxygen in blood.

par value The amount of money paid to a bond holder when the bond matures.
passbook A book to record deposits, withdrawals, and interest earned in a bank account.
payee The person to whom a check is payable.
percent (%) The ratio of a number to 100.
permutation An arrangement of objects in a set into a certain order.
personal loan A loan not requiring security, based on the borrower's good credit.
personal property The items owned by an individual.
piecework Work paid for at a fixed rate per unit.
plan A method or program for action.
possible outcomes All possible occurrences in a probability problem.
prediction Determining if the probability of an event occurring is greater than the probability of it not occurring.
premium Monthly, semiannual, or annual cost of an insurance policy.
principal An amount of money borrowed or loaned; amount upon which interest is computed.
probability The ratio of the number of times a certain outcome can occur to the number of total possible outcomes.
probability line A line showing where the probabilities fall between certainty and impossibility.
product The answer to a multiplication problem.
promissory note A written promise to repay a loan by a certain time.
pulse rate The number of heartbeats per minute.

quarterly Every 3 months, or 4 times a year.
quotient The answer to a division problem.

random Selected so that members of a set have an equal chance of being chosen.
rate A ratio used to compare two measures.
rate of assessment The percent used to calculate the assessed value of a house given the market value.
reciprocal Two numbers are *reciprocals* if their product is 1; inverse.
reconcile Procedure used to see that the checkbook balance and the bank statement agree.
remainder The number remaining when one number is divided by another; the answer to a subtraction problem.
renaming Expressing a fraction in lowest terms.
replacement cost Amount of money needed to replace an insured item.
respiration rate The total number of breaths taken every minute.
resume A statement of qualifications, education and experience, used when applying for a job.
retirement plan A plan to save funds toward retirement.

sales tax State or local tax on the sale of goods, usually a percent of the selling price.

sales tax schedule Table showing the amount of sales tax to be placed on the cost of certain items. See *sales tax*.

sample A part taken as a representative of a whole.

secured loan A loan in which a valuable item held as security can be taken by the lender if the loan is not repaid.

selling costs The sum of the base price plus all taxes and fees.

semiannually Every 6 months, or twice a year.

service charge Money charged by a bank for handling an account.

service job A job involving the sale or delivery of a product to the customer.

share One of the equal portions into which the capital of a company or corporation is divided.

shipping charges The cost of sending an item.

situation The position and condition of the problem.

social security The federal insurance program that provides retirement benefits. The amount of money received every month depends on average annual income and age at retirement.

square A quadrilateral with sides of equal length and four right angles; the product of a number multiplied by itself.

statistics Collection of data, usually numbers, relating to any topic, often shown in charts, tables, and graphs. See *data*.

sticker price The sum of the base price and the cost of any extra features.

stock Money invested in a company or a corporation.

straight commission Money earned by people in sales jobs who do not receive wages or tips; they are paid a straight percent of the sales.

strategy The course of action taken to answer a problem.

sum The answer to an addition problem.

table An arrangement of related facts, usually in rows and columns.

tally Recording an account or a score.

taxable income The amount of money that can be taxed after exemptions.

terminating debts Bills that have a final payment date.

total The answer to an addition problem.

trade-in allowance Amount of money allowed for a used article as part of the purchase price of a new item.

translate Put a word problem into number form.

tree diagram A diagram that shows all possible outcomes of an event or a set of permutations.

unit price The cost of one unit of an item.

utility costs Money spent on gas, oil, electricity, water, and telephone service.

vertical scale The scale used on the vertical (top to bottom) axis of a graph.

vital capacity The volume of air exhaled from the lungs after taking the deepest breath possible.

W-2 form Record of an employee's earnings, prepared by the employer, and attached to the employee's tax return when filing.

weekly gross income The amount of money earned every week before deductions.

weekly pay The money earned every week.

white collar job Clerical or professional job, not involving manual skills.

withholding tax Federal income taxes deducted from gross pay every pay period.

yearly pay Annual salary.

TRY THIS ANSWERS

1-1 **a.** exact **b.** estimate **c.** paper and pencil **d.** estimation

1-2 **a.** 9500 (9415) **b.** \$55 (\$60.32) **c.** 300 (232) **d.** 4.00 (4.09) **e.** 240 (244) **f.** 1600 (1603) **g.** 3900 (3940) **h.** 190 (187) **i.** \$100 (\$101.43) **j.** 100 (106)

1-3 **a.** 1163 **b.** 43,551 **c.** 18,766 **d.** \$120.70 **e.** 27.281 **f.** 1.349

1-4 **a.** 2256 **b.** 41,846 **c.** \$6.86 **d.** 26.99 **e.** 4.343 **f.** 0.2882

1-5 **a.** 137 **b.** 197 **c.** 82 **d.** 800 (797) **e.** \$15 (\$15.07)

1-6 **a.** 350 (322) **b.** \$350 (331.55) **c.** 120,000 (139,840) **d.** 1500 (1631.499) **e.** 10 (≈10.89) **f.** \$8 (\$8.266) **g.** 6 (≈6.08) **h.** 60 (≈64.95)

1-7 **a.** 3515 **b.** 57,072 **c.** \$66.50 **d.** \$3833.75 **e.** 36.75 **f.** 1.47

1-8 **a.** 708 **b.** 47 **c.** \$0.75 **d.** 1.48 **e.** 4.80 **f.** 50.00

1-9 **a.** 180 **b.** 36,189 **c.** \$4.20 **d.** \$32,760 **e.** 6530 **f.** 7004 **g.** 3009 **h.** 10,204

2-1 **a.** \$13,468 **b.** \$1,401

2-2 **a.** ≈1.4 million **b.** manufacturing **c.** 2 t **d.** 48 feedings

2-3 **a.** 300 calories **b.** utility company **c.** gardener

2-4 **a.** \$252 **b.** \$19.20

2-5 140

2-6 **a.** not reasonable—2.7 miles **b.** reasonable—18

2-7 4 sets of 3 pairs and one single pair

2-8 7 min

3-1 **a.** $\frac{2}{5}$ **b.** $\frac{2}{3}$ **c.** $\frac{2}{7}$ **d.** $\frac{1}{2}$ **e.** $\frac{5}{8}$ **f.** 9 **g.** 27 **h.** 30 **i.** 91 **j.** 72

3-2 **a.** $\frac{8}{5}$ **b.** $\frac{5}{2}$ **c.** $\frac{25}{6}$ **d.** $\frac{11}{7}$ **e.** $\frac{29}{8}$ **f.** $1\frac{4}{5}$ **g.** $1\frac{5}{7}$ **h.** 7 **i.** $1\frac{3}{5}$ **j.** 8

3-3 **a.** $\frac{2}{5}$ **b.** 6 **c.** $\frac{2}{5}$ **d.** $20\frac{1}{2}$ **e.** $6\frac{3}{8}$

3-4 **a.** 1 **b.** $2\frac{1}{2}$ **c.** $\frac{2}{7}$ **d.** $6\frac{3}{7}$ **e.** $1\frac{3}{11}$

3-5 **a.** $\frac{3}{4}$ **b.** $\frac{11}{12}$ **c.** $2\frac{1}{10}$ **d.** $5\frac{1}{16}$ **e.** $6\frac{17}{18}$

3-6 **a.** $\frac{1}{2}$ **b.** $\frac{4}{9}$ **c.** $2\frac{2}{3}$ **d.** $1\frac{3}{4}$ **e.** $1\frac{11}{15}$

3-7 **a.** 7 **b.** $8\frac{1}{2}$ **c.** $9\frac{13}{28}$ **d.** 26 **e.** 36 **f.** $16\frac{1}{4}$

3-8 **a.** $9\left(9\frac{9}{19}\right)$ **b.** \$2 (\$1.93) **c.** $6\left(5\frac{25}{31}\right)$ **d.** 20 (18.39) **e.** \$8 (\$7.66) **f.** $5\left(5\frac{1}{16}\right)$ **g.** $68\left(64\frac{2}{9}\right)$ **h.** 36 (35.7)

3-10 **a.** 1.7905 **b.** 0.5 **c.** 0.6 **d.** 2

4-1 **a.** 59% **b.** $83\frac{1}{3}\%$ **c.** 23.25% **d.** 58.4%

4-2 **a.** 0.51 **b.** 0.07 **c.** 2.75 **d.** 0.832 **e.** 0.001 **f.** $\frac{83}{100}$ **g.** $\frac{4}{25}$ **h.** $\frac{1}{500}$ **i.** $\frac{1}{6}$ **j.** $2\frac{3}{4}$

4-3 **a.** 90% **b.** 80% **c.** $66\frac{2}{3}\%$ **d.** $216\frac{2}{3}\%$ **e.** 200% **f.** 137.5% **g.** 94% **h.** 3% **i.** 220% **j.** 100% **k.** 3.25% **l.** $12\frac{1}{2}\%$

4-4 **a.** $<$ **b.** $>$ **c.** $<$ **d.** $>$ **e.** $=$ **f.** $<$ **g.** $<$ **h.** $>$

4-5 **a.** $105\% \times 50 = n$ **b.** $92 = n\% \times 200$ **c.** $14.7 = 7\% \times n$ **d.** $n = 33\frac{1}{3}\% \times 66$

e. $n = \frac{5}{6} \times 18$ **f.** $0.625 \times n = 12$
g. $\frac{1}{2} \times n = 9.37$ **h.** $\frac{2}{3} = \frac{1}{6} \times n$

4-6 **a.** 20.16 **b.** 1.47 **c.** $17.60
d. 1.68 **e.** 14.76 **f.** 0.0075
g. 27 **h.** $20.40 **i.** 80
j. 18 **k.** $6\frac{2}{3}$ **l.** 150
m. 10 **n.** 300 **o.** 5

4-7 **a.** 25% **b.** $33\frac{1}{3}$% **c.** 120%
d. $1\frac{1}{3}$% **e.** 80% (79.6%)
f. 50% (59%) **g.** $66\frac{2}{3}$% (71%)
h. 400% (404.2%)

4-8 **a.** $33\frac{1}{3}$% **b.** 60% **c.** 100%
d. 25% **e.** 37.5% **f.** 60%
g. 300% increase **h.** 62.5% decrease

4-9 **a.** $87\frac{1}{2}$% **b.** $112\frac{1}{2}$% **c.** 137.5%
d. $37\frac{1}{2}$%

5-2 **a.** B: $368 more per year
b. B: $3.12 more per week

5-3 **a.** $2028 **b.** $385 **c.** $150
d. $79.50

5-4 **a.** $277.20 **b.** $316.88 **c.** $401.63

5-5 **a.** $13.12 **b.** $1.43 **c.** $428
d. $210

5-6 **a.** $5214 **b.** $487.63 **c.** $472

5-7 **a.** $197.40 **b.** $64.50

5-8 **a.** $16,950 **b.** $16,688

5-10 The solution is $48.87 low.

6-1 $162.33

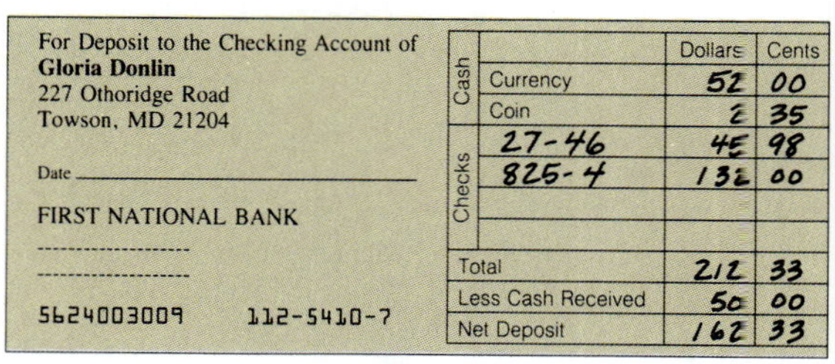
For Deposit to the Checking Account of
Gloria Donlin
227 Othoridge Road
Towson, MD 21204
Date
FIRST NATIONAL BANK
5624003009 112-5410-7

		Dollars	Cents
Cash	Currency	52	00
	Coin	2	35
Checks	27-46	45	98
	825-4	132	00
Total		212	33
Less Cash Received		50	00
Net Deposit		162	33

6-2 **a.**

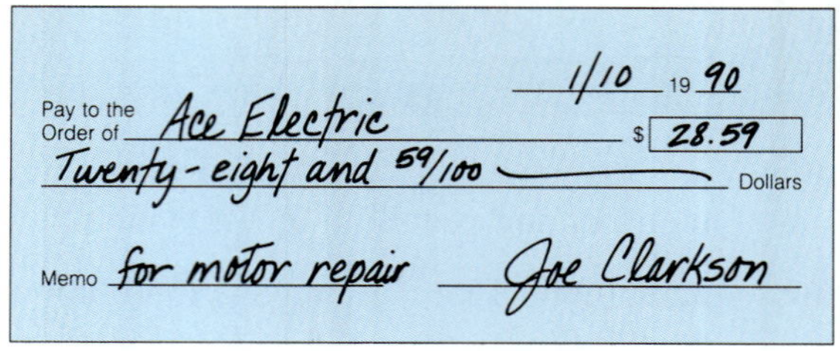
1/10 19 90
Pay to the Order of Ace Electric $ 28.59
Twenty-eight and 59/100 Dollars
Memo for motor repair Joe Clarkson

b.

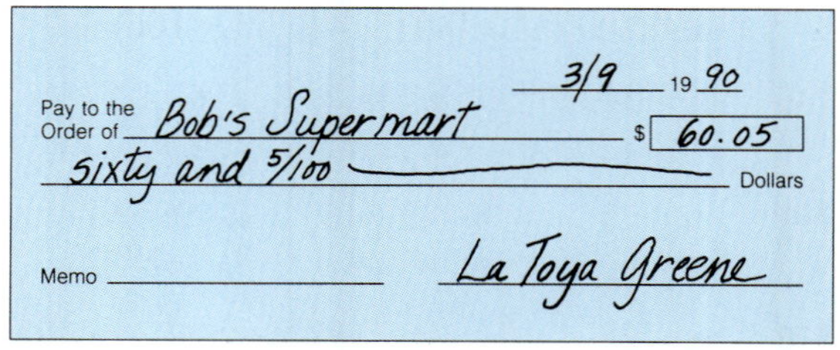
3/9 19 90
Pay to the Order of Bob's Supermart $ 60.05
sixty and 5/100 Dollars
Memo La Toya Greene

6-3 **a.** $468.94 **b.** $419.76 **c.** $341.51

6-4 **a.** disagree **b.** agree

6-5 **a.** $2640 **b.** $236.25 **c.** $163.50

6-6 **a.** $9520 **b.** $2893.75

6-7 **a.** $161.04 **b.** $206.08
c. 7.86 years or about
7 years 10 months

6-8 **a.** $456.56 **b.** $52.68 **c.** $70.24
d. $368.36 **e.** 8.2% **f.** 15.5%
g. 6.8% **h.** 10.9%

7-1 **a.** 1 lb 2 oz **b.** 98 sq ft

7-2 **a.** $47.96 **b.** $482.40 **c.** $1.67
d. $1.49

7-3 **a.** $691.70; $21.70
b. $1909.76; $409.76
c. $432.99; $7.99

7-4 **a.** $43.01 **b.** $59.53

7-5 **a.** $192.20 **b.** $220.43

7-6 **a.** $90 **b.** $20 **c.** $260
d. $29

7-7 **a.** $29.53 **b.** $18.10 **c.** $11.46

7-8 **a.** \$0.63 **b.** \$3.15 **c.** \$3.88
d. \$2.77

8-1

Interval	Tally	Frequency
16–18	IIII	4
19–21	~~IIII~~ I	6
22–24	IIII	4
25–27	II	2
28–30	~~IIII~~	5
31–33	I	1
34–36	IIII	4
37–39	II	2
40–42	II	2
43–45	IIII	4
46–48	I	1
49–51	I	1
52–54	II	2
55–57	I	1
58–60	I	1

8-2 **a.** \$105 **b.** 7A **c.** no mode

8-3 **a.** 72.5 **b.** \$14,760

8-4 **a.** $9\frac{1}{2}$ **b.** 2.79

8-7 **a.** Manuel's scores are the same on both graphs.
b. The vertical scales differ.
c. Graph A; it has a smoother incline than Graph B.

8-8 **a.** 10,200 **b.** Carlyle

9-1 **a.** $\frac{1}{7}$ **b.** 0 **c.** $\frac{2}{7}$ **d.** 1

e.

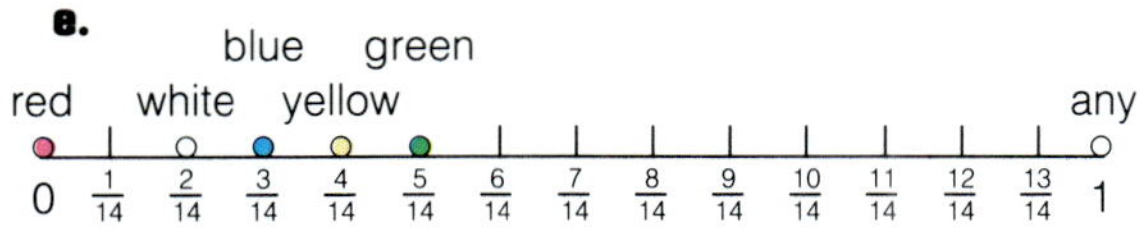

9-2 **a.** 5 to 1 **b.** 1 to 9 **c.** 2 to 9

9-3 **a.** $\frac{1}{12}$ **b.** $\frac{1}{16}$ **c.** 0.2116

9-4 **a.** about 3 days **b.** 2 prizes
c. 15 donations

9-5 **a.** $\frac{15}{250}$, or 6%, no
b. \$252 for > 50% **c.** 70%, yes
d. 50%, not predictable

9-6

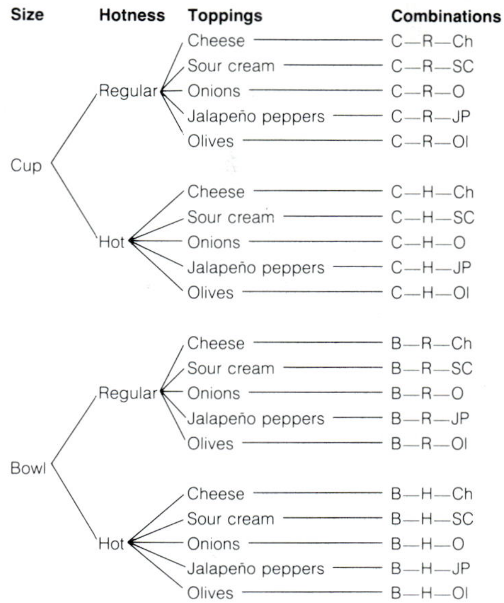

9-7 **a.** 12 **b.** 18

9-8 **a.** 24 different ways
b. 720 permutations

10-1 **a.** 3282 calories/day
b. 2373 calories/day
c. 2491 calories/day

10-2 **a.** 590 **b.** 1028.75
c. 95 more calories and 5 g more carbohydrates in rice

10-3 **a.** 575 calories **b.** 793 calories
c. 396 calories

10-4 **a.** 770 **b.** 31.5 g **c.** 0.6 mg
d. 0.028 mg **e.** less than 1.7 g

10-5 **a.** 100 cal; 30 mg Vit. C
b. 400–600 cal; 24–40 g of protein
c. 280 cal; 8 g of protein

10-6 **a.** \$219.17 **b.** \$898.33

10-7 **a.** save \$11.28 gardening
b. save \$7.35 at home

10-8 **a.** increase of 15 million acres
b. 5–10 quarts **c.** 12–18 quarts
d. 4–6 quarts

11-1 **a.** \$656.25; \$318.75
b. \$2168; \$592
c. \$1012.50; \$187.50
d. \$117.60; \$79.40

11-2 **a.** same
b. CC pays \$41 more.
c. WH pays \$112 more.
d. same

11-3 **a.** 17.71 mL; 17.974 mL
b. 19.37 mL; 19.655 mL
c. 14.69 mL; 14.927 mL

11-4 **a.** 49.8% **b.** \$6320
c. 50 births/month

11-5 **d.** 137 **e.** 166 **f.** 159
g. 3.5 liters **h.** 2.9 liters

11-6 **a.** 64 lb **b.** 152 lb

11-7 **a.** 171 lb **b.** 90 lb

11-8 **a.** \$3.70 **b.** \$18.90

12-1 **a.** \$0.35 **b.** \$0.04 **c.** \$1.20
d. \$1.05 **e.** \$87.07

12-2 **a.** \$23 **b.** \$1

12-3 **a.** \$27.56 **b.** \$58.83
c. \$175.93 **d.** \$131.25

12-4 **a.** \$9956; \$1493.40
b. \$41,067; \$7631.26
c. \$17,404; \$2610.60

12-5 **a.** \$285.66 **b.** \$875.15
c. \$329.50 **d.** \$411.48

12-6 **a.** \$9750; \$292.50
b. \$60,500; \$453.75

12-7 **a.** \$4548 **b.** \$51,280 **c.** \$91,030
d. \$0.56 **e.** \$13.79

12-8 **a.** \$0.08 **b.** \$0.07 **c.** \$0.09

12-12 no, $5 \times 3.14 \approx 15$

13-1 \$10,751.90

13-2 **a.** 19.27 mpg **b.** 26.43 mpg
c. \$653.50 **d.** \$577.38

13-3 **a.** \$5070 **b.** \$9530 **c.** 13%
d. 27%

13-4 **a.** \$2372 **b.** \$7600 **c.** \$2967.28

13-5 **a.** \$517.50 **b.** \$874 **c.** \$144
d. \$46

13-6 **a.** \$408.60; \$160.20; \$568.80
b. \$729.30; \$366.30; \$1095.60

13-7 **a.** \$350.60 **b.** \$411.25
c. \$212.20 **d.** \$321.73

13-8 \$74.79

14-1 **a.** 7:35 **b.** 8:01

14-2 **a.** \$2.25 **b.** \$7.75 **c.** \$2.75
d. \$6.25 **e.** $1\frac{1}{4}$ mi **f.** $4\frac{1}{2}$ mi
g. $1\frac{3}{4}$ mi

14-3 **a.** \$19.67; \$6.50; \$13.17
b. \$70.74; \$22.50; \$48.24

14-4 **a.** 1325 mi **b.** 50 mi/h
c. 30.9 mi/gal

14-5 **a.**

Road	Direction	Distance
State Route 3	North	12 miles
State Route 144	East	10 miles

b.

Road	Direction	Distance
State Route 148	South	11 miles
Interstate 57	South	2 miles
Interstate 24	Southeast	About 12 miles

14-6 **a.** 9:18 a.m. **b.** \$42.50

14-7 **a.** 1:35 p.m. **b.** 3 stops
c. 2 h 57 min

14-8 **a.** \$373.15; \$29.85
b. \$366.67; \$29.33
c. \$391.67; \$31.33
d. \$347.22; \$27.78
e. \$375; \$27.78
f. \$302.25; \$16.79
g. \$214.50; \$13.90

15-1 **a.** \$330

15-2 **a.** \$545 **b.** \$518 **c.** \$1180

15-3 **a.** \$11,484 **b.** \$13,500 **c.** no **d.** yes

15-4 **a.** \$356.29 **b.** \$879.65 **c.** \$922.40 **d.** \$812.60

15-5 **a.** \$3980
b. Down payment = \$13,235
Price = \$88,235
Cash Paid = \$17,215
Monthly Mortgage Payment = \$686.25

15-6 **a.** 59,589 **b.** \$4.76

15-7 **a.** \$132 **b.** \$2296 **c.** \$512.50

15-8 **a.** \$82.08 **b.** ≈5 yr 7 mo

16-1 **a.** $2\frac{3}{4}$ in. **b.** $1\frac{1}{2}$ in. **c.** $4\frac{3}{4}$ in. **d.** 156 in. **e.** $7\frac{2}{3}$ yd **f.** 288 in.

16-2 **a.** 4.5 gal, or 4 gal 2 qt **b.** 2.195 gal, or 2 gal 1 qt

16-3 **a.** 14 **b.** 19 **c.** \$1012.70

16-4 **a.** \$4608 **b.** ≈2 squares

16-5 **a.** \$69 **b.** \$115 **c.** \$107.65

16-6 \$820

16-7 **a.** \$1152 **b.** \$525

16-8 **a.** \$8.33 **b.** \$43.15

17-1 **a.** 7.9 h or 7 h 54 min; 11.5 gal

17-2 **a.** \$339 **b.** 15

17-3 **a.** \$498.96 **b.** \$336.96

17-4 **a.** \$937.65 **b.** \$177.68

17-5 **a.** 6.5 knots **b.** 13.9

17-6 **a.** 76 **b.** 71

17-7 **a.** 2.212 atm **b.** 2.765 atm **c.** 17 min

17-8 **a.** doubled **b.** one fourth as much **c.** open $\frac{1}{4}$ as long **d.** open 2 times as long

18-1 **a.** \$0.84 **b.** 570 **c.** \$15,650 **d.** \$107.25

18-2 **a.** 2000 **b.** \$817.50 **c.** \$67.50

18-3 **a.** \$4135.50 **b.** 327.26

18-4 **a.** 11% **b.** 789%

18-5 **a.** \$6000 (\$7592) **b.** \$40,000 (\$43,280) **c.** 2 (2.5397)

18-6 **a.** \$580 **b.** 13

18-7 **a.** \$13,065.12 **b.** \$30,860

18-8 **a.** \$21,146.40 **b.** This plan pays \$213.60 more.

INDEX

Extra TEAMS Practice

Practicing certain types of problems will help you with the Exit Level TEAMS Math test. The following pages provide additional practice using some of the skills you will need to:

- add, subtract, and multiply fractions and mixed numbers,
- solve word problems with multiple operations,
- solve word problems using proportions,
- solve word problems using percent,
- use either metric or customary measurements,
- use geometric properties to solve problems with shapes,
- solve word problems using formulas.

Extra TEAMS Practice

Fractions and Mixed Numbers

Solve and simplify. (Write the answer in lowest terms.)

1. $\frac{1}{2}\times\frac{1}{4}$	2. $\frac{3}{4}\times\frac{1}{3}$	3. $\frac{4}{5}\times\frac{1}{4}$	4. $\frac{5}{6}\times\frac{3}{5}$	5. $\frac{3}{7}\times\frac{7}{3}$
6. $\frac{4}{9}\times 3$	7. $\frac{5}{7}\times\frac{14}{5}$	8. $\frac{7}{8}\times 0$	9. $12\times\frac{1}{3}$	10. $\frac{3}{8}\times\frac{4}{11}$
11. $\frac{7}{6}\times\frac{9}{14}$	12. $\frac{1}{5}\times\frac{25}{23}$	13. $\frac{3}{5}\times\frac{3}{5}$	14. $32\times\frac{7}{16}$	15. $\frac{11}{24}\times\frac{6}{11}$
16. $\frac{22}{15}\times\frac{30}{11}$	17. $\frac{24}{45}\times\frac{25}{16}$	18. $\frac{16}{9}\times\frac{63}{12}$	19. $18\times\frac{5}{9}$	20. $\frac{5}{11}\times 1$
21. $\frac{14}{17}\times\frac{17}{2}$	22. $\frac{15}{21}\times\frac{14}{10}$	23. $\frac{3}{5}\times\frac{4}{7}$	24. $\frac{8}{9}\times\frac{12}{11}$	25. $\frac{6}{15}\times\frac{5}{3}$
26. $\frac{8}{11}\times\frac{33}{10}$	27. $\frac{15}{16}\times 64$	28. $\frac{8}{5}\times\frac{10}{6}$	29. $\frac{24}{21}\times 42$	30. $\frac{12}{13}\times 2$
31. $1\frac{2}{3}\times\frac{3}{4}$	32. $2\frac{1}{5}\times\frac{10}{11}$	33. $3\frac{1}{4}\times\frac{4}{5}$	34. $5\frac{2}{3}\times\frac{9}{17}$	35. $5\frac{1}{3}\times 1\frac{2}{13}$
36. $4\frac{1}{2}\times\frac{11}{18}$	37. $3\frac{3}{7}\times\frac{14}{3}$	38. $2\frac{1}{2}\times 1\frac{1}{5}$	39. $2\frac{2}{7}\times 1\frac{1}{6}$	40. $3\frac{5}{6}\times 1\frac{5}{7}$
41. $4\frac{2}{7}\times 1\frac{2}{5}$	42. $8\frac{1}{3}\times 6$	43. $5\frac{1}{5}\times 1\frac{2}{3}$	44. $9\frac{1}{3}\times 1\frac{1}{14}$	45. $4\frac{1}{2}\times 1\frac{2}{3}$
46. $7\frac{1}{2}\times 1\frac{3}{25}$	47. $5\frac{5}{6}\times 2\frac{4}{7}$	48. $4\frac{2}{11}\times 1\frac{10}{23}$	49. $3\frac{1}{8}\times 3\frac{1}{5}$	50. $2\frac{3}{5}\times 1\frac{1}{2}$
51. $\frac{2}{3}$ of 9	52. $\frac{4}{5}$ of 20	53. $\frac{4}{7}$ of $\frac{14}{3}$	54. $\frac{1}{7}$ of 8	55. $\frac{2}{3}$ of $4\frac{1}{5}$
56. $\frac{3}{4}$ of $1\frac{1}{15}$	57. $\frac{4}{7}$ of $2\frac{5}{8}$	58. $\frac{5}{3}$ of $1\frac{1}{2}$	59. $\frac{6}{5}$ of $2\frac{1}{12}$	60. $\frac{6}{11}$ of $2\frac{1}{5}$

Extra TEAMS Practice

Fractions and Mixed Numbers

Solve and simplify. (Write answers in lowest terms.)

1. $\frac{4}{9} + \frac{1}{9}$
2. $\frac{5}{11} + \frac{3}{11}$
3. $\frac{7}{10} - \frac{2}{10}$
4. $\frac{7}{8} - \frac{5}{8}$
5. $\frac{2}{9} + \frac{4}{9}$
6. $\frac{3}{7} + \frac{4}{7}$
7. $\frac{8}{15} + \frac{4}{15}$
8. $\frac{11}{13} - \frac{8}{13}$
9. $\frac{3}{4} - \frac{1}{4}$
10. $\frac{11}{18} + \frac{5}{18}$
11. $2\frac{6}{7} - 1\frac{3}{7}$
12. $\frac{8}{15} + 2\frac{3}{15}$
13. $6\frac{1}{2} - 3\frac{1}{2}$
14. $4\frac{2}{5} + \frac{3}{5}$
15. $11\frac{3}{8} - 4$
16. $4\frac{5}{9} - 2\frac{2}{9}$
17. $5\frac{3}{7} + 2\frac{5}{7}$
18. $9\frac{5}{12} - 3\frac{3}{12}$
19. $10\frac{5}{16} + 2\frac{7}{16}$
20. $3\frac{11}{18} + \frac{5}{18}$
21. $\frac{5}{12} - \frac{1}{3}$
22. $\frac{2}{5} + \frac{3}{10}$
23. $\frac{1}{2} + \frac{1}{4}$
24. $\frac{3}{7} - \frac{4}{14}$
25. $\frac{9}{16} + \frac{3}{8}$
26. $\frac{5}{6} + \frac{5}{18}$
27. $\frac{15}{24} - \frac{3}{8}$
28. $\frac{4}{21} + \frac{1}{7}$
29. $\frac{7}{30} - \frac{1}{10}$
30. $\frac{1}{2} + \frac{2}{3}$
31. $\frac{4}{21} + \frac{6}{7}$
32. $\frac{19}{24} - \frac{5}{8}$
33. $\frac{19}{30} - \frac{3}{10}$
34. $\frac{12}{33} + \frac{7}{11}$
35. $\frac{19}{40} + \frac{9}{10}$
36. $\frac{3}{10} + \frac{2}{15}$
37. $\frac{7}{8} - \frac{3}{10}$
38. $\frac{2}{9} + \frac{3}{4}$
39. $\frac{1}{3} + \frac{1}{2} + \frac{1}{4}$
40. $\frac{2}{5} + \frac{1}{10}$
41. $2\frac{3}{4} + \frac{1}{2}$
42. $5\frac{4}{5} - 2\frac{1}{10}$
43. $3\frac{3}{8} + 2\frac{1}{4}$
44. $2\frac{7}{12} - 1\frac{1}{3}$
45. $3\frac{4}{9} + 2\frac{2}{3}$
46. $\frac{5}{14} + 1\frac{2}{7}$
47. $5\frac{3}{4} - 2\frac{1}{5}$
48. $3\frac{15}{16} - 2\frac{1}{8}$
49. $10\frac{13}{24} - 7\frac{3}{8}$
50. $3\frac{9}{20} + 7\frac{4}{5}$
51. $8\frac{2}{9} + 5\frac{5}{6}$
52. $2\frac{4}{25} - \frac{1}{10}$
53. $4\frac{15}{16} - 3\frac{5}{12}$
54. $1\frac{2}{3} + \frac{5}{7}$
55. $8\frac{5}{6} - 3\frac{7}{15}$
56. $3\frac{7}{10} + 8\frac{3}{4}$
57. $1\frac{1}{3} + 5\frac{5}{8}$
58. $5\frac{9}{20} + 3\frac{4}{15}$
59. $1\frac{1}{3} + 3\frac{1}{2} + 5\frac{1}{6}$
60. $2\frac{2}{5} + 7\frac{1}{10}$

Multiple Operations

Solve.

1. Joan Jellison's house has a wooden deck that is a rectangle 6.5 m long and 4 m wide. Deck paint comes in 1-liter cans and cost $6.00 a can. Each liter of paint will cover about 8 m^2. How many cans of paint should Joan buy to paint the deck?

2. Joan needs 4 L of paint. She found paint sold in 1-liter cans for $6.00 per can and in 3-liter cans for $15.96 per can. What is the less expensive way to buy 4 L of paint? How much less expensive is it?

3. A deck that measures 6.5 m by 4.0 m needs 2 coats of paint. The first coat will use 1 L to cover 8 m^2, and the second coat will use 1 L to cover 10 m^2. How many liters of paint will be needed for the 2 coats? Round the answer to the nest largest number.

4. Joan bought 1 three-liter can of paint for $15.96 and and one 1-liter can of paint for $6.00. What is the cost, to the nearest cent, of the paint per liter?

5. A record store was open 8 hours on Friday and 9 hours on Saturday. The total sales for Friday $2720. The total sales for Saturday were $3249. How much greater were total sales per hour on Saturday than on Friday?

6. Each of the 45 members of the Modesto service club agreed to sell 2 books of 10 tickets each for the annual grape festival and dance. Each ticket cost $2. What was the total number of tickets they agreed to sell?

7. The Salazars have been saving to buy a car. They now have $8465 in a savings account. The cars they like cost between $6595 and $7599. What is the least amount that would be left in their account if they pay cash?

8. Rudy worked 3 jobs last summer. He made a total of $1350. One job earned him one half of that amount. The second job paid him $275. How much did Rudy earn from the third job?

9. The recorded high score on an electronic game is 9880 points. Harley's highest score is $\frac{7}{8}$ of the record. Nadine's highest score is 1456 points less than Harley's. What is her highest score?

10. The manager of Kenbrook Farms owns 15 riding horses. After taking out the cost of caring for the horses, he earned $11,500 from selling rides. He sold 3560 rides at $5 each. What was his cost to keep each horse?

Multiple Operations

Solve.

1. In a 4-H Club sheep auction, 199 sheep will be sold. 12 sheep are in each group. 16 groups of sheep will be shown. How many sheep will be left over?

2. Ms. Long plans to buy 20 gal of asphalt at $12 per gal. She finds asphalt on sale in 5-gal cans for $50 each. How much would she save by buying 5-gal cans?

3. Kaela works part-time in a shop. Her hourly pay is $5.15. Last week she worked 18 hours . So far this week she has worked 3 hours 15 minutes on Monday, 2 hours 45 minutes on Tuesday, 2 hours 45 minutes on Wednesday and Friday. How long does Kaela have to work on Saturday to equal last week's work time?

4. A bus driver drives from Davis to Nelson once a week. On the way to Nelson he stops in Santa Ana. The distance for this trip is 350 miles. The return trip without the stop is 310 miles. How many miles does he drive in 4 weeks?

5. A basketball arena holds 15,000 people. 11,630 spectators attend a game. Three rows of 25 seats each are removed for this game. How many seats will be left empty?

6. A railroad tank car carries 11,970 gallons of oil. Each barrel holds 42 gallons. The value of each barrel is $21. What is the total value of the oil?

7. The Holiday Skate Shop has 715 pairs of skates to rent. 70 new pairs of skates have been delivered. 58 pairs have been brought in broken. How many skates does the store have to rent now?

8. A yardage store owner orders 23 yards of lace, 47 yards of black ribbon, 16 yards of yellow ribbon, and 7 yards of red ribbon. How many yards of notions did the owner order?

9. The Alamo Brothers concert tickets sell for $20 for the main floor and $16 for balcony seats. There is a $3 fee for each order that is charged by phone. D'arcy bought and charged 7 balcony tickets by phone. How much did she spend?

10. The Daisy Farm day care center charges 75¢ per minute for late pick-up of children. One Friday, 3 parents were 1 minute late, 2 parents 2 minutes late, and 2 parents 5 minutes late. How much did the center earn in late charges that day?

Proportions

Solve each proportion.

1. $\frac{5}{6}=\frac{f}{12}$
2. $\frac{u}{40}=\frac{7}{8}$
3. $\frac{7}{g}=\frac{1}{4}$
4. $\frac{x}{18}=\frac{2}{3}$
5. $\frac{d}{40}=\frac{8}{12}$
6. $\frac{8}{t}=\frac{24}{30}$
7. $\frac{30}{k}=\frac{15}{18}$
8. $\frac{3}{6}=\frac{8}{m}$
9. $\frac{14}{35}=\frac{b}{20}$
10. $\frac{4}{5}=\frac{s}{25}$
11. $\frac{9}{12}=1\frac{10}{a}$
12. $\frac{h}{24}=\frac{6}{8}$
13. $\frac{3}{c}=\frac{7}{35}$
14. $\frac{9}{12}=\frac{15}{y}$
15. $\frac{4}{6}=\frac{n}{30}$
16. $\frac{y}{50}=\frac{12}{30}$
17. $\frac{q}{72}=\frac{7}{8}$
18. $\frac{7}{21}=\frac{8}{e}$
19. $\frac{8}{15}=\frac{r}{30}$
20. $\frac{5}{j}=\frac{135}{54}$

Solve.

21. One person in 20 has red hair. In a group of 60 people, how many would you expect to find with red hair?

22. Melissa drove her car 200 kilometers in 2.5 hours. How far can she drove in 6 hours at the same rate?

23. 3 apples cost 29 ¢, how much will 12 apples cost?
$\frac{3}{29}=\frac{12}{x}$, x=$1.16

24. Steve swam 200 meters in 2 minutes. How many meters is this per second?

25. Lee ran 400 meters in 50 seconds. How many meters is this per second.

26. On a map, 1 centimeter represents 70 kilometers. What distance does 6.2 centimeters represent?

27. Marie worked 7.5 hours and earned $50. How much would she earn working 30 hours at the same pay scale?

28. A gear rotates 60 times every 25 minutes. How many times will it rotate in 10 minutes?

Proportions

Each pair of triangles is similar. Find the missing side length.

1.
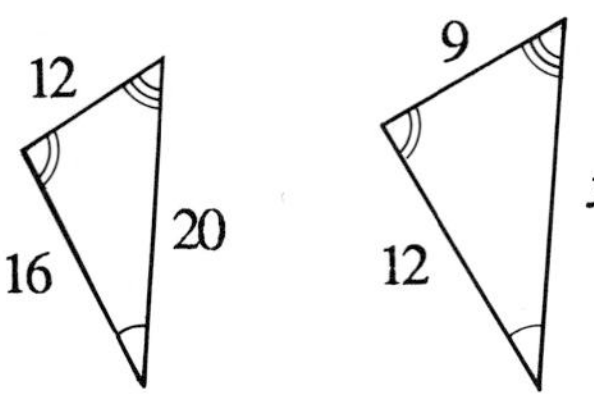

2.
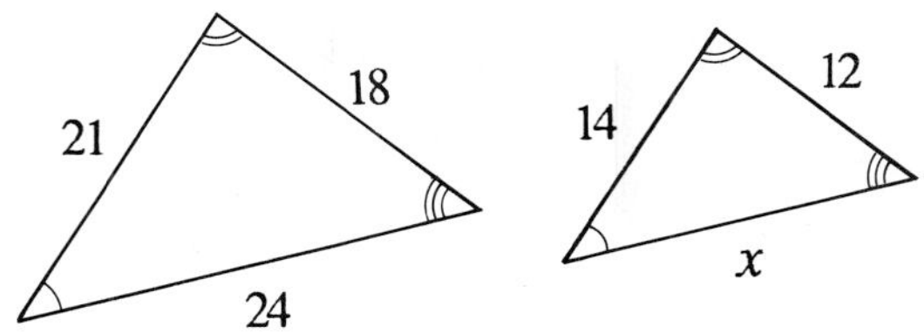

3.
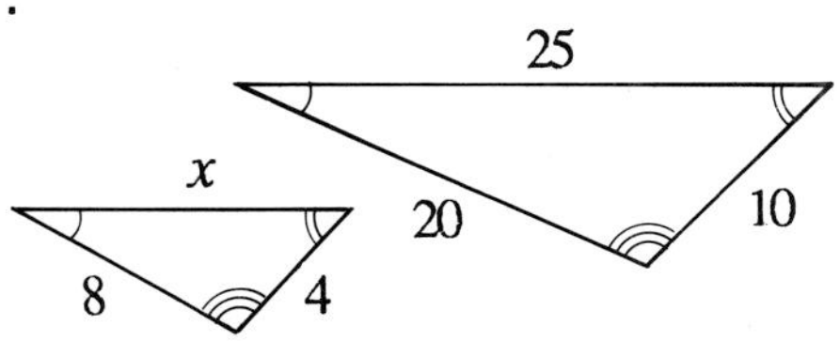

4.
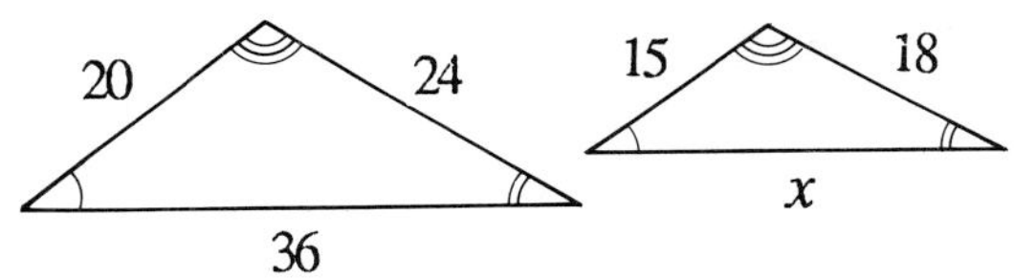

5.
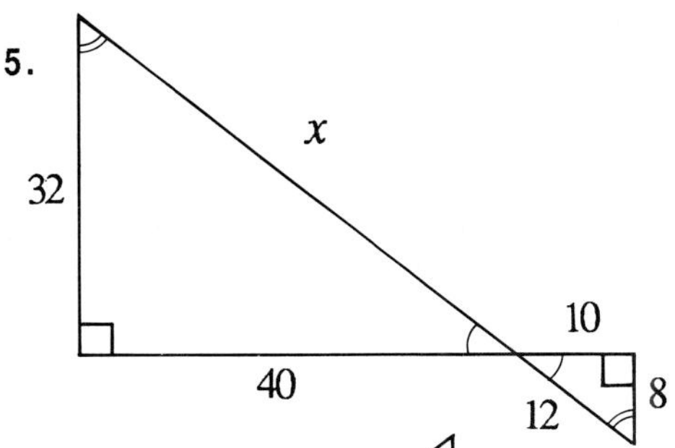

6.
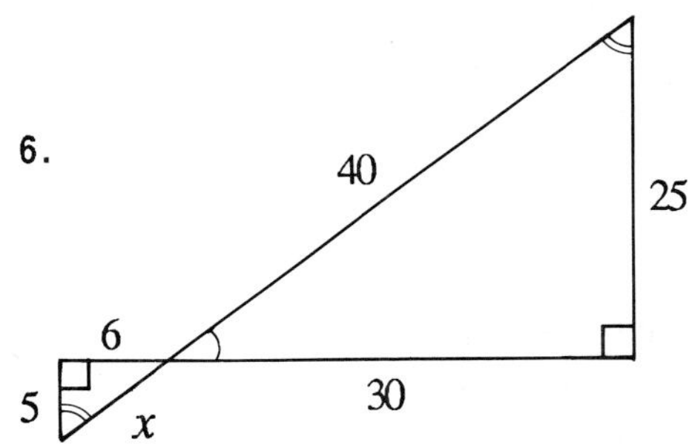

7.
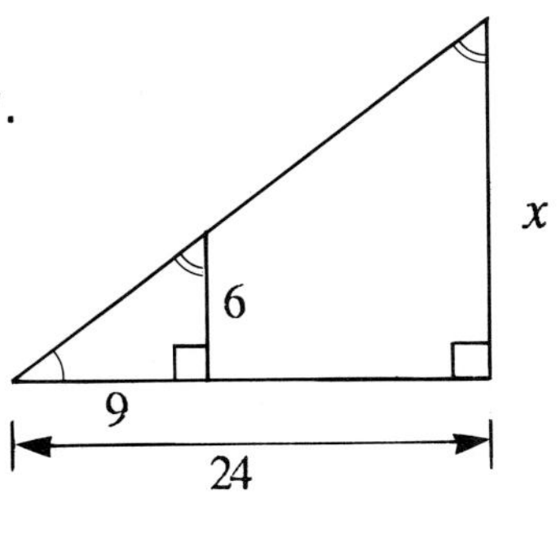

8.
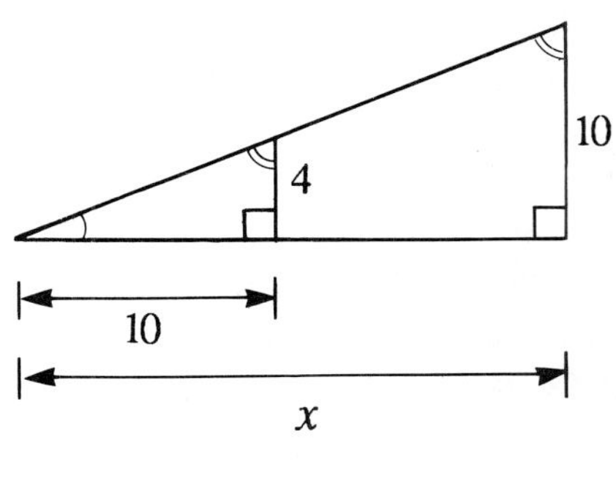

Percent

Solve.

1. What number is 20% of 25?
2. 50% of 64 is what number?
3. 33.3% of 93 is what number?
4. What number is 10% of 70?
5. What number is 25% of 12?
6. 30% of 70 is what number?
7. 15% of 160 is what number?
8. What number is 8% of 50?
9. What number is 12.5% of 48?
10. 7% of 30 is what number?
11. 150% of 62 is what number?
12. What number is 200% of 72?
13. 20 is what percent of 80?
14. What percent of 16 is 12?
15. What percent of 24 is 9?
16. 12 is what percent of 36?
17. What percent of 50 is 30?
18. 24 is what percent of 36?
19. What percent of 9 is 5.4?
20. 25 is what percent of 25?
21. 33 is what percent of 66?
22. What percent of 82 is 123?
23. 22 is what percent of 20?
24. What percent of 25 is 21?
25. 12 is 25% of what number?
26. 5% of what number is 7?
27. 13.5 is 50% of what number?
28. 12.5% of what number is 12?
29. 15 is 20% of what number?
30. 5.4 is 9% of what number?
31. 44 is 200% of what number?
32. 60% of what number is 12?
33. 15% of what number is 15?
34. 45 is 75% of what number?
35. 72 is 80% of what number?
36. 1.5 is 30% of what number?

Percent

Find the percent and direction of change.

1. 5 to 8
2. 20 to 30
3. $45 to $36
4. 12 to 15
5. 24 to 6
6. 40 to 44
7. $80 to $48
8. 15 to 30
9. 64 to 40
10. 90 to 60
11. 124 to 62
12. $25 to $45
13. 16 to 18
14. 20 to 19
15. 300 to 225
16. 30 to 75
17. 9 to 21
18. 55 to 44
19. 90 to 81
20. 120 to 144

Solve.

21. Four rooms in a 5-room apartment take up, respectively, 20%, 18%, 27%, and 30% of the space. Find the percent taken up by the fifth room.
22. Three rooms in a 4-room office take up, respectively, 47%, 13%, and 17% of the space. Find the percent taken up by the fourth room.
23. Four rooms in a 5-room building take up, respectively, 15%, 19%, 38%, and 9% of the space. Find the percent taken up by the fifth room.
24. Four rooms in a 5-room house take up, respectively, 19%, 35%, 9%, and 22% of the space. Find the percent taken up by the fifth room.
25. Three rooms in a 4-room office take up, respectively, 18.4%, 33.9%, and 44.6% of the space. Find the percent taken up by the fourth room.
26. Two rooms in a 3-room office take up, respectively, 14.5% and 55.25% of the space. Find the percent taken up by the third room.

Measurement

Solve.

1. The temperature at 3 p.m. on Tuesday was 68°F. The temperature at 9 p.m. was 0.75 the temperature at 3 p.m. What was the temperature at 9 p.m.?

2. Enrique bought 3 90-minute cassettes. He recorded 4.25 hours of music. How many minutes did he have left to record on tape? (1 h = 60 min)

3. Jean paid $0.20 for 0.5 pint of milk. Isabel paid $1.08 for 0.5 gallon of milk. How much more or less per quart did Isabel pay than Jean? (1 gal = 4 qt, 1 qt = 2 pt)

4. A raisin weighs about 500 milligrams. Suppose there are 600 raisins in a box. What is the total weight of the raisins in grams? (1 g = 1000 mg)

5. The Komodo dragon is about 3 meters long. The Gila monster is about 60 centimeters long. Which lizard is longer, and by how much? (1 m = 100 cm)

6. Juan figured he had watched about 24,000 hours of television. How many years and days is this? (1 yr = 365 d, 1 d = 24 h)

7 A wooden gift box contained 18 large apples. The box weighed 11 pounds when full. The empty box weighed 2 pounds. About how many ounces did one apple weigh? (1 lb = 16 oz)

8. Gina exercises 90 minutes each day of the week. How many hours does she spend exercising all week? (1 h = 60 min)

9. A family size package of ground beef weighs 1.5 kilograms. How many 100-gram patties can you make? (1 kg = 1000 g)

10. Butter melts at about 30°C. To bake a casserole the temperature should be about 7 times as hot. At what temperature should you bake a casserole?

11. Miguel worked at the hospital 3 75 hours in the morning and 3.5 hours in the afternoon. How many minutes did Miguel work each day? (1 h = 60 min)

12. An armadillo weighed 12.85 pounds one month and 13.1 pounds the next. How many ounces did the armadillo gain? (1 lb = 16 oz)

Measurement

Solve.

1. Bonnie figures that driving from Houston to Dallas is 1.5 times farther than driving the 160 miles from Houston to Austin. How longer in miles is the Houston-Dallas trip?

2. Destiny has to decide whether to buy potatoes at Store A at 18 pounds for $2.70 or at Store B at 24 pounds for $3.36. Which is the better buy and for by how much per pound?

3. Mark has 6 classes that are each 45 minutes long. How many hours does Mark spend in class? (1 h = 60 min)

4. A carton of orange juice contains 1.6 liters. How many 200-milliliter cups can you fill? (1 L = 1000 mL)

5. Nelly made 3 gallons of jelly. How many 1-pint jars can she fill? (1 qt = 2 pt, 1 gal = 4 qt)

6. The lowest recorded temperature in Rio Grande City is 21°F. The highest recorded temperature in Corpus Christi is 5 times as high. What is the record high in Corpus Christi?

7 A salesperson receives $0.20 per mile for traveling expenses. If she drives 270 miles from Fort Worth to San Antonio, how much will she receive for expenses?

8. To buy toothpaste, Helen compares prices at Store A (8 ounces for $2.80) and Store B (5 ounces for $1.25). Which is the better buy, and by how much?

9. The distance between bases in baseball is about 0.0274 kilometers. If Sean hits a homerun, how many meters will he travel to score? (1 km = 1000 meters)

10. Barry worked 6 hours and 40 minutes, 7 hours and 20 minutes, and 8 hours and 20 minutes on consecutive weekends for $5 per hour. How much did he earn?

11. Umberto ran 880 yards the first day, 1120 yards the second day, and 2000 yards the third day. How many miles and yards did he run altogether? (1 mi = 1760 yd)

12. A bantamweight boxer cannot weigh more that 118 pounds. A featherweight boxer cannot weigh more than 126 pounds. Skip weighs 1920 ounces. In which weight division would he box? (1 lb = 16 oz)

Geometric Properties

Find the missing angle.

1. 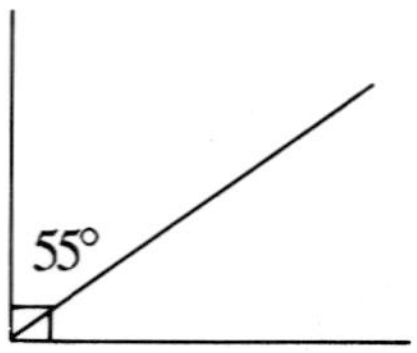

2. 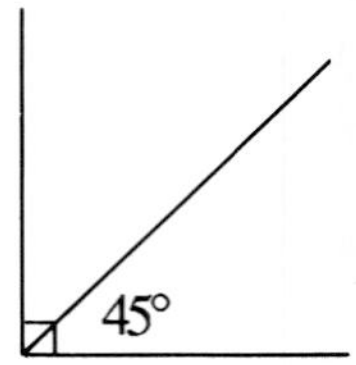

3. 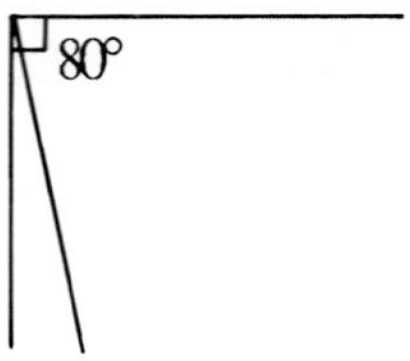

4. 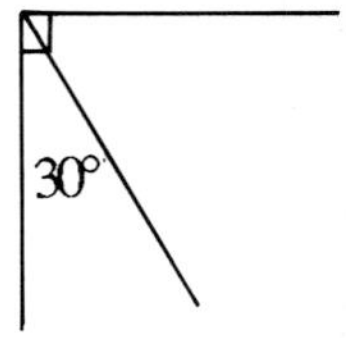

5. 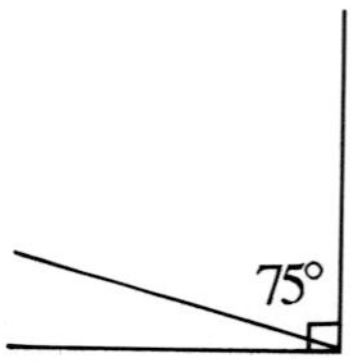

6. 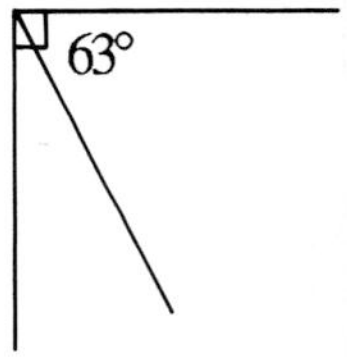

7. 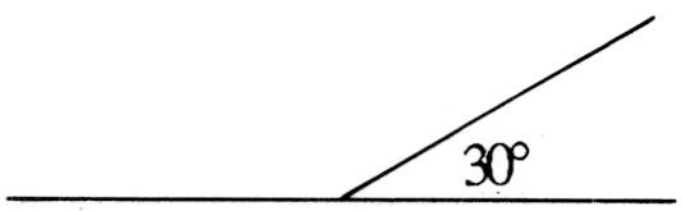

8. 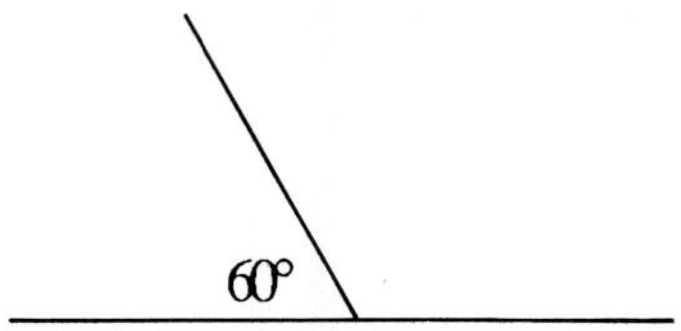

9. 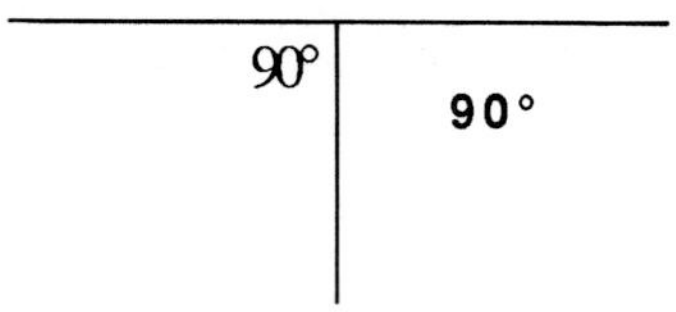

10. 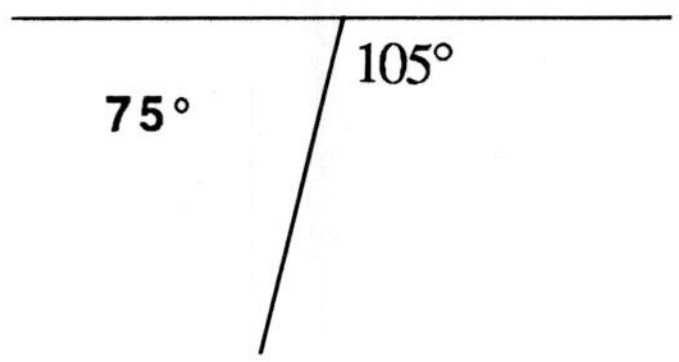

11. 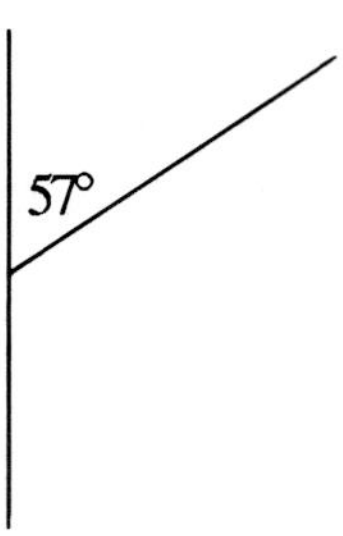

12. 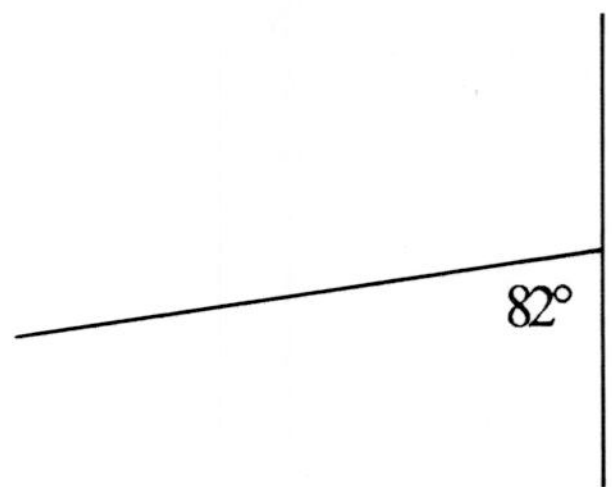

Extra TEAMS Practice

Geometric Properties

Find the missing angle.

1. 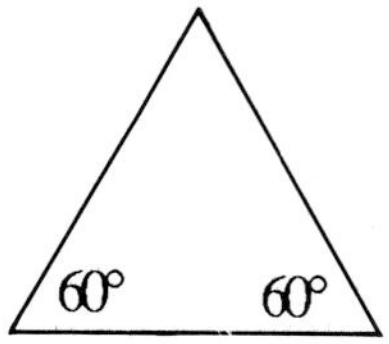

2. 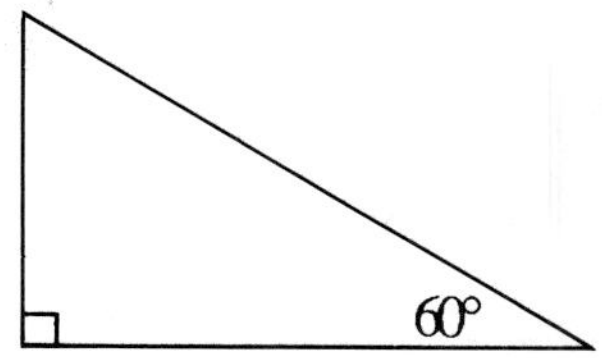

3. 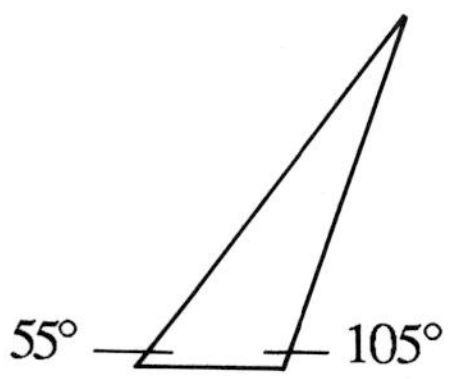

4. 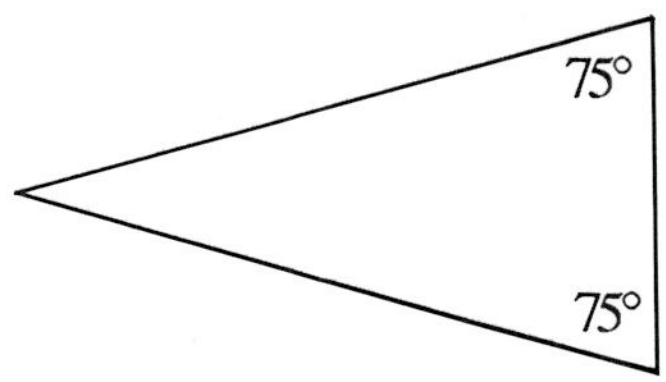

5. 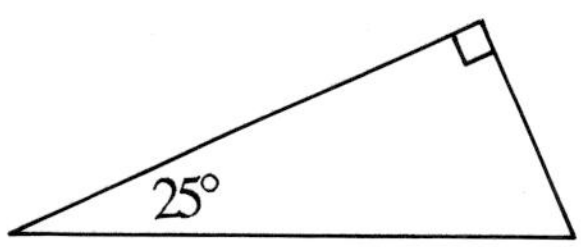

6. 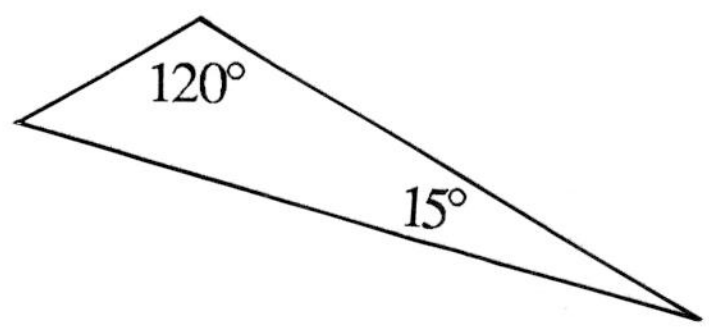

Find the missing side.

7. 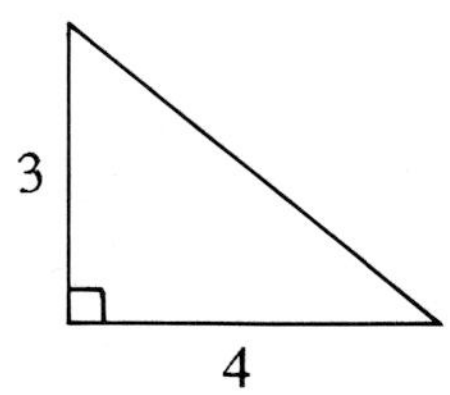

8. 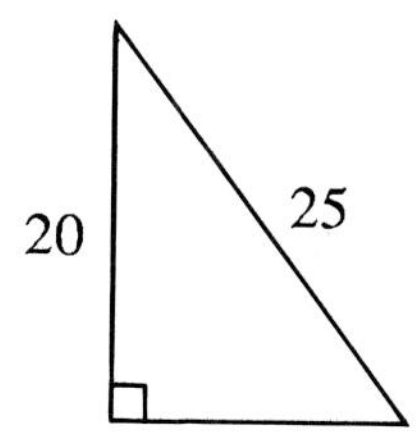

9. 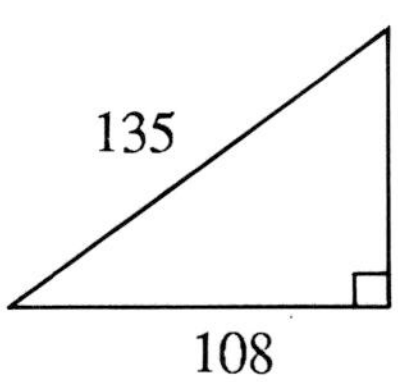

10. 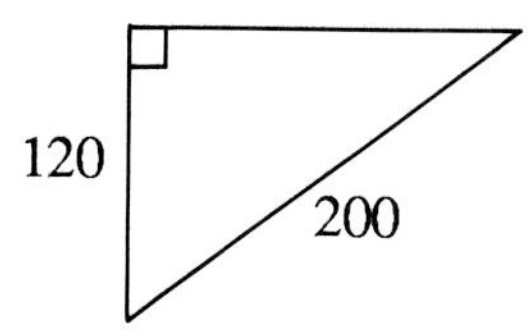

11. 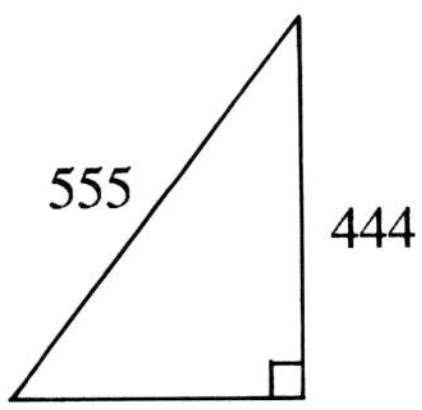

12. 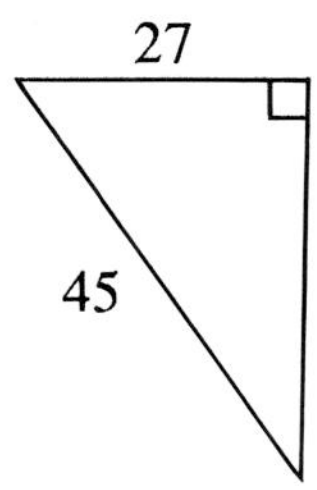

Formulas

Solve.

1. $t + 6 = 24$
2. $c + 52 = 52$
3. $g + 12 = 29$
4. $42 + e = 67$
5. $n + 27 = 69$
6. $p + 10 = 6$
7. $z + 26 = 61$
8. $e + 22 = {}^{-}3$
9. $h + 17 = 39$
10. $b + 11 = {}^{-}21$
11. $v + 13 = {}^{-}41$
12. $u + 29 = 17$
13. $a + 46 = 27$
14. $k + 12 = {}^{-}21$
15. $w + 21 = {}^{-}53$
16. $i + 28 = {}^{-}13$
17. $o + 32 = {}^{-}23$
18. $d + 19 = {}^{-}46$
19. $7 = q + 3$
20. $f + 9 = {}^{-}33$
21. $16 = p + 9$
22. ${}^{-}43 = 9 + s$
23. ${}^{-}27 = {}^{-}12 + p$
24. $42 = 56 + m$
25. $t - 5 = 11$
26. $p - 21 = 17$
27. $x - 13 = 42$
28. $y - 36 = 29$
29. $n - 21 = 61$
30. $r - 66 = 53$
31. $z - 14 = {}^{-}7$
32. $e - 20 = {}^{-}14$
33. $h - 17 = {}^{-}8$
34. $k - 39 = {}^{-}21$
35. $h - 20 = {}^{-}79$
36. $e - 54 = {}^{-}81$
37. $9 = q - 12$
38. ${}^{-}6 = l - 1$
39. $17 = a - 6$
40. $21 = m - 9$
41. ${}^{-}11 = c - 13$
42. ${}^{-}33 = n - 51$
43. ${}^{-}29 = b - 42$
44. $d - ({}^{-}10) = 23$
45. $i - ({}^{-}16) = 31$
46. $j - ({}^{-}33) = 19$
47. ${}^{-}8 = c - 3$
48. ${}^{-}21 = u - 7$
49. $4x = 20$
50. $2y = 10$
51. $5t = 30$
52. $8f = 48$
53. $11g = 99$
54. ${}^{-}7m = 42$
55. $-5n = 45$
56. $12r = 36$
57. $9p = {}^{-}72$
58. $-6j = 24$
59. $11m = 44$
60. $1k = {}^{-}72$

Formulas

Solve.

1. Find the number s that when multiplied by 9 gives a product of 63.

2. Find the number g that when divided by 3 gives a quotient of 29.

3. Find the number d that when multiplied by 12 gives a product of 132.

4. Find the number a that when divided by 15 gives a quotient of 4.

5. $2x + 4 = 36$

6. $6y + 10 = 46$

7. $7p + 7 = 63$

8. $3z + 26 = 59$

9. $57 = 9s + 3$

10. $42 = 4d + 6$

11. $105 = 8p + 9$

12. $^{-}3y + 3 = ^{-}3$

13. $^{-}6x + 16 = ^{-}8$

14. $17 = ^{-}5t + 2$

15. $76 = ^{-}9k + 4$

16. $55 = ^{-}7h + 68$

17. $^{-}8z + 12 = 60$

18. $6n - 10 = 32$

19. $22 = 4y - 2$

20. $41 = 6m - 7$

21. $\frac{1}{2}k + 9 = 30$

22. $\frac{1}{3}g - 7 = 15$

23. $\frac{1}{2}f - 22 = 48$

24. $\frac{1}{4}j - 7 = 15$

25. $0.5x - 0.6 = 8.4$

26. $0.5t - 0.9 = 13.1$

27. $-25.7 = 2.2i + 0.7$

28. $^{-}22.8 = 1.5l + 1.2$

Measurement

Use a formula to solve.

rectangle	cube	simple interest	circle	rectangular solid	triangle
$A = lw$	$V = s^3$	$i = prt$	$A = \pi r^2$	$V = lwh$	$A = \frac{1}{2}bh$

1. Find the annual interest on a loan to buy a $40,000 beach house at 12%.

2. Alex bought a $950 sailboard. Find the simple interest on his loan at 11% for two years.

3. Find the volume of a 1' × 3' × 1.5' picnic basket.

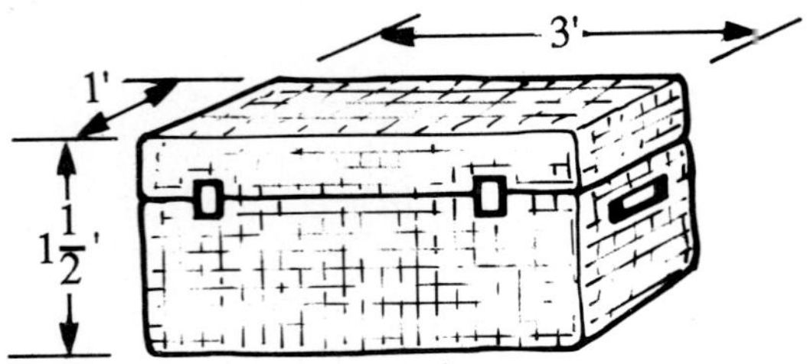

4. Find the area of an 11-inch diameter frisbee.

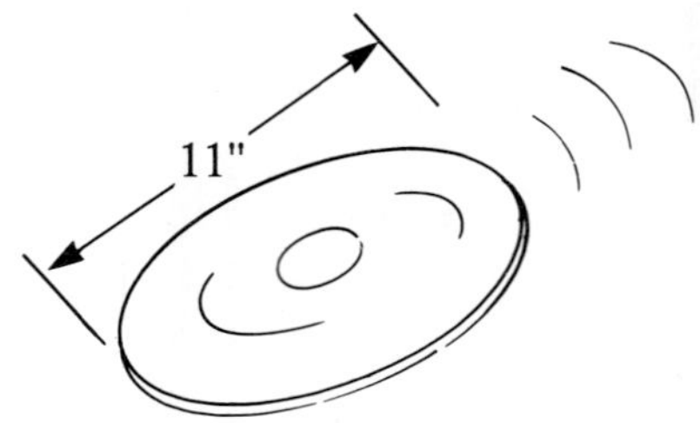

5. Find the approximate area of a pizza with a radius of 9 inches.

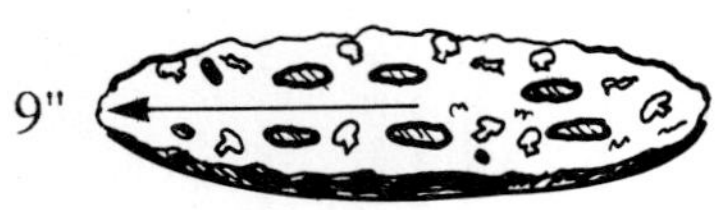

6. Find the volume of a 12" × 14" × 16" ice chest.

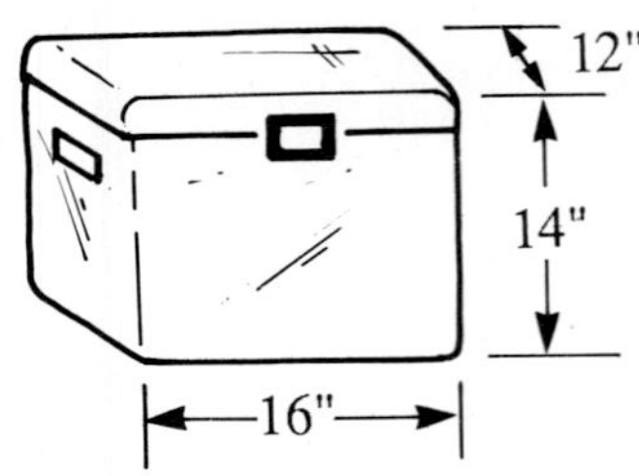

7. Find the area of one side of an 18' × 12' triangular sail.

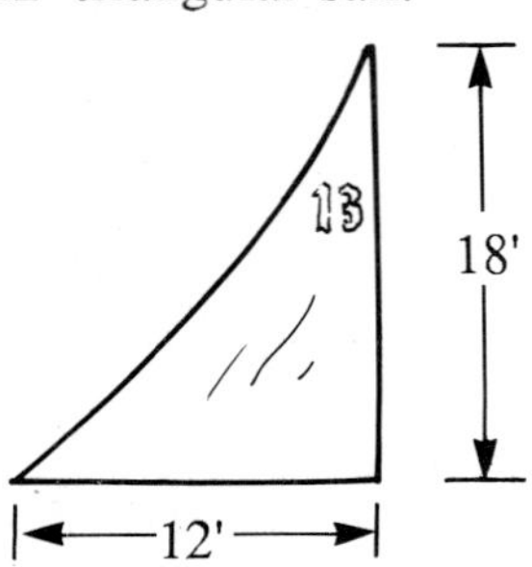

8. Find the area of the top surface of a 1.2-meter × 2-meter log raft.

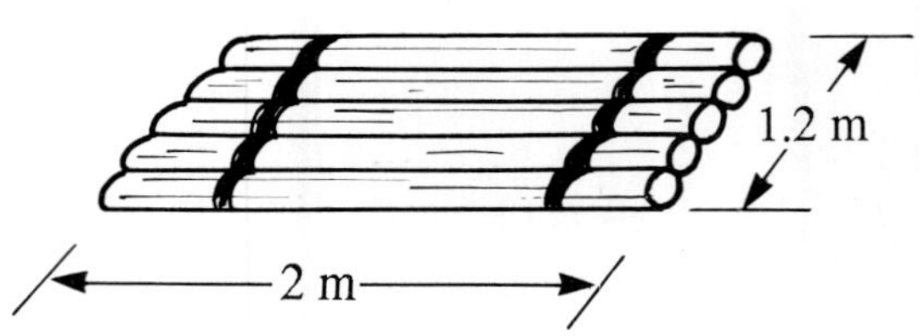